Introduction to Analysis

Arthur Mattuck

Massachusetts Institute of Technology

Prentice Hall
Upper Saddle River, New Jersey 07458

Library of Congress Cataloging-in-Publication Data

Mattuck, Arthur
    Introduction to Analysis / Arthur Mattuck
        p.   cm.
    Includes index.
    ISBN 0-13-081132-7
    1. Mathematical Analysis.   I. Title
  QA300.M23 1999 515–dc21   98–25850
                                              CIP

Cover art: **Sunrise** by Robert Kabak, 1956

Editorial director, Tim Bozik
Editor-in-chief, Jerome Grant
Acquisition editor, George Lobell
Executive managing editor, Kathleen Schiaparelli
Managing editor, Linda Behrens
Editorial assistants, Gale Epps, Nancy Bauer
Assistant VP production/manufacturing, David W. Riccardi
Manufacturing manager, Trudy Pisciotti
Manufacturing buyer, Alan Fischer
Creative director, Paula Maylahn
Art director, Jane Conte
Marketing manager, Melody Marcus

Printed in the United States of America

10 9 8 7 6 5 4 3

ISBN 0-13-081132-7

Prentice-Hall International (UK) Limited, *London*
Prentice-Hall of Australia Pty. Limited, *Sydney*
Prentice-Hall of Canada, Inc., *Toronto*
Prentice-Hall Hispanoamericana, S. A., *Mexico*
Prentice-Hall of India Private Limited, *New Delhi*
Prentice-Hall of Japan, Inc., *Tokyo*
Pearson Education Asia Pte. Ltd., *Singapore*
Editora Prentice-Hall do Brasil, Ltda., *Rio de Janeiro*

to the memory of my parents

Jacob A. Mattuck
Rae B. Mattuck

and my teacher

Arnold Dresden

Contents

Appendix

Preface

This book is for a one-semester undergraduate real analysis course, taught here at M.I.T. for about 25 years, and in its present form for about 15. It runs in parallel with a more difficult course based on point-set topology.

Its origins Many years ago, my older brother Dick, then a very smart graduate student in theoretical physics here, came in fuming.

"I'm taking this graduate complex variable course, and I can't understand a word of it. The book (it was Ahlfors) just goes on and on in English about open sets and closed sets, and then says 'This concludes the proof.' I was always great at mathematics. What am I doing wrong?"

As he was my chief tormenter in childhood, my advice was a smug, "Just keep on trying, I'm sure you'll get it."

A month later he came in again.

"Last night I suddenly saw it, and it's all completely trivial."

Years later, a senior physics undergraduate sat down before me and sighed.

"Well, this is the fourth time I'm dropping analysis. Each time I get a little further into the course, but the open sets always win out in the end. Isn't it possible to teach it so guys like me could understand it? We understand derivations, but they give us proofs instead. Inequalities are OK, as long as they look like equations, but this analysis doesn't look like the math we know — it's all in English instead of symbols. And as far as any of us can tell, the only thing any theorem is good for is proving the next theorem."

Kenneth Hoffman agreed to see what could be done; then Frank Morgan and Steven Robbins each taught it for a year. Afterwards, profiting from their experiences and the notes they generously left me, I took it over and developed the notes from which this book has evolved.

The course The course is basically one-variable analysis. The emphasis throughout is not on the algebraic or topological aspects of analysis, but on estimation and approximation: how analysis replaces the equalities of calculus with inequalities: certainty with uncertainty. This represents for students a step up in maturity.

To help, arguments use as little English as possible, and are formulated to look like successions of equations or inequalities: derivations, in other words.

Calculus is used freely from the beginning as a source of examples, so students can see how the ideas are used. The real numbers are discussed briefly in the first chapter, with most of the emphasis on the completeness property. The aim is to get to interesting things as quickly as possible. Several appendices present extended applications.

Needless to say, point-set topology, the *pons asinorum* of analysis courses, has been banished to near the end, and presented in abbreviated form, just before it is needed in the study of integrals depending on a parameter. By then, students can understand the arguments, and even enjoy them as something new-looking.

A few times I got through the whole book, but nowadays don't seem to get that far. There are different goals one can aim at, any of which provide a sense of fulfillment, since they justify results known from previous courses but unproved there, or generally provoke curiosity:

    ⋆ differentiating the Laplace transform under the integral sign (Section 27.5);

    ⋆ the existence and uniqueness theorem for first order ODE's (Appendix E);

    ⋆ differentiating a power series term-by-term (Chapter 22);

    ⋆ what the Lebesgue integral does for you (Chapter 23);

or just the theorems of calculus revisited (Chapter 20 or 21), with an easy chapter illustrating some new uses for them (Appendix C). For each of these goals, certain chapters can be skipped, which teachers will be able to identify without trouble.

Features of the book   Some account of what I take to be distinguishing features of the book may be helpful.

Questions   At the end of each short section (1-3 pages) are questions, with answers at the end of the chapter. (At the end of the chapter are the customary exercises, tied to the sections, and also somewhat harder problems, going with the chapter as a whole.) The questions have many functions.

(a) There are many worked-out examples in the book; if there were more, the main thread of the exposition would tend to get lost. But students always want more examples. The questions supply them. And because they have to turn over some pages to get the answers, there's at least some hope they will actually try them without first peeking. It's an unusual student who will do this with worked-out examples in the text.

(b) Answers are often complete proofs, written in proper style, that students can use as models.

(c) The questions act as traffic signals, not-so-gently prodding the students to stop and see if they understand what they have just read, instead of mindlessly continuing with their yellow highlighting. Some questions ask about specific things in the proofs. Occasionally a proof will deliberately gloss over some point, leaving the more detailed explanation as the answer to a question.

Writing Most of the students taking this course have not seriously studied proofs before, and haven't had to write any of their own. A major goal of the book is to get them to be able to do this. So there are remarks in the first few chapters (towards the ends of Chapters 3 and 5 for instance) about how to write up arguments, as well as warnings sprinkled here and there about common pitfalls to avoid. Some of these are in the answers to questions.

A lot of this was written in the wan hope of getting assignments that were easier to read and grade, I must confess.

Reading and Typography: a Diatribe The ability to write proofs goes hand-in-hand with the ability to read them. To facilitate this I have tried hard to make the book readable, for example by asking average students to note in the margins everything that puzzled them on a first reading. (I learned a lot about the placement of subordinate clauses from this, and recommend it heartily to fellow-sufferers in scrivening.)

In my opinion, many otherwise good books are spoiled by indifference to layout. How can one expect students to write arguments decently, when they study from books where a proof is presented as a solid block of type, equations broken in the middle, the final line and crux of a proof appearing all by itself on the other side of the page? Decent compositors don't do these things; alas, some TeX-happy authors do.

But TeX can be a blessing instead of a curse: it allows arguments and formulas to be set out in the clearest format; one can experiment on the spot, rewriting sentences to get a better layout, something the best compositor cannot do.

I've tried to avail myself of these possibilities: almost no proofs require a page turn; implication arrows are lined up for maximum clarity, spacing in formulas, mathematical phrases, and between paragraphs has been adjusted to correspond to the pauses one would make in reading aloud. I hope there are no frightening pages here.

Dept. of Fuller Explanation To make analysis more accessible, the book goes into more detail than usual about elementary things concerning functions, inequalities, and so on.

About all I can say in defense of this is that these are things which I find that many of my students don't seem to know, or don't know explicitly. They subtract inequalities, are vague about inverse functions, and not always sure just what functions are or how they should think about them.

Mathematical notation The book makes use of some non-traditional notation and terminology, which classes have received well, while (more to the point) their teachers have not objected too strenuously.

Proofs are allowed to end with 42ϵ instead of ϵ — in the book, this bears the name "K-ϵ principle", introduced explicitly in Chapter 3, and used throughout.

When the approximation is the main idea to be expressed, the estimate $|a - b| < \epsilon$ is often written $a \underset{\epsilon}{\approx} b$. This too is harmless.

More serious is the use of these terms borrowed from applied mathematics: "for $n \gg 1$" (for large n) and "for $x \approx a$" (for x sufficiently close to a).

The former for example is introduced at the end of Chapter 2 and used right away at the beginning of Chapter 3 in the definition of the limit of a sequence. By avoiding the explicit use of an N or δ, these terms provide a very gentle introduction to limits, and suppress a lot of unnecessary details in later arguments. I've used them with classes for thirty years, and would not go back to using N and δ routinely. (When needed for an argument, they can be introduced by a phrase such as "for $n \gg 1$, say for $n > N$".)

The main places where N or δ must be made explicit are:

$\star$ the very beginning, in proving that some expression really has a given number L as its limit;

$\star$ in the proof of limit theorems involving composite functions (there are only two or three; sequential continuity is one);

$\star$ in the definition of uniform continuity and uniform convergence; these occur later in the course, however, and students can by then handle the complexity of another quantifier;

$\star$ in forming the negations needed for negative arguments.

This last is serious. The average student cannot negate "$a_n > 0$ for all n" correctly. If you teach students how to do it (see Appendix B), they start making even the simplest argument negative, a terrible habit. Therefore, the book avoids negative arguments when possible, and handles the negations informally when they become necessary (principally in Chapter 13).

Appendix A discusses negative arguments, but warns against using them as first choice. Appendix B is the forbidden fruit — it discusses both quantifiers and negation explicitly. Classes that read it early are I think asking for trouble.

Acknowledgments

I thank first of all my students over the years, whose puzzlements in class and stumbles on their written work taught me a lot.

The mathematicians at Mt. Holyoke who have used the book for a few years — Donal O'Shea, Lester Senechal, and most recently Harriet Pollatsek — have offered helpful comments, as have teachers of the course at M.I.T., especially Sigurdur Helgason, Ali Nadim, and Mary Lou Zeeman; so have the anonymous reviewers.

Comments by David Eisenbud and Norton Starr improved the early sections.

Silvio Levy and Bonnie Friedman helped me through several TeX crises, and Nicholas Romanelli was a genial production editor.

To all of these, and to the senior physics undergraduate, wherever he may be, my thanks.

Arthur Mattuck
M.I.T.
apm@math.mit.edu
Comments, errata, etc. see: http://www-math.mit.edu/~apm

1

Real Numbers and Monotone Sequences

1.1 Introduction. Real numbers.

Mathematical analysis depends on the properties of the set $\mathbb{R}$ of real numbers, so we should begin by saying something about it.

There are two familiar ways to represent real numbers. Geometrically, they may be pictured as the points on a line, once the two reference points corresponding to 0 and 1 have been picked. For computation, however, we represent a real number as an infinite decimal, consisting of an integer part, followed by infinitely many decimal places:

$$3.14159\ldots, \qquad -.033333\ldots, \qquad 101.2300000\ldots.$$

There are difficulties with decimal representation which we need to think about. The first is that two different infinite decimals can represent the same real number, for according to well-known rules, a decimal having only 9's after some place represents the same real number as a different decimal ending with all 0's (we call such decimals *finite* or *terminating*):

$$26.67999\ldots = 26.68000\ldots = 26.68\ , \qquad -99.999\ldots = -100.$$

This ambiguity is a serious inconvenience in working theoretically with decimals.

Notice that when we write a finite decimal, in mathematics the infinite string of decimal place zeros is dropped, whereas in scientific work, some zeros are retained to indicate how accurately the number has been determined.

Another difficulty with infinite decimals is that it is not immediately obvious how to calculate with them. For finite decimals there is no problem; we just follow the usual rules—add or multiply starting at the right-hand end:

$$
\begin{array}{r}
2.389 \\
+\ 2.389 \\
\hline
\ldots 78
\end{array}
\qquad\qquad
\begin{array}{r}
2.849 \\
\times\ \ .09 \\
\hline
\ldots 41
\end{array}
$$

But an infinite decimal has no right-hand end...

To get around this, instead of calculating with the infinite decimal, we use its truncations to finite decimals, viewing these as approximations to the infinite decimal. For instance, the increasing sequence of finite decimals

(1) $$\qquad\qquad 3, \quad 3.1, \quad 3.14, \quad 3.141, \quad 3.1415, \quad \ldots$$

gives ever closer approximations to the infinite decimal $\pi = 3.1415926\ldots$; we say that π is the *limit* of this sequence (a definition of "limit" will come soon).

1

To see how this allows us to calculate with infinite decimals, suppose for instance we want to calculate

$$\pi + \sqrt[3]{2} \, .$$

We write the sequences of finite decimals which approximate these two numbers:

π	is the limit of	3,	3.1,	3.14,	3.141,	3.1415,	3.14159,...;
$\sqrt[3]{2}$	is the limit of	1,	1.2,	1.25,	1.259,	1.2599,	1.25992,...;

then we add together the successive decimal approximations:

$\pi + \sqrt[3]{2}$ is the limit of   4,   4.3,   4.39,   4.400,   4.4014,   4.40151,...,

obtaining a sequence of numbers which also increases.

The decimal representation of this increase isn't as simple as it was for the sequence representing π, since as each new decimal digit is added on, the earlier ones may change. For instance, in the fourth step of the last row, the first decimal place changes from 3 to 4. Nonetheless, as we compute to more and more places, the earlier part of the decimals in this sequence ultimately doesn't change any more, and in this way we get the decimal expansion of a new number; we then define the sum $\pi + \sqrt[3]{2}$ to be this number, 4.4015137 . . . .

We can define multiplication the same way. To get $\pi \times \sqrt[3]{2}$, for example, multiply the two sequences above for these numbers, getting the sequence

(2)   3,   3.72,   3.9250,   3.954519, ...

Here too as we use more decimal places in the computation, the earlier part of the numbers in the sequence (2) ultimately stops changing, and we define the number $\pi \times \sqrt[3]{2}$ to be the limit of the sequence (2).

As the above shows, even the simplest operations with real numbers require an understanding of sequences and their limits. These appear in analysis whenever you get an answer not at once, but rather by making closer and closer approximations to it. Since they give a quick insight into some of the most important ideas in analysis, they will be our starting point, beginning with the sequences whose terms keep increasing (as in (1) and (2) above), or keep decreasing. In some ways these are simpler than other types of sequences.

Appendix A.0 contains a brief review of set notation, and also describes the most essential things about the different number systems we will be using: the integers, rational numbers, and real numbers, as well as their relation to each other. Look through it now just to make sure you know these things.

Questions 1.1

(Answers to the Questions for each section of this book can be found at the end of the corresponding chapter.)

1.   In the sequence above for $\pi + \sqrt[3]{2}$, the first decimal place of the final answer is not correct until four steps have been performed. Give an example of addition where the first decimal place of the final answer is not correct until k steps have been performed. (Here k is a given positive integer.)

1.2   Increasing sequences.

By a **sequence** of numbers, we mean an infinite list of numbers, written in a definite order so that there is a first, a second, and so on; we write it either

(3) $a_0, \ a_1, \ a_2, \ a_3, \ldots, a_n, \ldots,$ or $\{a_n\}, \ n \geq 0.$

We call a_n the **n-th term** of the sequence; often there is an expression in n for it. Some simple examples of sequences written in both forms are:

(4)   1, 1/2, 1/3, 1/4, ... $\{1/n\}, \ n \geq 1$

(5)   1, −1, 1, −1, ... $\{(-1)^n\}, \ n \geq 0$

(6)   1, 4, 9, 16, ... $\{n^2\}, \ n \geq 1$

(7)   3, 3.1, 3.14, 3.141, 3.1415, ...

For the last sequence, there is no expression in n for the n-th term. In the other cases, the range of values of n is specified, though this can be omitted if it is the standard choice $n \geq 0$. As this book progresses, we will with increasing frequency omit the braces, referring to (5) for example simply as the sequence $(-1)^n$.

Definition 1.2   We say the sequence $\{a_n\}$ is

increasing if $a_n \leq a_{n+1}$ for all n;   **strictly increasing** if $a_n < a_{n+1}$ for all n;

decreasing if $a_n \geq a_{n+1}$ for all n;   **strictly decreasing** if $a_n > a_{n+1}$ for all n.

As examples, the sequences (6) and (7) above are increasing, and even strictly increasing, while (4) is strictly decreasing. The phrase "for all n" has the meaning "for all values of n for which a_n is defined"; this is usually $n \geq 0$ or $n \geq 1$ for the sequences in this chapter.

According to the definition, the sequence $2, 2, 2, \ldots$ has to be called increasing. This may seem strange, but remember that, like Humpty-Dumpty, mathematicians can define words to mean whatever they want them to mean. Here the mathematical world itself is split over what one should call these sequences. One possibility is on the right, but our choice is on the left — we have a dislike for negative-sounding words, since they point you in the non-right direction.

$$\begin{aligned} \text{increasing} \quad &= \quad \text{non-decreasing;} \\ \text{strictly increasing} \quad &= \quad \text{increasing.} \end{aligned}$$

Questions 1.2   (Answers at end of chapter)

1. Under the natural ordering, which of the following are sequences?

    (a) all integers   (b) all integers ≥ -100 (c) all integers ≤ 0

2. Give each sequence in the form $\{a_n\}, \ n \geq \ldots$, as in (4) or (5):

    (a) $0, 1, 0, -1, 0, 1, 0, -1, \ldots$ (use $\sin x$)   (b) $1/2, \ 2/3, \ 3/4, \ldots$

3.   For each of the following sequences, tell without proof whether it is increasing (strictly?), decreasing (strictly?) or neither.

(a) $\{(3/4)^n\}$, $n \geq 0$ (b) $\{\cos(1/n)\}$, $n \geq 1$ (c) $\left\{\dfrac{n-1}{n}\right\}$, $n \geq 1$

(d) $\{n^2 - n\}$, $n \geq 0$ (e) $\{n(n-2)\}$, $n \geq 0$ (f) $\{\ln(1/n)\}$, $n \geq 1$

1.3   The limit of an increasing sequence.

We now make our earlier observations about adding and multiplying reals more precise by giving a provisional definition for the limit of an increasing sequence. (A more widely applicable definition will be given in Chapter 3.)

In the definition, we assume for definiteness that none of the a_n ends with all 9's—i.e., they are written as terminating decimals, if possible. The limit L however might appear in either form (cf. Question 1.3/3 below); we will refer to the form in which it appears as a "suitable" decimal representation for L.

Definition 1.3A   A number L, in a suitable decimal representation, is the **limit** of the increasing sequence $\{a_n\}$ if, given any integer $k > 0$, all the a_n after some place in the sequence agree with L to k decimal places.

The two notations for limit are (often the braces are omitted):

$$\lim_{n \to \infty} \{a_n\} = L, \qquad \{a_n\} \to L \text{ as } n \to \infty.$$

If such an L exists, it must be unique, since its first k decimal places (for any given k) are the same as those of all the a_n sufficiently far out in the sequence.

On the other hand, such an L need not exist; the sequence $1, 2, \ldots, n, \ldots$ has no limit, for example. Here is the key hypothesis which is needed.

Definition 1.3B   A sequence $\{a_n\}$ is said to be **bounded above** if there is a number B such that $a_n \leq B$ for all n.

Any such B is called an **upper bound** for the sequence.

For example, the sequences (4), (5), and (7) are bounded above, while (6) is not. For (4) and (5), any number ≥ 1 is an upper bound.

Theorem 1.3

A positive increasing sequence $\{a_n\}$ which is bounded above has a limit.

We cannot give a formal proof but hope the argument below will seem plausible to those who have watched odometers on long car trips. (The theorem is also true for sequences with negative terms; these will be discussed in Section 1.6.)

Write out the decimal expansions of the numbers a_n and arrange them in a list, as illustrated at right.

$$a_0 = 15.34576\ldots$$
$$a_1 = 16.26745\ldots$$
$$a_2 = 16.33654\ldots$$
$$a_3 = 16.34722\ldots$$
$$a_4 = 16.34745\ldots$$
$$a_5 = 16.34747\ldots$$
$$a_6 = 16.34748\ldots$$

Look down the list of numbers. We claim that after a while the integer part and first k decimal places of the numbers on the list no longer change. Take these unchanging values to be the corresponding places of the decimal expansion of the limit L.

To see this in more detail, look first at the integer parts of the numbers in the list. They increase (in the sense of Definition 1.2), but they cannot strictly increase infinitely often, because the sequence formed by the integer parts is bounded above. So after some index $n = n_0$, the integer part never changes.

Starting from this term a_{n_0}, continue down the list, looking now just at the first decimal place. It increases (Definition 1.2), but if it ever got beyond 9, i.e., turned into 0, the integer part would have to change, and we just agreed it doesn't. So after some later index $n_1 \geq n_0$, the first decimal place will stay constant.

Continue down from the term a_{n_1}; after a while the second decimal place will stay constant, otherwise it would get beyond 9 and the first decimal place would have to change. Continuing in this way (or using mathematical induction — see Appendix A.4), we see that ultimately the integer part and first k decimal places remain constant, and these define the first k decimal places of L. Since k was arbitrary, we have defined L. $\square$

Questions 1.3

1. Which of these sequences is bounded above? For each that is, give an upper bound. (In each case use $n \geq 0$ if it makes sense, otherwise $n \geq 1$.)

    (a) $\{(-1)^n/n\}$ (b) $\{\sqrt{n}\}$ (c) $\{\sin n\}$ (d) $\{\ln n\}$

2. Which of these increasing sequences is bounded above? For each that is, give: (i) an upper bound; (ii) the limit.

    (a) $a_n = (n-1)/n$, $n \geq 1$ (b) $a_n = \cos(1/n)$, $n \geq 1$

    (c) $a_n = 2n/(n+1)$ (d) $a_n = 1 + \frac{1}{2} + \frac{1}{4} + \ldots + \frac{1}{2^n}$

3. Apply the method given in the argument for Theorem 1.3 to find the "suitable" decimal representation (cf. Definition 1.3A) of the limit L of the increasing sequence $a_n = 1 - 1/10^n$.

4. Where in the plausibility argument are we using the fact that the a_n are written in terminating form, if possible?

1.4   Example: the number e

We saw in Section 1.1 how the notion of limit lets us define addition and multiplication of positive real numbers. But it also gives us an important and powerful method for constructing particular real numbers. This section and the next give examples. They require some serious analytic thinking and give us our first proofs.

The aim in each proof is to present an uncluttered, clear, and convincing argument based upon what most readers already know or should be willing to accept as clearly true. The first proof for example refers explicitly to the binomial theorem

$$(8) \quad (1+x)^k = 1 + kx + \ldots + \binom{k}{i} x^i + \ldots + x^n, \qquad \binom{k}{i} = \frac{k(k-1)\cdots(k-i+1)}{i!},$$

which you should know. But it also uses without comment the result

$$1 + \frac{1}{2} + \frac{1}{4} + \frac{1}{8} + \ldots + \frac{1}{2^n} < 2,$$

which is "obvious" geometrically:

and also follows from the formula for the geometric sum (taking $r = 1/2$):

$$1 + r + r^2 + \ldots + r^n = \frac{1 - r^{n+1}}{1 - r}.$$

If you didn't think of the picture and didn't remember or think of using the formula, you will feel a step has been skipped. One person's meat is another person's gristle; just keep chewing and it will ultimately go down.

As motivation for this first example, we recall the compound interest formula: invest P dollars at the annual interest rate r, with the interest compounded at equal time intervals n times a year; by the end of the year it grows to the amount

$$A_n = P\left(1 + \frac{r}{n}\right)^n.$$

Thus if we invest one dollar at the rate $r = 1$ (i.e., 100% annual interest), and we keep recalculating the amount at the end of the year, each time doubling the frequency of compounding, we get a sequence beginning with

$$
\begin{aligned}
A_1 &= 1 + 1 & &= 2 & &\textit{simple interest;} \\
A_2 &= (1 + 1/2)^2 & &= 2.25 & &\textit{compounded semiannually;} \\
A_4 &= (1 + 1/4)^4 & &\approx 2.44 & &\textit{compounded quarterly.}
\end{aligned}
$$

Folk wisdom suggests that successive doubling of the frequency should steadily increase the amount at year's end, but within bounds, since banks do manage to stay in business even when offering daily compounding. This should make the following proposition plausible. (The limit is e.)

Proposition 1.4　*The sequence*　$a_n = \left(1 + \dfrac{1}{2^n}\right)^{2^n}$　*has a limit.*

Proof.

By Theorem 1.3, it suffices to prove $\{a_n\}$ is increasing and bounded above.

To show it is increasing, if $b \neq 0$ we have $b^2 > 0$, and therefore,

$$(1 + b)^2 > 1 + 2b ;$$

raising both sides to the 2^n power, we get

$$(1 + b)^{2 \cdot 2^n} > (1 + 2b)^{2^n}.$$

If we now put $b = 1/2^{n+1}$, this last inequality becomes $a_{n+1} > a_n$.　　　$\square$

To show that a_n is bounded above, we will prove a stronger statement ("stronger" because it implies that a_n is bounded above: cf. Appendix A):

(9) $$\left(1 + \frac{1}{k}\right)^k \leq 3 \qquad \text{for any integer } k \geq 1 .$$

To see this, we have by the binomial theorem (8),

(10) $$\left(1 + \frac{1}{k}\right)^k = 1 + k\left(\frac{1}{k}\right) + \ldots + \frac{k(k-1)\cdots(k-i+1)}{i!}\left(\frac{1}{k}\right)^i + \ldots + \frac{k!}{k!}\left(\frac{1}{k}\right)^k .$$

To estimate the terms in the sum on the right, we note that

$$k(k-1)\cdots(k-i+1) \leq k^i , \qquad i = 1, \ldots, k ,$$

since there are i factors on the left, each at most k; and by similar reasoning,

(11) $$\frac{1}{i!} = \frac{1}{i} \cdot \frac{1}{i-1} \cdot \ldots \cdot \frac{1}{2} \leq \left(\frac{1}{2}\right)^{i-1}, \quad i = 2, \ldots, k .$$

Therefore, for $i = 2, \ldots, k$ (and $i = 1$ also, as you can check),

(12) $$\frac{k(k-1)\cdots(k-i+1)}{i!} \cdot \left(\frac{1}{k}\right)^i \leq \frac{1}{2^{i-1}} .$$

Using (12) to estimate the terms on the right in (10), we get, for $k \geq 2$,

(13) $$\left(1 + \frac{1}{k}\right)^k \leq 1 + 1 + \frac{1}{2} + \frac{1}{4} + \ldots + \frac{1}{2^{k-1}} ;$$
$$\leq 1 + 2 ;$$

and this is true for $k = 1$ as well.   □□

Remarks.

1. Euler was the first to encounter the number $\lim a_n$; he named it e because of its significance for the exponential function (or maybe after himself).

2. In the proof that a_n is increasing, the b could have been dispensed with, and replaced from the start with $1/2^{n+1}$. But this makes the proof harder to read, and obscures the simple algebra. Also, for greater clarity the proof is presented (as are many proofs) backwards from the natural procedure by which it would have been discovered; cf. Question 1.4/1.

3. In the proof that a_n is bounded by 3, it is easy enough to guess from the form of a_n that one should try the binomial theorem. Subsequent success then depends on a good estimation like (12), which shows the terms of the sum (10) are small. In general, this estimating lies at the very heart of analysis; it's an art which you learn by studying examples and working problems.

4. Notice how the three inequalities after line (10) as well as the two in line (13) are lined up one under the other. This makes the proof much easier to read and understand. When you write up your arguments, do the same thing: use separate lines and line up the $=$ and $\leq$ symbols, so the proof can be read as successive transformations of the two sides of the equation or inequality.

Questions 1.4

1. Write down the proof that the sequence a_n is increasing as you think you would have discovered it. (In the Answers is one possibility, with a discussion of the problems of writing it up. Read it.)

2. Define $b_n = 1 + 1/1! + 1/2! + 1/3! + \ldots + 1/n!$; prove $\{b_n\}$ has a limit (it is e). (Hint: study the second half of the proof of Prop. 1.4.)

3. In the proof that $(1 + 1/k)^k$ is bounded above, the upper estimate 3 could be improved (i.e., lowered) by using more accurate estimates for the beginning terms of the sum on the right side of (10). If one only uses the estimate (11) when $i \geq 5$, what new upper bound does this give for $(1 + 1/k)^k$?

1.5   Example: the harmonic sum and Euler's number.

We consider an increasing sequence (its terms are called "harmonic sums") which does not have a limit. This somewhat subtle fact cannot even be guessed at by experimental calculation; it is only known because it can be proved. We will give two proofs for it.

Proposition 1.5A   *Let* $a_n = 1 + \dfrac{1}{2} + \dfrac{1}{3} + \ldots + \dfrac{1}{n}$, $n \geq 1$.

The sequence $\{a_n\}$ *is strictly increasing, but not bounded above.*

Proof 1.   We will show that the terms $a_1, a_2, a_4, a_8, a_{16}, \ldots$ become arbitrarily large. This will show that $\{a_n\}$ is not bounded above.

Consider the term a_n, where $n = 2^k$. We write it out as follows, grouping the terms after the first two into groups of increasing length: $2, 4, 8, \ldots, 2^{k-1}$:

$$a_{2^k} = 1 + \frac{1}{2} + \underbrace{\frac{1}{3} + \frac{1}{4}} + \underbrace{\frac{1}{5} + \ldots + \frac{1}{8}} + \underbrace{\frac{1}{9} + \ldots + \frac{1}{16}} + \ldots + \frac{1}{2^k} \; .$$

We have

$$\frac{1}{3} + \frac{1}{4} > \frac{1}{4} + \frac{1}{4} = \frac{1}{2} \, ,$$

$$\frac{1}{5} + \ldots + \frac{1}{8} > \frac{1}{8} + \ldots + \frac{1}{8} = \frac{1}{2} \, ,$$

and so on. Thus each of the groupings has a sum $> 1/2$. Since there are $k - 1$ such groupings, in addition to the beginning terms $1 + 1/2$, we get finally

$$a_{2^k} > 1 + \frac{1}{2} + (k - 1)\left(\frac{1}{2}\right) \, ,$$

which shows that a_{2^k} becomes arbitrarily large as k increases. $\square$

The next two proofs will use geometric facts about the graph of $1/x$ and the relation between areas and definite integrals. If you are after a completely logical, rigorous presentation of analysis, you can complain that these things haven't been defined yet. This is a valid objection, but we assume a reader who knows calculus already, wants to see how the ideas of analysis are used in familiar and unfamiliar settings, and is willing to wait for a rigorous presentation of the definite integral.

Proof 2.   Draw the curve $y = 1/x$, and put in the rectangles shown, of width 1, and of height respectively 1, 1/2, 1/3, ..., 1/n .

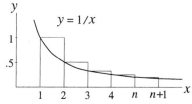

We compare the total area of the rectangles with the area under the curve between $x = 1$ and $x = n + 1$.

$$\text{total area of the rectangles} \ = \ 1 + \frac{1}{2} + \ldots + \frac{1}{n} \ = \ a_n \ ;$$

$$\text{area under curve and over } [1, n + 1] \ = \ \int_1^{n+1} \frac{dx}{x} \ = \ \ln(n + 1) \ .$$

Since their tops lie above the curve, the rectangles have greater total area:

$$a_n \ > \ \ln(n + 1) \ .$$

Since $\ln n$ increases without bound as n increases, so does a_n, and it follows that $\{a_n\}$ is not bounded above. $\qquad\square$

Though this second proof is less elementary, it has the advantage of giving more insight into the approximate size of a_n than the first proof does. The picture suggests that a_n increases at about the same rate as $\ln(n + 1)$. What can we say about the difference between them?

Proposition 1.5B   Let $b_n = 1 + \dfrac{1}{2} + \ldots + \dfrac{1}{n} \ - \ \ln(n + 1)$, $n \geq 1$.

Then $\{b_n\}$ has a limit (denoted by γ and called "Euler's number").

Proof.   It is sufficient to show $\{b_n\}$ is increasing and bounded above.

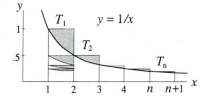

Referring to the picture at the right and the ideas of Proof 2 above, and letting T_i denote the area of the i-th shaded curvilinear triangle in the picture, we have

$$b_n = \text{ area of rectangles } - \text{ area under curve}$$
$$= \ T_1 + \ldots + T_n \ .$$

The sequence $\{b_n\}$ is increasing, since $b_{n+1} = b_n + T_{n+1}$.

The sequence $\{b_n\}$ has 1 as an upper bound, since all of the "triangles" can be moved horizontally without overlapping into the rectangle of area 1 lying over the interval $1 \leq x \leq 2$. $\qquad\square$

Remarks.   How big is Euler's number $\gamma = \lim b_n$? From the proof,

$$\gamma \ = \ T_1 + T_2 + T_3 + \ldots \ = \ \text{total area of the triangles.}$$

Each of the triangles has a curved hypotenuse; if this were replaced by a straight side, the total area of the resulting triangles would add up to exactly half the area inside the rectangle over $[1, 2]$, i.e., to 1/2. This shows that $1/2 < \gamma < 1$, but the picture suggests it is much closer to 1/2.

It turns out that $\gamma = .577$, to three decimal places.

One of the mysteries about γ is whether or not it can be expressed in terms of other known numbers, like e, π, $\ln 2$ or $\sin 3$. It is also not known whether γ is an algebraic number, i.e., a zero of some polynomial with integer coefficients. These problems have been open for over 200 years.

We will meet γ again toward the end of this book.

Questions 1.5

1. Estimate the size of $1 + 1/2 + 1/3 + \ldots + 1/999$, by using the ideas of Proof 2. The approximation $\ln 10 \approx 2.3$ will be helpful (and is useful enough to be worth memorizing).

2. To see that Proposition 1.5A is not an experimental fact, suppose a computer adds up in a second 1,000,000 terms of the harmonic sum $\sum 1/k$. Estimating there are roughly 100,000,000 seconds/year, what will be the approximate value of a_n after one year of calculation? After two years of calculation?

3. In the proof of Proposition 1.5B, why is the phrase "without overlapping" included in the next-to-last line?

If the curve $1/x$ were replaced by the graph of some other function, what property should its graph have to guarantee a similar argument can be made?

1.6   Decreasing sequences. The Completeness Property.

To conclude, we gain a little more flexibility by first extending the notion of limit to decreasing sequences. The definition is the same as the one in Section 1.3; as before we assume that none of the a_n end with all 9's.

Definition 1.6A   A number L, in a suitable decimal representation, is the **limit** of the decreasing sequence $\{a_n\}$ if, given any integer $k > 0$, all the a_n after some place in the sequence agree with L to k decimal places.

Definition 1.6B   A sequence $\{a_n\}$ is said to be **bounded below** if there is a number C such that $a_n \geq C$ for all n.

Any such B is called a **lower bound** for the sequence.

Theorem 1.6   *A positive decreasing sequence has a limit.*

The plausibility argument is similar to the one we gave before and is omitted (think of the odometer on a car running in reverse). Note that since all the terms are positive, the sequence is bounded below by zero.

Theorems 1.3 and 1.6 can be extended to include sequences all or some of whose terms are not positive. Consider for example an increasing sequence $\{a_n\}$ which has a term ≤ 0. There are three cases:

a) The sequence also contains a positive term a_N. In this case, all the terms after a_N will be positive, and the argument for Theorem 1.3 applies.

b) All the terms are negative. In this case, just change the sign of all the terms: the sequence $\{-a_n\}$ will be a positive decreasing sequence, so it will have a limit L by Theorem 1.6; then $-L$ is the limit of $\{a_n\}$. For, since the decimal places of L agree with those of the $\{-a_n\}$, the places of $-L$ agree with those of the $\{a_n\}$.

c) Neither of the above. Left for you to figure out (see the Questions).   □

Decreasing sequences with a non-positive term can be handled similarly.

We would now like to combine all these cases into a single concise statement about the existence of a limit; it will be one of the cornerstones of our work in this book. For this we need two more words.

Definition 1.6C   A sequence $\{a_n\}$ is **bounded** if it is bounded above and bounded below; i.e., there are constants B and C such that

$$C \leq a_n \leq B \qquad \text{for all } n \, .$$

Notice that an increasing sequence is always bounded below (by its first term), so that for an increasing sequence it makes no difference whether we say it is bounded or bounded above. Similarly, saying a decreasing sequence is bounded below is the same as saying it is bounded.

Definition 1.6D   A sequence is **monotone** if it is increasing for all n, or decreasing for all n.

> Humpty-Dumpty strikes again: in a rational world, "monotone" ought to be reserved for a sequence like $2, 2, 2, \ldots$, which is both increasing *and* decreasing. When you need a verbal macro, either you give a new meaning to an old word, or you coin a new one, like "scofflaw" for those who run speakeasies or red lights. Mathematicians do both.

We can summarize Theorems 1.3 and 1.6, allowing also non-positive terms, by the following statement; it is one form of what is called the *Completeness Property* of the real number system $\mathbb{R}$.

Completeness Property.   *A bounded monotone sequence has a limit.*

The word "completeness" is used because the property says that the real line is "complete"— it has no holes. The early Greeks thought all numbers were rational; their line contained only points corresponding to the rational numbers. The discovery by Pythagoras that $\sqrt{2}$ is irrational (cf. Appendix A.2) was a mathematical earthquake; it meant that on the Greek line, there would be no limit for the sequence of points 1, 1.4, 1.41, 1.414, ..., since the line had no point representing the number $\sqrt{2}$. Passing from the rationals to the reals can be thought of as filling in the holes in the pre-Pythagorean line—making it complete, in other words.

Our definition of the limit of a sequence is reasonably intuitive, but has two defects. It works only for monotone sequences, and it is wedded too closely to decimal notation. We shall free it from both limitations in Chapter 3, but to do this, we need some ideas of estimation and approximation which are fundamental to all of analysis. So we turn to these in the next chapter.

Questions 1.6

1. For each of the a_n below, tell if the sequence $\{a_n\}$ is bounded or monotone; if both, give its limit. (Use $n = 0$ as the starting point, or $n = 1$ if a_0 would be undefined.)

(a) $1/n$ (b) $\sin 1/n$ (c) $\sin 4/n$ (d) $(-1)^n$ (e) $\ln(1/n)$

2. What is case (c) in the extension of the limit definition to increasing sequences with some non-positive terms? How would the argument for Theorem 1.3 go for this case?

Exercises   (The exercises go with the indicated section of the chapter.)

1.2

1. For each of the a_n below, tell if the sequence $\{a_n\}$, $n \geq 1$, is increasing (strictly?), decreasing (strictly?), or neither; show reasoning.

(If simple inspection fails, try considering the difference $a_{n+1} - a_n$, or the ratio a_{n+1}/a_n, or relate the sequence to the values of a function $f(x)$ known to be increasing or decreasing.)

(a) $1 - \frac{1}{2} + \frac{1}{3} - \ldots + (-1)^{n-1}\frac{1}{n}$ (b) $n/(n+1)$

(c) $\sum_1^n \sin^2 k$ (d) $\sum_1^n \sin k$

(e) $\sin(1/n)$ (f) $\sqrt{1 + 1/n^2}$

1.3

1. Show increasing; find an upper bound, if it exists; give the limit if you can.

(a) $\dfrac{\sqrt{n^2 - 1}}{n}$ (b) $\left(2 - \dfrac{1}{n}\right)\left(2 + \dfrac{1}{n}\right)$ (c) $\sum_0^n \sin^2 k\pi$ (d) $\sum_0^n \sin^2 k\pi/2$

2. Let $a_n = \sum_1^n 1/10^i$. Apply the method in Theorem 1.3 to find the limit L in its "suitable" decimal form; also express it as a rational number a/b.

3. Let $\{a_n\}$ be increasing, and $\lim_{n\to\infty} a_n = L$, where L is a terminating decimal. Show that if $\{a_n\}$ is *strictly* increasing, the "suitable" decimal representation for L in Definition 1.3A is always the non-terminating form (ending with all 9's).

4. Read Section A.4 on mathematical induction (just the first page will be enough for now) and finish the argument for Theorem 1.3 by using induction.

1.4

1. Consider the sequence $\{a_n\}$, where
$$a_n = 1 + \frac{1}{1 \cdot 3} + \frac{1}{1 \cdot 3 \cdot 5} + \frac{1}{1 \cdot 3 \cdot 5 \cdot 7} + \ldots + \frac{1}{1 \cdot 3 \cdot \ldots \cdot (2n-1)} \,.$$
Decide whether $\{a_n\}$ is bounded above or not, and prove your answer is correct. (Hint: cf. Question 1.4/2 .)

2. Prove the sequence $a_n = n^n/n!$ is

(a) increasing;   (b) not bounded above (show $a_n > n$).

1.5

1. (a) Let $a_n = 1 + \dfrac{1}{1 \cdot 2} + \dfrac{1}{2 \cdot 3} + \ldots + \dfrac{1}{n(n+1)}$. Prove $\{a_n\}$ is bounded above. (Hint: $\dfrac{1}{2 \cdot 3} = \dfrac{1}{2} - \dfrac{1}{3}$.)

(b) Let $b_n = 1 + \dfrac{1}{4} + \dfrac{1}{9} + \ldots + \dfrac{1}{n^2}$. Prove $\{b_n\}$ is bounded above by comparing it to $\{a_n\}$. What upper bound does this give for $\{b_n\}$?

2. Prove the sequence $\{b_n\}$ of the preceding exercise is bounded, by expressing b_n as the area of a set of rectangles and comparing this with the area under a suitable curve. What upper bound does this give for $\{b_n\}$?

In fact, it is known that the limit of $\{b_n\}$ is $\pi^2/6$; how close is this to the bounds you got in this exercise and the preceding one?

3. Let $a_n = 1 + 1/\sqrt{2} + 1/\sqrt{3} + \ldots + 1/\sqrt{n}$. Prove $\{a_n\}$ is unbounded.

4. Let $b_n = a_n - 2\sqrt{n+1}$, where a_n is as in the previous exercise. Prove $\{b_n\}$ has a limit. (See Proposition 1.5B.)

1.6

1. Show the sequence $a_n = (n+1)/(n-1)$ is strictly decreasing and bounded below, and give its limit.

2. Show that $a_n = n/2^n$, $n \geq 1$, is a monotone sequence.

3. Define a sequence $\{a_n\}$ by: $a_{n+1} = 2a_n^2$; assume $0 < a_0 < 1/2$. Prove that a_n is strictly decreasing; is it bounded below?

4. Prove the sequence $a_n = \dfrac{1 \cdot 3 \cdots (2n-1)}{2 \cdot 4 \cdots 2n}$ has a limit.

Problems

1-1 Define a sequence by

$$a_{n+1} = \frac{a_n + 1}{2}, \quad n \geq 0; \quad a_0 \text{ arbitrary} .$$

(a) Prove that if $a_0 \leq 1$, the sequence is increasing and bounded above, and determine (without proof) its limit.

(b) Consider analogously the case $a_0 \geq 1$.

(c) Interpret the sequence geometrically as points on a line; this should make (a) and (b) intuitive.

1-2 Prove that $a_n = \left(1 + \frac{1}{2}\right)\left(1 + \frac{1}{3}\right) \cdots \left(1 + \frac{1}{n}\right)$ is strictly increasing, and not bounded above.

1-3 Prove that $a_n = \dfrac{1 \cdot 3 \cdot \cdots \cdot (2n+1)}{2 \cdot 4 \cdots (2n)}$ is strictly increasing and not bounded above.

1-4 Let A_n denote the area of the regular 2^n-sided polygon inscribed in a unit circle. (Assume $n \geq 2$.) Explain geometrically why the sequence $\{A_n\}$ is monotone and bounded above, and give its limit. Then use trigonometry to get an explicit expression for A_n, and prove the same facts analytically, using anything you know from calculus.

Answers to Questions

1.1

1. Example: $.099\ldots9 + .000\ldots1$ (both numbers have k decimal places). The sum sequence is $.0, .09, .099, \ldots, .099\ldots9$ ($k-1$ places), $.100\ldots0$ (k places), so the correct first decimal place of the sum—namely, 1—appears for the first time only in the k-th term of the sum sequence.

1.2

1. Only (b) is a sequence.

2. (a) $\{\sin n\pi/2\}$, $n \geq 0$ (b) $\left\{\dfrac{n}{n+1}\right\}$, $n \geq 1$, or $\left\{\dfrac{n+1}{n+2}\right\}$, $n \geq 0$.

3. (a) strictly decreasing (b) strictly increasing (c) strictly increasing

(d) increasing (e) neither (f) strictly decreasing

1.3

1. (a) 1/2 or anything larger;   (c) 1 or anything larger;

(b) and (d) are not bounded above.

2. The limits are: (a) 1 (b) 1 (c) 2 (d) 2.

These or anything larger are upper bounds for the respective sequences.

3. The sequence is $.9, .99, .999, \ldots$; following the method in Theorem 1.3 gives the limit $L = .999\ldots$, i.e., the non-terminating form of 1.

4. The argument says at various points: the integer parts increase; after a certain point, the first decimal place increases; after a later point, the second decimal place increases, and so on.

This would not necessarily be true if we did not require a uniform choice (terminating form) for the decimal expansions of the a_n. For instance, the constant sequence $1.000, .999\ldots, 1.000, \ldots$, whose representation alternates between the terminating and non-terminating forms of 1, is increasing and bounded, yet the integer part of its terms is not increasing.

Similarly, in the constant sequence $1.300, 1.299\ldots, 1.300$, the integer part is unchanging, but the sequence formed by the first decimal places $3, 2, 3, \ldots$ is not an increasing sequence.

1.4

1. The following is a fairly common method, both of discovering the argument that a_n is increasing, and writing it up.

$$a_{n+1} \geq a_n ; \tag{1}$$

$$\left(1 + \frac{1}{2^{n+1}}\right)^{2^{n+1}} \geq \left(1 + \frac{1}{2^n}\right)^{2^n} ; \tag{2}$$

$$\left(1 + \frac{1}{2^{n+1}}\right)^2 \geq \left(1 + \frac{1}{2^n}\right) ; \tag{3}$$

$$1 + \frac{2}{2^{n+1}} + \frac{1}{2^{2(n+1)}} \geq 1 + \frac{1}{2^n} ; \tag{4}$$

$$\frac{1}{2^{2(n+1)}} \geq 0 . \tag{5}$$

At this point (or at step (4)), the problem is considered solved.

The meaning of this falling-domino argument is presumably:

(1) is true if (2) is true: (2) $\Rightarrow$ (1);
(2) is true if (3) is true: (3) $\Rightarrow$ (2);
(3) is true if (4) is true: (4) $\Rightarrow$ (3);
(4) is true if (5) is true: (5) $\Rightarrow$ (4);
(5) is true. Therefore (1) is true.

The argument is written backwards. Students often try to express this by writing the first four inequalities using an invented symbol, such as $\geq?$. A formal argument would usually write the successive inequalities in the opposite order, without any ? symbols. But written in this way, the argument is unmotivated, and can be difficult to follow, since the readers can't tell where they are headed.

A reasonable compromise might be to use some symbol like $\geq?$, with the understanding that the next line then *must* imply the line containing the $\geq?$.

2. By Theorem 1.3 it suffices to show that $\{b_n\}$ is increasing and bounded above. (The limit is again e.)

It is strictly increasing since $b_{n+1} = b_n + 1/(n+1)!$.

It is bounded above since

$$b_n = 1 + 1/1! + 1/2! + 1/3! + \ldots$$
$$\leq 1 + 1 + 1/2 + 1/2^2 + 1/2^3 + \ldots, \qquad \text{by (11)};$$
$$\leq 3 \ .$$

3. Using the estimate (11) only starting with the fifth term, we have

$$\left(1 + \frac{1}{k}\right)^k = 1 + k\left(\frac{1}{k}\right) + \ldots + \frac{k(k-1)\cdots(k-i+1)}{i!}\left(\frac{1}{k}\right)^i + \ldots \ ;$$
$$\leq \ 1 + 1 + \frac{1}{2!} + \frac{1}{3!} + \frac{1}{2^3} + \frac{1}{2^4} + \ldots \ ;$$
$$\leq \frac{8}{3} + \frac{1}{2^3}\left(1 + \frac{1}{2} + \frac{1}{2^2} + \ldots\right);$$
$$\leq 2.67 + .25 = 2.92 \ .$$

1.5

1. $0 < a_{999} - \ln(1000) < 1$ by proofs 1.5A[2], 1.5B; $\ln(1000) \approx 6.9$.

2. $\ln(10^{14}) \approx 14(2.3) \approx 32.2$, after one year;

$\ln(2 \cdot 10^{14}) = \ln 2 + \ln 10^{14} \approx .7 + 32.2 \approx 32.9$, after two years.

3. The statement (total area of triangles) $\leq$ (area of rectangle) can only be made if the triangles do not overlap when fitted inside the rectangle.

If the function is positive and strictly increasing or strictly decreasing, the triangles will not overlap.

1.6

1. (a) both (decreasing); 0   (b) both (decreasing); 0

(c) bounded, not monotone   (d) bounded, not monotone

(e) monotone (decreasing), not bounded

2. Case (c): The increasing sequence contains no positive terms, but not all the terms are negative.

The sequence then contains the term 0, but no positive terms. Since it is increasing, all terms after 0 must also be 0, so the limit exists and is 0.

2

Estimations and Approximations

2.1 Introduction. Inequalities.

Our work with bounded sequences $\{a_n\}$ in Chapter 1 — the sequences whose limits were e and γ and the sequence of harmonic sums — depended in each case on estimating the size of a_n. In general, estimation is not formal or routine; it is an art, but one that can be learned by studying and imitating examples. It is at the core of analysis.

Two simple tools form the language of estimations: *inequalities* for making comparisons and *absolute values* for measuring size and distance. We begin in this section with inequalities.

First of all, in comparing numbers with 0 we will say:

a is *positive* if $a > 0$, a is *non-negative* if $a \geq 0$,

a is *negative* if $a < 0$, a is *non-positive* if $a \leq 0$.

The terms "non-negative" and "non-positive" point in the wrong direction and are awkward; fortunately, they are not so often needed.

INEQUALITY LAWS   The algebraic laws for working with inequalities are familiar, but we flag several which even good mathematicians sometimes trip over. We will use $<$ in the statements; the laws using $\leq$ are analogous.

Addition   You can only add inequalities in the same direction:

$$
\begin{array}{l}
a < b \\
\underline{c < d} \\
a + c < b + d
\end{array}
\qquad\qquad
\begin{array}{l}
a < b \\
\underline{c > d} \\
a + c \ ? \ b + d
\end{array}
$$

Subtraction   Don't even think of doing this; see discussion below.

Multiplication   This is legal provided that the inequalities are in the same direction and the numbers are *positive*:

$$a < b, \ c < d \ \Rightarrow \ ac < bd, \quad \text{if } a, b, c, d > 0.$$

Sign-change law   Changing signs reverses an inequality:

$$a < b \ \Rightarrow \ -a > -b; \qquad a < b \Rightarrow ka > kb, \quad \text{if } k < 0.$$

Reciprocal law   The inequality reverses, if both numbers are positive:

$$a < b \ \Rightarrow \ 1/a > 1/b, \quad \text{if } a, \ b > 0.$$

Remarks.

All of the preceding are common sources of error, but the worst is probably the sign-change law. The reason is that changing all signs in an *equality* (to get rid of excess minus signs) is legal, and so people tend to apply the same rule to inequalities, failing to reverse the inequality sign. Or they multiply both sides of an inequality by $\sin x$, failing to realize that the direction of the resulting inequality depends on whether $\sin x > 0$ or $\sin x < 0$.

After the above, the most common sin is trying to subtract inequalities in the same direction. To see why you can't do this, change the subtraction to addition, by using the sign-change law:

$$(a < b) - (c < d) \quad \text{is the same as} \quad (a < b) + (-c > -d):$$

you're trying to add inequalities in opposite directions.

Questions 2.1   (Answers at end of chapter)

1. Give a numerical example to show you cannot subtract inequalities in the same direction: $a < b$ and $c < d \Rightarrow a - c < b - d$ is false.

2. Show the reciprocal law is also true if both numbers are negative.
(Call them $-a$ and $-b$, and use the sign-change law.)

3. Prove if true; if false, give a counterexample, and if possible amend the statement so it becomes true and then prove it. (For some proofs, try proving the contrapositive instead. See Appendix A.3 for "counterexample", A.2 for "contrapositive".)

  (i) $a < b \Rightarrow \sqrt{a} < \sqrt{b}$ (assume the roots exist);

  (ii) $a^2 < b^2 \Rightarrow a < b$;

  (iii) $\{a_n\}$ increasing, $a_n \neq 0$ for all $n \Rightarrow \{1/a_n\}$ is decreasing.

4. If $a_n > 0$, under what hypotheses will $\{1/a_n\}$ be bounded above?

2.2   Estimations.

One of the major uses of inequalities is to make estimations. These have a language of their own that you need to get used to.

Definition 2.2   If c is a number we are estimating, and $K < c < M$, we say

  K is a **lower estimate** for c ; M is an **upper estimate** for c .

Sometimes the respective phrases *lower bound* or *upper bound* for c are used.

If two sets of upper and lower estimates satisfy the inequalities

$$K < K' < c < M' < M ,$$

we say K', M' are **stronger** or **sharper** estimates for c , while K, M are **weaker** estimates.

Estimates are obtained in many ways; the inequality laws of the preceding section usually play an important role. Here are some examples to illustrate.

Example 2.2A   Give upper and lower estimates for $\dfrac{1}{a^4 + 3a^2 + 1}$.

Solution.   We are not told what a is, so our estimates will have to be valid no matter what a is, i.e., valid for all a.

We first estimate the denominator. Since $a^2 \geq 0$, the polynomial has its smallest value 1 when $a = 0$, and it has no upper bound since a can be arbitrarily large. We express this symbolically by the inequalities

$$1 \leq a^4 + 3a^2 + 1 < \infty .$$

Using the reciprocal law for inequalities (and writing symbolically $1/\infty = 0$), the inequalities reverse and we get the upper and lower estimates

$$0 < \frac{1}{a^4 + 3a^2 + 1} \leq 1 .$$

These are the sharpest possible estimates valid for all a : the upper bound 1 is attained when $a = 0$; the lower bound is 0 since the fraction can be made arbitrarily close to 0 by taking a sufficiently large.

Example 2.2B   Give upper and lower estimates for $\dfrac{1 + \sin^2 n}{1 + \cos^2 n}$, for $n \geq 0$.

Solution.   Both the numerator and the denominator lie between 1 and 2.

To make the quotient biggest, we use the *upper* bound for the top and the *lower* bound for the bottom; to make it smallest, we use the *lower* bound for the top and the *upper* bound for the bottom. Thus we get the estimates

$$\frac{1}{2} \leq \frac{1 + \sin^2 n}{1 + \cos^2 n} \leq 2 , \quad \text{for } n \geq 0 .$$

The above argument is very informal, but this is the way most people mentally answer the question. You should also be able to give a formal argument using the inequality laws: see Question 1 below.

Example 2.2C   By interpreting the integral as the area under $1/x$ and over the interval $[1, 2]$, estimate $\ln 2 = \displaystyle\int_1^2 \frac{dx}{x}$.

Solution.   Referring to the picture, since the region under the curve lies between the rectangle (area $1/2$) and the trapezoid (area $3/4$), we get

$$.50 < \ln 2 < .75 .$$

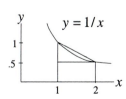

This is pretty crude; can we sharpen the estimates?

Referring to the new picture, we divide the interval in half, use two trapezoids for the upper bound and a trapezoid and rectangle for the lower bound. (Note that the slope at $x = 1$ is -1.) Calculating the area of the trapezoids and rectangle gives

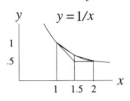

$$5/8 \; < \; \ln 2 \; < \; 17/24 \; ;$$
$$.63 \; < \; \ln 2 \; < \; .71 \; .$$

Since $\ln 2 \approx .69$, the upper estimate is pretty good, while the lower one is weaker; the picture suggests this, in fact.

Questions 2.2

1. In example 2.2B, estimates were given for a fraction of the form A/B; the argument was rather informal however. Assuming the estimates $1 \leq A \leq 2$ and $1 \leq B \leq 2$, complete the argument more formally, using the laws in 2.1.

2. Give upper and lower estimates for $\dfrac{4 + \cos na}{3 + \sin na}$, valid for all a and all integers $n > 0$. Then make a careful argument, using the inequality laws in 2.1 .

3. In example 2.2C, if the lower estimate is made by halving the interval and using two rectangles, instead of the trapezoid and rectangle, is the resulting estimate stronger or weaker than the one given?

2.3   Proving boundedness.

Rather than an upper and lower estimate, often you want an estimate just in one direction. This was the case with our work with sequences in Sections 1.4 and 1.5. Using the language of estimations, we summarize the principles used there; for simplicity, we assume the sequences are non-negative: $a_n \geq 0$.

To show $\{a_n\}$ is bounded above, get one upper estimate: $a_n \leq B$, for all n;

to show $\{a_n\}$ is not bounded above, get a lower estimate for each term: $a_n \geq B_n$, such that B_n tends to ∞ as $n \to \infty$.

The method for showing unboundedness is rather restricted, but it's good enough most of the time. We will give a precise definition of "tend to ∞" in the next chapter; for now we will use it intuitively. The examples we gave in Sections 1.4 and 1.5 illustrate these principles.

Example 1.4 (9) $b_k = \left(1 + \dfrac{1}{k}\right)^k$.

We showed $\{b_k\}$ bounded above by the upper estimate: $b_k < 3$ for all k.

Example 1.5A $a_n = 1 + \dfrac{1}{2} + \dfrac{1}{3} + \ldots + \dfrac{1}{n}$.

Here we showed $\{a_n\}$ was not bounded above, by proving the lower estimate $a_n > \ln(n + 1)$, and relying on the known fact that $\ln n$ tends to ∞ as $n \to \infty$.

These examples show that to estimate, you may have to first decide what kind of estimate you are looking for: upper or lower. If you aren't told this in advance, deciding may not be so easy. For example, by using the rectangles in Example 1.5A a bit differently, we could easily have obtained the *upper* estimate

$$a_n < 1 + \ln(n + 1)$$

for the terms of the sequence. But this would have been useless for showing the sequence was unbounded, since the estimate goes in the wrong direction. Nor does it show the sequence is bounded, since the estimate $1 + \ln(n + 1)$ depends on n and tends to ∞, instead of being the same for all the terms.

As another example of the difficulty in deciding which type of estimate to look for, consider the sequence formed like the one above in Example 1.5, but using only the prime numbers in the denominators (here p_n denotes the n-th prime):

$$a_n = 1/2 + 1/3 + 1/5 + 1/7 + 1/11 + \ldots + 1/p_n \; .$$

Should we look for an upper estimate that will show it is bounded, or a lower estimate that clearly tends to infinity? Anyone who can answer this without having studied number theory is a mathematical star of zeroth magnitude.

Questions 2.3

1. Let $a_n = 1 + \frac{1}{4} + \frac{1}{9} + \ldots + \frac{1}{n^2}$; is it bounded or unbounded; i.e., should one look for an upper or a lower estimate for the a_n? (cf. Example 1.5A)

2.4   Absolute values. Estimating size.

We turn now to the other basic tool of estimation, the absolute value. As with inequalities, its use requires both care and knowing some rules.

Definition 2.4   Absolute value.   Define $|a| = \begin{cases} a, & \text{if } a \geq 0; \\ -a, & \text{if } a < 0. \end{cases}$

The formal definition has been confusing students for centuries. Simply put: $|\;|$ erases the $-$ sign if a number is negative; it leaves positive numbers alone.

Here are two good ways to think about absolute value.

Absolute value measures magnitude: $|a|$ is the size of a.

Negative numbers can be big too: a million dollar loss is a big sum for a small business. One says $-N$ is "negatively large" or "large in absolute value".

Absolute value measures distance: $|a - b|$ is the distance between a and b.

This is often used to describe intervals on the x-axis. For instance the interval $[2, 4]$ can be desribed as $\{x : |x - 3| \leq 1\}$, i.e., the set of points whose distance from the point 3 is at most 1.

We give two simple but common and important uses of the absolute value.

(1) $|a|$ *is guaranteed to be non-negative.*

For instance, $a \leq b \Rightarrow |c|a \leq |c|b$ is true regardless of the sign of c.

The absolute value is also an efficient way to give **symmetric bounds**: the inequality on the left below says the same thing as the two on the right,

(2) $|a| \leq M \quad \Leftrightarrow \quad -M \leq a \leq M$,

and it is often more convenient to use. If the bounds are not symmetric to start with, they can be made so by using the following important fact:

(2a) $K \leq a \leq L \Rightarrow |a| \leq M$,   where $M = \max(|K|, |L|)$.

Proof of (2a).   Regardless of whether a, K, and L are positive or negative,
$$-M \leq -|K| \leq K \leq a \leq L \leq |L| \leq M ;$$
now apply (2) to the two ends.   □

ABSOLUTE VALUE LAWS   In working with absolute values we will make frequent use of the simple

(3)   **multiplication law** $|ab| = |a||b|;$ $|a/b| = |a|/|b|$, if $b \neq 0$;

and the law connecting absolute value with sums:

(4)   **triangle inequality** $|a + b| \leq |a| + |b|$;

(4a)   **extended triangle inequality** $|a_1 + \ldots + a_n| \leq |a_1| + \ldots + |a_n|$.

Proof of (4). We use (2) in the direction $\Rightarrow$ to transform the inequalities $|a| \leq |a|$ and $|b| \leq |b|$, and then we add them; the resulting inequalities are:
$$-|a| \leq a \leq |a| ;$$
$$-|b| \leq b \leq |b| ;$$
$$-|a| - |b| \leq a + b \leq |a| + |b| ;$$
and according to (2) again, this last line is the same as the inequality (4).   □

The extended triangle inequality (4a) follows by applying (4) repeatedly (i.e., using mathematical induction, Appendix A.4). We leave it for the Exercises. □□

The triangle inequality is used to get an *upper* estimate for a sum. Two equivalent but less familiar forms of (4) are sometimes useful for getting *lower* estimates of sums and differences:

(4b)   **difference forms of the triangle inequality** $|a - b| \geq |a| - |b|$; $|a + b| \geq |a| - |b|$.

The first form follows from (4) by changing a to $a - b$; the second form is just the first with b changed to $-b$. In using either form, a should be the larger-sized number (i.e., $|a| > |b|$), otherwise the inequality gives no information.

Most of the estimates in this book and in introductory analysis are estimates of *size*: they will use the absolute value laws combined with the inequality laws. In the hope of forestalling the two most common errors:

To estimate size, you have to use $|\ |$, and the estimations have to be in the right direction:

> *to show a is small in size, show $|a| <$ a small number;*
>
> *to show a is large in size, show $|a| >$ a large number.*

This hardly sounds like the sort of advice one would pay money for, yet one often sees on student papers inequalities obtained at great effort, but going in the wrong direction and therefore useless for estimating.

Even more common is the failure, largely through sloppiness, to use $|\ |$ in size estimations. If you want to show a_n is small, it does no good to show that

$$a_n < 1/n$$

unless you know in advance that a_n is non-negative, for otherwise a_n could be negatively large and still satisfy the above inequality; instead, show

$$|a_n| < 1/n .$$

Example 2.4A   In Fourier analysis, one uses trigonometric sums of the form
$$S_n = c_1 \cos t + c_2 \cos 2t + \ldots + c_n \cos nt .$$
If $c_i = 1/2^i$, give an upper estimate for the size of S_n.

Solution.   This is a typical use of the extended triangle inequality (4a), whose *raison d'etre* is to give upper estimates for sums. Note first that
$$|\cos A| \le 1 \quad \text{for all } A ;$$
thus by the extended triangle inequality and the multiplication law (3),
$$|S_n| \le |c_1||\cos t| + |c_2||\cos 2t| + \ldots + |c_n||\cos nt| ;$$
$$\le \frac{1}{2} \cdot 1 + \frac{1}{4} \cdot 1 + \ldots + \frac{1}{2^n} \cdot 1 \quad < \quad 1 .$$
Thus $|S_n| < 1$ for all n.   □

Example 2.4B   If $|a| \ge 2$ and $|b| \le 1/2$, estimate $|a + b|$ from below.

Solution   Since we want a lower estimate for the size of a sum, the triangle inequality in the difference form (4b) is the thing to use. It gives
$$|a + b| \ge |a| - |b| \ge 3/2 ,$$
since the first term is at least 2, and we subtract at most 1/2. (One can also argue more formally using the inequality laws; see question 4 below.)   □

Important Warning 2.4   You can't use the triangle inequality mechanically. For example, the above reasoning gives a silly lower estimate for $|\sin n - \cos n|$:
$$|\sin n - \cos n| \ge |\sin n| - |\cos n| \ge 0 - 1 = -1 .$$

As another example (one of the most common misuses, actually), if a and b are close, you won't see this using the triangle inequality:

$$|\pi - 3.14| \leq |\pi| + |3.14| \leq 6.3 .$$

Even worse, many try to do this estimation by using

$$|a - b| \leq |a| - |b|, \qquad\qquad \textit{(the fool's inequality)},$$

which is obvious nonsense (what if $|a| < |b|$?)—if you highlight your text, highlight that one in black.

As another example of size estimation, we show that $|\ |$ lets us prove boundedness by using just one inequality, instead of two. This proposition, though simple, is important; always formulate boundedness this way, unless there's some obvious reason not to (for instance, if a_n are known to be positive).

Proposition 2.4

(5) $\{a_n\}$ is bounded $\Leftrightarrow$ there is a B such that $|a_n| \leq B$ for all n .

Proof. $\Leftarrow$:   Using (2), $|a_n| \leq B \Rightarrow -B \leq a_n \leq B$, so $\{a_n\}$ is bounded.

$\Rightarrow$:   Using (2a), $K \leq a_n \leq L$ for all $n \Rightarrow |a_n| \leq \max(|K|, |L|)$ for all n. $\square$

Questions 2.4

1. Proof or counterexample: (i) $a < b \Rightarrow |a| < |b|$ (ii) $|a| < |b| \Rightarrow a < b$.

2. (a) Without looking it up, write the proof of the triangle inequality (4).
    (b) Using (4), prove the extended triangle inequality (4a) when $n = 3$.

3. In example 2.4A, estimate $|S_n|$ if $c_i = 1/i!$

4. Write out the end of example 2.4B formally, using the inequality laws.

5. Express the boundedness of the sequence $\{\frac{1}{n} \cos n\pi\}$, $n \geq 1$ by
    (a) giving upper and lower bounds (the sharpest possible);
    (b) one inequality (the sharpest possible), as in (5).

(Remember that to show boundedness, the bounds must be independent of n.)

2.5   Approximations.

A common way of estimating some quantity is to say that it is approximately equal to some number. In scientific work one often writes $a \approx b$ to mean that a and b are approximately equal. This statement has no exact mathematical meaning however, since it says that a and b are close, but not how close. After all, a short distance for a snail might seem an intolerable distance to an amoeba. We must be more precise.

Suppose that a and b are within ϵ of each other, where ϵ is some positive number (invariably thought of as small). The standard way of writing this says that the distance between a and b is less than ϵ:

(6) $|a - b| < \epsilon .$

While we shall use this notation frequently (and your teacher may insist that you use it always), we shall also use the non-standard notation

$$(6a) \qquad\qquad\qquad a \underset{\epsilon}{\approx} b$$

to express the same meaning as (6), when we particularly want to focus on the approximation point of view.

> Often a mathematical argument which depends on making different sorts of approximations can be worked out roughly by using $\approx$ everywhere, and then it can be made precise by putting in the ϵ's underneath. Lazy folk can even skip this last step and still feel they understand the argument more or less.

As an example of approximation, let $a > 0$ be written as an infinite decimal (if it terminates, assume it ends with all 0's, rather than all 9's):

$$a \;=\; a_0 \,.\, a_1 a_2 a_3 \ldots a_n \ldots; \qquad \begin{cases} a_0 = \text{integer part,} \\ a_i = i\text{-th decimal place,} \end{cases} \qquad a > 0 .$$

We define its *n-th truncation* to be:

$$a^{(n)} \;=\; a_0 \,.\, a_1 a_2 a_3 \ldots a_n\, 0\,0\ldots, \qquad a_i = 0,\; i > n.$$

Like any terminating decimal, $a^{(n)}$ is a rational number (e.g., $3.14 = 314/100$.) Thus it gives an approximation to a by a rational number; how close are they?

$$(7) \qquad\qquad\qquad a \underset{\epsilon}{\approx} a^{(n)}, \quad \text{where } \epsilon = \frac{1}{10^n}.$$

We leave the simple proof for Question 1b. In practice, the approximation may be much closer than what (7) guarantees. For example, since $\pi = 3.14159\ldots$,

$$\pi \underset{.1}{\approx} 3.1, \quad \pi \underset{.01}{\approx} 3.14 \quad \text{by (7)}, \quad \text{whereas actually} \quad \pi \underset{.05}{\approx} 3.1, \quad \pi \underset{.002}{\approx} 3.14 .$$

As an example of the use of the truncation approximation in analysis, we prove a theorem we will use occasionally. Some of the details of the proof are left as Questions, to give you practice with these ideas and with inequalities.

In what follows, the set of rational numbers is denoted by $\mathbb{Q}$; the sum, difference, product, and quotient of rational numbers is again a rational number. A number which is not rational is called *irrational*; for example, in Appendix A.2 it is proved that $\sqrt{2}$ is irrational, a fact we shall make use of in the proof.

(Appendix A.0 describes the notation and number facts that we use.)

Theorem 2.5   *For any two real numbers $a < b$, there is*

  (i)   *a rational number $r \in \mathbb{Q}$ between them: $a < r < b$;*

  (ii)   *an irrational number $s \notin \mathbb{Q}$ between them: $a < s < b$.*

Proof. (i) We prove (i); assume $b > 0$ and make two cases.

If b is rational choose n so large that $a < b - \frac{1}{10^n} < b$; then we are done since the middle number is also rational.

If b is not rational, choose n so large that $a < b^{(n)} < b$. (see the picture). To do this, by (7) we need only choose n so that $\frac{1}{10^n} < b - a$. (See Question 2a.)

If $b \leq 0$, then $b + N > 0$ for some integer N; this reduces it to the previous case, since $a + N < r < b + N \Rightarrow a < r - N < b$, and $r - N$ is still rational. $\square$

(ii) To prove this, from (i) we have $a < r < b$, where r is rational. Then if n is a sufficiently large integer, we have

$$a \;<\; r \;<\; r + \frac{\sqrt{2}}{n} \;<\; b\,,$$

and $r + \sqrt{2}/n$ is irrational (see Question 2b). $\square\square$

There are a set of laws for calculating with approximations, of which we single out the two most important; others are in the Questions and Exercises.

(8) **transitive law** $a \underset{\epsilon}{\approx} b$ and $b \underset{\epsilon'}{\approx} c \;\Rightarrow\; a \underset{\epsilon+\epsilon'}{\approx} c\,;$

(9) **addition law** $a \underset{\epsilon}{\approx} a'$ and $b \underset{\epsilon'}{\approx} b' \;\Rightarrow\; a + b \underset{\epsilon+\epsilon'}{\approx} a' + b'\,.$

Both are proved by the triangle inequality; for example, to prove (9),

$$|(a + b) - (a' + b')| \;\leq\; |a - a'| + |b - b'| < \epsilon + \epsilon'\,. \qquad \square$$

If we think of ϵ as being the possible error in a measurement, the addition law (9) tells how errors propagate under addition. Their behavior under multiplication and taking reciprocals is a bit less simple, and left for the exercises.

Because the notation $\approx$ suggests a sort of rough equality, it ought to obey the transitive law: $a \approx b$, $b \approx c \;\Rightarrow\; a \approx c$. While this is suggestive, it has no precise meaning. In arguments, it must be used in the form (8) with the ϵ's put in. Forget this and your teacher won't let you use $\approx$ (he or she might not, regardless).

Questions 2.5

1. (a) We have $e = 2.718\dots$. If we write $e \underset{\epsilon}{\approx} 2.72$, what can we take for ϵ?

   (b) Prove (7).

2. Fill in the details in Theorem 2.5 by

   (a) in part (i), proving the inequalities without referring to a picture;

   (b) in part (ii), giving an explicit expression for how large n must be, and proving $r + \sqrt{2}/n$ is irrational.

3. (a) Prove the transitive law (8).

   (b) If $a_1 \underset{\epsilon}{\approx} a_2$, $a_2 \underset{\epsilon}{\approx} a_3$, $\dots$, $a_{n-1} \underset{\epsilon}{\approx} a_n$, what accuracy can you give for the approximation $a_1 \approx a_n$? Prove it, either using (4a), or using (8) and induction (Appendix A.4.)

4. If $a \underset{\epsilon}{\approx} b$, what is the accuracy of

   (i) $a^2 \approx b^2$?   (ii) $1/a \approx 1/b$?   (assume $a, b \neq 0$)

2.6   The terminology "for n large".

In estimating or approximating the terms of a sequence $\{a_n\}$, sometimes the estimate is not valid for all terms of the sequence; for example, it might fail for the first few terms, but be valid for the later terms. In such a case, one has to specify the values of n for which the estimate holds.

Example 2.6A   Let $a_n = \dfrac{5n}{n^2 - 2}$, $n \geq 2$; for what n is $|a_n| < 1$?

Solution. For $n = 1$, the estimate is not valid; if $n > 1$, then a_n is positive, so we can drop the absolute value. Then we have

$$\frac{5n}{n^2 - 2} < 1 \quad \Leftrightarrow \quad 5n < n^2 - 2 \quad \Leftrightarrow \quad 5 < n - \frac{2}{n} ,$$

and by inspection, one sees this last inequality holds for all $n \geq 6$. $\qquad\square$

Example 2.6B   In the sequence $a_n = \dfrac{n^2 + 2n}{n^2 - 2}$, for what n is $a_n \underset{.1}{\approx} 1$?

Solution. $\left| \dfrac{n^2 + 2n}{n^2 - 2} - 1 \right| = \dfrac{2n + 2}{n^2 - 2}$,

$$\text{which is } < \frac{1}{10} \quad \Leftrightarrow \quad n(n - 20) > 22 ,$$

and by inspection, this last inequality holds for all $n \geq 22$. $\qquad\square$

The above is a good illustration of Warning 2.4. To show a_n and 1 are close, we get a small upper estimate of the difference by first transforming it algebraically; trying instead to use the triangle inequality, in either the sum or difference form, would produce nothing useful.

As the above examples illustrate, sometimes a property of a sequence a_n is not true for the first few terms, but only starts to hold after a certain place in the sequence. In this case, a special terminology is used.

Definition 2.6   The sequence $\{a_n\}$ has property $\mathcal{P}$ **for n large** if

(10)   there is a number N such that a_n has property $\mathcal{P}$ for all $n \geq N$.

One can say instead *for large n, for n sufficiently large*, etc. Most of the time we will use the symbolic notation *for $n \gg 1$* , which can be read in any of the above ways.

Note that in the definition, N need not be an integer.

To illustrate the definition, Examples 2.6A and 2.6B show, respectively:

$$\text{if } a_n = \frac{5n}{n^2 - 2}, \quad \text{then } |a_n| < 1 \quad \text{for } n \gg 1 ;$$

$$\text{if } a_n = \frac{n^2 + 2n}{n^2 - 2}, \quad \text{then } |a_n| \underset{.1}{\approx} 1 \quad \text{for } n \gg 1 .$$

In both examples, we gave the smallest integer value of N (it was 6 and 22, respectively), such that the stated property of a_n held for all $n \geq N$. In general, this is overkill: to show something is true for n large, one only has to give *some* number N which works, not the "best possible" N. And as we said, it need not be an integer.

The symbols $a \gg b$, with $a, b > 0$, have the meaning "a is relatively large compared with b", that is, a/b is large. Thus we do not write "for $n \gg 0$"; intuitively, every positive integer is relatively large compared with 0.

Example 2.6C   Show the sequence $\{\sin 10/n\}$ is decreasing for large n.

Proof.   The function $\sin x$ is increasing on the interval $0 < x < \pi/2$, i.e.,

$$a < b \;\Rightarrow\; \sin a < \sin b, \qquad \text{for } 0 < a < b < \pi/2 \;.$$

Thus

$$\sin \frac{10}{n+1} \;<\; \sin \frac{10}{n} \;, \qquad \text{if} \quad \frac{10}{n} < \frac{\pi}{2} \;, \quad \text{i.e., if} \quad n > \frac{20}{\pi} \;. \qquad \square$$

Remarks.   This completes the argument, since it shows we can use $N = 20/\pi$. If you prefer an integer value, take $N = 7$, the first integer after $20/\pi$.

In this example, it was no trouble to find the exact integer value $N = 7$ at which the sequence starts to be decreasing. However this is in general not necessary. Any value for N greater than 7 would do just as well in showing the sequence is decreasing for large n.

Thus, for example, if it turned out for some sequence a_n that

$$a_{n+1} < a_n \qquad \text{if } n^2 + n > 100 \;,$$

to show the sequence is decreasing for large n, it is a waste of time to solve the quadratic equation $n^2 + n = 100$; by inspection one sees that if $n \geq 10$, then $n^2 + n > 100$, so that one can take $N = 10$.

The final two examples illustrate how to use "for $n \gg 1$" in a formal argument; we give proofs as model arguments you can imitate. Watch carefully how the N is handled.

Example 2.6D   If $\{a_n\}$ is bounded above for $n \gg 1$, it is bounded above.

Proof. There is a B and an N (we may take it to be an integer) such that

$$a_n \leq B \qquad \text{for } n \geq N \;.$$

Let B' be a number greater than $a_0, a_1, \ldots, a_N$ and B. Then

$$a_n \leq B' \qquad \text{for } n = 0, 1, \ldots, N \;;$$
$$a_n \leq B < B' \qquad \text{for } n \geq N \;;$$
$$\text{therefore} \quad a_n \leq B' \qquad \text{for all } n \;. \qquad \square$$

Example 2.6E   Suppose that $\{a_n\}$ and $\{b_n\}$ are increasing for $n \gg 1$.

Prove that $\{a_n + b_n\}$ is increasing for $n \gg 1$.

Proof.   By hypothesis, there are numbers N_1 and N_2 such that

$$a_n \leq a_{n+1} \quad \text{for } n \geq N_1 ,$$
$$b_n \leq b_{n+1} \quad \text{for } n \geq N_2 .$$

Choose any $N \geq N_1, N_2$; for example, let $N = \max(N_1, N_2)$. Then

$$a_n \leq a_{n+1} \quad \text{for } n \geq N, \text{ since } N \geq N_1 ;$$
$$b_n \leq b_{n+1} \quad \text{for } n \geq N, \text{ since } N \geq N_2 ; \quad \text{therefore}$$
$$a_n + b_n \leq a_{n+1} + b_{n+1} \quad \text{for } n \geq N . \qquad \square$$

You should be very clear about why the proof does not consist simply of the last three lines (with the two "since" clauses deleted). When we start out, each sequence has its private N_1; the point of the proof is to get them to agree to use the same N, so that the inequalities can be added.

Remarks.   A couple of further remarks about "for $n \gg 1$" might be useful. In this phrase, an N is concealed. Why is it hidden, and when do we have to make it explicit?

We have to make the N explicit whenever we have to *prove* that some property holds for large n. To make such a proof, we have to go back to the definition of "for $n \gg 1$" and exhibit explicitly the N which works. We saw this in Examples 2.6A and 2.6B.

Most of the time however the N remains hidden, and hiding it is the main function of the phrase "for $n \gg 1$". We use it whenever we want to say a sequence behaves in a certain way when $n \geq N$, but we don't want to have to specify exactly what N is. Maybe it would be awkward to calculate a value for N. More likely, the value of N is of no importance, and we don't want to clutter up the definition or argument with a lot of irrelevant numbers or letters, just as a novelist does not give names to the characters who appear for a moment in the story, but will not be seen again.*

There will be many examples of "for n large" in the next chapter. For now, let's conclude by seeing how we can use the terminology to weaken slightly the hypothesis of the Completeness Property formulated at the end of the preceding chapter.

Suppose $\{a_n\}$ is increasing and bounded above. In finding its limit, we see that how the sequence behaves near its beginning is not important; the early terms

---

* There are, of course, exceptions: "With only two minutes left to reach the Bursar's office with the check for my first tuition installment, I sprinted the four blocks, took the five flights of stairs three steps at a time, and arrived gasping at the window just as Algernon Finch, Payments, with a thin-lipped smile, closed it."

could be changed, but the limit would stay the same. Indeed, if the sequence is bounded, but is increasing only after some term a_N :

$$a_n \leq a_{n+1}, \qquad \text{for } n \geq N,$$

it still will have a limit, and exactly the same procedure we used before (watching each decimal place until it no longer changes) will produce it. The same observation applies to decreasing bounded sequences, and we are led to a slightly more general form of the **Completeness Property**:

> *A sequence which is bounded and monotone for $n \gg 1$ has a limit.*

Questions 2.6

1. Show that each of the following sequences has the indicated property for $n \gg 1$. (This means you have to specify some number N and show that the sequence has the property when $n \geq N$ or $n > N$.)

    (a) $a_n = \cos 20\pi/n$ (increasing);   (b) $a_n = n^4 - 100n^3 - 1$ (positive).

2. Proof or counterexample: $a_n < k$ for $n \gg 1 \Rightarrow a_n/n < k$ for $n \gg 1$.

3. Prove $\dfrac{2n}{n-1} \underset{.1}{\approx} 2$,   for large n.

4. Prove: $\{a_n\}$ increasing and not bounded above $\Rightarrow a_n > 0$ for $n \gg 1$.

Exercises

2.1

1. Let $\{a_n\}$ and $\{b_n\}$ be increasing; are the following increasing? Proof or counterexample.   (i) $\{a_n + b_n\}$ (ii) $\{a_n - b_n\}$

2. Let $\{a_n\}$ and $\{b_n\}$ be bounded above. State hypotheses which guarantee $\{a_n b_n\}$ will also be bounded above, and prove it.

3. "If $\{a_n\}$ and $\{b_n\}$ are increasing, then $\{a_n b_n\}$ is increasing". Show this is false, change the hypothesis on $\{b_n\}$, and prove the amended statement.

4. Define a sequence by $a_{n+1} = -a_n{}^2$, with $-1 \leq a_0 < 0$. Prove that $\{a_n\}$ is increasing and bounded above.

5. Prove $\{\sqrt{n+1} - \sqrt{n}\}$, $n \geq 0$, is monotone, using just algebra. (You can use the true results in Question 2.1/3; cf. Ans. 1.4/1 for the write-up.)

2.2

1. Give upper and lower bounds for the sequences ($n \geq 0$):

    (a) $\dfrac{1}{3 + \cos n}$ (b) $\dfrac{\sin n}{n^2 + 1}$ (give the sharpest bounds)

2. Give an upper estimate for $\ln 3 = \displaystyle\int_1^3 \frac{dx}{x}$ as in example 2.2C, by using

    (a) one trapezoid;   (b) two trapezoids.

3. Let $\alpha = 1/\sqrt{3}$. Give a lower bound for $\tan^{-1}\alpha = \int_0^\alpha \dfrac{dx}{1+x^2}$, by using a single trapezoid; compare it with the exact value (use $\sqrt{3} \approx 1.73$).

(The first positive point of inflection of $1/(1+x^2)$ is α; why is this relevant?)

2.3

1. For each sequence, tell if it is bounded above or not (indicate a reason):

(a) $a_n = 1 + 1/4 + 1/7 + \ldots + 1/(3n+1)$;

(b) $a_n = 1 + \dfrac{1}{2\sqrt{2}} + \dfrac{1}{3\sqrt{3}} + \ldots + \dfrac{1}{n\sqrt{n}}$.

2.4

1. Prove: $\max(a,b) = \dfrac{a+b+|a-b|}{2}$, $\min(a,b) = \dfrac{a+b-|a-b|}{2}$,

where max and min denote respectively the greater and the lesser of the two numbers a and b.

2. If $|a_1 \sin b + a_2 \sin 2b + \ldots + a_n \sin nb| > n$, prove that $|a_i| > 1$ for at least one of the a_i. (Use contraposition: cf. Appendix A.2.)

3. Give upper and lower estimates (in terms of n alone) for

$$\left| \cos na + \frac{\sin nb}{n} \right| ,$$

valid for all a, b; show these estimates are the best possible, by giving values of a and b for which the bounds are actually attained.

4. Give the best upper and lower estimates of the form cA^n (c and A are fixed numbers), valid for all a and b, that you can for

$$\left| \frac{\sin^3 na - 2}{(1 + (\cos b)/2)^n} \right| ;$$

then tell by using your estimates for what values of n you can be sure that the expression is < 100, and for what values you can be sure it is $> 1/3$, regardless of what a and b are.

5. Consider the sum $S_n = \sum_1^n c_k \cos kt$, where $c_k = 1/2^k$ (cf. Ex. 2.4A). Prove the lower estimate $S_n \geq .7$, for $0 \leq t \leq .1$.

(Find a lower bound for the first three terms — use $\cos u > .95$, $0 \leq u \leq .3$; estimate the remaining terms, and then use the difference form of the triangle inequality.)

6. Give an alternate proof of the triangle inequality (4), based on showing that if both sides of (4) are squared, an inequality holds between them. Write up the proof carefully, citing a reference for any non-trivial step.

7. Prove the extended triangle inequality (4a) by induction.

2.5

1. If $a \underset{\epsilon}{\approx} b$, and $|a| \leq K$, $|b| \leq K$, give an estimate for the accuracy of the approximation $a^n \approx b^n$, where n is a positive integer.

2.   If $a \underset{\epsilon}{\approx} 1$ and $a \underset{\epsilon}{\approx} 2$, then $\epsilon > \frac{1}{2}$. This is obvious if you draw a picture, but prove it without a picture.

3.   Suppose that a and b are two numbers with this property: for any $\epsilon > 0$, it is true that $a \underset{\epsilon}{\approx} b$. Prove $a = b$. (Use contraposition, cf. Appendix A.2.)

4.   Prove Theorem 2.5(i) a different way using this suggestion.

We are looking for integers m and n such that $an < m < bn$ (why?) What property of two real numbers guarantees there is an integer between them?

5.   If $a \underset{\epsilon}{\approx} b$, $\epsilon = \frac{1}{10^n}$, state and prove the best estimate you can give for $a^{(n)} \approx b^{(n)}$. Show by example that it is the best possible. (Assume $a, b > 0$.)

2.6

1.   Prove that if $a > 0$, the sequence $a^n/n!$ is monotone for large n.

2.   Prove: if $a > 1$, the sequence n/a^n is monotone for n large.

3.   Prove that if $\epsilon > 0$ is given, then $\dfrac{n}{n+2} \underset{\epsilon}{\approx} 1$, for $n \gg 1$.

4.   Prove $\{a_n\}$ is decreasing for $n \gg 1$, if $a_0 = 1$ and

(a) $a_{n+1} = \dfrac{n-5}{(n+1)(n+2)} a_n$ (b) $a_{n+1} = \dfrac{n^2+15}{(n+1)(n+2)} a_n$

Problems

2-1   Let $\{a_n\}$ be a sequence. We construct from it another sequence $\{b_n\}$, its *sequence of averages*, defined by

$$b_n = \frac{a_1 + \ldots + a_n}{n} = \text{average of the first } n \text{ terms.}$$

(a) Prove that if $\{a_n\}$ is increasing, then $\{b_n\}$ is also increasing.

(b) Prove that if $\{a_n\}$ is bounded above, then $\{b_n\}$ is bounded above.

2-2   Prove that a bounded increasing sequence $\{a_n\}$ of integers is constant for $n \gg 1$.

(Do not use an indirect argument. Use a direct argument which, starting from the hypotheses, produces the constant value L; then prove that L has the desired property.)

2-3   *The arithmetic-geometric mean inequality.*

(a) Prove: for any $a, b \geq 0$, $\sqrt{ab} \leq \dfrac{a+b}{2}$,

with equality holding if and only if $a = b$.

(b) How does the figure illustrate the inequality?

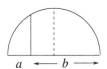

$a \longleftarrow b \longrightarrow$

2-4   A positive sequence is defined by $a_{n+1} = \sqrt{1 + a_n^2/4}$, $0 \leq a_0 < 2/\sqrt{3}$.

(a) Prove the sequence is strictly increasing.

(b) Prove the sequence is bounded above.

Answers to Questions

2.1 1. $1 < 2$ and $-3 < 1$, for example.

2.

$$-a < -b \;\Rightarrow\; a > b \qquad \text{(sign-change law)}$$
$$\Rightarrow\; 1/a < 1/b \quad \text{(reciprocal law; since } a, b > 0)$$
$$\Rightarrow\; 1/(-a) > 1/(-b) \quad \text{(sign-change law)}$$

3. (i) true. Note that $a, b \geq 0$ since their square roots exist. By contraposition: $\sqrt{a} \geq \sqrt{b} \;\Rightarrow\; a \geq b$ (multiplication law).

(ii) false: $1 < 4$, but $-1 > -2$. If $a, b \geq 0$, it follows from (i).

(iii) false: $\{-1, 1, 2, 3, \ldots\}$ is increasing, but $\{-1, 1, 1/2, 1/3, \ldots\}$ is not decreasing. True if all $a_n > 0$ or all $a_n < 0$, by the reciprocal law and 2.1/2 .

4. $a_n > c > 0$ for all $n \;\Rightarrow\; 1/a_n < 1/c$ for all n.

(The hypothesis is usually phrased: "a_n is bounded away from 0".)

2.2 1.   We have

$$1 \leq B \leq 2 \;\Rightarrow\; 1/2 \leq 1/B \leq 1 \quad \text{(reciprocal law)};$$
$$1 \leq A \leq 2 ;$$

therefore, by the multiplication law,

$$1/2 \leq A/B \leq 2, \quad \text{since everything is } > 0.$$

2.   Call it A/B.
$$2 \leq B \leq 4 \;\Rightarrow\; 1/4 \leq 1/B \leq 1/2 \quad \text{by the reciprocal law};$$
$$3 \leq A \leq 5 \;\Rightarrow\; 3/4 \leq A/B \leq 5/2 \quad \text{by the mutliplication law}.$$

3.   Area of left rectangle $= 1/3$, area of left trapezoid $= 3/8$. Since $1/3 < 3/8$, the rectangles give a weaker lower estimate of the area under the curve.

2.3 1.   Get an upper estimate by comparing the terms to the area under the curve $1/x^2$ and lying over the interval $1 \leq x < \infty$.

2.4 1.   Both false. (i) $-3 < -2$ (ii) $|-2| < |-3|$

2.   (b) Using the triangle inequality (4) twice:
$$|a + b + c| = |(a + b) + c| \leq |a + b| + |c|$$
$$\leq |a| + |b| + |c|.$$

3.   We have
$$|S_n| \leq |c_1||\cos t| + |c_2||\cos 2t| + \ldots + |c_n||\cos nt|$$
$$\leq (1/1!) \cdot 1 + (1/2!) \cdot 1 + \ldots + (1/n!) \cdot 1 \; < \; e - 1, \text{ by Ques. } 1.4/2 .$$

4. $|a| \geq 2$; $-|b| \geq -1/2$ (sign-change law); add the inequalities.

5.   (a) $-1 \leq a_n \leq 1/2$ (b) $|a_n| \leq 1$

2.5

1. (a) any $\epsilon > .002$, based on the given information. (If we knew that $e > 2.718$, then we could also take $\epsilon = .002$.)

(b) $\quad a - a^{(n)} = .0\ldots0\,a_{n+1}a_{n+2}\ldots \geq 0 \quad (n \text{ zeros})$;

Therefore, $\quad 0 \leq a - a^{(n)} < .00\ldots01 \quad (n \text{ places})$

$$= 1/10^n \; .$$

2. (a) To save space we write horizontally; it would be clearer vertically.

First part: $n > -\ln(b-a)/\ln 10 \;\Rightarrow\; \frac{1}{10^n} < b - a \;\Rightarrow\; a < b - \frac{1}{10^n} < b$.

Second part: by (7), $b - b^{(n)} < \frac{1}{10^n} \;\Rightarrow\; b - \frac{1}{10^n} < b^{(n)} \;\Rightarrow\; a < b - \frac{1}{10^n} < b^{(n)} < b$.

(b) We have $\quad r + \sqrt{2}/n < b \quad\Leftrightarrow\quad n > \dfrac{\sqrt{2}}{b - r}$.

Indirect proof of irrationality (cf. appendix A.2):

If $r + \sqrt{2}/n = s$, a rational number, then $\sqrt{2} = n(s - r)$, contradicting the fact that $\sqrt{2}$ is irrational.

3. (a) $|a - c| = |(a - b) + (b - c)| \leq |a - b| + |b - c| < \epsilon + \epsilon'$

(b)
$$|a_1 - a_n| = |(a_1 - a_2) + (a_2 - a_3) + \ldots + (a_{n-1} - a_n)|$$
$$\leq |a_1 - a_2| + |a_2 - a_3| + \ldots + |a_{n-1} - a_n|$$
$$< (n-1)\epsilon \; .$$

4. (i) $|a^2 - b^2| = |a - b||a + b| < \epsilon \cdot |a + b|$, i.e., $a^2 \underset{|a+b|\epsilon}{\approx} b^2$.

(ii) $\left|\dfrac{1}{a} - \dfrac{1}{b}\right| = \dfrac{|b - a|}{|ab|} < \dfrac{\epsilon}{|ab|}$, thus $\dfrac{1}{a} \underset{\epsilon'}{\approx} \dfrac{1}{b}$, where $\epsilon' = \epsilon/|ab|$.

2.6

1. (a) $\cos x$ is decreasing for $0 \leq x \leq \pi$. So the sequence is increasing if $20\pi/n \leq \pi$, i.e., if $n \geq 20$.

(b) $n^4 - 100n^3 - 1 = n^3(n - 100) - 1$, which is > 0 if $n > 100$.

2. Counterexample: $a_n = -2$ for all n, $k = -1$.

3. $\left|\dfrac{2n}{n-1} - 2\right| = \left|\dfrac{2}{n-1}\right| < .1$ if $n - 1 > 20$, or $n > 21$.

4. Some $a_N > 0$, since otherwise all $a_n \leq 0$, and the sequence would be bounded above. Then $a_n > 0$ for all $n \geq N$, since $\{a_n\}$ is increasing.

3

The Limit of a Sequence

3.1 Definition of limit.

In Chapter 1 we discussed the limit of sequences that were monotone; this restriction allowed some short-cuts and gave a quick introduction to the concept. But many important sequences are not monotone—numerical methods, for instance, often lead to sequences which approach the desired answer alternately from above and below. For such sequences, the methods we used in Chapter 1 won't work. For instance, the sequence

$$1.1, \ .9, \ 1.01, \ .99, \ 1.001, \ .999, \ \ldots$$

has 1 as its limit, yet neither the integer part nor any of the decimal places of the numbers in the sequence eventually becomes constant. We need a more generally applicable definition of the limit.

We abandon therefore the decimal expansions, and replace them by the approximation viewpoint, in which "the limit of $\{a_n\}$ is L" means roughly

a_n *is a good approximation to* L , *when* n *is large.*

The following definition makes this precise. After the definition, most of the rest of the chapter will consist of examples in which the limit of a sequence is calculated directly from this definition. There are "limit theorems" which help in determining a limit; we will present some in Chapter 5. Even if you know them, don't use them yet, since the purpose here is to get familiar with the definition.

Definition 3.1 The number L is the **limit** of the sequence $\{a_n\}$ if

(1) $$\text{given } \epsilon > 0, \qquad a_n \underset{\epsilon}{\approx} L \quad \text{for } n \gg 1.$$

If such an L exists, we say $\{a_n\}$ **converges**, or *is convergent*; if not, $\{a_n\}$ **diverges**, or *is divergent*. The two notations for the limit of a sequence are:

$$\lim_{n \to \infty} \{a_n\} = L \ ; \qquad a_n \to L \ \text{ as } n \to \infty \ .$$

These are often abbreviated to: $\lim a_n = L$ or $a_n \to L$.

Statement (1) looks short, but it is actually fairly complicated, and a few remarks about it may be helpful. We repeat the definition, then build it in three stages, listed in order of increasing complexity; with each, we give its translation into English.

Definition 3.1 $\lim a_n = L$ if: given $\epsilon > 0$, $a_n \underset{\epsilon}{\approx} L$ for $n \gg 1$.

Building this up in three succesive stages:

(i) $a_n \underset{\epsilon}{\approx} L$ (a_n approximates L to within ϵ);

(ii) $a_n \underset{\epsilon}{\approx} L$ for $n \gg 1$ $\begin{pmatrix}\text{the approximation holds for all } a_n \\ \text{far enough out in the sequence;}\end{pmatrix}$;

(iii) given $\epsilon > 0$, $a_n \underset{\epsilon}{\approx} L$ for $n \gg 1$

                      (the approximation can be made as close as desired, provided we go far enough out in the sequence—the smaller ϵ is, the farther out we must go, in general).

The heart of the limit definition is the approximation (i); the rest consists of the if's, and's, and but's. First we give an example.

Example 3.1A   Show $\lim\limits_{n \to \infty} \dfrac{n-1}{n+1} = 1$, directly from definition 3.1.

Solution.   According to definition 3.1, we must show:

(2)   given $\epsilon > 0$, $\dfrac{n-1}{n+1} \underset{\epsilon}{\approx} 1$ for $n \gg 1$.

We begin by examining the size of the difference, and simplifying it:

$$\left| \frac{n-1}{n+1} - 1 \right| = \left| \frac{-2}{n+1} \right| = \frac{2}{n+1} .$$

We want to show this difference is small if $n \gg 1$. Use the inequality laws:

$$\frac{2}{n+1} < \epsilon \ \text{ if } \ n+1 > \frac{2}{\epsilon} , \quad \text{i.e., if } \ n > N, \text{ where } N = \frac{2}{\epsilon} - 1 ;$$

this proves (2), in view of the definition (2.6) of "for $n \gg 1$".   □

The argument can be written on one line (it's ungrammatical, but easier to write, print, and read this way):

Solution.   Given $\epsilon > 0$, $\left| \dfrac{n-1}{n+1} - 1 \right| = \dfrac{2}{n+1} < \epsilon$, if $n > \dfrac{2}{\epsilon} - 1$.   □

Remarks on limit proofs.

1. The heart of a limit proof is in in the approximation statement, i.e., in getting a small upper estimate for $|a_n - L|$. Often most of the work will consist in showing how to rewrite this difference so that a good upper estimate can be made. (The triangle inequality may or may not be helpful here.)

Note that in doing this, you *must* use $| \ |$; you can drop the absolute value signs only if it is clear that the quantity you are estimating is non-negative.

2. In giving the proof, you must exhibit a value for the N which is lurking in the phrase "for $n \gg 1$". You need not give the smallest possible N; in example 3.1A, it was $2/\epsilon - 1$, but any bigger number would do, for example $N = 2/\epsilon$.

Note that N depends on ϵ: in general, the smaller ϵ is, the bigger N is, i.e., the further out you must go for the approximation to be valid within ϵ .

3. In Definition 3.1 of limit, the phrase "given $\epsilon > 0$" has at least five equivalent forms; by convention, all have the same meaning, and any of them can be used. They are:

$$\text{for all } \epsilon > 0 \ , \quad \text{for every } \epsilon > 0 \ , \quad \text{for any } \epsilon > 0 \ ;$$
$$\text{given } \epsilon > 0 \ , \quad \text{given any } \epsilon > 0 \ .$$

The most standard of these phrases is "for all $\epsilon > 0$", but we feel that if you are meeting (1) for the first time, the phrases in the second line more nearly capture the psychological meaning. Think of a **limit demon** whose only purpose in life is to make it hard for you to show that limits exist; it always picks unpleasantly small values for ϵ. Your task is, given any ϵ the limit demon hands you, to find a corresponding N (depending on ϵ) such that $a_n \underset{\epsilon}{\approx} L$ for $n > N$.

Remember: the limit demon supplies the ϵ; you cannot choose it yourself.

In writing up the proof, good mathematical grammar requires that you write "given $\epsilon > 0$" (or one of its equivalents) *at the beginning*; get in the habit now of doing it. We will discuss this later in more detail; briefly, the reason is that the N depends on ϵ, which means ϵ must be named first.

4. It is not hard to show (see Problem 3-3) that if a monotone sequence $\{a_n\}$ has the limit L in the sense of Chapter 1—higher and higher decimal place agreement—then L is also its limit in the sense of Definition 3.1. (The converse is also true, but more trouble to show because of the difficulties with decimal notation.) Thus the limit results of Chapter 1, the Completeness Property in particular, are still valid when our new definition of limit is used. From now on, "limit" will always refer to Definition 3.1.

Here is another example of a limit proof, more tricky than the first one.

Example 3.1B Show $\displaystyle \lim_{n \to \infty} (\sqrt{n+1} - \sqrt{n}) = 0$.

Solution.   We use the identity $A - B = \dfrac{A^2 - B^2}{A + B}$, which tells us that

(3) $$\left| (\sqrt{n+1} - \sqrt{n}) \right| \ = \ \frac{1}{\sqrt{n+1} + \sqrt{n}} \ < \ \frac{1}{2\sqrt{n}} \ ;$$

given $\epsilon > 0$, $\qquad \dfrac{1}{2\sqrt{n}} < \epsilon$ if $\dfrac{1}{4n} < \epsilon^2$, i.e., if $n > \dfrac{1}{4\epsilon^2}$. $\square$

Note that here we need not use absolute values since all the quantities are positive.

It is not at all clear how to estimate the size of $\sqrt{n+1} - \sqrt{n}$; the triangle inequality is useless. Line (3) is thus the key step in the argument: the expression must first be transformed by using the identity. Even after doing this, line (3) gives a further simplifying inequality to make finding an N easier; just try getting an N without this step! The simplification means we don't get the smallest possible N; who cares?

Questions 3.1

1. Directly from the definition of limit (i.e., without using theorems about limits you learned in calculus), prove that

(a) $\dfrac{n}{n+1} \to 1$ 　　　　　　　　　(b) $\dfrac{\cos na}{n} \to 0$ 　　(a is a fixed number)

(c) $\dfrac{n^2+1}{n^2-1} \to 1$ 　　　　　　　(d) $\dfrac{n^2}{n^3+1} \to 0$ 　$\left(\begin{array}{l}\text{cf. Example 3.1B: make}\\ \text{a simplifying inequality}\end{array}\right)$

2. Prove that, for any sequence $\{a_n\}$, $\quad \lim a_n = 0 \quad \Leftrightarrow \quad \lim |a_n| = 0$. (This is a simple but important fact you can use from now on.)

3. Why does the definition of limit say $\epsilon > 0$, rather than $\epsilon \geq 0$?

3.2　The uniqueness of limits.　The K-ϵ principle.

Can a sequence have more than one limit? Common sense says no: if there were two different limits L and L', the a_n could not be arbitrarily close to both, since L and L' themselves are at a fixed distance from each other.. This is the idea behind the proof of our first theorem about limits. The theorem shows that if $\{a_n\}$ is convergent, the notation $\lim a_n$ makes sense; there's no ambiguity about the value of the limit. The proof is a good exercise in using the definition of limit in a theoretical argument. Try proving it yourself first.

Theorem 3.2A　Uniqueness theorem for limits.

A sequence a_n has at most one limit:　$a_n \to L$ and $a_n \to L' \Rightarrow L = L'$.

Proof.　By hypothesis, given $\epsilon > 0$,
$$a_n \underset{\epsilon}{\approx} L \ \text{ for } n \gg 1, \quad \text{and} \quad a_n \underset{\epsilon}{\approx} L' \ \text{ for } n \gg 1.$$
Therefore, given $\epsilon > 0$, we can choose some large number k such that
$$L \underset{\epsilon}{\approx} a_k \underset{\epsilon}{\approx} L' \, .$$
By the transitive law of approximation (2.5 (8)), it follows that

(4) 　　　　　　　　　　　given $\epsilon > 0, \quad L \underset{2\epsilon}{\approx} L' \, .$

To conclude that $L = L'$. we reason indirectly (cf. Appendix A.2).

Suppose $L \neq L'$; choose $\epsilon = \frac{1}{2}|L - L'|$. We then have
$$|L - L'| < 2\epsilon, \qquad \text{by (4);} \quad \text{i.e.,}$$
$$|L - L'| < |L - L'|, \quad \text{a contradiction.} \qquad \square$$

Remarks.

1. The line (4) says that the two numbers L and L' are arbitrarily close. The rest of the argument says that this is nonsense if $L \neq L'$, since they cannot be closer than $|L - L'|$.

2. Before, we emphasized that the limit demon chooses the ϵ; you cannot choose it yourself. Yet in the proof we chose $\epsilon = \frac{1}{2}|L - L'|$. Are we blowing hot and cold?

The difference is this. Earlier, we were trying to prove a limit existed, i.e., were trying to prove a statement of the form:

given $\epsilon > 0$,   some statement involving ϵ is true.

To do this, you must be able to prove the truth no matter what ϵ you are given.

Here on the other hand, we don't have to prove (4)—we already deduced it from the hypothesis. It's a true statement. That means we're allowed to *use* it, and since it says something is true for every $\epsilon > 0$, we can choose a particular value of ϵ and make use of its truth for that particular value.

To reinforce these ideas and give more practice, here is a second theorem which makes use of the same principle, also in an indirect proof. The theorem is "obvious" using the definition of limit we started with in Chapter 1, but we are committed now and for the rest of the book to using the newer Definition 3.1 of limit, and therefore the theorem requires proof.

Theorem 3.2B $\{a_n\}$ *increasing,* $L = \lim a_n \Rightarrow a_n \leq L$ *for all* n;

$\{a_n\}$ *decreasing,* $L = \lim a_n \Rightarrow a_n \geq L$ *for all* n.

Proof.   Both cases are handled similarly; we do the first.

Reasoning indirectly, suppose there were a term a_N of the sequence such that $a_N > L$. Choose $\epsilon = \frac{1}{2}(a_N - L)$. Then since $\{a_n\}$ is increasing,

$$a_n - L \ \geq \ a_N - L \ > \ \epsilon, \quad \text{for all } n \geq N,$$

contradicting the Definition 3.1 of $L = \lim a_n$. $\square$

The K-ϵ principle.

In the proof of Theorem 3.2A, note the appearance of 2ϵ in line (4). It often happens in analysis that arguments turn out to involve not just ϵ but a constant multiple of it. This may occur for instance when the limit involves a sum or several arithmetic processes. Here is a typical example.

Example 3.2   Let $a_n = \dfrac{1}{n} + \dfrac{\sin n}{n + 1}$.   Show $a_n \to 0$, from the definition.

Solution   To show a_n is small in size, use the triangle inequality:

$$\left| \frac{1}{n} + \frac{\sin n}{n + 1} \right| \ \leq \ \left| \frac{1}{n} \right| + \left| \frac{\sin n}{n + 1} \right| .$$

At this point, the natural thing to do is to make the separate estimations

$$\left| \frac{1}{n} \right| < \epsilon, \ \text{ for } n > \frac{1}{\epsilon} \ ; \qquad \left| \frac{\sin n}{n + 1} \right| < \epsilon, \ \text{ for } n > \frac{1}{\epsilon} - 1 \ ;$$

so that, given $\epsilon > 0$,

$$\left| \frac{1}{n} + \frac{\sin n}{n + 1} \right| \ < \ 2\epsilon \ , \quad \text{for } n > \frac{1}{\epsilon} \ .$$

This is close, but we were supposed to show $|a_n| < \epsilon$. Is 2ϵ just as good?

The usual way of handling this would be to start with the given ϵ, then put $\epsilon' = \epsilon/2$, and give the same proof, but working always with ϵ' instead of ϵ. At the end, the proof shows

$$\left| \frac{1}{n} + \frac{\sin n}{n+1} \right| < 2\epsilon', \quad \text{for } n > \frac{1}{\epsilon'} \; ;$$

and since $2\epsilon' = \epsilon$, the limit definition is satisfied.

Instead of doing this, let's once and for all agree that if you come out in the end with 2ϵ, or 22ϵ, that's just as good as coming out with ϵ. If ϵ is an arbitrary small number, so is 22ϵ. Therefore, if you can prove something is less than 22ϵ, you have shown that it can be made as small as desired.

We formulate this as a general principle, the "K-ϵ principle". This isn't a standard term in analysis, so don't use it when you go to your next mathematics congress, but it is useful to name an idea that will recur often.

Principle 3.2   The K-ϵ principle.

Suppose that $\{a_n\}$ is a given sequence, and you can prove that

(5)   *given any $\epsilon > 0$, $a_n \underset{K\epsilon}{\approx} L$ for $n \gg 1$,*

where $K > 0$ is a fixed constant, i.e., a number not depending on n or ϵ.

Then $\lim\limits_{n \to \infty} a_n = L$.

The K-ϵ principle is here formulated for sequences, but we will use it for a variety of other limits as well. In all of these uses, the essential point is that K must truly be a constant, and not depend on any of the variables or parameters.

Questions 3.2

1. In the last (indirect) part of the proof of the Uniqueness Theorem, where did we use the hypothesis $L \neq L'$?

2. Show from the definition of limit that if $a_n \to L$, then $ca_n \to cL$, where c is a fixed non-zero constant. Do it both with and without the K-ϵ principle.

3. Show from the definition of limit that $\lim\left(\dfrac{1}{n+1} - \dfrac{2}{n-1}\right) = 0$.

3.3   Infinite limits.

Even though ∞ is not a number, it is convenient to allow it as a sort of "limit" in describing sequences which become and remain arbitrarily large as n increases. The definition is like the one for the ordinary limit.

Definition 3.3   We say the sequence $\{a_n\}$ *tends to infinity* if

(6)   given any $M > 0$, $a_n > M$ for $n \gg 1$.

In symbols: $\lim\limits_{n \to \infty} \{a_n\} = \infty$,   or $a_n \to \infty$ as $n \to \infty$.

As for regular limits, to establish that $\lim\{a_n\} = \infty$, what you have to do is give an explicit value for the N concealed in "for $n \gg 1$", and prove that it does the job, i.e., prove that $a_n > M$ when $n \geq N$. In general, this N will depend on M: the bigger the M, the further out in the sequence you will have to go for the inequality $a_n > M$ to hold.

As before, it is not you who chooses the M; the limit demon does that, and you have to prove the inequality in (6) for whatever positive M it gives you.

Note also that even though we are dealing with size, we do not need absolute values, since $a_n > M$ means the a_n are all positive for $n \gg 1$.

One should not think that infinite limits are associated only with increasing sequences. Consider these examples, neither of which is an increasing sequence.

Examples 3.3A   Do the following sequences tend to ∞? Give reasoning.

   (i) $\{a_n\} = 1, 10, 2, 20, 3, 30, 4, 40, \ldots, k, 10k, \ldots, \quad (k \geq 1)$;

   (ii) $\{a_n\} = 1, 2, 1, 3, \ldots, 1, k, \ldots, \quad (k \geq 1)$.

Solution.   (i) A formula for the n-th term is $a_n = \begin{cases} 5n, & n \text{ even}; \\ (n+1)/2, & n \text{ odd}. \end{cases}$

This shows the sequence tends to ∞ since (6) is satisfied: given $M > 0$,

$$a_n > M \quad \text{if } (n+1)/2 > M; \text{ i.e., if } n > 2M - 1 \,.$$

   (ii) The second sequence does not tend to ∞, since (6) is not satisfied for every given M: if we take $M = 10$, for example, it is not true that after some point in the sequence all $a_n > 10$, since the term 1 occurs at every odd position in the sequence.

Example 3.3B   Show that $\{\ln n\} \to \infty$.

Solution.   We use the fact that $\ln x$ is an increasing function, that is,

$$\ln a > \ln b \quad \text{if } a > b;$$

therefore,   given $M > 0$, $\ln n > \ln(e^M) = M$ if $n > e^M$.

Questions 3.3

   1. (a) Formulate a definition for $\lim_{n \to \infty} a_n = -\infty$:   "a_n tends to $-\infty$".

      (b) Prove $\ln(1/n) \to -\infty$.

   2. Which of these sequences tend to ∞? For those that do, prove it.

        (a) $(-1)^n n$ (b) $n|\sin n\pi/2|$ (c) $\sqrt{n}$ (d) $n + 10\cos n$

   3. Prove: if $a_n \to \infty$, then a_n is positive for large n.

3.4   An important limit.

As a good opportunity to practice with inequalities and the limit definition, we prove an important limit that will be used constantly later on.

Theorem 3.4   The limit of a^n .

(7)
$$\lim_{n\to\infty} a^n = \begin{cases} \infty, & \text{if } a > 1; \\ 1, & \text{if } a = 1; \\ 0, & \text{if } |a| < 1. \end{cases}$$

Proof.   We consider the case $a > 1$ first. Since $a > 1$, we can write

$$a = 1 + k, \quad k > 0.$$

Thus $a^n = (1+k)^n$,   which by the binomial theorem

$$= 1 + nk + \frac{n(n-1)}{2!}k^2 + \frac{n(n-1)(n-2)}{3!}k^3 + \ldots + k^n.$$

Since all the terms on the right are positive,

(5) $a^n > 1 + nk$;

     $> M$,   for any given $M > 0$, if $n > M/k$,   say.

This proves that $\lim a^n = \infty$ if $a > 1$, according to Definition 3.3.   □

The second case $a = 1$ is obvious. For the third, in outline the proof is:

$$|a| < 1 \quad\Rightarrow\quad \frac{1}{|a|} > 1 \quad\Rightarrow\quad \left(\frac{1}{|a|}\right)^n \to \infty \quad\Rightarrow\quad a^n \to 0 .$$

Here the middle implication follows from the first case of the theorem. The last implication uses the definition of limit; namely, by hypothesis,

$$\text{given } \epsilon > 0, \quad \left(\frac{1}{|a|}\right)^n > \frac{1}{\epsilon} \quad \text{for } n \text{ large;}$$

by the reciprocal law of inequalities (2.1) and the multiplication law for $|\ |$,

$$|a^n| < \epsilon \quad \text{for } n \text{ large.} \qquad \square\square$$

Why did we begin by writing $a = 1 + k$? Experimentally, you can see that when $a > 1$, but very close to 1 (like $a = 1.001$), a increases very slowly at first when raised to powers. This is the worst case, therefore, and it suggests writing a in a form which shows how far it deviates from 1.

The case $a \leq -1$ is not included in the theorem; here the a^n alternate in sign without getting smaller, and the sequence has no limit. A formal proof of this directly from the definition of limit is awkward; instead we will prove it at the end of Chapter 5, when we have more technique.

Questions 3.4

1. Find (a) $\lim_{n\to\infty} \cos^n a$;   (b) $\lim_{n\to\infty} \ln^n a$, for $a \geq 1$.

2. Suppose one tries to prove the theorem for the case $0 < a < 1$ directly, by writing $a = 1 - k$, where $0 < k < 1$ and imitating the argument given for the first

case $a > 1$. Where does the argument break down? Can one prove a^n is small by dropping terms and estimating? Could one use the triangle inequality?

3.5   Writing limit proofs.

Get in the habit of writing your limit proofs using correct mathematical grammar. The proofs in the body of the text and some of the answers to questions (those answers which aren't just brief indications) are meant to serve as models. But it may also help to point out some common errors.

One frequently sees the following usages involving "for n large" on student papers. Your teacher may know what you mean, but the mathematical grammar is wrong, and technically, they make no sense; avoid them.

Wrong	*Right*
$a_n \to 0$ for $n \gg 1$;	$a_n \to 0$ as $n \to \infty$;
$\lim 2^n = \infty$ for $n \gg 1$;	$\lim 2^n = \infty$;
$\lim(1/n) = 0$ if $n > 1/\epsilon$;	$\begin{cases} \lim(1/n) = 0; \\ (1/n) \underset{\epsilon}{\approx} 0 \quad \text{if } n > 1/\epsilon. \end{cases}$

In the first two, the limit statement applies to the sequence as a whole, whereas "for $n >$ some N" can only apply to individual terms of the sequence. The third is just a general mess; two alternatives are offered, depending on what was originally meant.

As we said earlier, "given $\epsilon > 0$" or "given $M > 0$" must come first:

Poor: $1/n < \epsilon$, for $n \gg 1$ (what is ϵ, and who picked it?)

Wrong: For $n \gg 1$, given $\epsilon > 0$, $1/n < \epsilon$.

This latter statement is wrong, because according to mathematical conventions, it would mean that the N concealed in "for $n \gg 1$" should not depend on ϵ. This point is more fully explained in Appendix B; rather than try to study it there at this point, you will be better off for now just remembering to first present ϵ or M, and then write the rest of the statement.

Another point: write up your arguments using plenty of space on your paper (sorry, clarity is worth a tree). Often in the book's examples and proofs, the inequality and equality signs are lined up below each other, rather than strung out on one line; it is like properly-written computer code. See how it makes the argument clearer, and imitate it in your own work. If equalities and inequalities both occur, the convention we will follow is:

$$
\begin{aligned}
A \; &< \; n(n+1) \\
&= \; n^2 + n \, ,
\end{aligned}
\qquad \text{rather than} \qquad
\begin{aligned}
A \; &< \; n(n+1) \\
&< \; n^2 + n \, .
\end{aligned}
$$

The form on the right doesn't tell you explicitly where the second line came from; in the form on the left, the desired conclusion $A < n^2 + n$ isn't explicitly stated, but it is easily inferred.

3.6   Some limits involving integrals.

To broaden the range of applications and get you thinking in some new directions, we look at a different type of limit which involves definite integrals.

Example 3.6A   Let $a_n = \int_0^1 (x^2 + 2)^n dx$. Show that $\lim_{n \to \infty} a_n = \infty$.

The way *not* to do this is to try to evaluate the integral, which would just produce an unwieldy expression in n that would be hard to interpret and estimate. To show that the integral tends to infinity, all we have to do is get a lower estimate for it that tends to infinity.

Solution.   We estimate the integral by estimating the integrand.
$$x^2 + 2 \geq 2 \qquad \text{for all } x;$$
therefore,
$$(x^2 + 2)^n \geq 2^n \qquad \text{for all } x \text{ and all } n \geq 0.$$

Thus
$$\int_0^1 (x^2 + 2)^n dx \geq \int_0^1 2^n dx = 2^n .$$

Since $\lim 2^n = \infty$ by Theorem 3.4, the definite integral must tend to ∞ also:
$$\text{given } M > 0, \quad \int_0^1 (x^2 + 2)^n dx \geq 2^n \geq M, \text{ for } n > \log_2 M. \qquad \square$$

Example 3.6B   Show $\lim_{n \to \infty} \int_0^1 (x^2 + 1)^n dx = \infty$.

Solution.   Once again, we need a lower estimate for the integral that is large. The previous argument gives the estimate $(x^2 + 1)^n \geq 1^n = 1$, which is useless. However, it may be modified as follows.

Since $x^2 + 1$ is an increasing function which has the value $A = 5/4$ at the point $x = .5$ (any other point on $(0, 1)$ would do just as well), we can say
$$x^2 + 1 \geq A > 1 \qquad \text{for } .5 \leq x \leq 1;$$
therefore,
$$(x^2 + 1)^n \geq A^n \qquad \text{for } .5 \leq x \leq 1;$$
since $\lim A^n = \infty$ by Theorem 3.4, the definite integral must tend to ∞ also:
$$\text{given } M > 0, \quad \int_0^1 (x^2 + 1)^n dx \geq \int_{.5}^1 A^n dx = \frac{A^n}{2} \geq M, \text{ for } n \text{ large.} \quad \square$$

Questions 3.6

1. By estimating the integrand, show that: $\dfrac{1}{3} \leq \int_0^1 \dfrac{x^2 + 1}{x^4 + 2} dx \leq 1$.

2. Show without integrating that $\lim_{n \to \infty} \int_0^1 x^n (1 - x)^n dx = 0$.

3.7   Another limit involving an integral.

We give a third example, this one a little more sophisticated than the other two. You can skip it, but it is worth studying and understanding.

Example 3.7   Let $a_n = \displaystyle\int_0^{\frac{\pi}{2}} \sin^n x \, dx$. Determine $\lim a_n$.

Remarks.   As before, attempting to evaluate the integral will lead to an expression in n whose limit is not so easy to determine.

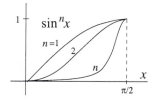

The integral represents an area, so it helps to have some idea of how the curves $\sin^n x$ look. Since $\sin x < 1$ on the interval $0 \le x < \frac{\pi}{2}$, by Theorem 3.4 the powers $\sin^n x \to 0$. Thus as n increases, the successive curves get closer and closer to the x-axis, except that the right-hand end always passes through the point $(\frac{\pi}{2}, 1)$.

The area under the curve seems to get small, as n increases, so the limit should be 0. But how do we prove this?

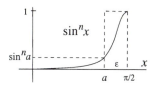

For a big value of n, the area really is composed of two parts, both of which are small, but for different reasons. The vertical strip on the right is small in area because it is thin. The horizontal strip below is small in area because the curve is near the x-axis.

This suggests the following procedure.

Solution.   Given $\epsilon > 0$, we show:

area under $\sin^n x$ and over $[0, \frac{\pi}{2}]\ <\ 2\epsilon$, for $n \gg 1$.

(This suffices, by the K-ϵ principle of section 3.2.)

Mark off the the point a shown, where $a = \frac{\pi}{2} - \epsilon$. Divide up the area under the curve into two pieces as shown, and draw in the two rectangles.

right-hand area $<$ area of right-hand rectangle $= \epsilon$;

left-hand area $<$ area of horizontal rectangle $= a \sin^n a$

$<\ \epsilon$,   for $n \gg 1$,

since $|\sin a| < 1$ (cf. Theorem 3.4). Therefore,

$$\int_0^{\frac{\pi}{2}} \sin^n x \, dx = \begin{matrix}\text{total area} \\ \text{under curve}\end{matrix} < \begin{matrix}\text{area of the} \\ \text{two rectangles}\end{matrix} < 2\epsilon, \quad \text{for } n \gg 1. \qquad \square$$

It would be easy to get rid of the pictures and just use integral signs everywhere, but it wouldn't make the argument more rigorous, just more obscure.

Exercises

3.1

1. Show that the following sequences have the indicated limits, directly from the definition of limit.

(a) $\lim\limits_{n\to\infty} \dfrac{\sin n - \cos n}{n} = 0$ 　　(b) $\lim\limits_{n\to\infty} \dfrac{2n-1}{n+2} = 2$

(c) $\lim\limits_{n\to\infty} \dfrac{n}{n^2 + 3n + 1} = 0$ 　　(d) $\lim\limits_{n\to\infty} \dfrac{n}{n^3 - 1} = 0$

(e) $\lim\limits_{n\to\infty} \sqrt{n^2 + 2} - n = 0$

2. Prove that if a_n is a non-negative sequence, $\lim a_n = 0 \Rightarrow \lim \sqrt{a_n} = 0$.

3.2

1. Prove that if $a_n \to L$ and $b_n \to M$, then $a_n + b_n \to L + M$.

Do this directly from the definition 3.1 of limit.

2. Suppose $\{a_n\}$ is a convergent increasing sequence, and $\lim a_n = L$.

Let $\{b_n\}$ be another sequence "interwoven" with the first, i.e., such that
$$a_n < b_n < a_{n+1} \quad \text{for all } n.$$
Prove from the definition of limit that $\lim b_n = L$ also.

3. (a) Prove the sequence $a_n = \dfrac{1}{n+1} + \dfrac{1}{n+2} + \ldots + \dfrac{1}{2n}$ has a limit.

(b) Criticize the following "proof" that its limit is 0:

Given $\epsilon > 0$, then for $i = 1, 2, 3, \ldots$, we have
$$\frac{1}{n+i} < \epsilon, \quad \text{if } \frac{1}{n} < \epsilon, \text{ i.e., if } n > 1/\epsilon .$$
Adding up these inequalities for $i = 1, \ldots, n$ gives
$$0 < a_n < n\epsilon, \quad \text{for } n > 1/\epsilon ;$$
therefore, 　　　　　　$a_n \underset{n\epsilon}{\approx} 0, \quad \text{for } n \gg 1$.

By the definition of limit and the K-ϵ principle, $\lim a_n = 0$. 　　　　□

4. Prove that $\lim\limits_{n\to\infty} \left(\dfrac{1}{n^2+1} + \dfrac{1}{n^2+2} + \ldots + \dfrac{1}{n^2+n} \right) = 0$.

(Modify the incorrect argument in the preceding exercise.)

5. Let $\{a_n\}$ be a convergent sequence of integers, having the limit L. Prove that it is "eventually constant", that is, $a_n = L$ for large n.

(Apply the limit definition, taking $\epsilon = 1/4$, say. Why is it legal for you to choose the ϵ in this case?)

3.3

1.   For each sequence $\{a_n\}$, tell whether or not $a_n \to \infty$; if so, prove it directly from Definition 3.3 of infinite limit.

(a) $a_n = \dfrac{n^2}{n-1}$ (b) $a_n = n^2|\cos n\pi|$ (c) $a_n = \dfrac{n^3}{n^2+2}$

(d) $a_n = 1,\ 2,\ 3;\ 2,\ 3,\ 4;\ 3,\ 4,\ 5;\ \ldots$ (e) $\sqrt{n}$ (f) $\ln\ln(n)$

2. Prove: if $a_n < b_n$ for $n \gg 1$, and $a_n \to \infty$, then $b_n \to \infty$. Base the proof on Definition 3.3.

3. Prove: if $\{a_n\} \to \infty$, then $\{a_n\}$ is not bounded above. (This is "obvious"; the point is to get practice in using the definitions to construct arguments. Give an indirect proof; see Appendix A.2.)

3.4

1. Define a_n recursively by $a_{n+1} = ca_n$, where $|c| < 1$. Prove $\lim a_n = 0$, using Theorem 3.4 and Definition 3.1.

2. Let $a_n = r^n/n!$. Prove that for all $r \in \mathbb{R}$, we have $a_n \to 0$.

(If $|r| \le 1$, this is easy. If $|r| > 1$, it is more subtle. Compare two successive terms of the sequence, and show that if $n \gg 1$, then $|a_{n+1}|$ is less than half of $|a_n|$. Then complete the argument.)

3. Prove that if $a > 1$, then $a^n/n \to \infty$.

(Hint: imitate the proof in theorem 3.4, but use a different term in the binomial expansion.)

4.   Prove that $na^n \to 0$ if $0 < a < 1$, using the result in the preceding exercise.

5. Prove that $\lim a^{1/n} = 1$, if $a > 0$.

(Here $a^{1/n}$ means the real positive n-th root of a. Nothing is said about $a < 0$, since then $a^{1/n}$ is not a real number if n is even.

Note how $\{a^{1/n}\}$ and $\{a^n\}$ have opposite behavior as sequences. As we take successive n-th roots, all positive numbers *approach* 1; in contrast, as we take successive n-th powers, all positive numbers $\ne 1$ *recede from* 1.)

(a) Consider first the case $a > 1$, fix a value of n, put $a^{1/n} = 1 + h_n$, and show that $h_n \to 0$, by reasoning like that in Theorem 3.4.

(b) If $a < 1$, then $1/a > 1$; use this and the definition of limit to deduce this case from the previous one. (Use only the definition, not other "obvious" facts about limits.)

3.6

1.   Modeling your arguments on the two examples given in this section, prove the following without attempting to evaluate the integrals explicitly.

(a) $\displaystyle\lim_{n\to\infty} \int_1^2 \ln^n x\, dx = 0$ (b) $\displaystyle\lim_{n\to\infty} \int_2^3 \ln^n x\, dx = \infty$

3.7

   1. Show that $\lim\limits_{n\to\infty} \int_0^1 (1-x^2)^n dx = 0$, without attempting to evaluate the integral explicitly.

Problems

   3-1 Let $\{a_n\}$ be a sequence and $\{b_n\}$ be its sequence of averages:
$$b_n = (a_1 + \ldots + a_n)/n \qquad \text{(cf. Problem 2-1).}$$

   (a) Prove that if $a_n \to 0$, then $b_n \to 0$.

(Hint: this uses the same ideas as example 3.7. Given $\epsilon > 0$, show how to break up the expression for b_n into two pieces, both of which are small, but for different reasons.)

   (b) Deduce from part (a) in a few lines without repeating the reasoning that if $a_n \to L$, then also $b_n \to L$.

   3-2 To prove a^n was large if $a > 1$, we used "Bernouilli's inequality":
$$(1+h)^n \geq 1+nh, \quad \text{if } h \geq 0 .$$
We deduced it from the binomial theorem. This inequality is actually valid for other values of h however. A sketch of the proof starts:
$$(1+h)^2 = 1 + 2h + h^2 \geq 1 + 2h, \qquad \text{since } h^2 \geq 0 \text{ for all } h;$$
$$(1+h)^3 = (1+h)^2(1+h) \geq (1+2h)(1+h), \quad \text{by the previous case,}$$
$$= 1 + 3h + 2h^2,$$
$$\geq 1 + 3h .$$

   (a) Show in the same way that the truth of the inequality for the case n implies its truth for the case $n+1$. (This proves the inequality for all n by mathematical induction, since it is trivially true for $n = 1$.)

   (b) For what h is the inequality valid? (Try it when $h = -3$, $n = 5$.) Reconcile this with part (a).

   3-3 Prove that if a_n is a bounded increasing sequence and $\lim a_n = L$ in the sense of Definition 1.3A, then $\lim a_n = L$ in the sense of Definition 3.1.

   3-4 Prove that a convergent sequence $\{a_n\}$ is bounded.

Answers

3.1

1. We write these up in four slightly different styles; take your pick.

(a) Given $\epsilon > 0$, $\left| \dfrac{n}{n+1} - 1 \right| = \dfrac{1}{n+1}$,

which is $< \epsilon$ if $n > 1/\epsilon - 1$, or $n > 1/\epsilon$.

(b) Given $\epsilon > 0$, $\left| \dfrac{\cos na}{n} \right| \leq \dfrac{1}{n}$, since $|\cos x| \leq 1$ for all x;

and $1/n < \epsilon$ if $n > 1/\epsilon$.

(c) Given $\epsilon > 0$, $\left| \dfrac{n^2 + 1}{n^2 - 1} - 1 \right| = \left| \dfrac{2}{n^2 - 1} \right|$,

$< \epsilon$ if $n^2 - 1 > 2/\epsilon$, or $n > \sqrt{2/\epsilon + 1}$.

(d) Given $\epsilon > 0$, $\dfrac{n^2}{n^3 + 1} < \dfrac{1}{n} < \epsilon$ if $n > 1/\epsilon$.

2. $\lim a_n = 0$ means: given $\epsilon > 0$, $|a_n - 0| < \epsilon$ for $n \gg 1$.

$\lim |a_n| = 0$ means: given $\epsilon > 0$, $\big||a_n| - 0\big| < \epsilon$ for $n \gg 1$.

But these two statements are the same, since

$$\big||a_n| - 0\big| \;=\; \big||a_n|\big| \;=\; |a_n| \;=\; |a_n - 0| \,.$$

3.   If the limit demon were allowed to give you $\epsilon = 0$, then since $\underset{0}{\approx}$ is the same as equality $=$, it would have to be true that $a_n = L$ for $n \gg 1$; in other words, any sequence which had the limit L would from some point on have to be constant and equal to L. This would be too restricted a notion of limit. (The sequences which do behave this way are said to be "eventually constant"; cf. Exercise 3.2/5.)

3.2

1. $L \neq L' \Rightarrow \epsilon = |L - L'|/2 > 0$, which is essential (cf. Question 3.1/3).

2. (a) Given $\epsilon > 0$, $|a_n - L| < \epsilon$ for $n > N$, say. Therefore,

$$|ca_n - cL| \;=\; |c||a_n - L| \;<\; |c|\epsilon \text{ for } n > N,$$

so we're done by the K-ϵ principle.

(b) Given $\epsilon > 0$, $|a_n - L| < \epsilon/|c|$ for $n > N_1$, say. Therefore,

$$|ca_n - cL| \;=\; |c||a_n - L| \;<\; |c|\epsilon/|c| = \epsilon \text{ for } n > N_1.$$

3. Given $\epsilon > 0$,

$$\left| \frac{1}{n+1} - \frac{2}{n-1} \right| \;\leq\; \left| \frac{1}{n+1} \right| + \left| \frac{2}{n-1} \right| \qquad \text{(triangle inequality)}$$

$< \epsilon + \epsilon$,   if $n + 1 > 1/\epsilon$, and $n - 1 > 2/\epsilon$;

$< 2\epsilon$,   if $n > 1 + 2/\epsilon$;

so we're done by the K-ϵ principle.

3.3

1. (a) Given $-M < 0$, $a_n < -M$ for $n \gg 1$. (The $-$ signs can be omitted.)

(b) $\ln(1/n) = -\ln n < -M$ if $\ln n > M$, i.e., if $n > e^M$.

2. (a) no; alternate terms are negative;

(b) no; alternate terms are 0;

(c) yes; given $M > 0$, $\sqrt{n} > M$ if $n > M^2$;

(d) yes; given $M > 0$, $n + 10 \cos n > n - 10 > M$, if $n > M + 10$.

3. Given M, $a_n > M$ for $n \gg 1$. Take $M = 0$: $a_n > 0$ for $n \gg 1$.

3.4

1. (a) limit is 0, except: limit is 1 if $a = 2n\pi$, no limit if $a = (2n+1)\pi$.

(b) limit is: $\begin{cases} 0, & \text{if } 1 \le a < e; \\ 1, & \text{if } a = e; \\ \infty, & \text{if } a > e. \end{cases}$

2. Since the terms alternate in sign, one cannot get an inequality after dropping most of the terms to simplify the expression. Basically, a^n is small not because the individual terms are small, but because they cancel each other out. Thus the triangle inequality cannot help either. So this approach doesn't lead to a usable estimation that would show a_n is small.

3.6

1. Over the interval $0 \le x \le 1$,

$$1 \le x^2 + 1 \le 2; \qquad 2 \le x^4 + 2 \le 3.$$

By standard reasoning (see Example 2.2B for instance),

$$\frac{1}{3} \le \frac{x^2 + 1}{x^4 + 2} \le \frac{2}{2};$$

therefore its integral over the interval $[0, 1]$ of length 1 also lies between these bounds.

2. On $[0, 1]$, the maximum of $x(1 - x)$ occurs at $x = 1/2$, therefore

$$0 \le x(1 - x) \le 1/4 \quad \text{on } [0, 1];$$

$$\text{given } \epsilon > 0, \quad 0 \le \int_0^1 x^n(1 - x)^n dx \le (1/4)^n < \epsilon, \text{ if } 4^n > 1/\epsilon ;$$

therefore, denoting the integral by I_n, we see that

$$\text{given } \epsilon > 0, \quad |I_n| < \epsilon \quad \text{if } n > \ln(1/\epsilon)/\ln 4 ,$$

which proves that $\lim I_n = 0$.

4

Error Term Analysis

4.1   The error term.

It is an important practical (and often theoretical) matter to know not just that a sequence $\{a_n\}$ converges to a limit L, but also to have some idea of how rapidly it converges to L. For instance, it can be proved that both of the following sequences converge to $\ln 2$:

$$a_n = 1 - \frac{1}{2} + \frac{1}{3} - \frac{1}{4} + \ldots + \frac{(-1)^{n-1}}{n} \; ;$$

$$b_n = \frac{2}{1 \cdot 3} + \frac{2}{3 \cdot 3^3} + \frac{2}{5 \cdot 3^5} + \ldots + \frac{2}{(2n-1) \cdot 3^{2n-1}} \; .$$

The first however is useless for computing $\ln 2$, because it converges so slowly. The formula for the n-th term shows for instance that

$$a_{100} = a_{99} - \frac{1}{100} \; ,$$

so that at the 100-th term of the sequence, the second decimal place is still changing. By contrast, the other sequence b_n converges rapidly; the term b_3 already gives $\ln 2$ to three decimal places.

To think about questions of this kind, we want to change our point of view about limits in a slight, but important way. Instead of looking at the approximation itself, $a_n \underset{\epsilon}{\approx} L$, we focus our attention on the **error term**

$$e_n = a_n - L \; ,$$

which measures how far away a_n is from its limit. The sequence, its limit, and the error term are related by the easy

Theorem 4.1   Error-form Principle.   Let $a_n = L + e_n$;   then

(1) $a_n \to L \; \Leftrightarrow \; e_n \to 0 \; .$

This change in viewpoint is simple, but useful. Calculations and proofs involving limits are often simpler or more intuitive when your attention is focused on the error term. The rest of this chapter will illustrate.

Questions 4.1

1. Prove Theorem 4.1, from Definition 3.1 of limit.

2. If $a_n = 4n^2/(n^2 - 1)$, find an N such that $|e_n| < .01$ for $n > N$.

4.2   The error in the geometric series. Application.

To illustrate the use of the error term, consider the high-school formula

$$(2) \qquad\qquad 1 + a + a^2 + \ldots + a^n + \ldots = \frac{1}{1-a} , \quad |a| < 1 .$$

This is just an expressive way of writing the limit of a certain sequence; its meaning is given by

Proposition 4.2   The geometric sum limit.   Consider the geometric sum

$$(3) \qquad\qquad a_n = 1 + a + a^2 + \ldots + a^n ;$$

then

$$\lim_{n \to \infty} a_n = \frac{1}{1-a} , \quad \text{if } |a| < 1 .$$

Proof.   The compact formula for a_n is (check it by cross-multiplying):

$$(4) \quad 1 + a + a^2 + \ldots + a^n = \frac{1 - a^{n+1}}{1-a} = \frac{1}{1-a} - \frac{a^{n+1}}{1-a} , \quad a \neq 1.$$

Comparing (4) with (1), we see that the error term for the desired limit is

$$(5) \qquad\qquad e_n = \frac{-a^{n+1}}{1-a} , \qquad a \neq 1 .$$

By Theorem 3.4, if $|a| < 1$, then $a^n \to 0$. It follows that $e_n \to 0$ (see Question 3.2/2); by the error-form principle, this proves Proposition 4.2.   □

We can make similar limit proofs and study convergence rate for sequences related to (3). Here is the one that started off this chapter.

Example 4.2   Let $b_n = 1 - \dfrac{1}{2} + \dfrac{1}{3} - \dfrac{1}{4} + \ldots + \dfrac{(-1)^{n-1}}{n}$; show $\lim_{n \to \infty} b_n = \ln 2$.

Solution.   We substitute $-u$ for a into (4), getting

$$1 - u + u^2 - \ldots + (-1)^{n-1} u^{n-1} = \frac{1}{1+u} - (-1)^n \frac{u^n}{1+u} , \quad u \neq -1.$$

If we integrate both sides from 0 to 1, we get (check this):

$$(6) \quad 1 - \frac{1}{2} + \frac{1}{3} - \frac{1}{4} + \ldots + \frac{(-1)^{n-1}}{n} = \ln 2 \pm \int_0^1 \frac{u^n}{1+u} \, du .$$

The limit we want is $\ln 2$; comparing (6) with the error form (1) shows

$$(7) \qquad\qquad e_n = \pm \int_0^1 \frac{u^n}{1+u} \, du .$$

According to the error-form principle 4.1, we have to show $e_n \to 0$, and to do this we need an upper estimate for $|e_n|$ which shows it is small. Since e_n is the definite integral of a rational function, we could try evaluating the integral exactly, but that would turn out to be worse than useless. (Try it and see.)

To get a small upper estimate for (7), we estimate the integrand:

$$\frac{u^n}{1+u} \leq u^n , \qquad 0 \leq u \leq 1 .$$

Integrating both sides of the inequality from 0 to 1 preserves it; by (7),

$$(8) \qquad |e_n| = \int_0^1 \frac{u^n}{1+u}\, du \leq \int_0^1 u^n du = \frac{1}{n+1} ,$$

which shows (by Definition 3.1 of limit) that $e_n \to 0$; it therefore proves by (6) and the error-form principle (1) that

$$(9) \qquad \lim_{n \to \infty} 1 - \frac{1}{2} + \frac{1}{3} - \frac{1}{4} + \ldots + \frac{(-1)^{n-1}}{n} = \ln 2 . \qquad \square$$

The estimate (8) for the error term shows that $|e_n|$ is bounded by the size of the next term of the sum; in fact, this turns out to be roughly the size of the error. As we pointed out at the beginning of this chapter, this slow rate of convergence makes the sequence (9) a poor way to calculate $\ln 2$. On the other hand, one could hardly wish for a more elegant result!

It is like the hands of Marie Antoinette—reputed to be beautiful, but useless for cutting her meat.

Questions 4.2

1. Go from 0 to 1 on the x-axis, then back half-way to $\frac{1}{2}$, then forward half as far to $\frac{3}{4}$, then back half as far to $\frac{5}{8}$, then forward half as far, and so on.

   (a) Give an explicit formula for your position after the n-th step.

   (b) Where do you end up, and what does "end up" mean?

2. Taking a_n as in (3), with $a = \frac{1}{4}$, what is $\lim a_n$, and how large must n be for a_n to agree with its limit to two decimal places? (Study the error term.)

3. If you try to study $1 + \frac{1}{2} + \frac{1}{3} + \frac{1}{4} + \ldots + \frac{1}{n}$ by using the argument of Example 4.2, replacing a in (4) by u (instead of $-u$), at what two places does the resulting argument break down?

4.3   A sequence converging to $\sqrt{2}$:   Newton's method.

In the rest of this chapter, we illustrate the use of the error form on sequences whose general term a_n is not given directly in terms of n, but instead is given recursively by a formula involving a_{n-1} and perhaps previous terms as well. Such sequences are the normal thing one encounters in numerical analysis and computation, for example. Establishing the convergence of such sequences often requires some thought.

Consider for example Newton's method, which is a numerical method for locating a zero α of a given function $f(x)$ to any accuracy desired. It produces a recursively defined sequence $\{a_n\}$ which converges to α (if all goes well).

The method is shown at the right. We start with some approximation a_0 to the zero α, then get a_1 as shown, by seeing where the tangent to the graph of $f(x)$ at P_0 intersects the x-axis. We then repeat the process starting with a_1; this gives us a_2, and so on.

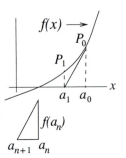

The analytic formula connecting a_n and a_{n+1} is obtained by equating two expressions for the slope of the hypotenuse of the triangle shown. This gives

$$f'(a_n) = \frac{f(a_n)}{a_n - a_{n+1}} \ ;$$

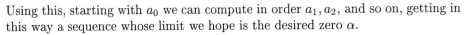

solving for a_{n+1} gives the desired formula:

$$(10) \qquad a_{n+1} = a_n - \frac{f(a_n)}{f'(a_n)} \ .$$

Using this, starting with a_0 we can compute in order a_1, a_2, and so on, getting in this way a sequence whose limit we hope is the desired zero α.

Note that the sequence is calculated recursively: each term is calculated from the preceding one. There is in general no formula for a_n purely in terms of n.

Example 4.3   Find a sequence $\{a_n\}$ such that $a_n \to \sqrt{2}$, by using Newton's method; prove the sequence converges to $\sqrt{2}$, and study its rate of convergence.

Solution.   We use Newton's method to find the positive zero of $f(x) = x^2 - 2$. By using (10) and algebra, we see the sequence $\{a_n\}$ is given recursively by

$$(11) \qquad a_{n+1} = \frac{1}{2}\left(a_n + \frac{2}{a_n}\right) .$$

The picture suggests that any starting a_0 close enough to $\sqrt{2}$ will generate a sequence converging to $\sqrt{2}$. To prove this, by 4.1 we should prove $e_n \to 0$, where

$$e_n = a_n - \sqrt{2}.$$

Since we have no explicit formula for a_n, we have none for e_n either; the best we can do is use (11) to relate e_{n+1} to e_n, and hope that somehow out of the resulting formula we will be able to see that $e_n \to 0$.

Replacing n by $n + 1$ in the above expression for e_n and using (11), we get

$$e_{n+1} = \frac{1}{2}\left(a_n + \frac{2}{a_n}\right) - \sqrt{2} \ ;$$

using $a_n = \sqrt{2} + e_n$, this becomes

$$e_{n+1} = \frac{1}{2}\left((\sqrt{2} + e_n) + \frac{2}{\sqrt{2} + e_n}\right) - \sqrt{2} \ .$$

Things look bleak, but if we put all the terms on the right side over a common denominator, we get a dramatic simplification:

$$(12) \qquad e_{n+1} = \frac{e_n^2}{2(\sqrt{2} + e_n)} \ .$$

To show that e_{n+1} gets small, we must show that the denominator is *not* small, i.e., we need a lower estimate for it. This is just the sort of situation which calls for the difference form of the triangle inequality 2.4 (4b); it gives

$$|\sqrt{2} + e_n| \geq \sqrt{2} - |e_n|$$
$$> 1.4 - .9 = .5 ,$$

provided $|e_n| < .9$.

This estimation shows that as soon as $|e_n| < .9$, the denominator of (12) is greater than 1, and then (12) tells us

$$|e_{n+1}| < e_n^2 .$$

If we choose a starting value a_0 for which $|e_0| < .9$, we see that

$$|e_1| < e_0^2, \quad |e_2| < e_1^2 < e_0^4, \quad ..., \quad |e_n| < e_0^{2^n} ;$$

this shows that all subsequent $|e_n| < .9$; moreover it shows by Theorem 3.4 that $e_n \to 0$, and the proof is complete. $\square$

Since $|e_{n+1}| < e_n^2$, the size of the error term is at least squared with each successive step, which shows it tends very rapidly to 0. Indeed, once it gets less than .1, it will be at most .01 at the next step, then .0001, so that in general, at each successive step the number of correct decimal places should double.

Questions 4.3

1. (a) Assuming $a_n > 0$, show that if a_n is on one side of $\sqrt{2}$, then $2/a_n$ is on the other. (So it makes sense in (11) to take a_{n+1} to be their average.)

(b) For which a_0 will the sequence defined by (11) converge to $\sqrt{2}$? (Draw the graph of $x^2 - 2$ and reason geometrically.)

(c) Starting with $a_0 = 1$, verify by calculator that the number of correct decimal places doubles at each step. (Use 8 decimal places.)

2. Just for practice: what sequence does Newton's method produce for finding the zero of $f(x) = x^2$?

Starting with $a_0 = 1$, how many steps will it take to get within .001 of the zero?

Why does the method converge so much more slowly than it did for $\sqrt{2}$? (Look at the graph.)

4.4   The sequence of Fibonacci fractions.

The *Fibonacci sequence* 0, 1, 1, 2, 3, 5, 8, 13, ... is the sequence which starts with 0, 1 and for which each succeeding term is the sum of the preceding two terms. The corresponding *sequence of Fibonacci fractions* is formed by taking the ratios of successive terms of the Fibonacci sequence:

(13)
$$\frac{1}{1}, \quad \frac{1}{2}, \quad \frac{2}{3}, \quad \frac{3}{5}, \quad \frac{5}{8}, \quad \frac{8}{13}, \quad \frac{13}{21}, \quad \ldots ;$$
$$1.000, \ .500, \ .667, \ .600, \ .625, \ .614, \ .619, \ \ldots , \text{ to three places.}$$

If you continue calculating, experimentally the sequence (13) seems to have a limit $M = .618\ldots$. We would like to determine this number M exactly, and prove it is indeed the limit of the sequence.

Note first that the general term a_n is not given by an explicit formula in n but rather by a *recursion formula* which tells you how to get a_{n+1} from a_n. We begin by making this recursion formula explicit. Writing a_n as the quotient of two integers C and D, according to its rule of formation this part of the Fibonacci sequence is $\ldots, C, D, C + D, \ldots$, so we have

$$a_n = \frac{C}{D}, \qquad a_{n+1} = \frac{D}{C+D},$$

from which, dividing numerator and denominator by D, we get the formula

(14)
$$a_{n+1} = \frac{1}{a_n + 1}.$$

If we assume that $M = \lim a_n$ exists, it is surprisingly easy to see what its value must be. Namely, since M is the limit, for large n both a_{n+1} and a_n should be essentially indistinguishable from M, so (14) tells us

(15)
$$M = \frac{1}{M+1}, \qquad \text{or} \qquad M^2 + M - 1 = 0.$$

This quadratic equation has the unique positive root

$$M = \frac{\sqrt{5}-1}{2} = .618\ldots.$$

To see that this number M actually is the limit of $\{a_n\}$, we examine the error term $e_n = a_n - M$, and try to show $e_n \to 0$. Since we have no explicit expression for a_n, we have none for e_n either, and so as in 4.3, the best we can do is to use (14) to get a recursion formula expressing e_{n+1} in terms of e_n and hope it tells us something. We have

$$e_{n+1} = a_{n+1} - M = \frac{1}{a_n + 1} - M, \qquad \text{by (1) and (14);}$$

$$= \frac{1}{e_n + M + 1} - M, \qquad \text{by (1);}$$

$$= \frac{1 - M - M^2 - Me_n}{e_n + M + 1}, \qquad \text{by algebra.}$$

This last expression simplifies by using (15), and we get finally

(16) $\qquad e_{n+1} = -\dfrac{M}{e_n + M + 1}\, e_n = -\dfrac{\sqrt{5}-1}{2e_n + \sqrt{5}+1}\, e_n\ .$

We want to show e_n gets small, so we want an upper estimate for the coefficient; for this we need a lower estimate for the denominator. Using the difference form of the triangle inequality (2.4b), and $\sqrt{5} \approx 2.23$, we get

$$|\sqrt{5}+1+2e_n| \ \geq\ 2.2+1-2(.2)\ =\ 2.8, \qquad\qquad \text{if } |e_n| \leq .2\ ;$$

and so from (16) we get the upper estimate (for the top we use $\sqrt{5} < 2.3$):

(17) $\qquad\qquad |e_{n+1}| \ <\ \dfrac{1.3}{2.8}\,|e_n| \ <\ \dfrac{1}{2}\,|e_n|\ , \qquad\qquad \text{if } |e_n| \leq .2\ .$

Since (13) shows that $|e_2| \approx .12$, it follows recursively from (17) that for all $n \geq 2$, we have $|e_n| < .2$; therefore (17) is valid for all $n \geq 2$.

Since (17) shows the error $|e_n|$ is at least halved at each step, it follows that $e_n \to 0$; consequently $a_n \to M$, by the error-form principle 4.1 .

To check this numerically, call the first error term e_1 and calculate the sequence of e_n from the original sequence (13) (remember that $M = .618\ldots$):

$$.382, \quad -.118, \quad .049, \quad -.018, \quad .007, \quad -.004, \quad .001, \quad \ldots\ .$$

As (16) predicts, the error terms alternate in sign, and as (17) shows, each is in size less than half the preceding one (up to round-off error).

The ancient Greeks thought the most beautiful rectangle was one for which

$$\frac{\text{width}}{\text{length}} = \frac{\text{length}}{\text{length} + \text{width}}\ .$$

Let $R =$ width/length, and on the right, divide numerator and denominator by length. The equation becomes $R = 1/(1 + R)$, which means $R = M$, by (15) above.

Thus M is the ratio of width/length for the most beautiful rectangle; the Greeks called it the "golden mean". Since the sequence (13) has M as its limit, its successive terms give the width/length ratio for presumably ever more beautiful rectangles.

Questions 4.4

1. Consider the sequence $\{a_n\}$ defined recursively by

$$a_{n+1} = 1/a_n, \qquad a_0 = k\ .$$

If this sequence is used in place of the sequence defined by (14) and the rest of the argument is followed through (equation (15)), we conclude that the limit of the sequence is $L = \pm 1$. But this is not correct. Where is the error?

Exercises

4.1

1. If $a_n \to L$ and $b_n \to M$, then

(i) $a_n + b_n \to L + M$;   (ii) $ca_n \to cL$.

Prove, using the error form for the limits and the K-ϵ principle (3.2).

4.2

1. By mimicking the argument in 4.2, prove Leibniz' famous result

$$\frac{\pi}{4} = \lim_{n\to\infty} 1 - \frac{1}{3} + \frac{1}{5} - \frac{1}{7} + \ldots + \frac{(-1)^n}{2n+1} \quad :$$

(a) derive a formula $1 - u^2 + u^4 - \ldots + (-1)^n u^{2n} = \dfrac{1}{1+u^2} + e_n(u)$,

where the error term $e_n(u)$ depends on u;

(b) integrate every term from $u = 0$ to $u = 1$, and show the integral of the error term tends to 0 as $n \to \infty$. (Compare it with another definite integral.)

2. Proceed as in the preceding exercise, with the following two changes:

(a) substitute u^2 for a in (4);   (b) integrate from 0 to 1/3.
What sequence do you get? What is its limit? Prove it converges to this limit.

4.3

1. (a) Suppose Newton's method is used to find $\sqrt{3}$, by finding the positive zero of $x^2 - 3$. Give the recursion formula for a_{n+1} in terms of a_n.

(b) Obtain a recursion formula for the error term $e_n = a_n - \sqrt{3}$. Use it to prove that $a_n \to \sqrt{3}$, if the starting value a_0 lies in a suitable interval.

2. Sketch the graph of $f(x) = x^3 - x - 1$ (sketch $x^3 - x$ first). This function has a unique positive zero; call it K. Show by intuitive geometric reasoning based on the graph of $f(x)$ that if you start with $a_0 > 1/\sqrt{3}$, and use Newton's method, the resulting sequence converges to K. (Use calculus to find out what you need to know about the graph of $f(x)$.)

3. Using Newton's method to find M, the unique positive zero of $x^2 + x - 1$:

(i) Give the recursion formula for a_{n+1} in terms of a_n.

(ii) Obtain a recursion formula for the error term $e_n = a_n - M$. Use it to prove $a_n \to M$, if a_0 lies in a suitable interval. (Hint: in the algebraic calculations, it is best to leave M as a letter; at a certain point you can simplify the expressions by using $M^2 + M - 1 = 0$.)

(iii) Section 4.4 describes another sequence that converges to M. Read the conclusion of section 4.4; which sequence converges faster to M, and why?

4.4

1. Start with any positive rational number A/B (here A and B are positive integers). Form a new one having numerator $2A$ and denominator $A + B$. Repeat this process on the new number. Continuing, you get a sequence of rational numbers. Guess what its limit L is (try an example; cf. (15), 4.4). Then by finding the recursion formula for the error term e_n, prove that the sequence converges to L if   (a) $A > B$ (b) $A < B$.

2. Proceed as in the preceding exercise, but take $3A$ as the new numerator each time. What is the limit of the new sequence? Prove it.

3. Define a sequence recursively by $a_{n+1} = 2a_n^2$, $a_0 > 0$.

(a) As in 4.4, show that if $L = \lim a_n$ exists, then $L = 0$ or $L = 1/2$.

(b) Show that the limit is in general not $1/2$ by proving that:
$$a_0 < 1/2 \Rightarrow \lim a_n = 0; \qquad a_0 > 1/2 \Rightarrow \lim a_n = \infty.$$

Problems

4-1 Prove that $n^{1/n} \to 1$ as $n \to \infty$.
(Use the error form principle 4.1; cf. the hint for Exercise 3.4/3.)

4-2 Pick a positive number between 0 and $\pi/2$, take its cosine, then take the cosine of that number, and keep on taking cosines. You get a sequence $\{a_n\}$ given by $a_{n+1} = \cos a_n$.

(a) Try it on a calculator a few times. What eight place decimal number L do you end up with? What equation is it a root of?

(b) Prove the sequence converges to this limit, by studying the error term and showing it tends to 0 in the limit; cf. the examples in this chapter.
(Use the estimations $|1 - \cos x| \leq x^2/2$ and $|\sin x| \leq |x|$, valid for all x.)

4-3 *A sequence for calculating the cube root of a number:* $A = \sqrt[3]{B}$.

(a) If $a > 0$, show A always lies between a and B/a^2 (if $a < A$, then $B/a^2 > A$, etc.).

(b) The preceding suggests that, starting with a, a *weighted average* of the two numbers a and B/a^2 will be a good next approximation to A. So we write
$$a_{n+1} = r\,a_n + s\,B/a_n^2 , \qquad r + s = 1, \ r > 0, \ s > 0;$$
we try to fix r and s so the error $e_n = a_n - A$ goes to 0 most rapidly.
Show the corresponding expression for e_{n+1} in terms of e_n is
$$e_{n+1} = r(A + e_n) + \frac{sA}{(1 + e_n/A)^2} - A .$$

(c) Using the approximation: $\dfrac{1}{(1 + \epsilon)^2} \approx (1 - \epsilon + \epsilon^2)^2 \approx 1 - 2\epsilon + 3\epsilon^2$,
show the above becomes, when written in powers of e_n, approximately
$$e_{n+1} \approx (r - 2s)e_n + \frac{3s}{A} \cdot e_n^2 .$$

(d) What values of r and s will give most rapid convergence of $e_n \to 0$? Show that with this choice, the resulting recursion formula for a_n is the same as the one produced by Newton's method (take $f(x) = x^3 - B$).

Answers to Questions

4.1

1. Since (1) shows that $|a_n - L| = |e_n|$, we have:

given $\epsilon > 0$, $\quad |a_n - L| < \epsilon \quad$ for $n \gg 1 \quad \Leftrightarrow \quad |e_n| < \epsilon \quad$ for $n \gg 1$.

2. We have $a_n = \dfrac{4}{1 - 1/n^2}$, which shows that $a_n \to 4$ as $n \to \infty$; thus

$$|e_n| = \left| \frac{4n^2}{n^2 - 1} - 4 \right| = \left| \frac{4}{n^2 - 1} \right| ,$$
$$< .01 \quad \text{if } n^2 - 1 > 400, \text{ or } n \geq 21 .$$

4.2

1. (a) $1 - 1/2 + 1/4 - \ldots \pm 1/2^{n-1} = \frac{2}{3}(1 - (-1/2)^n)$, by formula (2);

(b) $\lim a_n = 2/3$.

2. $\lim a_n = \dfrac{1}{1 - 1/4} = \dfrac{4}{3}; \quad |e_n| = \left(\dfrac{1}{4}\right)^{n+1} \cdot \dfrac{4}{3}$, which is $< .005$, if $n \geq 4$.

3. The right side of (6) becomes $\displaystyle\int_0^1 \frac{du}{1-u} - \int_0^1 \frac{u^n \, du}{1 - u}$.

The first integral is infinite; for the integrand of the second, there is no easy analog of the estimate u^n, since the denominator becomes arbitrarily small as $u \to 0$ through positive values.

4.3

1. (a) $0 < a_n < \sqrt{2} \Rightarrow \dfrac{2}{a_n} > \dfrac{2}{\sqrt{2}} = \sqrt{2} \quad$ and $\quad a_n > \sqrt{2} \Rightarrow \dfrac{2}{a_n} < \sqrt{2}.$

(b) For all $a_0 > 0$.

(c) To eight decimal places the sequence is:

1., 1.5, 1.41666667, 1.41421569, 1.41423562 $(= \sqrt{2})$,

which shows the sequence converges very rapidly, and the number of correct decimal places doubles at each step.

2. The sequence is $a_{n+1} = a_n/2$. Since the zero of x^2 is at $x = 0$, in this case $a_n = e_n$. The error halves at each step, so if $a_0 = 1$, we get $a_n = 1/2^n$, and it therefore takes 10 steps ($a_{10} = 1/2^{10} = 1/1028$) to get $|a_n| < .001$. The convergence is slow since the graph is tangent to the x-axis at the zero.

4.4

1. The argument only shows that *if* the limit exists, it is ± 1; but the limit actually does not exist.

5

The Limit Theorems

5.1 Limits of sums, products, and quotients.

This chapter is devoted to a group of theorems which are of great help in finding the limits of sequences: they use known limits to find new limits. Their proofs will give good practice in working with the definition of limit.

The first theorem tells how to find the limit of the new sequence formed when you combine two sequences $\{a_n\}$ and $\{b_n\}$ algebraically.

Theorem 5.1 *Assume that* $a_n \to L$ *and* $b_n \to M$, *as* $n \to \infty$.

(1) **Linearity Theorem** $\qquad ra_n + sb_n \;\to\; rL + sM, \quad$ *for any* $r, s \in \mathbb{R}$;

(2) **Product Theorem** $\qquad\qquad a_n b_n \;\to\; L \cdot M$;

(3) **Quotient Theorem** $\qquad\quad b_n / a_n \;\to\; M/L$, *if* $L, a_n \neq 0$ *for all* n.

Before giving the proofs, we illustrate with an example.

Example 5.1A By using the limit theorems, show that
$$\lim_{n \to \infty} \frac{3n^2 - 2n - 1}{n^2 + 1} \;=\; 3 \;.$$

Solution. We divide numerator and denominator by n^2, getting
$$\frac{3 - 2/n - 1/n^2}{1 + 1/n^2} \;.$$

Since $1/n \to 0$, the Product Theorem (2) shows $1/n^2 \to 0$ and $2/n \to 0$. Then the Linearity Theorem (extended to several terms) shows the numerator $\to$ 3 and the denominator $\to 1$, so by the Quotient Theorem, the limit is $3/1$. $\qquad\square$

Observe that the limit theorems actually give the limit L of the sequence, whereas the definition of limit (3.1) only allows you to verify that a given L is indeed the limit — it doesn't tell you what L is if you don't already know.

In addition, the limit theorems give a quick way of proving convergence without using ϵ-N arguments. To do Example 5.1A just from the definition of limit, we would have to show that
$$\text{given } \epsilon > 0, \quad \left| \frac{3n^2 - 2n - 1}{n^2 + 1} - 3 \right| < \epsilon, \quad \text{for } n \gg 1 \;.$$

We can do this with some algebra and estimations, but it is tedious. The limit theorems make these error estimations for us once and for all: they occur in the proofs of the limit theorems, and then need not be repeated.

Proofs. The Linearity and Product Theorems are proved most straightfor-wardly by using the error form of the limit (4.1). We prove the Product Theorem here; the proof of the Linearity Theorem goes the same way, but is even easier, and left as Question 5.1/3.

For all three theorems, the hypotheses say

(4) $\quad a_n = L + e_n, \quad e_n \to 0 \quad$ and $\quad b_n = M + e'_n, \quad e'_n \to 0 \; ; \quad$ i.e.,

(5) $\qquad$ given $\epsilon > 0, \quad |e_n| < \epsilon$ and $|e'_n| < \epsilon \quad$ for $n \gg 1$.

Proof of the Product Theorem. We multiply the equations in (4), getting

(6) $\qquad a_n b_n = L \cdot M + (e_n M + e'_n L + e_n e'_n)$.

The error term is the one in parentheses, and we want to show its limit is 0.

To get a small upper estimate for this error term $e_n M + e'_n L + e_n e'_n$, we use the triangle inequality. Given $\epsilon > 0$,

$$
\begin{aligned}
|e_n M + e'_n L + e_n e'_n| &\leq |e_n||M| + |e'_n||L| + |e_n||e'_n| \\
&< \epsilon|M| + \epsilon|L| + \epsilon \cdot \epsilon \quad \text{by (5)} \\
&< \big(|M| + |L| + 1\big)\,\epsilon, \quad \text{since we may assume } \epsilon < 1.
\end{aligned}
$$

So we are done by the K-ϵ principle 3.2 (taking $K = |M| + |L| + 1$). $\qquad \square$

To prove the Product Theorem directly, without using the error form, we have to give a small upper estimate for $|a_n b_n - LM|$. But it is not clear at first how to do this. There is however a fairly straightforward method, explained in Exercise 5.1/3, which you should study.

Proof of the Quotient Theorem. It will be enough to prove that

(7) $\qquad\qquad\qquad\qquad \dfrac{1}{a_n} \to \dfrac{1}{L} , \quad L \neq 0,$

since the Product Theorem (2) will then show that

$$
\frac{b_n}{a_n} = b_n\Big(\frac{1}{a_n}\Big) \to M\Big(\frac{1}{L}\Big) = \frac{M}{L} .
$$

In proving (7), the error form at first offers no advantages, so we proceed directly. By algebra, and the multiplication law for absolute values,

(8) $\qquad\qquad\qquad \left|\dfrac{1}{a_n} - \dfrac{1}{L}\right| = \dfrac{|L - a_n|}{|a_n||L|} \; ;$

to show the quotient on the right is small, we must estimate the numerator from above, which is easy, and $|a_n|$ from below (to show the denominator is not too small). To get this lower estimate, use the error form (4) and (5): since $a_n \to L$,

given $\epsilon > 0, \quad a_n = L + e_n, \quad$ where $|e_n| < \epsilon$ for $n \gg 1$.

Using the difference form of the triangle inequality (2.4b) on $|a_n| = |L + e_n|$,

$$|a_n| \geq |L| - |e_n|;$$
$$> |L| - \epsilon, \qquad \text{since } |e_n| < \epsilon;$$
$$> |L|/2, \qquad \text{if } \epsilon < |L|/2 ,$$

which we may assume since $L \neq 0$.

With this lower estimate, we can now estimate the right side of (8): for the numerator we have, by (4) and (5),

$$|L - a_n| = |e_n| < \epsilon ;$$

using our lower estimate for the denominator, we get

$$\left| \frac{1}{a_n} - \frac{1}{L} \right| = \frac{|L - a_n|}{|a_n||L|} < \frac{\epsilon}{|L|/2 \cdot |L|} , \qquad \text{for } n \gg 1,$$

which proves (7), in view of the K-ϵ principle, taking $K = \dfrac{1}{|L|^2/2}$. $\square$

Theorem 5.1∞   Algebraic operations for infinite limits.

(9) $a_n \to \infty,$ $\begin{cases} b_n \to \infty, \ b_n \to L > 0, \text{ or} \\ b_n \text{ bounded below} \end{cases}$ $\Rightarrow$ $a_n + b_n \to \infty ;$

(10) $a_n \to \infty,$ $\begin{cases} b_n \to \infty, \ b_n \to L > 0, \text{ or} \\ b_n \geq K > 0 \text{ for } n \gg 1 \end{cases}$ $\Rightarrow$ $a_n b_n \to \infty ;$

(11) $a_n \to \infty \ \Rightarrow \ 1/a_n \to 0 ;$

(12) $a_n \to 0, \ a_n > 0 \text{ for all } n \ \Rightarrow \ 1/a_n \to \infty .$

Proofs.   Left as exercises, based on the definition of infinite limit (3.3).

In (9) and (10) the first two of the alternative hypotheses are stronger, but often more convenient to use. The converse to (11) is false (Question 5.1/3). Occasionally we will use analogous results for sequences which tend to $-\infty$.

Example 5.1B   Find $\lim\limits_{n \to \infty} n(a + \cos n\pi)$, for different values of a.

Solution.   We have $\cos n\pi = (-1)^n$.

If $a > 1$, then $a + \cos n\pi \geq a - 1 > 0$, so the limit is ∞, by (10).

If $a < -1$, then $a + \cos n\pi \leq a + 1 < 0$, so the limit is $-\infty$, by the analog to (10) for negatively infinite limits.

If $a = \pm 1$, the terms alternate between 0 and $\pm 2n$; there is no limit.

If $|a| < 1$, the terms alternate in sign, but tend to ∞ in size, so again there is no limit. (We accept for now these last two cases as intuitive; the end of the chapter will justify them more rigorously.) $\square$

Questions 5.1

1. Evaluate, using the limit theorems.

(a) $\lim_{n\to\infty} \dfrac{n^2+n}{1-2n^2}$ (b) $\lim_{n\to\infty} \left(\dfrac{n}{1-n}\right)\left(\dfrac{1-2n^2}{n^2}\right)$

2. How can you assume $\epsilon < 1$ in the proof of the Product Theorem, or assume $\epsilon < |L|/2$ in the proof of the Quotient Theorem? Don't you have to work with whatever ϵ the limit demon gives you?

3. Prove the Linearity Theorem (1); follow in outline the proof of the Product Theorem.

4. Prove (11); give a counterexample to the converse.

5. Prove (9), using the last of the alternative hypotheses on b_n.

5.2   Comparison theorems.

Another type of limit theorem is somewhat different from the ones in the previous section. It uses inequalities to compare a given sequence with other sequences whose limits we already know. The most widely used has several forms and goes under various picturesque names; here it's the "Squeeze Theorem".

Theorem 5.2   Squeeze Theorem for limits of sequences.

We are given three sequences $\{a_n\}$, $\{b_n\}$, and $\{c_n\}$, such that

$$a_n \leq b_n \leq c_n \quad \text{for } n \gg 1 .$$

Suppose that $a_n \to L$ and $c_n \to L$; then $b_n \to L$ also.

In words, if the two outside sequences both converge, and to the same limit L, then the one caught in the middle also converges, and to this same limit. To the physically adept, this is the "flycatching theorem", others call it the "funnel theorem", and still others the "sandwich theorem", which at least suggests the hypotheses, if not the conclusion.

Proof.   We will need the two standard and equivalent versions of the approximation statement:

(13) $a_n \underset{\epsilon}{\approx} L \quad \Leftrightarrow \quad L-\epsilon < a_n < L+\epsilon .$

To prove the Squeeze Theorem, the hypotheses say that, given $\epsilon > 0$,

$$a_n \underset{\epsilon}{\approx} L \quad \text{and} \quad c_n \underset{\epsilon}{\approx} L \quad \text{for } n \gg 1 .$$

We use (13) to turn these approximations into inequalities; combining them with the inequalities given in the hypotheses, we get

$$L-\epsilon < a_n \leq b_n \leq c_n < L+\epsilon , \quad \text{for } n \gg 1 .$$

By (13) again, this shows $b_n \underset{\epsilon}{\approx} L$ for $n \gg 1$, as was to be proven. $\square$

Example 5.2A   Show that $\sqrt[n]{2 + \cos na} \to 1$, for any fixed number a.

Solution.   We need the implication: for $a, b > 0$,

(14) $$a \le b \;\Rightarrow\; \sqrt[n]{a} \le \sqrt[n]{b} \,,$$

whose proof by contraposition is immediate: $\sqrt[n]{a} > \sqrt[n]{b} \;\Rightarrow\; a > b$.

We will also use the result of Exercise 3.4/5: $a > 0 \;\Rightarrow\; \lim\limits_{n \to \infty} \sqrt[n]{a} = 1$.

We turn now to the solution. Since $-1 \le \cos na \le 1$, we have

$$1 \le 2 + \cos na \le 3 \quad \text{for all } n \,;$$

applying (14) then gives the line of inequalities below, and the proof is concluded by applying the Squeeze Theorem; note that the right-hand sequence tends to the limit 1 by the result in Exercise 3.4/5 quoted above.

$$
\begin{array}{ccccc}
1 & \le & \sqrt[n]{2 + \cos na} & \le & \sqrt[n]{3} & \quad \text{for all } n; \\
\downarrow & & \Downarrow & & \downarrow \\
1 & & 1 & & 1 & \quad \text{as } n \to \infty \,.
\end{array}
$$

The format is an expressive one we will use to summarize the squeeze argument. It says that the two outside sequences tend to the limit 1 (single arrows); therefore (double arrow), the middle sequence also tends to the limit 1. This format is not standard notation; you won't see it outside this book.

Those who clap flies between their palms refer to the above variant of the squeeze argument (in which one of the outside sequences is a constant sequence) as the "fly-swatting theorem".

Before giving another example, we remark that there is an analogue of the Squeeze Theorem involving infinite limits, but not much squeezing is going on: it is more like a "flotation theorem". The proof is fairly immediate, and left as a Question.

Theorem 5.2∞ Squeeze Theorem for infinite limits.

$$a_n \to \infty, \quad b_n \ge a_n \;\Rightarrow\; b_n \to \infty \,.$$

We have been using this without comment since Chapter 1, to prove that sequences were unbounded. For example, we proved in Section 1.5 that

$$1 + 1/2 + 1/3 + \ldots + 1/n > \ln n,$$

so that by Theorem 5.2∞, the sequence on the left tends to infinity, since $\ln n$ does.

Example 5.2B   Prove: $a > 1 \;\Rightarrow\; a^n \to \infty$.

Proof.   This is Theorem 3.4; the proof there can now be shortened to:

$$a > 1 \;\Rightarrow\; a = 1 + k, \quad k > 0;$$
$$\Rightarrow\; a^n > 1 + nk, \quad \text{by the binomial theorem;}$$
$$\Rightarrow\; a^n \to \infty, \qquad \text{by Theorem 5.2}\infty,$$

since $\lim\limits_{n \to \infty} 1 + nk = \infty$ if $k > 0$. $\qquad\qquad\qquad\qquad\qquad\square$

There are many applications of the Squeeze Theorem which involve the definite integral and the sums associated with it. Here is a typical one, closely related to a well-known estimation of $n!$ known as Stirling's Formula (see Chapter 20).

Example 5.2C Show that $\lim\limits_{n \to \infty} \dfrac{\ln n!}{n \ln n} = 1$.

Remark. This result is usually written more expressively as

$$\ln n! \sim n \ln n,$$

where the symbol $\sim$, read "is asymptotic to", means exactly that the ratio of the two sides tends to 1 as $n \to \infty$. The result gives a rough estimate of the size of $\ln n!$, and therefore also of $n!$, if n is large. We will get a better estimate later on, but it will cost us more effort.

Solution. The trick is to observe that

$$\ln n! = \ln 1 + \ln 2 + \ldots + \ln n .$$

Referring to the picture, the sum is the total area of the rectangles shown. We compare it with the area under the curve $\ln x$ and over the interval $[1, n]$; using integration by parts, this area

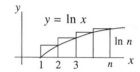

$$= \int_1^n \ln x \, dx ;$$

$$= x \ln x - x \Big|_1^n ;$$

$$= n \ln n - n + 1 .$$

Since the curved area lies below the rectangles, we get the first inequality below:

(*) $\underset{\text{area under curve}}{n \ln n - n + 1} \; \le \; \underset{\text{total area of rectangles}}{\ln 1 + \ln 2 + \ldots + \ln n} \; \le \; n \ln n$.

The right-hand inequality holds since the sum has n terms, each $\le \ln n$. Divide the chain of inequalities (*) through by $n \ln n$, then use the Squeeze Theorem:

$$1 - \frac{1}{\ln n} + \frac{1}{n \ln n} \;\le\; \frac{\ln(n!)}{n \ln n} \;\le\; 1 \;;$$

$$\quad\;\; \downarrow \qquad\qquad\qquad \Downarrow \qquad\;\; \downarrow$$

$$1 + 0 + 0 \qquad\qquad\quad 1 \qquad\;\; 1 \;\; .$$

Questions 5.2

1. Evaluate the limits and prove you are correct by the Squeeze Theorem.

   (a) $\lim\limits_{n \to \infty} \dfrac{\cos n}{n}$ (b) $\lim\limits_{n \to \infty} \dfrac{n - \sin n}{1 + 3n}$ (c) $\lim\limits_{n \to \infty} \left(1 + \dfrac{1}{n}\right)^{3/2}$

2. Prove $\lim\limits_{n \to \infty} |a_n| = 0 \Rightarrow \lim\limits_{n \to \infty} a_n = 0$, by using the Squeeze Theorem.

3. Write out the proof of Theorem 5.2∞ .

4. Show that in Example 5.2C the right-hand inequality in (*) can also be interpreted as a comparison of areas.

5.3 Location theorems.

Another group of elementary theorems tell how the location of the terms of a convergent sequence are related to the location of their limit. They are almost equivalent to each other; we will use the first most often.

Theorem 5.3A   Limit location theorem.   *If $\{a_n\}$ is convergent, then*

(15a) $\qquad\qquad a_n \leq M \ \text{ for } n \gg 1 \ \Rightarrow \ \lim a_n \leq M \ ;$

(15b) $\qquad\qquad a_n \geq M \ \text{ for } n \gg 1 \ \Rightarrow \ \lim a_n \geq M \ .$

For example, the Location Theorem says that the limit of an eventually non-negative sequence is non-negative:

$$a_n \geq 0 \quad \text{for } n \gg 1 \ \Rightarrow \ \lim a_n \geq 0 \ .$$

Note that the $\geq$ in the above cannot be replaced by $>$ everywhere, since the limit of a positive sequence can certainly be 0.

On the other hand, the hypothesis "$a_n \leq M$ for all $n \gg 1$", while convenient to state, is a little stronger than necessary (see the Exercises).

Proof.   We prove the first statement (15a), which we may write

(15a) $\qquad\qquad a_n \ \leq \ M \ \text{ for } n \gg 1 \ , \quad a_n \to L \ \Rightarrow \ L \ \leq \ M \ .$

By hypothesis, given $\epsilon > 0$, $\ a_n \underset{\epsilon}{\approx} L \ $ for $n \gg 1$; that is,

$$L - \epsilon \ < \ a_n \ < \ L + \epsilon, \qquad \text{for } n \gg 1 \ .$$

Combining the lower inequality with our hypothesis $a_n \leq M$ gives us

(*) $\qquad\qquad\qquad L - \epsilon \ < \ M \ , \quad \text{for any } \epsilon > 0 \ .$

This implies $L \leq M$; for if $L > M$, choosing $\epsilon = L - M$ would give a positive value for ϵ that contradicts (*) .   □

The statement in (15b) may be proved similarly, using the upper inequality $a_n < L + \epsilon$ in place of the lower inequality.   □□

A variant of the Limit Location Theorem is sometimes useful:

(15c) $\qquad \{a_n\}, \ \{b_n\} \text{ convergent}, \ \ a_n \leq b_n \text{ for } n \gg 1 \ \Rightarrow \ \lim a_n \leq \lim b_n \ .$

Proof. $\quad a_n - b_n \leq 0 \ \text{ for } n \gg 1 \ \Rightarrow \ \lim (a_n - b_n) \leq 0, \quad \text{by (15a)};$

$\qquad\qquad\qquad\qquad \Rightarrow \ \lim a_n - \lim b_n \leq 0, \quad \text{by linearity (1)};$

$\qquad\qquad\qquad\qquad \Rightarrow \ \lim a_n \leq \lim b_n \ . \quad \square$

Theorem 5.3B   Sequence location theorem. *Assuming* $\{a_n\}$ *converges,*

$$(16a) \qquad\qquad \lim a_n < M \;\Rightarrow\; a_n < M \text{ for } n \gg 1 \;;$$

$$(16b) \qquad\qquad \lim a_n > M \;\Rightarrow\; a_n > M \text{ for } n \gg 1 \;.$$

Proof.   Let $L = \lim a_n$. Then, given $\epsilon > 0$,

$$L - \epsilon \;<\; a_n \;<\; L + \epsilon \qquad \text{for } n \gg 1 \;.$$

For statement (16a), the hypothesis is $L < M$. To prove the conclusion, choose $\epsilon = M - L$ in the second inequality above; this gives

$$a_n < M \quad \text{for } n \gg 1. \qquad\qquad\qquad \square$$

Statement (16b) is proved similarly, and left as an exercise.

Here the hypotheses $<$ or $>$ cannot be replaced throughout by $\leq$ or $\geq$, as easy examples show.

Questions 5.3

1. By choosing one of the two words within the parentheses, four statements can be made from the following. Mark each true or false; if false, give a counterexample.

If a convergent sequence is (positive, non-negative), then its limit is (positive, non-negative).

2. Prove (15b) without looking at the book: $a_n \geq M,\; a_n \to L \;\Rightarrow\; L \geq M$.

3. Show that (15a) can be deduced from (15c).

4. Prove that a convergent sequence is bounded, by using the theorems of this section..

5.4   Subsequences.   Non-existence of limits.

Sometimes one wants to form a new sequence by picking out some of the terms of the old one. The next notion captures this idea.

Definition 5.4   A **subsequence** of $\{a_n\}$ is a sequence composed of terms of $\{a_n\}$ and having the form $a_{n_1}, a_{n_2}, \ldots, a_{n_i}, \ldots$, where $n_1 < n_2 < \ldots$.

What the definition says is that in forming the subsequence, the order of the terms in the subsequence must be the same as their order in the original sequence. Thus if the sequence is

$$1,\ 2,\ 1,\ 3,\ 1,\ 4,\ 1,\ 5,\ \ldots,$$

then the first line below gives subsequences, while the second line does not:

$$1,\ 1,\ 1,\ \ldots, \qquad 1,\ 2,\ 3,\ 4,\ \ldots;$$

$$2,\ 2,\ 2,\ \ldots, \qquad 3,\ 2,\ 5,\ 4,\ 7,\ 6,\ \ldots\ .$$

Theorem 5.4   Subsequence theorem.

If $\{a_n\}$ converges, every subsequence also converges, and to the same limit:

$$\lim_{n \to \infty} a_n = L \;\Rightarrow\; \lim_{i \to \infty} a_{n_i} = L, \quad \text{for every subsequence } \{a_{n_i}\} \;.$$

Proof.   By hypothesis, given $\epsilon > 0$,

$$a_n \underset{\epsilon}{\approx} L \quad \text{for } n \gg 1.$$

That is, there is an N (depending on ϵ) such that

(17) $$a_n \underset{\epsilon}{\approx} L \quad \text{for } n > N.$$

We look now at the indices $n_1, n_2, \ldots, n_i, \ldots$ of the terms in the subsequence; since they are increasing, $n_1 < n_2 < n_3 < \ldots$, eventually they will get and stay bigger than N, that is,

(18) $$n_i \; > \; N \quad \text{for } i \gg 1.$$

Combining (17) and (18), we see that, given $\epsilon > 0$,

$$a_{n_i} \underset{\epsilon}{\approx} L \quad \text{for } i \gg 1.$$

This proves, by the definition of limit, that

$$\lim_{i \to \infty} a_{n_i} \; = \; L \;. \qquad \qquad \square$$

Notice in the preceding proof of the Subsequence Theorem the explicit need for the N which is always hiding in the shadows when you use the phrase "for $n \gg 1$". It is not often needed after the first few exercises practicing with the definition of limit, but this is one of those times.

Subsequences play an important role in showing limits do not exist. But first let us look more generally at negative statements involving limits. Such statements are often quite tricky to handle, and one should avoid dealing with them, if possible. Here is a typical error that students make.

Example 5.4A   Prove that $\lim a_n^2 = 0 \;\Rightarrow\; \lim a_n = 0$.

"Solution".   By contraposition; we will show $\lim a_n \neq 0 \;\Rightarrow\; \lim a_n^2 \neq 0$.

Suppose therefore that $\lim a_n = L$, where $L \neq 0$. Then by the Product Theorem for limits (2),

$$\lim a_n^2 = \lim a_n \cdot \lim a_n = L^2 \neq 0 \;. \quad \square?$$

Stop and try to find the mistake, before proceeding. There is nothing wrong with proof by contraposition (see Appendix A.2): you prove $A \Rightarrow B$ by proving instead that not-$B \Rightarrow$ not-A.

The error lies in the successive forms used for the negative hypothesis:

it is not true that $\lim a_n = 0$;

$$\lim a_n \neq 0 \; ;$$

$$\lim a_n = L, \quad \text{where } L \neq 0 \, .$$

The first two forms are arguably equivalent, though the second is a little ambiguous. But the third is definitely different; for suppose the sequence has no limit: then the first two statements are correct, but the third is not. The real villain is the second statement, which is technically equivalent to the first, but misleading, since it suggests that the limit exists.

In the preceding example, there is no difficulty proving the result by a direct argument, instead of trying to do it indirectly, by contraposition. But the question still remains, how *does* one prove a sequence has no limit? This may not be so easy; however, some general advice can be given.

Always try for positive statements. If $\{a_n\}$ has no limit because it tends to ∞, one of the infinite limit theorems will often work, or a direct demonstration that $a_n > M$ for n large. On the other hand, if $\{a_n\}$ has no limit because it jumps around, often one can use the Subsequence Theorem, as the next two examples illustrate.

Example 5.4B   Prove $\displaystyle\lim_{n\to\infty} \sin \frac{n\pi}{2}$ does not exist.

Proof.   There are two subsequences which converge to different limits; they correspond to selecting the terms where $n = 2k$ and those where $n = 4k + 1$:

$$\sin \frac{2k\pi}{2} \to 0, \qquad \sin \frac{(4k+1)\pi}{2} \to 1 \, .$$

If therefore the original sequence had a limit, the above would contradict the Subsequence Theorem 5.4.   □

Example 5.4C   Prove $\lim \sin n$ does not exist.

This is harder than the preceding example, since we don't know the exact values of $\sin n$. But a proof requires only a crude estimate of these values. Here is one possible argument.

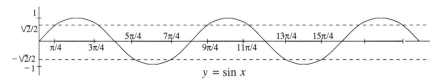

$y = \sin x$

Proof.   From the graph, we see there are an infinity of intervals of length $\pi/2$ on which $\sin x \geq \sqrt{2}/2$. Since each of these intervals has length > 1, in each of them we can choose an integer; let k_i be the integer chosen from the i-th interval. This gives us a subsequence $\{\sin k_i\}$ such that

(19) $$\sin k_i \geq \sqrt{2}/2 \, .$$

Similarly, we can choose an integer m_i from each of the successive intervals of length $\pi/2$ on which $\sin x \leq -\sqrt{2}/2$, giving a subsequence $\{\sin m_i\}$ such that

(20) $$\sin m_i \leq -\sqrt{2}/2 .$$

Reasoning indirectly now, suppose $\{\sin n\}$ converges; call its limit L. Then the Subsequence Theorem 5.4 says that the two subsequences also converge and to this same limit:

$$\lim \, \sin k_i \; = \; L \; ; \qquad \lim \, \sin m_i \; = \; L \, .$$

But since the two subsequences both converge, the Limit Location Theorem 5.3A, taken together with (19) and (20), shows that

$$\lim \, \sin k_i \; \geq \; \frac{\sqrt{2}}{2} \, , \qquad \lim \, \sin m_i \; \leq \; -\frac{\sqrt{2}}{2} \; ; \quad \text{i.e.,}$$

$$L \; \geq \; \frac{\sqrt{2}}{2} \, , \qquad L \; \leq \; -\frac{\sqrt{2}}{2} \, ,$$

a contradiction; thus our assumption that $\{\sin n\}$ has a limit L is false. $\qquad \square$

Questions 5.4

1. Define a "scrambled subsequence" of $\{a_n\}$ to be a sequence consisting of an infinity of terms of $\{a_n\}$, but not necessarily in their original order. How would you alter Definition 5.4 so it defines a scrambled subsequence?

2. Can $\{|\sin k\pi/2|\}$ be viewed as a subsequence of $\{\sin n\pi/2\}$?

3. Prove $\{(-1)^n + 1/n\}$ has no limit.

5.5   Two common mistakes.

By now you have had a little experience in reading and writing arguments, so perhaps it is a good place to signal two mistakes students often make. Some examples will illustrate. Try doing the first yourself before you read the discussion.

Example 5.5A   Prove: $a_n \to 0$, b_n bounded $\Rightarrow a_n b_n \to 0$.

Discussion.   Since we are taking the product of two sequences, many students first think of the Product Theorem 5.1. But this can't be used, because we don't know that b_n has a limit.

The natural thing to do next is write down the definition of boundedness and see if can be used. This often leads to the following argument.

Proof (?) We use the Squeeze Theorem 5.2. Since b_n is bounded, we have

$$L \quad \le \quad b_n \quad \le \quad M \; ;$$

$$a_n L \quad \le \quad a_n b_n \quad \le \quad a_n M$$

$$\downarrow \qquad \quad \Downarrow \qquad \quad \downarrow$$

$$0 \qquad \quad 0 \qquad \quad 0 \; ;$$

in the last line, the two outside limits follow from the Linearity Theorem 5.1. This shows that $a_n b_n \to 0$.   □

(Stop and see if you can find the error.)

The mistake in the preceding argument occurs in the second line: since we do not know if a_n is positive or negative, we don't know which way the inequalities should go. "No problem, we'll just make two cases: if $a_n \ge 0$, the argument is what we gave above; if $a_n \le 0$, just reverse the inequalities in the second line, and then the rest of the argument is exactly the same."

Alas, this is also wrong, since for example a_n might alternate between positive and negative, so that neither case holds for $n \gg 1$, which means the Squeeze Theorem cannot be used.

The way out of this difficulty is to use absolute values, which guarantee non-negativity; then you need not worry about the inequality signs. Write the hypothesis that $\{b_n\}$ is bounded in the form

$$|b_n| \le K \; ;$$

then the argument can be written in either of the following forms:

★   using the Squeeze Theorem, with the top line $0 \le |a_n||b_n| \le |a_n|K$;

the squeeze argument then shows $|a_n b_n| \to 0$, from which $a_n b_n \to 0$ follows (cf. Question 3.1/2);

★   given $\epsilon > 0$, $|a_n| \le \epsilon$ for $n \gg 1$;   therefore, $|a_n b_n| \le K\epsilon$ for $n \gg 1$, which shows directly that $a_n b_n \to 0$, by the K-ϵ principle 3.2.

> *Inequalities are tricky; use | | to avoid trouble.*

The second common failing is not really an error in the technical sense, but it is certainly a mistake: in their work, students endlessly repeat the proofs of theorems, instead of just quoting the theorems.

> *USE theorems by CITING them, not by repeating their proofs.*

For example, in the proof of the Quotient Theorem for limits in 5.1, we repeated part of the proof of the Sequence Location Theorem 5.3B, since it wasn't available then. Let's redo the proof, this time making use of Theorem 5.3B.

Example 5.5B   Prove: $a_n \to L$, $L \neq 0$ $\Rightarrow$ $1/a_n \to 1/L$.

Solution.   Our hypothesis says: given $\epsilon > 0$, $|a_n - L| < \epsilon$ for $n \gg 1$.

Assume $L > 0$. According to the Sequence Location Theorem 5.3B,

$$a_n > L/2 \quad \text{for } n \gg 1 .$$

Using this, (we drop the $|\ |$ in the denominator since the factors are positive),

$$\left| \frac{1}{a_n} - \frac{1}{L} \right| = \frac{|L - a_n|}{a_n L} < \frac{\epsilon}{(L/2)L} \quad \text{for } n \gg 1,$$

which proves the desired limit, according to the K-ϵ principle. $\square$

If $L < 0$, we can reduce to the above case by changing the signs:

$$a_n \to L \quad \Rightarrow \quad -a_n \to -L \quad \Rightarrow \quad \frac{1}{-a_n} \to \frac{1}{-L} \quad \Rightarrow \quad \frac{1}{a_n} \to \frac{1}{L}. \qquad \square\square$$

Exercises

5.1

1. Find the limit, with proof, using the limit theorems.

(a) $\displaystyle \lim_{n\to\infty} \frac{n^3 - n^2 - 1}{2n^3 + 1}$ (b) $\displaystyle \lim_{n\to\infty} \left(\frac{n-3}{2n-1} \right) 2^{-n}$

2. Without looking, prove directly from the definition of limit.

(a) $a_n \to L$, $b_n \to M$ $\Rightarrow$ $a_n + b_n \to L + M$

(b) $a_n \to L$, $b_n \to 0$ $\Rightarrow$ $a_n b_n \to 0$

3. Prove the Product Theorem: $a_n \to L$, $b_n \to M$ $\Rightarrow$ $a_n b_n \to LM$,
by following through these ideas.

The problem is how to get a small upper estimate for $|a_n b_n - LM|$. The idea is to move from $a_n b_n$ to LM in two steps, in each of which only one of the quantities is changing. Represent this symbolically by using "harpoons":

one step: $a_n \cdot b_n \rightharpoonup L \cdot M$;   two steps: $a_n \cdot b_n \rightharpoonup L \cdot b_n \rightharpoonup L \cdot M$;

algebraically, this corresponds to breaking up the difference into a sum:

$$a_n b_n - L \cdot M = (a_n - L) \cdot b_n + L \cdot (b_n - M) .$$

Now estimate the difference by using the triangle inequality and Problem 3-4. (Students angrily call the preceding equation a trick, but it's really a method based on a principle: change one thing at a time. ["method":= a trick used twice])

4. Prove: if $a_n/b_n \to L$, $b_n \neq 0$ for any n, and $b_n \to 0$, then $a_n \to 0$. (Here, as always, L represents a finite number, not ∞.)

5. Prove in Theorem 5.1∞: (a) (12);   (b) (10) (use the last hypothesis).

5.2

1. Evaluate with proof, using the Squeeze Theorem: $\displaystyle\lim_{n\to\infty}\frac{\sqrt{n}+\cos n}{\sqrt{n+1}}$.
(You may use the result in Problem 5-1.)

2. If a_n is increasing and bounded and b_n lies between a_n and a_{n+1}, prove $\lim a_n = \lim b_n$. (Same as Exercise 3.2/2, but now use the ideas of this section.)

3. Let $a_n = \sqrt{1}+\sqrt{2}+\ldots+\sqrt{n}$. Prove $a_n \sim \frac{2}{3}n^{\frac{3}{2}}$; i.e., the ratio has limit 1 as $n\to\infty$ (cf. the Remark in Example 5.2C).

(Model the argument on Example 5.2C, using the Squeeze Theorem. Move the rectangles one unit to the right or left to get the two inequalities. Question 5.2/1c may be helpful.)

4. Determine $\displaystyle\lim\left(\frac{1}{n+1}+\frac{1}{n+2}+\ldots+\frac{1}{2n}\right)$, and prove it.

Hint: use the Squeeze Theorem, modeling your work on Example 5.2C. You may assume: if $a_n \to L$, with $a_n > 0$ and $L > 0$, then $\ln a_n \to \ln L$.

5.3

1. By choosing one word within each parentheses, four statements can be made from the following. Which are false (give a counterexample) and which true?

If $\lim a_n$ is (positive, non-negative), then for large n the individual terms a_n are (positive, non-negative).

2. Prove (16b): $\lim a_n > M \Rightarrow a_n > M$ for $n \gg 1$.

3. (a) Prove that if $a_n \to \infty$ and $b_n \to L > 0$, then $a_n b_n \to \infty$.

(b) Suppose $L \geq 0$ is substituted in part (a). Find a sequence for which the resulting statement is false, and a sequence for which it is true.

4. Suppose a_n is increasing, $a_n \leq b_n$ for all n, and $b_n \to M$.

(a) Prove that a_n converges and $\lim a_n \leq M$. (Use Problem 3-4.)

(b) Show that in the above $\leq$ cannot be replaced throughout by $<$.

5. Prove that $a_n \to L \Rightarrow |a_n| \to |L|$. (Make cases and use Theorem 5.3B.)

6. Prove that $\lim |a_{n+1}/a_n| = 0 \Rightarrow \lim a_n = 0$.
(Show that $|a_{n+1}| < \frac{1}{2}|a_n|$ for $n \gg 1$.)

5.4

1. Suppose the terms of the sequence $\{a_n\}$ are colored, using k different colors (red, blue, yellow, etc.), and using each color infinitely often. Then we get subsequences: $\{a_{r_i}\} = $ all the red terms; $\{a_{b_i}\} = $ all the blue terms; etc.

Prove: if each of these k subsequences converges to the same limit L, then $\{a_n\}$ converges to L. (This is a partial converse to the Subsequence Theorem.)

2. Let $s(n)$ be the sum of the prime factors of the integer n. For example, $6 = 2\cdot 3$, so $s(6) = 5$; similarly, $8 = 2^3$, so $s(8) = 6$.

Prove that $\lim_{n \to \infty} \left(\dfrac{s(n)}{n} \right)$ does not exist. (Use Thm 5.4 and Exer. 3.4/4.)

5.5

1. Prove Example 5.5A by making two cases, as in the first attempt at fixing the proof, and then using Exercise 5.4/1 .

Problems

5-1 (a) If $a_n \geq 0$ for all n and $a_n \to L$, then $\sqrt{a_n} \to \sqrt{L}$.

Criticize the following "proof" of this result.

Let $a_n \to M$. Then by the Product Theorem for limits, $a_n^2 \to M^2$, so that $M^2 = L$. Therefore $M = \sqrt{L}$ or $M = -\sqrt{L}$. Since $\sqrt{a_n} \geq 0$, the Limit Location Theorem 5.3A shows that $M \geq 0$; therefore $M = \sqrt{L}$.

(b) Give a correct proof of the theorem stated in part (a).

5-2   Assume $a_{n+1}/a_n \to L$, where $L < 1$ and $a_n > 0$. Prove that

(a) $\{a_n\}$ is decreasing for $n \gg 1$;

(b) $a_n \to 0$.   (Give two proofs: an indirect one using (a), and a direct one.)

5-3 Show by counterexample that the partial converse to the Subsequence Theorem given in Exercise 5.4/1 would be false if we allowed k to be ∞. That is, if we divide up $\{a_n\}$ into an infinity of infinite subsequences, each of which converges to L, it does not follow that the sequence $\{a_n\}$ itself converges to L.

5-4 Let $\nu(n)$ be the number of prime factors of the integer n. For example, $\nu(8) = 3$, $\nu(5) = 1$. Prove: $\lim \nu(n)/n = 0$.

(You may use familiar limits from calculus in your argument.)

5-5 If $a_n \to L$ and b_n lies between a_n and a_{n+1}, prove $b_n \to L$. (Unlike 5.2/2, "between" does not tell you in which direction the inequalities go.)

5-6 (a) By analyzing the proof of the Limit Location Theorem 5.3A (15a), show that the hypothesis "$a_n \leq M$ for $n \gg 1$" can be weakened.

(b) Show that this strengthened version of the theorem is the contrapositive of the Sequence Location Theorem 5.3B (16b).

(By Appendix A.1, this shows the two theorems are equivalent. Note that to weaken the hypothesis of a theorem is to strengthen the theorem, i.e., to make a stronger assertion; again, see Appendix A.1).

5-7 Define a sequence recursively by $a_{n+1} = \sqrt{2a_n}$, $a_0 > 0$.

(a) Prove that for any choice of $a_0 > 0$, the sequence a_n is monotone and bounded.

(b) Part (a) shows the limit L exists; determine L, prove it is the limit, and is independent of the choice of $a_0 > 0$.

Answers to Questions

5.1

1. (a) Briefly, $\dfrac{n^2+n}{1-2n^2} = \dfrac{1+1/n}{1/n^2-2} \to \dfrac{1+0}{0-2}$, using in turn $1/n \to 0$, then $1/n^2 \to 0$ (by the Product Theorem), then the Sum and Quotient Theorems for limits.

(b) Briefly, $\dfrac{n}{1-n} \cdot \dfrac{1-2n^2}{n^2} = \dfrac{1}{1/n-1} \cdot \dfrac{1/n^2-2}{1} \to \dfrac{1}{-1} \cdot \dfrac{-2}{1} = 2$, using in turn $1/n \to 0$, then $1/n^2 \to 0$ by the Product Theorem, then the Sum, Quotient, and Product Theorems for limits.

2. If you can prove $a_n b_n \underset{\epsilon}{\approx} L \cdot M$ for $n \gg 1$ when $\epsilon < 1$, say, then it is automatically true for all larger $\epsilon' \geq 1$, since
$$|a_n b_n - L \cdot M| < \epsilon \Rightarrow |a_n b_n - L \cdot M| < \epsilon' .$$

3. To prove the Linearity Theorem (1), we add the equations in (4), getting
$$ra_n + sb_n = rL + sM + (re_n + se'_n) .$$

This is the error form for the limit (1); it suffices to show $re_n + se'_n \to 0$. To show this, we go back to Definition 3.1 of limit. Given $\epsilon > 0$,
$$|re_n + se'_n| \leq |r||e_n| + |s||e'_n|, \qquad \text{by the triangle inequality;}$$
$$< (|r| + |s|)\,\epsilon, \text{ for } n \gg 1, \quad \text{by (5).}$$
By the K-ϵ principle (taking $K = |r| + |s|$), this proves $\lim (re_n + se'_n) = 0$.

4. By going back to the definitions of $a_n \to \infty$ and $1/a_n \to 0$, we have
$$\text{given } \epsilon > 0, \quad a_n > 1/\epsilon \text{ for } n \gg 1, \text{ say for } n > N;$$
$$\Rightarrow 1/a_n < \epsilon \text{ for } n > N, \text{ therefore for } n \gg 1.$$
The sequence $a_n = (-1)^n n$ is a counterexample to the converse.

5. $b_n > B$ for all n,   by hypothesis;
given M, $a_n > M - B$ for $n \gg 1$;
adding, $a_n + b_n > M$ for $n \gg 1$,
which shows $a_n + b_n \to \infty$.

5.2

1. In outline, the arguments are as follows.

(a)
$$-\dfrac{1}{n} \;\leq\; \dfrac{\cos n}{n} \;\leq\; \dfrac{1}{n}$$
$$\downarrow \qquad\quad \Downarrow \qquad\quad \downarrow$$
$$0 \qquad\qquad 0 \qquad\qquad 0$$

(b)
$$\dfrac{n-1}{1+3n} \;\leq\; \dfrac{n-\sin n}{1+3n} \;\leq\; \dfrac{n+1}{1+3n}$$
$$\downarrow \qquad\qquad \Downarrow \qquad\qquad \downarrow$$
$$1/3 \qquad\qquad 1/3 \qquad\qquad 1/3$$

(c) $1 \;\leq\; \left(1+\dfrac{1}{n}\right)^{3/2} \;<\; \left(1+\dfrac{1}{n}\right)^{2}$

$\quad\;\;\downarrow\qquad\quad\;\; \Downarrow \qquad\qquad\quad\; \downarrow$

$\quad\;\;1\qquad\qquad 1 \qquad\qquad\quad\; 1$

2. Apply the Squeeze Theorem to the inequalities $-|a_n| \;\leq\; a_n \;\leq\; |a_n|$.

3. Using the definitions of $a_n \to \infty$ and $b_n \to \infty$, we have:

$$\text{given } M, \quad a_n > M \;\text{ for } n \gg 1$$
$$\Rightarrow\; b_n \geq a_n > M \;\text{ for } n \gg 1.$$

4. We note that $n \ln n$ is the area of the rectangle with base $[0,n]$ and height $\ln n$; it contains all of the smaller rectangles, so one gets the inequality.

5.3

1. TRUE: a positive sequence has a non-negative limit; a non-negative sequence has a non-negative limit;

FALSE: a positive sequence has a positive limit; a non-negative sequence has a positive limit.

Counterexamples to the two false statements are, respectively, $\{1/n\}$ and the constant sequence $\{0\}$.

2. The statement we wish to prove is

(15b) $\qquad a_n \geq M \;\text{ for } n \gg 1, \quad a_n \to L \;\Rightarrow\; L \geq M.$

By hypothesis, given $\epsilon > 0$, $\;a_n \underset{\epsilon}{\approx} L$ for $n \gg 1$; this gives us the inequalities, valid for any $\epsilon > 0$,

$$M \;\leq\; a_n \;<\; L+\epsilon, \qquad \text{for } n \gg 1\,; \quad \text{therefore,}$$

(∗) $\qquad\qquad\qquad M \;<\; L+\epsilon, \;\text{ for any } \epsilon > 0\,.$

This implies $M \leq L$, for if $M > L$, choosing $\epsilon = M - L$ would give a positive value for ϵ that contradicts (∗).

3. Apply (15c) to $\{a_n\}$ and the constant sequence $\{M\}$:

$$a_n \leq M \;\text{ for all } n \;\Rightarrow\; \lim a_n \leq \lim M \;\Rightarrow\; L \leq M.$$

4. Let $a_n \to L$; then according to the Sequence Location Theorem 5.3B,

$$L - 1 \;<\; \lim a_n \;<\; L+1 \;\Rightarrow\; L-1 \;<\; a_n \;<\; L+1 \;\text{ for } n \gg 1.$$

This shows a_n is bounded for $n \gg 1$; therefore it is bounded for all n (see 2.6).

5.4

1. Replace $n_1 < n_2 < \dots$ by: $n_i \neq n_j$, if $i \neq j$.

2. Yes: $0,1,0,1,0,1\dots$ is a subsequence of $0,1,0,-1,0,1,0,-1\dots$; it is the subsequence $a_0, a_1, a_4, a_5, \dots, a_{4k}, a_{4k+1}, \dots$.

3. It has two subsequences converging to different limits, which would violate the Subsequence Theorem if the whole sequence had a limit; namely,

$$(-1)^{2k} + \frac{1}{2k} \;\to\; 1; \qquad (-1)^{2k+1} + \frac{1}{2k+1} \;\to\; -1\,.$$

6

The Completeness Property

6.1 Introduction. Nested intervals.

So far, we know from the definition of limit in Chapter 3 how to recognize if a given number L is the limit of a given sequence a_n, and we know from Chapter 5 how to use known limits to produce new limits. But if the sequence itself is really new, unrelated to other sequences whose limits we already know, the only tool we have so far for showing it has a limit is the Completeness Property:

A bounded monotone sequence converges to a limit.

This chapter describes some other methods that can be used to construct or prove the existence of a limit. They apply to sequences which are not monotone; nevertheless, they are based on the Completeness Property, and for the most part are equivalent to it, though they may look rather different. In describing these methods, the main tools will be the Limit Location Theorem 5.3A and the notion of subsequence.

The first new method is the one given by the Nested Intervals Theorem.

Recall that a **closed interval** $[a, b]$ is one which contains its endpoints:

$$[a, b] = \{x : a \leq x \leq b\} \,.$$

Definition 6.1 Suppose we have a sequence of closed intervals

$$I_n = [a_n, b_n], \qquad n = 0, 1, 2, \ldots,$$

having the property that each interval lies inside the previous one, i.e.,

(1) $$a_n \leq a_{n+1} \leq b_{n+1} \leq b_n \,.$$

Such a sequence of intervals is said to be **nested** (like a set of bowls).

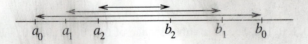

Theorem 6.1   The Nested Intervals Theorem.

Suppose $[a_n, b_n]$ is an infinite sequence of nested intervals, whose lengths tend to 0, i.e., $\lim (b_n - a_n) = 0$. Then there is one and only one number L in all the intervals; moreover, $a_n \to L$ and $b_n \to L$.

Proof. There are five steps; (B) and (E) use the Completeness Property.

(A)   *Each a_n lies to the left of all the b's: $a_n \le b_i$ for all i and n.* To prove this, we have two cases: $i \le n$ and $i \ge n$. As the picture shows,

$$\underset{a_i \quad a_n \quad b_n \quad b_i}{\rule{4cm}{0.4pt}} \qquad \underset{a_n \quad a_i \quad b_i \quad b_n}{\rule{4cm}{0.4pt}}$$

$$i \le n \;\Rightarrow\; a_n \le b_n \le b_i \;; \qquad i \ge n \;\Rightarrow\; a_n \le a_i \le b_i \;.$$

(B)   *The sequence $\{a_n\}$ is increasing and convergent; let $L = \lim a_n$.*
It is increasing by (1) since the intervals are nested, and (A) shows it is bounded above by any b_i. So by the Completeness Property, it converges.

(C)   *The number L is inside every interval: $a_n \le L \le b_n$ for all n.*
The first inequality follows from (B), by Theorem 3.2B. To prove the second inequality, we have $a_k \le b_n$ for all k, by (A); therefore $L = \lim a_k \le b_n$, by the Limit Location Theorem 5.3A.

(D)   *L is the only number inside every interval.*
Namely, let M also be inside every interval; we show $|L - M| = 0$. We have

$$|L - M| \;\le\; b_n - a_n \quad \text{for all } n;$$
$$|L - M| \;\le\; \lim(b_n - a_n), \quad \text{by the Limit Location Theorem 5.3A};$$
$$|L - M| \;\le\; 0, \text{ since we assumed } \lim(b_n - a_n) = 0 \;.$$

(E)   *$L = \lim b_n$.*
Since $\{b_n\}$ is a decreasing sequence, bounded below by L according to (C), it has a limit; set $\lim b_n = M$. We now have the following chain of inequalities:

$$a_n \le L \le M \le b_n \quad \text{for all } n;$$

the first inequality by (C), the third inequality by Theorem 3.2B, and the middle one since (C) tells us that $b_n \ge L$ for all n and therefore $\lim b_n \ge L$ by the Limit Location Theorem 5.3A.

The above chain of inequalities shows according to (D) that $M = L$, and therefore $\lim b_n = L$. $\qquad\qquad\qquad\qquad\qquad\qquad\qquad\qquad\qquad \square$

Example 6.1   Let $a_n = 1 - \frac{1}{2} + \frac{1}{3} - \ldots + (-1)^{n-1}\frac{1}{n}$, for $n > 0$; $a_0 = 0$.
   Show $\{a_n\}$ converges.

The sequence is not monotonic. In Section 4.2, we proved it converges to $\ln 2$ by using error-term analysis. Here is an elementary proof of just its convergence.

Solution.

$$\underset{\substack{a_0 \\ 0}}{\rule{2cm}{0.4pt}} \qquad \underset{\substack{a_2 \; a_4 \\ 1/2 \; 7/12}}{\rule{2cm}{0.4pt}} \quad \underset{\substack{a_5 \, a_3 \qquad a_1 \\ 5/6 \qquad 1}}{\rule{2cm}{0.4pt}}$$

From the picture,

(2) $[a_0, a_1], \quad [a_2, a_3], \quad \ldots, \quad [a_{2k}, a_{2k+1}], \quad [a_{2k+2}, a_{2k+3}], \quad \ldots$

is a sequence of nested intervals. (A formal proof will be given in Section 7.6 for a general class of sequences which includes this one.)

The lengths of the nested intervals (2) tend to 0 as $k \to \infty$, since

$$(3) \qquad\qquad |a_{2k+1} - a_{2k}| = \frac{1}{2k+1} \, .$$

Thus the hypotheses of the Nested Interval Theorem 6.1 are satisfied; it follows that there exists a unique L inside all the intervals, and since

$$a_{2k} \leq L \leq a_{2k+1} \qquad \text{for all } k,$$

it follows by (3) that regardless of whether n is even or odd,

$$\text{given } \epsilon > 0, \quad |a_n - L| \leq \frac{1}{n} \, , \quad \text{which is } \leq \epsilon \text{ for } n \gg 1;$$

this shows that $a_n \to L$. $\qquad\qquad\qquad\qquad\qquad\qquad\qquad\qquad\qquad\quad \Box$

Locating a real number L by constructing a set of nested intervals which contract to L is an effective process. At each stage, you know that L lies inside $[a_n, b_n]$, so that you have an estimate $b_n - a_n$ of the possible error.

By contrast, locating L by constructing an increasing sequence $\{a_n\}$ having L as its limit is not so effective, since at each stage you may have no way of estimating how far away $\{a_n\}$ is from L.

Questions 6.1

1. Show that given any real number a, there is a sequence of nested intervals of positive length having a as the unique point inside.

2. How would the conclusion of Theorem 6.1 be modified if the hypothesis $\lim(b_n - a_n) = 0$ were omitted?

3. Generalize Example 6.1; that is, what hypotheses on a sequence $\{c_n\}$ would enable you to use the same method as in Example 6.1 to prove that the sequence converges? (You need not give the proof.)

6.2   Cluster points of sequences.

Before discussing other forms of the completeness principle, we need to say something about cluster points. These are numbers that the sequence gets arbitrarily close to, infinitely often. Unlike the limit, which is unique if it exists, a sequence can have many cluster points.

Definition 6.2. K is a **cluster point** of the sequence $\{a_n\}$ if

$$(4) \qquad\qquad \text{given } \epsilon > 0, \quad a_n \underset{\epsilon}{\approx} K \quad \text{for infinitely many } n.$$

Remarks.   For both a limit L and a cluster point K of a sequence $\{a_n\}$, the a_n must get arbitrarily close. But the a_n must stay close to a limit L, whereas they need only visit the vicinity of a cluster point K infinitely often; the difference is expressed by the distinction between "for all $n \gg 1$" and "for infinitely many n".

Every limit L is automatically a cluster point, but there are cluster points which are not limits, as the following examples show.

Examples 6.2A

(a)   1; 1, 2; 1, 2, 3; 1, 2, 3, 4; 1, 2, 3, 4, 5; ... ;

every positive integer is a cluster point, but the sequence has no limit.

(b)   1, 1/2, 1, 1/3, 1, 1/4, ... ;

here 1 and 0 are cluster points, but the sequence has no limit.

(c) $\sin 1$, $\sin 2$, $\sin 3$, ..., $\sin n$, ... ;

every real number in the interval $[-1, 1]$ is a cluster point of this sequence, though this is hardly obvious! The sequence has no limit (see Example 5.4C).

Theorem 6.2. Cluster point theorem.

K *is a cluster point of* $\{a_n\}$ $\Leftrightarrow$ K *is the limit of some subsequence* $\{a_{n_i}\}$.

Proof. The backward direction is easy; the forward requires more thought.

$\Leftarrow$:   Given $\epsilon > 0$, $a_{n_i} \underset{\epsilon}{\approx} K$ for $i \gg 1$,   say for all $i \geq I$.   Then the approximation holds for an infinity of a_n, namely $a_{n_I}, a_{n_{I+1}}, a_{n_{I+2}}, \ldots$.

$\Rightarrow$:   Choose step by step a sequence $a_{n_1}, a_{n_2}, a_{n_3}, \ldots$, so that

$$a_{n_1} \underset{1}{\approx} K,$$

(5) $$a_{n_2} \underset{1/2}{\approx} K, \quad \text{and} \quad n_2 > n_1,$$

$$a_{n_3} \underset{1/3}{\approx} K, \quad \text{and} \quad n_3 > n_2, \quad \text{and so on.}$$

To see that this is possible, we note that at the i-th step, there are by hypothesis infinitely many a_n such that $a_n \underset{1/i}{\approx} K$. This means that we can choose one of them—call it a_{n_i}—for which $n_i > n_{i-1}$.

Since $n_i > n_{i-1}$, the a_{n_i} have the same order they have in $\{a_n\}$, so they form a subsequence of $\{a_n\}$. Moreover, $\{a_{n_i}\} \to K$, since the approximations (5) show that

$$\text{given } \epsilon > 0, \quad a_{n_i} \underset{\epsilon}{\approx} K \quad \text{for } 1/i < \epsilon, \quad \text{i.e., for } i > 1/\epsilon \,. \qquad \square$$

Example 6.2B   Let $a_n = \dfrac{1}{n} + (-1)^n$. Show $\{a_n\}$ has -1 and 1 as cluster points, but no limit.

Solution. We have the two convergent subsequences formed from the odd terms and the even terms ($n = 2k + 1$ and $n = 2k$):

$$\frac{1}{2k + 1} - 1 \;\to\; -1, \qquad \frac{1}{2k} + 1 \;\to\; 1.$$

This shows by the Cluster Point Theorem that -1 and 1 are cluster points. Since there are subsequences converging to different limits, the Subsequence Theorem 5.4 shows the sequence $\{a_n\}$ cannot have a limit.

Other names for cluster point are "point of accumulation" and "limit point". The latter must be carefully distinguished from "limit". A sequence can have at most one limit, but it can have many limit points.

Questions 6.2

1. Find the cluster points of $\{\sin n\pi/2\}$. For each, find a subsequence converging to it.

2. (a) Prove that if a sequence is convergent, it has only one cluster point. Use the Cluster Point and Subsequence Theorems to do this.

   (b) Find a sequence having only one cluster point, yet not convergent.

3. Show the following definition of cluster point is equivalent to the one in the text: L is a cluster point of $\{a_n\}$ if

(4')   given $\epsilon > 0$ and any N, $a_n \underset{\epsilon}{\approx} L$ for some $n > N$.

That is (cf. Appendix A.1), show: (4) is true $\Leftrightarrow$ (4') is true.

6.3   The Bolzano-Weierstrass theorem.

Sequences in general do not converge, but they often have subsequences which do. The theorem which asserts this is one of the cornerstones of analysis. The proof given here uses a method involving nested intervals known as the **bisection method**, since each new nested interval is created by halving the preceding one and choosing one of the halves.

Theorem 6.3.   Bolzano-Weierstrass

A bounded sequence $\{x_n\}$ has a convergent subsequence.

Proof.   By the Cluster Point Theorem 6.2, it suffices to show the bounded sequence $\{a_n\}$ has a cluster point, since it then has a subsequence which converges to this cluster point.

We find this cluster point by nested intervals. Since $\{x_n\}$ is bounded, there are points a_0 and b_0 such that

$$a_0 \leq x_n \leq b_0 \quad \text{for all } n.$$

We set

$$\text{length } [a_0, b_0] = d.$$

We can assume $d > 0$, since if $d = 0$, the sequence x_n is constant.

Divide $[a_0, b_0]$ in half at its midpoint c. Then since $[a_0, b_0]$ contains infinitely many x_n (all of them, in fact), we know that at least one of the half-intervals $[a_0, c], [c, b_0]$ contains infinitely many x_n. Call this half-interval $[a_1, b_1]$; if both do, use the left-hand one. We now have

$$[a_0, b_0] \supset [a_1, b_1], \qquad \text{length } [a_1, b_1] = d/2 ,$$

$$[a_1, b_1] \text{ contains an infinity of } x_n.$$

Similarly, by dividing $[a_1, b_1]$ in half, we get an $[a_2, b_2]$ such that
$$[a_1, b_1] \supset [a_2, b_2], \qquad \text{length } [a_2, b_2] = d/4 \ ,$$
$$[a_2, b_2] \text{ contains an infinity of } x_n.$$

Continuing, we get a sequence of nested intervals
$$[a_0, b_0] \supset [a_1, b_1] \supset [a_2, b_2] \supset \ldots \supset [a_n, b_n] \supset \ldots$$
such that
$$[a_n, b_n] \text{ contains an infinity of } x_n, \qquad \text{length } [a_n, b_n] = d/2^n \ .$$

By Theorem 6.1, there is a unique point L inside all these intervals. We claim this point L is a cluster point of $\{x_n\}$.

Namely, given any $\epsilon > 0$, choose an i so that length $[a_i, b_i] = d/2^i < \epsilon$; then $x_n \underset{\epsilon}{\approx} L$ for the infinity of x_n guaranteed to lie inside $[a_i, b_i]$. $\square$

> If you have not seen the proof, the Bolzano-Weierstrass theorem is far from obvious. For example, the sequence $\{\sin n\}$ is bounded between -1 and 1, so by the theorem, it must have a convergent subsequence. Indeed, the remarks in Example 6.2A show it has a huge number of convergent subsequences, since for any real number a in the interval $[-1, 1]$, there is a subsequence converging to a. But just try to find one!

Questions 6.3

1. Let $\{a_n\}$ be an arbitrary sequence. Which of the following sequences $\{b_n\}$ will always have a convergent subsequence, regardless of the choice of $\{a_n\}$? Indicate reason.

$$(a) \quad b_n = \sin a_n \qquad (b) \quad b_n = \frac{1}{1 + a_n} \qquad (c) \quad b_n = \frac{1}{1 + a_n^2}$$

2. Prove that every bounded sequence has a cluster point; use theorems.

6.4   Cauchy sequences.

The Completeness Property tells us that a bounded monotone sequence converges, without telling us or requiring us to know what the limit is. In this section we present another such convergence criterion; it is a bit less simple to state, but it applies to sequences which are not monotone.

We call the sequence $\{a_n\}$ a **Cauchy sequence** if it has this property:

(6)   given $\epsilon > 0, \quad a_m \underset{\epsilon}{\approx} a_n \quad$ for $m, n \gg 1$.

Theorem 6.4   The Cauchy criterion for convergence.

If $\{a_n\}$ is a Cauchy sequence, then $\{a_n\}$ converges.

Proof.   Let $\{a_n\}$ be a Cauchy sequence. There are three steps.

(A) $\{a_n\}$ *is bounded.*

To prove this, take $\epsilon = 1$ say. By (6), there is an N such that

$$a_n \underset{1}{\approx} a_m \qquad \text{for all } m, n \geq N;$$

therefore, $a_n \underset{1}{\approx} a_N \qquad \text{for all } n \geq N, \qquad \text{i.e.,}$

$$a_N - 1 \ \leq\ a_n \ \leq\ a_N + 1 \quad \text{for all } n \geq N.$$

This shows $\{a_n\}$ is bounded for $n \gg 1$; hence it is bounded for all n.

(B) $\{a_n\}$ *has a convergent subsequence* $\{a_{n_i}\}$.

This follows from (A) by the Bolzano-Weierstrass Theorem 6.3.

(C)   *Let* $L = \lim\{a_{n_i}\}$;   *then* $\{a_n\} \to L$.

Namely, since L is the limit of the subsequence, this means

$$\text{given } \epsilon > 0, \quad a_{n_i} \underset{\epsilon}{\approx} L \qquad \text{for } i \gg 1, \ \text{i.e., for } n_i \gg 1.$$

$$\text{But} \quad a_n \underset{\epsilon}{\approx} a_{n_i} \qquad \text{for } n, n_i \gg 1, \ \text{since } \{a_n\} \text{ is Cauchy.}$$

$$\text{Therefore} \quad a_n \underset{2\epsilon}{\approx} L \qquad \text{for } n \gg 1. \hspace{2cm} \square$$

Remarks.   Study this argument; it is a typical use of the Bolzano-Weierstrass Theorem. First one uses the theorem to find a convergent subsequence; call its limit L. Then one shows that the rest of the sequence adheres closely to this subsequence, which means the whole sequence must converge to L.

Think of the sequence as a herd of goats; if the sub-herd consisting of all goats carrying bells is converging on some pasture, then the other goats, trained to stay close to the bell-goats, will go there too.

The converse—that every convergent sequence is a Cauchy sequence—is true and left for the Exercises.

A common error is to write the definition of Cauchy sequence incorrectly as

$$\text{given } \epsilon > 0, \quad a_{n+1} \underset{\epsilon}{\approx} a_n \qquad \text{for } n \gg 1.$$

This hypothesis is too weak to allow one to conclude that $\{a_n\}$ converges. Counterexamples are not hard to give. In other words, you must know that after a certain point in the sequence, any two terms will be close, not merely that successive terms will be close.

Example 6.4   The sequence of Fibonacci fractions is defined recursively by

(7) $a_{n+1} = \dfrac{1}{a_n + 1}$, $n \geq 1$; $a_1 = 1$;   it starts: $\frac{1}{1}, \frac{1}{2}, \frac{2}{3}, \frac{3}{5}, \frac{5}{8}, \frac{8}{13}, \ldots$.

Prove it converges, and determine its limit.

In Section 4.3 we guessed the limit, then showed the error term $e_n \to 0$. Here the approach is different. Note that the sequence is not monotone.

Solution.   To prove convergence, we show the sequence is a Cauchy sequence.

To see that $|a_n - a_m|$ is small, we first use (7) to get an estimate of how rapidly the difference $|a_{n-1} - a_n|$ decreases as you move from one pair of successive terms to the next. We have by (7):

(8)
$$|a_n - a_{n+1}| = \left| \frac{1}{a_{n-1}+1} - \frac{1}{a_n + 1} \right|$$

$$= \frac{|a_{n-1} - a_n|}{(a_n + 1)(a_{n-1} + 1)} \ .$$

To estimate this, it looks from (7) as if $a_n \geq \frac{1}{2}$ for all n; assuming this for the moment, we get for the denominator

$$(a_n + 1)(a_{n-1} + 1) \ \geq \ \tfrac{3}{2} \cdot \tfrac{3}{2} \ > \ 2 \ ,$$

so that from (8) we get

(9)
$$|a_n - a_{n+1}| \ \leq \ \tfrac{1}{2}|a_{n-1} - a_n| \ , \quad \text{for all } n \geq 2.$$

We now use recursion, applying (9) in turn to $|a_{n-1} - a_n|$; this gives

$$|a_n - a_{n+1}| \ \leq \ \tfrac{1}{4}|a_{n-2} - a_{n-1}| \ ,$$

and continuing in this way, we get finally

(10)
$$|a_n - a_{n+1}| \ \leq \ \frac{1}{2^{n-1}}|a_1 - a_2| \ \leq \ \frac{1}{2^n} \ .$$

Now we use (10) to estimate $|a_n - a_m|$, for $m > n$; first we use the extended triangle inequality to break up $|a_n - a_m|$ into smaller steps. We have

$$|a_n - a_m| \ \leq \ |a_n - a_{n+1}| + |a_{n+1} - a_{n+2}| + \ldots + |a_{m-1} - a_m| \ ;$$

$$\leq \ \frac{1}{2^n} + \frac{1}{2^{n+1}} + \ldots + \frac{1}{2^{m-1}} \ , \quad \text{using (10);}$$

$$\leq \ \frac{1}{2^n} \left(1 + \frac{1}{2} + \frac{1}{4} + \ldots \right)$$

$$\leq \ \frac{1}{2^{n-1}} \ < \ \epsilon, \quad \text{for } m > n \gg 1.$$

This proves $\{a_n\}$ is a Cauchy sequence, which converges therefore to some limit L that can be easily found by using the limit theorems. Namely, since $a_n \to L$ and $a_{n+1} \to L$, the Algebraic Limit Theorems 5.1 applied to the terms of equation (7) show that

$$L = \frac{1}{L+1} \ ;$$

cross-multiplying turns this last into a quadratic equation whose unique positive root is

$$L = \frac{\sqrt{5} - 1}{2} \ .$$

The proof used $a_n \geq 1/2$ for all n; to show this, use recursion: $a_1 = 1$, and

$$\tfrac{1}{2} \leq a_n \leq 1 \;\Rightarrow\; \tfrac{3}{2} \leq a_n + 1 \leq 2$$

$$\Rightarrow\; \tfrac{1}{2} \leq \frac{1}{a_n + 1} \leq \tfrac{2}{3}, \qquad \text{by the inequality laws;}$$

$$\Rightarrow\; \tfrac{1}{2} \leq a_{n+1} \leq 1, \qquad \text{by (7).}$$

Questions 6.4

1. *"Given any $\epsilon > 0$, $a_{n+1} \underset{\epsilon}{\approx} a_n$ for $n \gg 1$."*

Give an example of an increasing sequence which has this property, but is not a Cauchy sequence. How do you know it is not a Cauchy sequence?

2. With the book closed, write the definition of Cauchy sequence, and prove a Cauchy sequence is bounded.

3. A sequence has the property: $|a_m - a_n| \leq 1/(m+n)$. Prove it is Cauchy.

4. A sequence has the property: $|a_n - a_{n+1}| \leq 1/2^n$. Prove it is Cauchy.

5. A convergent sequence is given recursively by $a_{n+1} = a_n/(1 + a_n)$.

Find its limit L. (Assume $a_n \neq -1$ for all n.)

6.5 The Completeness Property for sets.

Up to now we have worked with *sequences* of numbers, i.e., numbers ordered in a list. However, one can also discuss the Completeness Property taking as the primary notion a *set* of numbers, i.e., an unordered collection.

We give here a brief exposition of this *set-theoretic* point of view, as it is called, since we shall occasionally find use for it, and because it is standard in the literature. It is not difficult.

Definition 6.5A Let S be a set of real numbers: $S \subseteq \mathbb{R}$.

An **upper bound** for S is a number b such that $x \leq b$ for all $x \in S$.

S is said to be **bounded above** if S has an upper bound.

A number m is the **maximum** of S if m is an upper bound for S and $m \in S$ (notation: $m = \max S$).

For example, the interval $S = [0, 1]$ is bounded above, and $\max S = 1$.

The open interval $U = (0, 1) = \{x : 0 < x < 1\}$ is bounded above, but has no maximum: the number 1 is a sharp upper bound for U, but not the maximum, since it is not an element of U. We call it the "supremum" of U; like an exiled monarch, it rules from without.

Definition 6.5B   Let $S \subseteq \mathbb{R}$. The **supremum** of S ($\sup S$) is a number $\overline{m}$ satisfying:

sup-1: $\overline{m}$ is an *upper bound* for S: $x \le \overline{m}$ for all $x \in S$;

sup-2: $\overline{m}$ is the *least upper bound* for S, that is,

$$x \le b \quad \text{for all } x \in S \quad \Rightarrow \quad \overline{m} \le b.$$

You can use either " supremum" or " least upper bound" (notation: lub S); the first is preferred since it is Latin and therefore acceptable to the global market: sup S is universal, whereas lub S is provincial.

Example 6.5A   Find upper bounds, $\sup S$, and $\max S$ of

(a) $S = \{1 - \dfrac{1}{n} \; : \; n = 1, 2, \ldots\};$ (b) $S = \{1 + \dfrac{1}{n} \; : \; n = 1, 2, \ldots\}.$

Solution.

(a) Any $b \ge 1$ is an upper bound; $\sup S = 1$, and $\max S$ does not exist.

(b) Any $b \ge 2$ is an upper bound; $\sup S = 2$ and $\max S = 2$.

Proposition 6.5A   *If* $\max S$ *exists, then* $\sup S$ *exists, and* $\sup S = \max S$. *The numbers* $\sup S$ *and* $\max S$ *are unique, if they exist.*

The proof is not difficult, and left for the Questions.

Theorem 6.5A.   Completeness Property for sets.

If S is non-empty and bounded above, $\sup S$ exists.

Some like to say "If a set has an upper bound, it has a least upper bound," but the empty set is a counterexample to this catchy formulation (the only one, to be sure).

Proof.   We locate $\sup S$ by nested intervals, using bisection, then use the Limit Location Theorem 5.3A to show the point we have found really is $\sup S$.

Let b_0 be an upper bound for S, and choose an $a_0 \in S$.

Bisect the interval $[a_0, b_0]$, and choose the half-interval $[a_1, b_1]$ satisfying:

b_1 is an upper bound for S; $[a_1, b_1]$ contains a point of S.

(Let c be the midpoint of $[a_0, b_0]$; choose the half-interval $[a_0, c]$ if c is an upper bound for S, otherwise choose $[c, b_0]$. This works; check it out.)

Repeat this halving process with $[a_1, b_1]$ and continue. You get a sequence of nested intervals

$$[a_0, b_0] \supseteq [a_1, b_1] \supseteq \ldots \supseteq [a_n, b_n] \supseteq \ldots$$

such that

b_n is an upper bound for S; $[a_n, b_n]$ contains a point of S;

$$\text{length } [a_n, b_n] \to 0.$$

By the Nested Intervals Theorem 6.1, there is a point $\overline{m}$ which is in every interval, and

$$\lim a_n = \overline{m}, \qquad \lim b_n = \overline{m}.$$

We show now that $\overline{m} = \sup S$, by showing it satisfies the two conditions of Definition 6.5B.

    sup-1: $\overline{m}$ is an upper bound for S:

    $x \in S \Rightarrow x \leq b_n$ for all n,   since each b_n is an upper bound for S;

          $\Rightarrow x \leq \lim b_n = \overline{m}$,   by the Limit Location Theorem 5.3A.

    sup-2: $\overline{m}$ is the least upper bound of S:

    $x \leq b$ for all $x \in S$ (b is an upper bound for S)

          $\Rightarrow a_n \leq b$ for all n,   since $[a_n, b_n]$ contains a point of S;

          $\Rightarrow \lim a_n \leq b$,   by the Limit Location Theorem 5.3A;

          $\Rightarrow \overline{m} \leq b$. $\square$

We turn now to the analogous notions of lower bound, minimum, and infimum for a set S. They are formulated exactly the same, except the inequality signs are reversed. Try writing them down by yourself.

Definition 6.5C   Let S be a set of real numbers.
    A **lower bound** for S is a number b such that $x \geq b$ for all all $x \in S$.
    We say that S is **bounded below** if S has a lower bound.
    A number m is the **minimum** of S if m is a lower bound for S and $m \in S$ (notation: $m = \min S$).

Definition 6.5D   Let S be a set of real numbers. By the **infimum** of S (in symbols, $\inf S$), we mean a number $\underline{m}$ having these two properties:

    inf-1: $\underline{m}$ is a *lower bound* for S: $x \geq \underline{m}$ for all $x \in S$;

    inf-2: $\underline{m}$ is the *greatest lower bound* for S (in symbols, $\underline{m} = \text{glb } S$):

$$x \geq b \quad \text{for all } x \in S \quad \Rightarrow \quad \underline{m} \geq b.$$

Example 6.5B   Find lower bounds, $\inf S$, and $\min S$.

    (a) $S = \{1 - \dfrac{1}{n} : n = 1, 2, \ldots\}$ (b) $S = \{1 + \dfrac{1}{n} : n = 1, 2, \ldots\}$

Solution.
    (a) Any $b \leq 0$ is a lower bound; $\inf S = 0$, and $\min S = 0$.
    (b) Any $b \leq 1$ is a lower bound; $\inf S = 1$ and $\min S$ does not exist.

The analogues for min and inf of Proposition 6.5A and Theorem 6.5A are valid:

Proposition 6.5B   For a set S, $\min S$ and $\inf S$ are unique if they exist; if $\min S$ exists, they are equal.

Theorem 6.5B   *If S is non-empty and bounded below, $\inf S$ exists.*

This could be proved from scratch, using bisection as before, or it could be deduced as a corollary of Theorem 6.5A, by observing that if we define

$$-S = \{-x : x \in S\} \, ,$$

then

$$-\inf S = \sup(-S) \, .$$

The details are left for the Exercises.

Questions 6.5

1. For each of the following sets, determine the sup, inf, max, and min, if they exist; ($\mathbb{Z} = $ all integers, $\mathbb{N} = $ the positive integers, $\mathbb{R} = $ the real numbers).

(a) $\{\frac{1}{n} : n \in \mathbb{N}\}$ (b) $\mathbb{N}$ (c) $\mathbb{Z}$ (d) $\{x \in \mathbb{R} : |x| < 1\}$
(e) $\{n \in \mathbb{N} : n^2 = 2\}$

2. Prove $\max S$ is unique, if it exists.

3. Prove $\sup S$ is unique, if it exists (use sup-1 and sup-2).

Exercises

6.1

1. (a) In Example 6.1, give another proof that $c_n \to L$ by using the result of Exercise 5.4/1.

(b) Given a set of nested intervals $[a_n, b_n]$, $n = 0, 1, 2, \ldots$, for which $b_n - a_n \to 0$, prove the sequence $a_0, b_0, a_1, b_1, a_2, b_2, \ldots, a_n, b_n, \ldots$ converges.
(This generalizes part (a) and can be done the same way.)

2. Let α be a real number between 0 and 1 written in binary: *e.g.*,

$$\alpha = .1011001\ldots \qquad \text{means} \qquad \alpha = 1/2 + 1/2^3 + 1/2^4 + 1/2^7 + \ldots \, .$$

Make a set of nested intervals by starting with $I_0 = [0, 1]$, then defining recursively I_n to be the (closed) left half of I_{n-1} if the n-th place of α is 0, and the (closed) right half if the n-th place is 1.

Prove the resulting sequence of nested intervals converges to α, i.e., α is the unique number inside all the intervals.

6.2

1. Find the cluster points of: (a) $\{\sin(\frac{n+1}{n}\frac{\pi}{2})\}$ (b) $\{\sin(n + \frac{1}{n})\frac{\pi}{2}\}$.
For each cluster point, find a subsequence converging to it.

2. The terms of a sequence $\{x_n\}$ take on only finitely many values $a_1, \ldots, a_k$. That is, for every n, $x_n = a_i$ for some i (the index i will depend on n). Prove that $\{x_n\}$ has a cluster point.

6.3

1. Which of these sequences $\{b_n\}$ always has a convergent subsequence, regardless of what $\{a_n\}$ is? Indicate reasoning.

For those sequences $\{b_n\}$ which do not always have a convergent subsequence, give an example; that is, produce a sequence $\{a_n\}$ such that the corresponding $\{b_n\}$ has no convergent subsequence.

(a) $b_n = \cos^2 a_n$ (b) $b_n = \dfrac{a_n}{1 + a_n}$ (assume $a_n \neq -1$)   (c) $b_n = \dfrac{1}{1 + |a_n|}$

6.4

1. Prove that every convergent sequence is a Cauchy sequence.

2. Suppose a sequence $\{a_n\}$ has this property: there exist constants C and K, with $0 < K < 1$, such that

$$|a_n - a_{n+1}| < CK^n, \qquad \text{for} \quad n \gg 1.$$

Prove that $\{a_n\}$ is a Cauchy sequence.

3. Show that $\sqrt{n}$ is another example which illustrates Question 6.4/1: a sequence whose successive terms get arbitrarily close is not necessarily a Cauchy sequence..

6.5

1. For each of the following sets, determine the sup, inf, max, min if they exist. (Here $\mathbb{Z} = $ all integers, $\mathbb{N} = $ positive integers.)

(a) $\{\sin n\pi/6 \ : \ n \in \mathbb{Z}\}$ (b) $\{(\cos n\pi)/n \ : \ n \in \mathbb{N}\}$

(c) $\{1/n \ + \ \cos n\pi/2 \ : \ n \in \mathbb{N}\}$ (d) $\{n2^{-n} \ : \ n \in \mathbb{N}\}$

2. Let S be a non-empty bounded set of real numbers, and $\overline{m} = \sup S$. Prove
$$\inf\{\overline{m} - x \ : \ x \in S\} = 0 \ .$$

3. If A and B are two subsets of $\mathbb{R}$, and $c \in \mathbb{R}$, let

$\qquad cA = \{ca \ : \ a \in A\} \ , \qquad\qquad A + B = \{a + b \ : \ a \in A, \ b \in B\} \ .$

Assume A and B are bounded and non-empty subsets of $\mathbb{R}$.. Prove the following (for each part, you can assume any of the preceding parts).

(a) $A \subseteq B \Rightarrow \sup A \leq \sup B$ (b) $A \subseteq B \Rightarrow \inf A \geq \inf B$

(c) if $c > 0$, $\sup cA = c \sup A$ (d) if $c > 0$, $\inf cA = c \inf A$

(e) $\sup(-A) = -\inf A$ (f) $\inf(-A) = -\sup A$

(g) $\sup(A + B) \leq \sup A + \sup B$ (h) $\inf(A + B) \geq \inf A + \inf B$

(Actually, equality holds in (g) and (h); see Problem 6-2.)

4. Let S and T be non-empty subsets of $\mathbb{R}$, and suppose that for all $s \in S$ and $t \in T$, we have $s \leq t$. Prove that $\sup S \ \leq \ \inf T$.

Problems

6-1   Select two numbers a and b, and let $x_0 = a$, $x_1 = b$. Then continue the sequence by letting each new term be the average of the preceding two:
$$x_n = \frac{x_{n-1} + x_{n-2}}{2}, \quad n \geq 2 .$$

(a) Prove $\{x_n\}$ is a Cauchy sequence. (Hint: cf. Example 6.4: relate $x_n - x_{n-1}$ to $x_{n-1} - x_{n-2}$.)

(b) Find $\lim x_n$ in terms of a and b.

(Hint: $x_n = (x_n - x_{n-1}) + (x_{n-1} - x_{n-2}) + \ldots .$)

6-2 (a) Let S be a bounded non-empty subset of $\mathbb{R}$, and $\overline{m} = \sup S$. Prove there is a sequence $\{a_n\}$ such that $a_n \in S$ for all n, and $a_n \to \overline{m}$.

(You must show how to construct the sequence a_n. Use the properties sup-1 and sup-2 which characterize $\overline{m}$.)

(b) Let A and B be bounded non-empty subsets of $\mathbb{R}$. Prove the equality $\sup(A + B) = \sup A + \sup B$. (cf. Exercise 6.5/3g; use part (a).)

6-3   Suppose $f(x)$ is continuous and decreasing on $[0, \infty]$, and $f(n) \to 0$. Define $\{a_n\}$ by
$$a_n = f(0) + f(1) + \ldots + f(n-1) - \int_0^n f(x)dx .$$

(a) Prove $\{a_n\}$ is a Cauchy sequence directly from the definition.

(b) Evaluate $\lim a_n$ if $f(x) = e^{-x}$.

6-4   Prove the following two-dimensional form of the Bolzano-Weierstrass Theorem:

If $\{(x_n, y_n)\}$ is a sequence of points in the xy-plane, all of which lie in a rectangle (cf. Appendix A.0 for the notation $\times$ if unfamiliar),
$$R = [a, b] \times [c, d] = \{(x, y) : a \leq x \leq b, \ c \leq y \leq d\} ,$$
then there is a subsequence $\{(x_{n_i}, y_{n_i})\}$ which converges (i.e., the x's and y's each form a convergent sequence.)

(Do not use bisection; deduce it instead from Theorem 6.3, but be careful!)

6-5   Here is a proof of the Bolzano-Weierstrass Theorem which produces a convergent subsequence directly, without first having to find a cluster point. But the first part is a bit subtle.

(a) Prove that every sequence $\{a_n\}$ has a monotone subsequence.

Method: let $T_n = \{a_i : i \geq n\}$ denote the n-th "tail" of the sequence.

Case 1. If some tail has no maximum, deduce that $\{a_n\}$ has an increasing subsequence.

Case 2. If every tail has a maximum, deduce that $\{a_n\}$ has a decreasing subsequence, by selecting carefully from $\{\max T_n \ : \ n \in N\}$.)

(b) Deduce the Bolzano-Weierstrass Theorem from part (a).

6-6   lim sup and **lim inf**   For bounded sequences which do not converge, these are notions which sometimes can substitute for the non-existent limit.

Let $\{a_n\}$ be a bounded sequence. Define, for T_n as in Problem 6-5,

$$\bar{b}_n = \sup T_n \qquad \text{and} \qquad \underline{b}_n = \inf T_n.$$

(a) Prove the sequences $\{\bar{b}_n\}$ and $\{\underline{b}_n\}$ both converge.

We now define:

$$\limsup a_n = \lim \bar{b}_n, \qquad\qquad \liminf a_n = \lim \underline{b}_n.$$

(Other notation for these are $\overline{\lim}\, a_n$ and $\underline{\lim}\, a_n$, respectively.)

(b) Find lim sup and lim inf for the sequence $a_n = \dfrac{1}{n} + (-1)^n$.

(c) Prove that $\liminf a_n \leq \limsup a_n$.

(d) Prove that: $\lim a_n$ exists $\Leftrightarrow$ $\liminf a_n = \limsup a_n$.

6-7   Continuing 6-6, let S denote the set of cluster points of $\{a_n\}$.

Prove: $\limsup a_n = \max S, \qquad \liminf a_n = \min S.$

It is not obvious that $\min S$ and $\max S$ even exist. The problem requires you to show, among other things, that $\limsup a_n$ and $\liminf a_n$ are actually themselves cluster points of the sequence.

The result of this problem is probably the most intuitive way to think about lim sup and lim inf —they are respectively the highest and lowest cluster points of the sequence. Such cluster points exist by the Cluster Point Theorem and the Bolzano-Weierstrass Theorem, since $\{a_n\}$ is bounded.

Answers to Questions

6.1

1. $\{a, a + \frac{1}{n}\}$, for example.

2. There is a unique non-empty longest interval $[L, M] = \bigcap [a_n, b_n]$ inside all the $[a_n, b_n]$, and $a_n \to L$, $b_n \to M$.

3. $c_n = d_0 - d_1 + d_2 - \ldots + (-1)^n d_n$, where $\{d_i\}$ is a positive (or non-negative) decreasing sequence, and $\lim d_i = 0$. (Since the terms alternate in sign and decrease in size, the intervals are nested; $\lim d_i = 0$ implies the length of the nested intervals tends to 0.)

6.2

1. The sequence is $0, 1, 0, -1; 0, 1, 0, -1; \ldots$. Cluster points: $0, 1, -1$.

2. (a) Say $a_n \to L$. Then L is a cluster point (see remarks after (4)).

If K is also a cluster point, there is an $\{a_{n_i}\} \to K$. But $\{a_{n_i}\} \to L$ by the Subsequence Theorem. Hence $K = L$, since a limit is unique (Exercises 3.1/1 and 5.1/5).

(b) $1, 2, 1, 3, 1, 4, \ldots$, for instance.

3. $(4) \Rightarrow (4')$: Since $a_n \underset{\epsilon}{\approx} K$ is true for infinitely many n, you can find an $n > N$ for which it is true.

$(4') \Rightarrow (4)$: It is true for some $n_1 > 1$, then for some $n_2 > n_1$, then for some $n_3 > n_2$, etc. (taking successively $N = 1, n_1, n_2, \dots$); thus it is true for the infinitely many values $n_1, n_2, n_3, \dots$.

6.3

1. Use the B-W Theorem; the bounded ones are: (a), since $|b_n| \leq 1$, and (c), since $0 < b_n \leq 1$.

2. By the B-W Theorem, a bounded sequence $\{a_n\}$ has a convergent subsequence $\{a_{n_i}\}$; its limit K is a cluster point of $\{a_n\}$ by the Cluster Point Theorem 6.2.

6.4

1. $a_n = 1 + 1/2 + 1/3 + \dots + 1/n$; it is not convergent by Prop. 1.5A, therefore by theorem 6.4, it is not a Cauchy sequence.

3.   Given $\epsilon > 0$, $|a_m - a_n| \leq \dfrac{1}{m+n} < \epsilon$, if $m, n > 1/\epsilon$.

4.   See the lines following (9).

5.   Since $a_{n+1}(1 + a_n) = a_n$ for all n, taking the limit of both sides gives $L(1 + L) = L$, from which we deduce $L = 0$.

6.5

1. sup, inf, max min: (a) 1, 0, 1, none   (b) none, 1, none, 1
(c) all: none   (d) 1, -1, none, none   (e) all: none

2. Let $m_1 = \max S$ and $m_2 = \max S$; then $m_2 \leq m_1$ by the definition of m_1 and similarly $m_1 \leq m_2$, so $m_1 = m_2$.

3. Let $s_1 = \sup S$ and $s_2 = \sup S$. Then s_1 is an upper bound, so $s_2 \leq s_1$, by sup-2 for s_2. Similarly, $s_1 \leq s_2$. Therefore $s_1 = s_2$.

7

Infinite Series

7.1 Series and sequences.

An infinite series is a special kind of sequence.

The best way to begin their study is to look at examples. Here are some from previous chapters (looking ahead, we write $\{s_n\}$ instead of $\{a_n\}$).

$$s_n = 1 + \frac{1}{2} + \frac{1}{4} + \ldots + \frac{1}{2^n}, \qquad s_n \to 2; \qquad \textit{geometric sum, (4.2)};$$

$$s_n = 1 + \frac{1}{1!} + \frac{1}{2!} + \ldots + \frac{1}{n!}, \qquad s_n \to e; \qquad \textit{exponential sum, (Quest. 1.4/2)};$$

$$s_n = 1 + \frac{1}{2} + \frac{1}{3} + \ldots + \frac{1}{n}, \qquad s_n \to \infty; \qquad \textit{harmonic sum, (1.5)}.$$

What makes these sequences special is that to get from s_n to s_{n+1}, you add to s_n a simple expression in n. For the sequences above, we have respectively

$$s_{n+1} = s_n + \frac{1}{2^{n+1}}, \qquad s_{n+1} = s_n + \frac{1}{(n+1)!}, \qquad s_{n+1} = s_n + \frac{1}{(n+1)}.$$

By contrast, the recursive sequence $s_{n+1} = \frac{1}{2}(s_n + 2/s_n)$ for finding $\sqrt{2}$ by Newton's method (see 4.3), though simple, is not of this special type.

Definition 7.1 By an **infinite series**, we mean an expression

(1) $\qquad a_0 + a_1 + a_2 + \ldots + a_n + \ldots$, $\qquad$ (a_n is called the **n-th term**);

it is just a suggestive notation for the sequence $\{s_n\}$ defined by

(2) $\qquad s_n = a_0 + a_1 + a_2 + \ldots + a_n,$ $\qquad$ (s_n is the **n-th partial sum**);

or recursively by

(2a) $\qquad s_0 = a_0; \qquad s_{n+1} = s_n + a_{n+1}, \quad \text{for } n = 0, 1, 2, \ldots .$

If the sequence $\{s_n\}$ converges, with $\lim s_n = S$, we write symbolically

(3) $\qquad a_0 + a_1 + a_2 + \ldots + a_n + \ldots = S ,$

and say the series **converges** to the sum S; if not, we say the series **diverges**.

We write (1) in summation notation as $\sum_0^\infty a_n$, $\sum_0 a_n$, or just $\sum a_n$, if the lower limit of summation is either clear, or of no importance.

By virtue of (3), these symbols are used for both the series and its numerical sum, as the following examples illustrate; this double duty, while mildly confusing at first, spares us a lot of useless naming.

Examples 7.1A   Our earlier examples may now be written:

$$\sum_0^\infty \frac{1}{2^n} = 1 + \frac{1}{2} + \frac{1}{4} + \ldots + \frac{1}{2^n} + \ldots \ = 2 \qquad \textit{geometric series;}$$

$$\sum_0^\infty \frac{1}{n!} = 1 + \frac{1}{1!} + \frac{1}{2!} + \ldots + \frac{1}{n!} + \ldots \ = e \qquad \textit{exponential series;}$$

$$\sum_1^\infty \frac{1}{n} = 1 + \frac{1}{2} + \frac{1}{3} + \ldots + \frac{1}{n} + \ldots \quad \text{diverges} \qquad \textit{harmonic series.}$$

Example 7.1B   The geometric series.   We proved in Section 4.2 that

$$(4) \qquad\qquad \sum_0^\infty r^n = \frac{1}{1-r} \ , \quad \text{if } |r| < 1; \text{ otherwise it diverges .}$$

Example 7.1C   Find the limit of   0, 1, 1/2, 3/4, 5/8, 11/16, ... , the sequence formed by starting at 0, going right to 1, then going on the line alternately left and right, each time for half the preceding distance.

Solution.   According to its description, the sequence $\{s_n\}$ is given by

$$s_{n+1} = s_n + (-1)^n/2^n; \qquad s_0 = 0.$$

This law of formation shows that the sequence is of the infinite series type. That is, the s_n are the partial sums of the infinite series

$$0 + 1 - 1/2 + 1/4 - 1/8 + \ldots + (-1)^n/2^n + \ldots ,$$

a geometric series whose sum by (4) is seen to be 2/3. $\square$

Turning sequences into infinite series.

In Example 7.1C, we started with a sequence, and showed it could be viewed as the partial sums of an infinite series. In fact, we can start with any $\{s_n\}$, and by writing (2a) as

$$(5) \qquad\qquad a_0 = s_0, \qquad a_n = s_n - s_{n-1}, \quad n \geq 1 ,$$

view $\{s_n\}$ as the n-th partial sum of the infinite series

$$(6) \qquad \sum_0^\infty a_n \ = \ s_0 + (s_1 - s_0) + (s_2 - s_1) + \ldots + e(s_n - s_{n-1}) + \ldots \ .$$

When a series is written in this form (which makes self-evident the value of its n-th partial sum s_n), it is often called a *telescoping series*. For this form to be useful, however, the terms $s_n - s_{n-1}$ must be simple expressions in n.

Here is another example (from Prop. 1.5B) of changing a sequence to a series.

Example 7.1D   Let $s_n = 1 + 1/2 + 1/3 + \ldots + 1/n - \ln(n+1)$, $n \geq 1$. Convert the sequence into an infinite series.

Solution. This sequence does not seem to come from an infinite series; however by (5), it becomes the sequence of partial sums of $\sum a_n$ by defining

$$a_n = s_n - s_{n-1} = \frac{1}{n} - \ln(n+1) + \ln n$$

$$= \frac{1}{n} - \ln \frac{n+1}{n}, \quad n \geq 1.$$

By 1.5B, the resulting infinite series converges to Euler's constant:

$$\sum_{1}^{\infty} \left(\frac{1}{n} - \ln \frac{n+1}{n} \right) = \gamma.$$

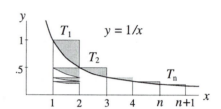

Here the infinite series $\sum a_n$ does not look like an improvement over the original sequence $\{s_n\}$. But the picture makes the series more natural by showing a_n may be interpreted as the area of the "triangle" T_n.

In the development of analysis, infinite series have been the most important type of sequence: they are used to represent functions, they arise naturally in the solution of differential equations, and it is much easier to study the convergence of infinite series than of more general sequences.

On the other hand, computers have given new life to iterative processes, which usually produce the general type of sequence, rather than the particular type called a series. Numerical methods for example (like Newton's method) give rise to iterative processes; so do discrete dynamical systems. So infinite series aren't the only kind of sequence in town.

Students often are confused about the difference between sequences and series because ordinary English uses the two words interchangeably. You won't have trouble if you remember that almost every October, two North American baseball teams play in the World Sequence.*

Questions 7.1

1. Express an arbitrary real number—say π—as the sum of an infinite series of rational numbers.

2. Convert these to telescoping series (6), find the partial sums $\{s_n\}$, and find the sum (if the series converges):

(a) $\displaystyle\sum_{1}^{\infty} \frac{1}{n(n+1)}$ $\left(\text{Hint} : \frac{1}{n(n+1)} = \frac{a}{n} - \frac{b}{n+1}\right)$ (b) $\displaystyle\sum_{1}^{\infty} \ln \frac{n}{n+1}$

3. If you remove the first N terms from $\sum_0 a_n$, show that this does not affect whether it converges or diverges.

---

* World Finite Sequence, really, since our sequences are always infinite.

7.2 Elementary convergence tests.

The infinite series convergence story is now briefly this.

First come the classical *tests for convergence* (or divergence). They are all based on estimating the size of the terms a_n of the series.

Except in some special cases, related to the geometric series or to formulas of calculus, *finding the sum* of a convergent series in terms of known numbers is generally impossible—there are so few numbers that can be named. So the computer just adds up the terms, which means you might want to know the *rate of convergence* of the series.

Finally, if the series is divergent but the sequence of partial sums is bounded, analysts have developed "summation methods" which sometimes permit the series to be summed in a weaker sense than that demanded by Definition 7.1. For instance, they assign the sum 1/2 to the divergent series $1 - 1 + 1 - 1 + 1 - 1 + \ldots$.

This chapter is mainly about the first topic mentioned above: some of the most important classical tests for convergence and divergence. They are based on our work with sequences, and on the geometric series (4) above. Remember: a series is just a special type of sequence, so our theorems about limits of sequences ought to tell us something about series. We begin in this section with some simple results about convergence and diveregence.

Theorem 7.2A   The n-th term test for divergence.
$$\sum a_n \text{ converges } \Rightarrow \quad \lim a_n = 0.$$

Proof.   Let s_n be the partial sums and $S = \sum a_n = \lim s_n$. We have

$$a_n = s_n - s_{n-1}, \qquad \text{by (5)};$$
$$\lim a_n = \lim s_n - \lim s_{n-1}, \quad \text{by the Linearity Theorem 5.1A};$$
$$= S - S = 0. \qquad \qquad \square$$

The most useful form for Theorem 7.2A is the contrapositive (see Appendix A.2) which turns it into the *n-th term divergence test*:

$$(7) \qquad\qquad \lim a_n \neq 0 \quad \Rightarrow \quad \sum a_n \text{ diverges.}$$

Example 7.2   Show these diverge: (a) $\sum n/(n+1)$ (b) $\sum (-1)^n$.

Solution. Using (7):   (a) $\lim a_n = 1$;   (b) $\lim a_n$ does not exist. $\square$

Warning 7.2   The most common mistake with infinite series.
$$\text{``}a_n \to 0 \quad \Rightarrow \quad \sum a_n \text{ converges''} \qquad \textit{is false.}$$

The quoted statement is the converse to Theorem 7.2A, but it is not true. The most familiar counterexample is the harmonic series:

$$1/n \to 0, \quad \text{but } \sum 1/n \text{ diverges.}$$

To put it another way, Theorem 7.2A says that the terms of a convergent series must get small, but this is not enough to ensure that a series will converge. In the language we do not use here, having the terms get small is a necessary but not a sufficient condition for the convergence of a sequence.

Alas, the false converse given in the Warning is an article of faith with many analysis students, to judge by the frequency with which it appears in exam answer booklets. Remember the harmonic series!

Theorem 7.2B   Tail-convergence theorem.

$$\sum_{N_0}^{\infty} a_n \quad \begin{array}{l}\text{converges} \\ \text{for some } N_0\end{array} \quad \Rightarrow \quad \sum_{0}^{\infty} a_n \quad \text{converges} \quad \Rightarrow \quad \sum_{N}^{\infty} a_n \quad \begin{array}{l}\text{converges} \\ \text{for all } N.\end{array}$$

Proof.   Easy; see Question 7.2/3.   □

Remarks.   The partial series starting with the N-th term is often called a *tail* of the original series. Theorem 7.2B says that if one tail converges, then they all do, including the original series. It expresses the fact that the early terms of a series don't matter if you just want to decide if it converges or not. (But they matter a great deal if you want to find the sum!)

The contrapositive of the Tail-convergence Theorem is the same statement about divergence: *if one tail diverges, then the series diverges and all its tails diverge.* To see this is the contrapositive, we negate both sides of the Tail-convergence Theorem and reverse the arrow:

$$\sum_{N}^{\infty} a_n \quad \begin{array}{l}\text{converges} \\ \text{for all } N\end{array} \qquad \text{negates to:} \qquad \sum_{N_0}^{\infty} a_n \quad \begin{array}{l}\text{diverges} \\ \text{for some } N_0\end{array} \ ;$$

$$\sum_{N_0}^{\infty} a_n \quad \begin{array}{l}\text{converges} \\ \text{for some } N_0.\end{array} \qquad \text{negates to:} \qquad \sum_{N}^{\infty} a_n \quad \begin{array}{l}\text{diverges} \\ \text{for all } N\end{array} \ .$$

Theorem 7.2C   Linearity theorem.   *Let p and q be real numbers; then*

$$\sum a_n \text{ and } \sum b_n \text{ converge} \quad \Rightarrow \quad \sum (pa_n + qb_n) \text{ converges,} \quad \text{and}$$

(8) $$\sum (pa_n + qb_n) = p\sum a_n + q\sum b_n.$$

Proof.   This is just a translation of the Linearity Theorem 5.1(i) for limits of sequences into the language of infinite series. In detail, let

$$s'_k = \sum_0^k a_n \qquad \text{and} \qquad s''_k = \sum_0^k b_n$$

be the respective partial sums. Then the partial sums of $\sum (pa_n + qb_n)$ are

$$s_k = \sum_0^k (pa_n + qb_n) = p\sum_0^k a_n + q\sum_0^k b_n = ps'_k + qs''_k.$$

Therefore by the Linearity Theorem 5.1(i), the sequence s_k converges and

$$\lim s_k = \lim(ps'_k + qs''_k) = p\lim s'_k + q\lim s''_k,$$

which is exactly (8), in view of the definition (3) of the sum of a series.   □

Corollary 7.2   *If convergent series are added, subtracted, or multiplied by a constant factor, the resulting series converge and to the corresponding sums:*

(9) $$\sum(a_n \pm b_n) \;=\; \sum a_n \pm \sum b_n; \qquad \sum(ca_n) \;=\; c\sum a_n.$$

By contrast, if two convergent series are multiplied or divided term-by-term, the resulting series will not necessarily converge, and even if it does, it will certainly not converge to the product or quotient of the two sums: the limit theorems about products and quotients of sequences are useless, since there is no relation between the partial sums of $\sum a_n b_n$ and $\sum a_n/b_n$ and those of $\sum a_n$ and $\sum b_n$.

Note also the double use of the summation symbol we called attention to earlier: in (8) and (9), $\sum$ refers to the numerical sum of the series, while in the line above (8), it refers to the series themselves. This double use avoids having to introduce new letters for the numerical sums of the series.

Theorem 7.2D   Comparison theorem for positive series. *Assume that*

(*) $$0 \le a_n \le a_n' \quad \text{for all } n. \qquad Then$$

$$\sum a_n' \text{ converges} \;\Rightarrow\; \sum a_n \text{ converges,} \quad \text{and} \quad \sum a_n \le \sum a_n' \;;$$

$$\sum a_n \text{ diverges} \;\Rightarrow\; \sum a_n' \text{ diverges.}$$

If (*) holds only for large n, the convergence and divergence statements are still valid, but not the inequality on the sums.

Proof.   It suffices to prove the convergence statement, since the divergence statement is just its contrapositive, and the remark then follows easily by applying the Tail-convergence Theorem 7.2B.

So consider the partial sums of the two series:
$$s_k = \sum_0^k a_n \qquad \text{and} \qquad s_k' = \sum_0^k a_n'.$$
We know the sequences $\{s_k\}$ and $\{s_k'\}$ are increasing, since $a_n \ge 0$ and $a_n' \ge 0$. By hypothesis, $\{s_k'\}$ converges to a limit S'; since $\{s_k'\}$ is increasing, its limit S' is an upper bound for the sequence by Theorem 3.2B, i.e.,
$$s_k' \;\le\; S' \quad \text{for all } k.$$
Since $a_n \le a_n'$ for all n, by hypothesis, it follows that
$$s_k \;\le\; s_k' \quad \text{for all } k.$$
Combining this with the earlier inequality shows that
$$s_k \;\le\; S' \quad \text{for all } k.$$
Thus $\{s_k\}$ is bounded above; since it is also increasing, the Completeness Property says it has a limit S, and by the Limit Location Theorem 5.3A,
$$S \;=\; \lim s_k \le S'.$$
This shows that $\sum a_n$ converges and that $\sum a_n \le \sum a_n'$.   □

Simple as it is—"obvious", if you like—the Comparison Theorem is the basis for most of the other convergence tests. It in turn is based on the Completeness Property.

Questions 7.2

1. Prove the following diverge: (a) $\sum \cos 1/n$;   (b) $\sum \sin n$.

2. (a) Find another example of a divergent series whose n-th term $\to 0$; include a proof of the divergence.

(b) Show the series $\sum (\sqrt{n+1} - \sqrt{n})$ is another example. (I assume you didn't think of this for part (a). Hint: Example 3.1B.)

3. Prove the Tail-convergence Theorem 7.2B, to get practice with notation.

4. Give a formal proof of the remark following the statement of the Comparison Theorem, i.e.:

if $0 \leq a_n \leq b_n$ for $n \gg 1$, then $\sum b_n$ converges $\Rightarrow \sum a_n$ converges.

(You can use the Comparison Theorem.)

5. Using theorems in this section, prove that the following diverge.

(a) $1/2 + 1/4 + \ldots + 1/2n + \ldots$

(b) $1 + 1/3 + 1/5 + \ldots + 1/(2n+1) + \ldots$

6. (a) $\sum 1/n^2$ is known to converge; using this, prove that $\sum 1/n^p$ converges whenever the constant $p > 2$.

(b) Discuss the convergence of $\sum 1/\sqrt{n}$.

7.3   The convergence of series with negative terms.

The comparison test applies only to series whose terms are positive. But many of the important series of analysis have both positive and negative terms. We handle this in a rather brutal fashion by changing all the $-$ signs to $+$ signs (i.e., replacing a_n by $|a_n|$) and comparison-testing the new series instead. Some convergent series don't object to this, but others do.

Definition 7.3 $\sum a_n$ is **absolutely convergent** if $\sum |a_n|$ converges;

$\sum a_n$ is **conditionally convergent** if $\sum a_n$ converges, but $\sum |a_n|$ diverges.

Notice that the definition of absolute convergence doesn't say the original series has to converge. But it always will; we'll prove this in a moment.

Examples 7.3

(a) $\sum (-1)^n/2^n$ and $\sum (-1)^n/n!$ are absolutely convergent, since the corresponding positive series are convergent: $\sum 1/2^n$ and $\sum 1/n!$.

(b) Any convergent series of positive terms is absolutely convergent.

(c) $\sum (-1)^n/n$ is conditionally convergent, since it converges (Example 6.1 or 4.2), but the corresponding positive series $\sum 1/n$ diverges (Example 1.5A).

A conditionally convergent series converges only because the negative terms partly cancel the positive ones, and so keep the total sum down. An absolutely convergent series is one whose terms are so small in size that even if you make them all positive, the resulting series converges. This makes the following theorem plausible.

Theorem 7.3   Absolute convergence theorem.

$$\sum |a_n| \text{ converges} \quad \Rightarrow \quad \sum a_n \text{ converges.}$$

Remarks. To make the proof clearer, we illustrate the idea on the second series in Example 7.3(a) above, which is absolutely convergent. To show it converges, write it as the difference of two non-negative series, one constructed from the positive terms, the other from the negative terms:

$$
\begin{array}{ccccccccc}
 & 1 & + & 0 & +1/2! & + & 0 & + & 1/4! + \ldots \\
- & 0 & + & 1/1! & +0 & + & 1/3! & + & 0 \quad + \ldots \\
= & 1 & - & 1/1! & +1/2! & - & 1/3! & + & 1/4! - \ldots
\end{array}
$$

Both of the top two series converge by the Comparison Theorem 7.2D, being non-negative and term-by-term not greater than the convergent series $\sum 1/n!$. Therefore their difference also converges, by the Linearity Theorem 7.2C.

Proof.   Write the given series as the difference of two series:

$$(10) \qquad\qquad \sum a_n = \sum (a_n^+ - a_n^-) = \sum a_n^+ - \sum a_n^-,$$

where $\sum a_n^+$ and $\sum a_n^-$ are the series defined by

$$a_n^+ = \begin{cases} |a_n|, & \text{if } a_n > 0; \\ 0, & \text{otherwise;} \end{cases} \qquad a_n^- = \begin{cases} |a_n|, & \text{if } a_n < 0; \\ 0, & \text{otherwise.} \end{cases}$$

Both series have non-negative terms, and according to their definition,

$$0 \le a_n^+ \le |a_n|, \qquad 0 \le a_n^- \le |a_n| \qquad \text{for all } n.$$

Since by hypothesis $\sum |a_n|$ converges, the comparison test 7.2D shows that

$$\sum a_n^+ \text{ converges} \qquad \text{and} \qquad \sum a_n^- \text{ converges.}$$

Since the difference of two convergent series is convergent (see (9)), equation (10) above shows that $\sum a_n$ converges.   □

Questions 7.3

1. True or false? A convergent series of negative terms is absolutely convergent.

2. Which converge absolutely?   (a) $\sum (-1)^n/n^2$ (b) $\sum (-1)^n/\sqrt{n}$

3. Why not simplify the proof of the Absolute Convergence Theorem in this way? Use the Comparison Theorem 7.2D:

$$a_n \le |a_n| \text{ for all } n, \quad \sum |a_n| \text{ converges} \quad \Rightarrow \quad \sum a_n \text{ converges.}$$

4. Given a conditionally convergent series, prove that the two series formed respectively from its positive and from its negative terms both diverge. (This will be used in Section 7.7. Use contraposition.)

7.4   Convergence tests: ratio and n-th root tests.

There are books devoted to convergence tests, but they are for specialists. The standard tests given in calculus are good enough for the series one ordinarily meets. Since they show the terms of the series get very small in size, they work only on absolutely convergent series. We begin with the one test everybody should know: it's easy to remember and use, a coarse test designed for the rapidly convergent series used to solve differential equations and in numerical calculation.

Theorem 7.4A   The ratio test.

Suppose $a_n \neq 0$ for $n \gg 1$, and $\displaystyle\lim_{n\to\infty}\left|\frac{a_{n+1}}{a_n}\right| = L.$ Then

$$L < 1 \quad\Rightarrow\quad \sum a_n \text{ converges absolutely ;}$$
$$L > 1 \quad\Rightarrow\quad \sum a_n \text{ diverges.}$$

(If $L = 1$ or there is no limit, the test fails and there is no conclusion.)

Discussion.   The proof should be studied carefully. The idea is to compare $\sum |a_n|$ with a geometric series. Since by hypothesis

$$\left|\frac{a_{n+1}}{a_n}\right| \approx L \qquad \text{for } n \gg 1,$$

we expect that, starting from some N, we have approximately

$$|a_{n+1}| \approx |a_n|L \quad \text{for } n \geq N;$$

using this over and over, we get

$$|a_{N+1}| \approx |a_N|L, \qquad |a_{N+2}| \approx |a_{N+1}|L \approx |a_N|L^2,$$

and similarly,

$$|a_{N+k}| \approx |a_N|L^k, \qquad k \geq 0.$$

Thus

$$(11) \qquad\qquad \sum |a_{N+k}| \approx |a_N| \sum L^k.$$

The series on the right is a geometric series (multiplied by a constant); as we know from 4.2, it converges if $L < 1$ and diverges if $L > 1$. This shows at least that $\sum a_n$ should be absolutely convergent if $L < 1$.

To make this argument rigorous, we have to replace in (11) the vague $\approx$ by the symbol $\leq$, so that we can use the Comparison Theorem 7.2D. As it stands, however, this replacement would not be valid. The trick therefore is to replace L by a slightly larger number M—big enough to make the inequality true, but still < 1 so that the geometric series converges.

Proof.   Assume $L < 1$, and choose a number M so that $L < M < 1$. Then by the Sequence Location Theorem 5.3B,

$$\left|\frac{a_{n+1}}{a_n}\right| \to L \quad\Rightarrow\quad \left|\frac{a_{n+1}}{a_n}\right| < M \text{ for } n \gg 1, \quad \text{say for } n \geq N.$$

The argument now continues as in the discussion, with inequalities involving M replacing the earlier approximations involving L. We have

$$\left|\frac{a_{n+1}}{a_n}\right| < M \quad \Rightarrow \quad |a_{n+1}| < |a_n|M, \qquad \text{for } n \geq N;$$

$$\Rightarrow \quad |a_{N+k}| < |a_N|M^k, \qquad \text{for } k \geq 1.$$

Therefore, term-by-term we have

$$\sum_1^\infty |a_{N+k}| \; < \; |a_N| \sum_1^\infty M^k.$$

The series on the right is a convergent geometric series, since $M < 1$. Thus by the Comparison Theorem 7.2D, the series on the left converges. Since it is a tail of $\sum |a_n|$, the latter also converges, by the Tail-convergence Theorem 7.2B.   □

The case $L > 1$ is simpler, since it does not require use of an M; divergence is proved by the n-th term test 7.2A. Since it is a good and non-tricky use of inequalities and comparison, we leave it for the Exercises.   □□

The ratio test is used all the time; the following test is less often needed. However, it is occasionally useful; we include it without giving its proof, which is similar to that of the ratio test and a good test of your understanding of that argument.

Theorem 7.4B The n-th root test.   *Suppose* $\lim\limits_{n\to\infty} |a_n|^{1/n} = L$. *Then*

$$L < 1 \quad \Rightarrow \quad \sum a_n \;\; \textit{converges absolutely;}$$

$$L > 1 \quad \Rightarrow \quad \sum a_n \;\textit{diverges.}$$

(*If* $L = 1$ *or there is no limit, the test fails and there is no conclusion.*)

Proof.   Similar to that of the ratio test, and left as an Exercise.

Example 7.4A   Test for convergence.   (a) $\sum(-1)^n n/2^n$ (b) $\sum 1/n^2$

Solution.   (a) Using the ratio test

$$\left|\frac{a_{n+1}}{a_n}\right| = \frac{n+1}{2^{n+1}} \cdot \frac{2^n}{n} = \frac{n+1}{2n} \to \frac{1}{2};$$

this shows it is absolutely convergent, and hence convergent, by Theorem 7.3.

(b) Using the ratio test,

$$\left|\frac{a_{n+1}}{a_n}\right| = \frac{n^2}{(n+1)^2} = \left(\frac{n}{n+1}\right)^2 \to 1^2 = 1,$$

so the test fails. The n-th root test also fails, for by Problem 4-1,

$$\lim n^{-2/n} = \lim (n^{1/n})^{-2} = 1^{-2} = 1.$$

This shows the ratio and n-th root tests are both too crude for this series—they only work for rapidly converging series, like the geometric series, or the exponential series; a series like the one in (b) requires more delicate tests, discussed in the next section.

Questions 7.4   1. Test for convergence, using the ratio or the n-th root test.

(a) $\sum \frac{(-1)^n}{3^n}$ (b) $\sum \frac{(-2)^n}{n!}$ (c) $\sum \frac{1}{\sqrt{n}}$ (d) $\sum \frac{2^n}{n^2}$ (e) $\sum \left(\frac{n+2}{2n} \right)^n$

7.5   The integral and asymptotic comparison tests.

In this section we consider tests which work on series like $\sum 1/n^2$, for which the ratio and n-th root tests fail. For the first test, we go back to the ideas in Section 1.5, where we compared the series with the area under a curve. We assume the relevant notions from calculus in the statement and the argument (it shouldn't be called a proof); they will all be justified later on when we study integration.

Theorem 7.5A.   The integral test.

Suppose $f(x) \geq 0$ and decreasing, for $x \geq$ some positive integer N.

Then $\sum f(n)$ converges if the area under $f(x)$ and over $[N, \infty)$ is finite, and diverges if the area is infinite.

Proof.   Assume first the area is finite. Referring to the picture,

(12) $0 \;\leq\; f(n+1) \;\leq\; A_n, \qquad$ for $n \geq N$.

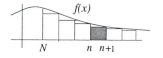

area of shaded   area under f(x)
rectangle   and over [n, n+1]

We claim that $\sum_N^\infty A_n$ converges: namely, its sequence of partial sums is given by

$$s_k = \sum_N^{N+k} A_n = \text{area over } [N,\ N+k+1],$$

so that $\lim s_k$ is the area over $[N, \infty)$, which we are assuming finite.

It follows from (12) and the Comparison Theorem that $\sum_N^\infty f(n)$ converges; then the Tail-convergence Theorem tells us $\sum f(n)$ converges as well.   □

If the area is infinite, shifting the rectangles one unit to the right gives circumscribed rectangles; similar reasoning shows $\sum f(n)$ diverges.   □□

Example 7.5A.   Prove that

(13) $\sum 1/n^p$ *converges if $p > 1$,   diverges if $p \leq 1$.*

Proof.   If $p \geq 0$, then $f(x) = 1/x^p$ is decreasing for $x > 0$.

We can therefore apply the integral test, studying whether the area integral on the left below is finite or not. If $p \neq 1$,

$$\int_1^\infty \frac{1}{x^p}\,dx = \lim_{r \to \infty} \frac{x^{1-p}}{1-p} \bigg]_1^r = \lim_{r \to \infty} \frac{r^{1-p} - 1}{1 - p}.$$

If $p > 1$, then $\lim_{r \to \infty} r^{1-p} = 0$, so the area is finite; this proves the convergence statement.

For the divergence statement, it's conceptually simplest to compare the series when $p < 1$ with the known-to-be-divergent harmonic series ($p = 1$). However, to get practice with the integral test, let's continue with that.

If $p = 1$, the integral evaluates to $\lim_{r \to \infty} \ln r$, so the area is infinite, and the series diverges.

If $0 \leq p < 1$, $\lim_{r \to \infty} r^{1-p} = \infty$, so the area is again infinite, and the series diverges.

If $p < 0$, the function is no longer decreasing, but the series is divergent since its individual terms tend to ∞. $\square$

The convergence result in this example can also be obtained by comparisons not involving integrals. However, the integral is the simplest, clearest, and most memorable way of seeing it. Remember the result, since these series are among the most important ones used in comparison tests.

Outside of the above example, the integral test is not often applied, since there are so few decreasing functions $f(x)$ that one would care to integrate. A substitute often used is the following convenient form of the comparison test.

Theorem 7.5B   Asymptotic comparison test.   *If*

$$|a_n| \sim |b_n|, \qquad (meaning: \quad \lim |a_n|/|b_n| = 1),$$

then

$$\sum |a_n| \ converges \quad \Leftrightarrow \quad \sum |b_n| \ converges.$$

Proof.   Left as a good and very reasonable exercise. $\square$

Examples 7.5B.   Do these converge or diverge?

$$(a) \quad \sum_{2}^{\infty} \frac{1}{n^3 - 2n + 1} \qquad\qquad (b) \quad \sum \sqrt{\frac{4n}{n^2 + 1}}$$

Solution.   (a) converges, by asymptotic comparison with $\sum 1/n^3$;

(b) diverges, by asymptotic comparison with $\sum 2/\sqrt{n}$.

If we had tried the integral test instead, we would have had to determine whether or not the integrals

$$\int_2^\infty \frac{dx}{x^3 - 2x + 1} \qquad and \qquad \int_0^\infty \sqrt{\frac{4x}{x^2 + 1}} \, dx \ .$$

were finite. This would not be difficult, provided you realize the way to do this is by estimating the integrands. Unfortunately many students try to do it by evaluating such integrals exactly, which is at best a waste of time, and at worst impossible.

Questions 7.5

1. Test the following for convergence, using comparison—either with the area of some infinite region, by using the integral test—or with other series known to converge or diverge, by using the asymptotic comparison test.

(a) $\sum \dfrac{n}{n^2+1}$ (both ways) (b) $\sum \sqrt{\dfrac{n}{n^4-4}}$ (c) $\sum \dfrac{1}{n^2+1}$ (both ways)

7.6   Series with alternating signs: Cauchy's test.

So far, all the tests we have given have been tests for absolute convergence. In general, the convergence of a conditionally convergent series is difficult to establish. But there is one simple and intuitively appealing test. The basic idea was given in Example 6.1, when we proved $\sum(-1)^n/n$ converged by using nested intervals.

Theorem 7.6   Cauchy's test for alternating series.

If $\{a_n\}$ is positive and strictly decreasing, and $\lim a_n = 0$, then

$$\sum(-1)^n a_n \quad \text{converges.}$$

The picture shows four successive partial sums of such a series; it should make the convergence very plausible.

Proof.   Since the signs alternate, we get for the even and odd partial sums of the series (recall that by hypothesis, $a_n > 0$):

(14) $$s_{2k} = s_{2k-1} + a_{2k}, \qquad s_{2k+1} = s_{2k} - a_{2k+1}.$$

Since $a_{2k} > 0$, this gives us a sequence of intervals

$$[s_{2k-1},\ s_{2k}], \quad k \geq 1; \qquad \text{length } [s_{2k-1},\ s_{2k}] = a_{2k} \to 0.$$

These intervals are nested, as the picture suggests; to see this formally, we combine the equations in (14), getting

$$s_{2k+1} = s_{2k-1} + (a_{2k} - a_{2k+1}),$$

and similarly,

$$s_{2k+2} = s_{2k} - (a_{2k+1} - a_{2k+2});$$

the quantities in parentheses are > 0 since $\{a_n\}$ is strictly decreasing, so that the two equations above imply respectively the first and third inequalities in

$$s_{2k-1} < s_{2k+1} < s_{2k+2} < s_{2k}.$$

The middle inequality in the chain comes from the left equation in (14), replacing k by $k+1$. This shows the intervals are nested, as in the picture.

The Nested Intervals Theorem 6.1 now says there is a unique real S in all of these intervals. To see that $s_n \to S$, the picture shows that, regardless of whether n is odd or even,

$$(15) \qquad\qquad\qquad |s_n - S| < a_{n+1} \, .$$

Since our hypothesis says that $a_n \to 0$, it follows by the Squeeze Theorem 5.2 that $|s_n - S| \to 0$, which shows the series converges to the sum S. $\qquad\square$

Questions 7.6

1. Test for convergence.

(a) $\sum \dfrac{(-1)^n}{\sqrt{n}}$ (b) $\sum(-1)^n \dfrac{n}{n+2}$ (c) $\sum(-1)^n \dfrac{\cos n\pi}{n}$

2. Let $\sum(-1)^n a_n$ be an alternating series satisfying the hypotheses of Cauchy's test; let S be its sum.

(a) If you stop summing at the term a_n, the size of the error is at most the size of the next term. What line of the proof of Cauchy's test shows this?

(b) According to part (a), how many terms of the alternating harmonic series should you add up to get its sum $\ln 2$ to three decimal places?

3. Is Cauchy's test for alternating series still valid if

(a) "strictly" is omitted?

(b) "positive and strictly decreasing" is true only for n large? Give a brief reason if true, or a counterexample if false.

7.7   Rearranging the terms of a series.

A natural question to ask is what happens to the sum of a convergent series if we add up its terms in a different order. Does it still converge? If so, do we get the same sum? The answer depends on whether the series is absolutely or conditionally convergent.

Theorem 7.7   Rearrangement Theorem.

If the terms of an absolutely convergent series are rearranged, the new series is still absolutely convergent, and has the same sum as the old one.

If the series is conditionally convergent, by rearranging its terms one can get a new series which will converge to any prescribed real number, or if one wishes, diverge to ∞ or $-\infty$.

Proof. We will not prove the second statement formally; the line of argument is sufficiently illustrated by an example. Let us rearrange the terms of the alternating harmonic series $\sum(-1)^{n-1}/n$ so that it converges to π, say.

As shown in Question 7.3/4, the series of positive terms and the series of negative terms both diverge:

$$1 + 1/3 + 1/5 + \ldots + 1/(2n+1) + \ldots \quad \text{diverges to } \infty;$$
$$-1/2 - 1/4 - 1/6 - \ldots - 1/2n - \ldots \quad \text{diverges to } -\infty.$$

(Briefly: the second diverges since it is a constant multiple of the divergent harmonic series; the first diverges since it is term-by-term greater than the negative of the second series.)

Now, add up the terms of the positive series until the sum S_0 first exceeds π, then add on terms of the negative series until the new total sum S_1 first becomes less than π, then add on more positive terms until the total sum S_2 again first exceeds π, then continue with more negative terms, and so on.

Let $\sum b_n$ be this rearranged series. Its sequence s_n of partial sums has S_i as a subsequence (that is, $S_i = s_{n_i}$ for some n_i). The construction shows that S_i cannot differ from π by more than the last added-on term, i.e.

$$|S_i - \pi| \leq |b_{n_i}|,$$

which shows that $\lim S_i = \pi$, since one can see that no matter how the original series is rearranged, $\lim b_{n_i} = 0$. Since by construction the other partial sums s_n always lie between two successive S_i, it is not hard to see that $\lim s_n = \pi$ also. Thus the rearranged series has π as its sum.

The argument depends on the fact that the sums of the positive and negative terms separately grow arbitrarily large. This is true of any conditionally convergent series; see Question 7.3/4.

To prove the Rearrangement Theorem for absolutely convergent series, we prove it first for a series $\sum a_n$ with positive terms: $a_n \geq 0$. Suppose the rearranged series is $\sum a_n'$, and the respective partial sums are s_n and s_n'.

Given a partial sum s_k', all its terms will occur in some partial sum s_{n_k} if you go far enough out in the series. Since all terms are positive,

$$s_k' \leq s_{n_k}.$$

But we also know

$$s_{n_k} \to S = \sum a_n,$$

since by the Subsequence Theorem 5.3, the subsequence $\{s_{n_k}\}$ has the same limit S as $\{s_n\}$. The s_k' thus form an increasing sequence, which is bounded above by S. By the Completeness Property, this sequence has a limit S', and by the Limit Location Theorem 5.3A,

$$S' \leq S.$$

This shows that $\sum a_n'$ converges, and to a sum $S' \leq S$. By symmetry, since $\sum a_n$ can be viewed as a rearrangement of $\sum a_n'$, we must also have $S \leq S'$. Therefore $S = S'$, and the two series have the same sum.   □

If the absolutely convergent series has both positive and negative terms, we form as in the proof of the Absolute Convergence Theorem 7.3 the two series of the positive and the negative terms. We have as in that proof,

$$\sum a_n = \sum a_n^+ - \sum a_n^- = S^+ - S^- ,$$

since both series converge. But we have also

$$\sum a_n' = \sum a_n'^+ - \sum a_n'^- = S^+ - S^- ;$$

namely, since the positive series $\sum a_n'^+$ is a rearrangement of the positive series $\sum a_n$, it has the same sum S^+, and similarly for the two negative series.

The two equations above show that $\sum a_n$ and $\sum a_n'$ have the same sum; this finishes the proof that absolutely convergent series can be rearranged without altering the sum. $\square\square$

Exercises

7.1

1. Show that the following series can be rewritten as telescoping series, and use this to prove they converge and evaluate their sum.

(a) $\displaystyle\sum_2^\infty \frac{1}{n^2 - 1}$ (b) $\displaystyle\sum_1^\infty \frac{(-1)^{n-1}}{n(n+2)}$ (c) $\displaystyle\sum_1^\infty \frac{1}{n(n+k)}$, k integer > 0

2. Translate the Cauchy criterion for convergence of a sequence (6.4) into a criterion for the convergence of an infinite series. (Interpret the sequence as the sequence of partial sums; express your criterion directly in terms of the series.)

3. An infinite product is represented as $\prod_0^\infty (1 + a_n)$, where $\prod_0^\infty$ means, "take the product of the terms as n goes from 0 to ∞". How would you define convergence for an infinite product, and how would you define its value if it converges?

(This value should not be allowed to be 0, since one wants to be able to convert the product to an infinite series by taking its natural logarithm.)

7.2

1. Given that $\displaystyle\sum_1^\infty \frac{1}{n^2} = \frac{\pi^2}{6}$, evaluate $\displaystyle\sum_0^\infty \frac{1}{(2n+1)^2}$; cite theorems used.

2. Prove that if $a_n \geq 0$ and $\sum a_n$ converges, then $\sum a_n^2$ converges. (Use comparison.)

3. Generalize the preceding exercise: let $\sum a_n$ and $\sum b_n$ be two convergent series with non-negative terms. Prove that $\sum a_n b_n$ converges. Do this two ways:

(a) prove and use an inequality relating the partial sums of the three series (call them s_n, s_n', and s_n'', respectively; you'll use limit theorems);

(b) use the Comparison Theorem.

4. Prove that if a_n and b_n are non-negative, and $\sum a_n^2$ and $\sum b_n^2$ converge, then $\sum a_n b_n$ converges. Is this a stronger or weaker theorem than the one in the preceding exercise; why? (Hint: $(a_n - b_n)^2 \geq 0$.)

5. Let $\sum a_n$ be a convergent series having the sum S. Let $\sum b_k$ be a new series, whose terms are formed by grouping the terms of $\sum a_n$ in pairs, and adding them. That is,

$$b_0 = a_0 + a_1, \quad b_1 = a_2 + a_3, \quad \ldots, \quad b_k = a_{2k} + a_{2k+1}, \quad \ldots .$$

Prove that $\sum b_k$ also converges, and to the same limit S.

(Hint: use the Subsequence Theorem 5.4.)

7.3

1. Theorem. If $\sum a_n$ is absolutely convergent, then

$$\left|\sum_1^\infty a_n\right| \leq \sum_1^\infty |a_n|, \qquad \textit{(infinite triangle inequality)}.$$

(a) Critique the following proof (supply missing steps and reasons):

$$\left|\sum_1^k a_n\right| \leq \sum_1^k |a_n|, \qquad \textit{(finite triangle inequality)}.$$

If we take the limit of both sides as $k \to 0$, we get the infinite triangle inequality.

(b) Give a slightly different proof, by imitating the proof of the usual triangle inequality (Section 2.4).

2. Prove: if $\sum a_n$ is absolutely convergent and $\{b_n\}$ is bounded, then $\sum a_n b_n$ is convergent.

3. (a) Let $\{a_n\}$ be a sequence, and $\{a_{n_i}\}$ be any subsequence. Prove that if $\sum_{n=0}^\infty a_n$ is absolutely convergent, then $\sum_{i=0}^\infty a_{n_i}$ is absolutely convergent.

(b) Show by counterexample that the above is false if the word "absolutely" is dropped everywhere.

4. (a) Prove: $\sum a_n$ absolutely convergent $\Rightarrow \sum a_n^2$ convergent.

(b) Show that "absolutely" cannot be dropped in part (a).

5. (a) How would you define "absolute divergence" of a series $\sum a_n$?

(b) "If $\sum a_n$ diverges absolutely, then it diverges."

Prove if true, give a counterexample if false.

6. Prove that a conditionally convergent series has an infinity of positive terms and an infinity of negative terms.

7.4-7.5

1. Test each of the following series for convergence; for two of them it helps to know that $\lim_{n\to\infty}(1 + r/n)^n = e^r$.

(a) $\displaystyle\sum_1 \frac{\sqrt{n}}{n^2 + 1}$ (b) $\displaystyle\sum_1 \frac{n^2}{2^n}$ (c) $\displaystyle\sum_1 \frac{\cos n}{n^2}$ (d) $\displaystyle\sum_0 \frac{(n!)^2}{(2n)!}$

(e) $\displaystyle\sum_1 \left(\frac{n+1}{2n+1}\right)^n$ (f) $\displaystyle\sum_2 \frac{1}{n\ln n}$ (g) $\displaystyle\sum_1 \sin(1/n)$ (h) $\displaystyle\sum_1 \frac{2^n n!}{n^n}$

(i) $\displaystyle\sum_0 \left(\frac{n}{n+2}\right)^{n^2}$ (j) $\displaystyle\sum_2 \frac{1}{n(\ln n)^p}$

2. Prove the case $L > 1$ of the ratio test.

3. (a) Prove the n-th root test.

(b) Give two examples of series $\sum a_n$ for which $(a_n)^{1/n} \to 1$: one convergent, the other divergent.

4. Prove the asymptotic comparison test 7.5B.

5. Prove the divergence statement in the integral test.

7.6

1. Test for conditional convergence.

(a) $\sum \dfrac{(-1)^n}{n^{1/3}}$ (b) $\sum \dfrac{(-1)^n}{n^2 + 1}$ (c) $\sum \dfrac{(-1)^n}{\tan^{-1} n}$ (d) $\sum (-1)^n \dfrac{n^5}{2^n}$

2. Is Theorem 7.6 still valid if "strictly decreasing" is omitted? Give a proof or counterexample.

7.7

1. Simpler than rearrangement is *grouping*: if the series is $\sum a_n$, we group its terms into a new series $\sum b_k$ by putting

$$b_0 = a_0 + \ldots + a_{n_0}, \quad b_1 = a_{n_0+1} + \ldots + a_{n_1}, \quad b_2 = a_{n_1+1} + \ldots + a_{n_2},$$

and so on. Prove that if $\sum a_n$ converges, so does $\sum b_k$, and to the same sum.

Problems

7-1 Here is a refinement of the n-th term test for divergence (7.2A).

Theorem.   *If $\{a_n\}$ is non-negative, decreasing, and $\sum a_n$ converges, then*

$$na_n \to 0.$$

(a) Illustrate the use of this theorem on the series $\sum 1/n^p$, for $p > 0$. What does it tell you about the convergence or divergence of this series for different values of p? Equally important, what does it *not* tell you?

(b) A converse is: $na_n \to 0$, $a_n \geq 0$, a_n decreasing $\Rightarrow \sum a_n$ converges. This converse is *not* true. Show this by giving a counterexample.

(c) Prove the theorem. Show first that

$$\text{given } \epsilon > 0, \qquad na_{2n} \leq a_{n+1} + \ldots + a_{2n} < \epsilon, \quad \text{for } n \gg 1.$$

Then handle the odd terms somehow, and complete the proof. (Hint: Exercises 7.1/2 and 6.4/1 may be helpful in this problem.)

7-2 Prove that if $|a_{n+1}/a_n| \leq |b_{n+1}/b_n|$ for $n \gg 1$, and $\sum b_n$ is absolutely convergent, then $\sum a_n$ is absolutely convergent.

7-3 Prove that if $|a_{n+1}/a_n| \leq r < 1$, for $n \gg 1$, then $\sum a_n$ converges. Show this is a stronger result than the convergence statement in the ratio test.

7-4 Prove: $\lim \left| \dfrac{a_{n+1}}{a_n} \right| = L \Rightarrow \lim |a_n|^{1/n} = L.$

Deduce that the n-th root test is stronger than the ratio test, in the sense that if a series satisfies the hypothesis for the ratio test, it will also satisfy the hypothesis of the n-th root test.

7-5 Give a different proof of the Absolute Convergence Theorem by using the infinite series form of the Cauchy criterion for convergence (6.4).

(Hint: cf. Exercise 7.1/2. Estimate $| \sum_n^m a_k | .$)

7-6 (a) Show the interval $[2k\pi + \pi/4, \ 2k\pi + 3\pi/4]$ contains an integer.

(b) Using this, or otherwise, prove that $\sum (\sin n)/n$ is not absolutely convergent.

7-7 A *summation method* is a way of assigning a sum to a series which diverges, but whose partial sums remain bounded. (Such a series must contain infinitely many positive and negative terms.) One of the best-known methods is *Cesaro summation*, defined as follows.

Given the series $\sum a_n$, let

$$c_n = (s_1 + \ldots + s_n)/n, \qquad \lim c_n = C.$$

If this limit C exists, the series $\sum a_n$ is said to be *Cesaro summable*, and C is called its *Cesaro sum*.

(a) As an example, find the Cesaro sum of $1 - 1 + 1 - 1 + \ldots$.

(b) Notice that c_n is the average of the first n partial sums of the series. To get a further feeling for c_n, express it directly in terms of the a_k.

(c) Show that if $\sum a_n$ converges, it is Cesaro summable and its Cesaro sum is the same as its ordinary sum. (A summation method with this desirable property is called *regular*.)

(Hint: cf. Problem 3-1b).

Answers to Questions

7.1

1. For example, $\pi = 3 + .1 + .04 + .001 + .0005 + \ldots$.

2. (a) $a = b = 1$, $s_n = 1 - 1/(n+1)$; convergent; sum $= \lim s_n = 1$.
(b) $s_n = -\ln(n+1)$, so series diverges.

3. Let s'_k be the k-th partial sum of the new series $\sum_N a_n$; then $s_{N+k} = s_{N-1} + s'_k$, for $k \geq 0$. Thus $\lim s'_k$ exists $\Leftrightarrow \lim_{k \to \infty} s_{N+k}$ exists.

7.2

1. Use the n-th term test: (a) $\lim a_n = 1$; (b) cf. Example 5.4C.

2. (a) $\sum 1/\sqrt{n} > \sum 1/n$, therefore it diverges (Comparison Theorem).

(b) $\lim a_n = 0$ (Example 3.1B); diverges since $s_n = \sqrt{n+1}$ (this is a telescoping series; see (6))

3. This is essentially Question 7.1/3.

4. Say $\leq$ holds for $n \geq N$; using in turn 7.2D and 7.2B, we have:
$$\sum_N b_n \text{ converges} \Rightarrow \sum_N a_n \text{ converges} \Rightarrow \sum a_n \text{ converges}.$$

5. (a) By 7.2C: $\sum 1/2n = (1/2) \sum 1/n$.
(b) By 7.2D; comparison with the series in part (a).

6. (a) By 7.2D: $\sum 1/n^p < \sum 1/n^2$ termwise.   (b) See 7.2/2a above.

7.3

1. True, by Linearity Theorem 7.2C, since $\sum a_n = \sum -|a_n| = -\sum |a_n|$.

2. (a), but not (b): see Question 7.2/6.

3. The Comparison Theorem requires $a_n \geq 0$, which is not assumed.

4. If say $\sum a_n^+$ converges, then $\sum a_n^- = -(\sum a_n - \sum a_n^+)$ also converges (by linearity, 7.2C); therefore $\sum |a_n| = \sum a_n^+ + \sum a_n^-$ converges.
This shows $\sum a_n$ is absolutely convergent, which is a contradiction.

7.4

1. (In each case, $|a_{n+1}/a_n|$ or $|a_n|^{1/n}$ is given, with its limit.)

(a) either test; $1/3 \to 1/3$, converges

(b) ratio; $2/(n+1) \to 0$; converges

(c) ratio; $(n/(n+1))^{1/2} \to 1$; test fails

(d) ratio; $2(n/(n+1))^2 \to 2$; diverges

(e) n-th root test; $(n+2)/2n \to 1/2$; converges

7.5

1. (a) diverges: by asymptotic comparison with $\sum 1/n$; or integral test:
$$\int_0^\infty \frac{x}{x^2+1}\, dx = \frac{1}{2}\ln(x^2+1)\Big]_0^\infty = \infty$$

(b) converges: asymptotic comparison with $\sum 1/n^{3/2}$

(c) converges; asym. comp. with $\sum 1/n^2$; or int. test: $\int_0^\infty \frac{dx}{x^2+1} = \frac{\pi}{2}$

7.6

1. (a) converges, by Cauchy's test   (b) diverges, by n-th term test
(c) diverges (the harmonic series)

2. (a) This is what (15) says.
(b) by convention, "3 decimal place accuracy" means:
error $< .0005 = 1/2000$. Since this is a_{2000}, we need to sum 1999 terms to get this accuracy.

3. (a) yes; change all $<$ in the proof to $\leq$; the intervals are still nested.
(b) yes; a tail of the series satisfies Cauchy's test and therefore converges, so the whole series converges (by the Tail-convergence Theorem).

8

Power Series

8.1 Introduction. Radius of convergence.

Classically, the heavily studied infinite series have been the *power series*: those series having the form

$$(1) \qquad \sum_0^\infty a_n x^n = a_0 + a_1 x + a_2 x^2 + \ldots + + a_n x^n + \ldots .$$

Power series are important because they are used to represent functions. Solutions to differential equations can often be expressed as power series when no other representation for them is available, and many new functions first arose in this way. The series are also useful in calculating values of the functions they represent, since the first few terms of the power series give a good approximation to the function if x is small.

How shall we think of the x in a power series? In this first section we will adopt the viewpoint of elementary algebra: that x represents some unspecified real number. Looked at this way, the expression (1) really represents not just one infinite series, but a whole family of them, one for each value of x. The first question therefore is, for which values of x does the series (1) converge? For the power series in daily commerce, the ratio test usually answers this question.

In using the ratio test and other results of the previous chapter, bear in mind that a_n there denoted the n-th term of the series, whereas in this chapter, the n-th term is written $a_n x^n$.

Example 8.1 For which values of x does $\sum_1^\infty \dfrac{x^{2n}}{2^n n}$ converge?

Solution. We use the ratio test: taking the limit of the ratio of two successive terms, we get

$$\lim_{n\to\infty} \left| \frac{x^{2(n+1)}}{2^{n+1}(n+1)} \cdot \frac{2^n n}{x^{2n}} \right| = \lim_{n\to\infty} \frac{|x|^2 n}{2(n+1)} = \frac{|x|^2}{2} ;$$

now since

$$\frac{|x|^2}{2} \le 1 \quad \Leftrightarrow \quad |x| < \sqrt{2} ,$$

the series converges for $|x| < \sqrt{2}$ and diverges for $|x| > \sqrt{2}$.

(What happens if $x = \pm\sqrt{2}$ is easily seen, but doesn't concern us just now.)

The series in Example 8.1 illustrates a basic property of power series: they converge in some interval $|x| < R$ symmetric about the origin, and diverge for $|x| > R$. (For $x = \pm R$, they may converge or diverge.) We prove this now; it is the main theorem about power series in this chapter, and a good exercise in using the Comparison Theorem for series (7.2D) and the Completeness Property for sets (6.5).

Theorem-Definition 8.1   Radius of convergence.

For each power series $\sum a_n x^n$ there is a unique $R \geq 0$ such that

$$\sum a_n x^n \text{ converges absolutely for } |x| < R, \text{ diverges for } |x| > R.$$

The number R is called the **radius of convergence** of the series. (By convention, we say $R = \infty$ if the series is absolutely convergent for all x.)

Proof.   The first part of the proof is devoted to showing:

(2) $\sum a_n x^n$ converges for $x = c$ $\Rightarrow$ $\sum |a_n x^n|$ converges for $|x| < |c|$.

For maximum clarity, we do this in two steps.

Case 1. $c = 1$. In this case, our hypothesis says: $\sum a_n$ converges.

$$\sum a_n \text{ converges} \quad \Rightarrow \quad a_n \to 0, \qquad \text{by the } n\text{-th term test 7.2A;}$$
$$\Rightarrow \quad |a_n| \to 0, \qquad \text{by Question 3.1/2;}$$
$$\Rightarrow \quad |a_n| < 1, \qquad \text{for } n \geq \text{some } N,$$

by the Sequence Location Theorem 5.3B. Using this, we show the tail $\sum_N |a_n x^n|$ converges for $|x| < 1$. Namely,

$$\sum_N |a_n x^n| \quad = \quad \sum_N |a_n||x|^n$$
$$< \quad \sum_N |x|^n, \quad \text{term by term, since } a_n < 1 \text{ for } n \geq N.$$

Since the series on the right converges for $|x| < 1$, so does the series on the left, by the Comparison Theorem 7.2D. This proves (2) when $c = 1$.

Case 2.   If $c \neq 0$ (for $c = 0$ there is nothing to prove), we reduce to the previous case by *scaling*, i.e., introducing the new variable $u = x/c$ so that $u = 1$ when $x = c$. Since $x = cu$, we have

$$\sum a_n x^n \text{ converges for } x = c$$
$$\Rightarrow \quad \sum a_n c^n u^n \text{ converges for } u = 1;$$
$$\Rightarrow \quad \sum |a_n c^n u^n| \text{ converges for } |u| < 1 \text{ (by case 1);}$$
$$\Rightarrow \quad \sum |a_n x^n| \text{ converges for } \left|\frac{x}{c}\right| < 1, \text{ or } |x| < |c|. \qquad \square$$

We now use the Completeness Property for sets (6.5) to locate R and prove that it separates the convergent sheep from the divergent goats.

Let A be the set of points where $\sum a_n x^n$ converges absolutely:

(3)
$$A = \{x \ : \ \sum |a_n x^n| \text{ converges}\} \ .$$

Our strategy is to define R to be sup A. For this to be valid, we need to know that sup A exists, which will follow from the Completeness Property if we can show that A is bounded above (note that A is non-empty since it contains 0). This and everything else we need to know about A will be based on the following property of A, which is based on the first half of the proof.

(4)
$$c > 0, \quad c \in A \quad \Rightarrow \quad (-c, c) \subset A \ .$$

Proof of (4). $c \in A \quad \Rightarrow \quad \sum |a_n c^n| \text{ converges, by (3)};$

$\qquad\qquad\qquad \Rightarrow \quad \sum a_n c^n \text{ converges, by Theorem 7.3};$

$\qquad\qquad\qquad \Rightarrow \quad \sum |a_n x^n| \text{ converges for all } |x| < c, \text{ by (2)};$

$\qquad\qquad\qquad \Rightarrow \quad (-c, c) \subset A, \text{ by (3).} \qquad\qquad \square$

In order now to define R as sup A, we distinguish two cases.

(i). $A = (-\infty, \infty)$. Then we set $R = \infty$.

(ii). $A \neq (-\infty, \infty)$ Then there is a point $b \notin A$; since (3) then shows that also $-b \notin A$, we may take $b > 0$.

We claim that this b is an upper bound for A. Reasoning indirectly, suppose A contained a point $c > b$; then according to (4),

$$[-b, b] \subset (-c, c) \subset A,$$

which contradicts $b \notin A$. Thus b is an upper bound for A.

It follows by the Completeness Property for sets (6.5) that sup A exists, and we define $R = \sup A$.

To see that R is the radius of convergence, we prove (5) and (6) below.

(5)
$$|x| < R \quad \Rightarrow \quad \sum |a_n x^n| \text{ converges, (i.e., } x \in A.)$$

Proof of (5).   Since R is the least upper bound of A,

$|x| < R \quad \Rightarrow \quad |x| \text{ is not an upper bound for } A;$

$\qquad\qquad \Rightarrow \quad |x| < c \leq R \text{ for some } c \in A;$

$\qquad\qquad \Rightarrow \quad x \in (-c, c) \subset A, \quad \text{by (4).} \qquad \square$

(6)
$$|x| > R \quad \Rightarrow \quad \sum a_n x^n \text{ diverges} \ .$$

Namely, choose any c such that $R < c < |x|$. Reasoning indirectly, if $\sum a_n x^n$ converged at x, by (2) it would converge absolutely at c, which would show $c \in A$; but this contradicts the fact that R is an upper bound for A. $\qquad\qquad \square\square$

Questions 8.1

1. For each of these power series, find the radius of convergence R, by using the ratio test. (The sums start at $n = 0$.)

$$\text{(a)} \sum \frac{x^n}{3^{2n+1}} \qquad \text{(b)} \sum n! x^n \qquad \text{(c)} \sum n^2 x^n \qquad \text{(d)} \sum \frac{x^n}{n!}$$

2. Is the radius of convergence of $\sum a_n x^n$ changed by

(a) dropping the first N terms from the series?

(b) altering the signs of the a_n at random?

3. Show that (4) can be strengthened to: $c > 0$, $c \in A \implies [-c, c] \subseteq A$.

8.2 Convergence at the endpoints. Abel summation.

If R is the radius of convergence of a power series, then we know the series converges for $|x| < R$ and diverges for $|x| > R$. But what about convergence at the two endpoints $x = R$ and $x = -R$? This is often hard to determine; in particular, if you determined R by using the ratio test, it is guaranteed not to work on the numerical series obtained by putting $x = \pm R$. However, one of the other tests may work, particularly Cauchy's test (7.6) for alternating series.

Example 8.2A   The following series all have radius of convergence 1; test them for convergence at 1 and -1.

$$\text{(a)} \sum x^n \qquad \text{(b)} \sum \frac{x^n}{n} \qquad \text{(c)} \sum \frac{x^n}{n^2}$$

Solution.

(a) diverges at 1 and -1, for example by the n-th term test (7.2A);

(b) diverges at 1 (see 7.5), but converges at -1 by Cauchy's test (7.6);

(c) absolutely convergent (see 7.5) at 1 and -1, so it converges (7.3).

Is there any way of predicting the radius of convergence in advance? Sometimes this is possible if we can calculate the sum of the series explicitly. For example consider the formula for the sum of the geometric series (4.2),

$$(7) \qquad\qquad \sum_{0}^{\infty} x^n = \frac{1}{1-x}, \qquad |x| < 1 .$$

Since the function on the right first becomes undefined when $x = 1$, it is reasonable to expect the series to have radius of convergence $R = 1$, as indeed it does. However this idea does not work if we start at 0 and move to the left, since the function is defined for all $x < 0$. But we know from Theorem 8.1 that the open interval of convergence has the form $|x| < R$; so since the series diverges at $R = 1$, it cannot converge for $x < -1$.

Unfortunately, even if we know the sum of the series explicitly, the preceding line of thought does not always lead to the correct R, as one can see by substituting $-x^2$ for x in (7); this gives

(9) $$\sum_0^\infty (-1)^n x^{2n} = \frac{1}{1+x^2}, \qquad |x| < 1;$$

(note that $|x^2| < 1$ and $|x| < 1$ are equivalent). Here the radius of convergence is still 1, yet the function on the right is defined for all x.

To understand this, one must replace in (9) the real variable x by the complex variable z; the function on the right is not defined when $z = \pm i$, so the complex series can only converge for $|z| < 1$, which means in turn that the real series can only converge when $|x| < 1$. Like shadow puppets on a wall produced by hands in a third dimension, the behavior of our real series on the line is controlled by its two-dimensional extension into the complex plane.

This confrontation—the power series diverges at a point a, yet the explicit sum of the series is defined at a — was used by Abel as a method of summing a divergent series. (Summing divergent series is an advanced topic, and we won't use it later in this book, but it is interesting.) For Abel's method, you need to know what a continuous function is, or at least be able to recognize one when you see it (a formal definition will be given in Chapter 11).

Definition 8.2   Abel summation.   Suppose
$$\sum a_n x^n = f(x), \qquad \text{for } |x| < 1,$$
where $f(x)$ is defined and continuous at $x = 1$, but the series diverges at 1. Then we say the series $\sum a_n$ is *Abel-summable* to $f(1)$, and write
$$\sum a_n = f(1) \qquad \text{(Abel summation)}.$$

Example 8.2B   Find the Abel sum of $1 - 1 + 1 - 1 + \ldots + (-1)^n + \ldots$.

Solution.   We form the corresponding power series and sum it:
$$1 - x + x^2 - x^3 + \ldots = \frac{1}{1+x}, \qquad |x| < 1.$$
The hypotheses are satisfied, so $\sum (-1)^n = 1/2$ (Abel summation).

Questions 8.2

1. Find R and determine convergence at $x = R$, $x = -R$.

(a) $\sum \dfrac{x^n}{\sqrt{n}}$ (b) $\sum \dfrac{x^{2n}}{n \, 4^n}$ (c) $\sum n \, x^n$

2. What is the Abel sum of $1 - 2 + 4 - 8 + 16 - \ldots$?

8.3   Operations on power series: addition.

Power series can be formally manipulated by the operations of algebra or calculus—added, multiplied, divided, differentiated, or integrated. The question is whether the sum of the resulting series is what you expect it to be. For instance, if we we formally differentiate both sides of

$$1 + x + x^2 + x^3 + \ldots + x^n + \ldots = \frac{1}{1-x} ,$$

we get

$$1 + 2x + 3x^2 + \ldots + nx^{n-1} + \ldots = \frac{1}{(1-x)^2} ,$$

but is this true? The standard theorem about differentiating sums only applies to finite sums; whether it extends to infinite sums is a subtle question we will have to leave for later, when we have more technique. We can however handle addition and multiplication now.

Theorem 8.3   Linearity theorem for power series.

If $\sum a_n x^n = f(x)$ and $\sum b_n x^n = g(x)$, for $|x| < K$, then for any constants p and q,

$$\sum (pa_n + qb_n)x^n = p\,f(x) + q\,g(x), \quad \text{for } |x| < K.$$

That is, if the each of the first two series converges to the indicated sum when $|x| < K$, so does the third series.

Proof.   By the Linearity Theorem for infinite series (7.2C), for each value of x such that $|x| < K$, we have

$$\sum (pa_n + qb_n)x^n = \sum (pa_n x^n + qb_n x^n)$$
$$= p\sum a_n x^n + q\sum b_n x^n = p\,f(x) + q\,g(x). \qquad \square$$

The theorem says one can add two power series term-by-term within an interval $|x| < K$ on which they both converge. If they have different radii of convergence, then K can be the smaller of the two radii.

Example 8.3   Check the validity of the theorem on the sum of the two series
$$1 + x + x^2 + x^3 + \ldots \quad \text{and} \quad 1 - x + x^2 - x^3 + \ldots .$$

Solution.   To get the second series, put $-x$ for x in the first one. So

$$1 + x + x^2 + x^3 + x^4 + x^5 + \ldots = \frac{1}{1-x}, \quad |x| < 1 ;$$

$$1 - x + x^2 - x^3 + x^4 - x^5 + \ldots = \frac{1}{1+x}, \quad |x| < 1 .$$

According to the Linearity Theorem, if we add the series, we get

$$2\left(1 + x^2 + x^4 + \ldots\right) = \frac{1}{1-x} + \frac{1}{1+x} = \frac{2}{1-x^2} ,$$

which we can confirm by substituting x^2 for x in the first series. $\qquad \square$

Questions 8.3

1. Using the power series representation $\sum \dfrac{x^n}{n!} = e^x$, and the Linearity Theorem, find the power series representation for

    (a) $\cosh x = \frac{1}{2}(e^x + e^{-x})$;   (b) $\sinh x = \frac{1}{2}(e^x - e^{-x})$.

8.4   Multiplication of power series.

Multiplication of power series is a bit more subtle, since in the product series, terms containing the same powers of x have to be grouped together. Series are written starting with the lowest powers, and that is how they must be multiplied: the method is adequately indicated by

$$a_0 + a_1 x + a_2 x^2 + a_3 x^3 + \ldots$$
$$b_0 + b_1 x + b_2 x^2 + b_3 x^3 + \ldots$$
$$\overline{a_0 b_0 + (a_0 b_1 + a_1 b_0)x + (a_0 b_2 + a_1 b_1 + a_2 b_0)x^2 + \ldots}$$

Study the way in which the coefficients are formed. The coefficient of x^3 would be, for example, $a_0 b_3 + a_1 b_2 + a_2 b_1 + a_3 b_0$.

Example 8.4   Multiply the two power series in example 8.3, and verify that multiplying their two sums gives you the sum of the product series.

Solution. $1 + x + x^2 + x^3 + x^4 + x^5 + \ldots \; = \; \dfrac{1}{1-x}$, $|x| < 1$;

$$1 - x + x^2 - x^3 + x^4 - x^5 + \ldots \; = \; \dfrac{1}{1+x} , \quad |x| < 1 .$$

Multiplying the series and their sums gives

$$1 + (1-1)x + (1-1+1)x^2 + \ldots = \dfrac{1}{1-x} \cdot \dfrac{1}{1+x} , \quad |x| < 1;$$

$$1 + x^2 + x^4 + \ldots = \dfrac{1}{1-x^2} , \quad |x| < 1 ,$$

which checks, as you can see by substituting x^2 for x in the first series.   □

Theorem 8.4A   Multiplication of power series.

$$\sum a_n x^n = f(x) \quad \text{and} \quad \sum b_n x^n = g(x) \;\; \Rightarrow \;\; \sum c_n x^n = f(x)g(x),$$

$$\text{where} \quad c_n = a_0 b_n + a_1 b_{n-1} + \ldots + a_n b_0 = \sum_{i+j=n} a_i b_j \; ;$$

in the sense that if the first two series converge for $|x| < K$, with the indicated sums, the same is true for the product series.

The formula for the coefficients c_n of the product series generalizes the previous formulas giving the first few coefficients. It would seem like a strange expression if one did not know its source. Of course, for a computer, no expression is strange.

Just as the Linearity Theorem for power series followed immediately from the same theorem for ordinary series, Theorem 8.4A follows immediately from

Theorem 8.4B   Multiplication theorem for series.

Suppose $\sum a_n$ and $\sum b_n$ converge absolutely, to the sums A and B respectively. Then if we put

$$c_n = a_0 b_n + a_1 b_{n-1} + \ldots + a_n b_0 = \sum_{i+j=n} a_i b_j \ ,$$

the series $\sum c_n$ converges absolutely to the sum $A \cdot B$.

Remarks.   In the statement, a_n and b_n are numbers; in the application to proving Theorem 6.4A, they represent the tems $a_n x^n$ and $b_n x^n$. Unlike the Linearity Theorem for series 7.2C, here we must assume the series converge absolutely. For the application to power series, this is no restriction, since by Theorem 8.1, a power series always converges absolutely inside its radius of convergence.

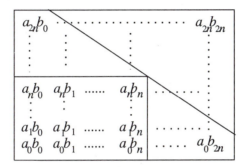

Proof.   Assume first all the coefficients a_n and b_n are non-negative. The matrix array shows all the possible products $a_i b_j$. If we denote the n-th partial sums of the three series respectivey by A_n, B_n, and C_n, then by multiplying these partial sums together, one can see that (check this):

$$
\begin{array}{rcl}
\text{the small square} & = & \text{all } a_i b_j \text{ occurring in } A_n B_n \ ; \\
\text{the lower triangle} & = & \text{all } a_i b_j \text{ occurring in } C_{2n} \ ; \\
\text{the big square} & = & \text{all } a_i b_j \text{ occurring in } A_{2n} B_{2n} \ .
\end{array}
$$

The argument now runs in outline:

$$
\begin{array}{ccccc}
\text{small square} & \subseteq & \text{triangle} & \subseteq & \text{big square;} \\
A_n B_n & \leq & C_{2n} & \leq & A_{2n} B_{2n} \ ; \\
\downarrow & & \Downarrow & & \downarrow \\
A \cdot B & & A \cdot B & & A \cdot B \ .
\end{array}
$$

The second line follows from the first, since all the $a_i b_j \geq 0$; for the rest, we let $n \to \infty$ and apply the Squeeze Theorem (5.2); note that $A_n \to A$, and $B_n \to B$, as $n \to \infty$.

Since the subsequence $C_{2n} \to A \cdot B$, and the whole sequence C_n is increasing, it follows that $C_n \to A \cdot B$ also, in other words, $\sum c_n = A \cdot B$. $\square$

The preceding argument assumed all a_n, $b_n \geq 0$; if all are negative, the same conclusion holds, by multiplying everything through by -1.

If the series contains both positive and negative terms, we proceed as in Section 7.3, writing

$$a_n = a_n^+ - a_n^-, \qquad\qquad b_n = b_n^+ - b_n^-,$$

$$\sum a_n = \sum a_n^+ - \sum a_n^-, \qquad \sum b_n = \sum b_n^+ - \sum b_n^-;$$

the second line follows from the first by the Linearity Theorem 7.2C, since our hypothesis of absolute convergence guarantees that the series $\sum a_n^+$, $\sum a_n^-$, etc. are all convergent (by the comparison test; cf. 7.3). Now

$$c_n = \sum_{i+j=n} (a_i^+ - a_i^-)(b_j^+ - b_j^-)$$

$$= \sum_{i+j=n} (a_i^+ b_j^+ + a_i^- b_j^-) - \sum_{i+j=n} (a_i^- b_j^+ + a_i^+ b_j^-)$$

$$= c_n^+ - c_n^-.$$

By the first part of the proof (the Multiplication Theorem for series with all positive or all negative terms), together with the Linearity Theorem, it follows from the above formulas for c_n^+ and c_n^- that

$$\sum c_n^+ = A^+ B^+ + A^- B^-, \qquad \sum c_n^- = A^- B^+ + A^+ B^-,$$

where $A^+ = \sum a_n^+$, $A^- = \sum a_n^-$, and so on.

Finally, since $c_n = c_n^+ - c_n^-$, the Linearity Theorem and the above calculations show that

$$\sum c_n = \sum c_n^+ - \sum c_n^-;$$

$$= A^+ B^+ + A^- B^- - A^- B^+ - A^+ B^-;$$

$$= (A^+ - A^-)(B^+ - B^-);$$

$$= A \cdot B.$$ ☐☐

Questions 8.4

1. By squaring the power series for $1/(1 - x)$, find the power series for the function $1/(1 - x)^2$. Compare your answer with what you get by formally differentiating term-by-term the power series whose sum is $1/(1 - x)$.

Exercises

8.1

1. Find the radius of convergence for the following power series.

(a) $\displaystyle\sum_1 \frac{x^n}{2^n\sqrt{n}}$ (b) $\displaystyle\sum_1 \frac{(n!)^2}{(2n)!}x^n$ (c) $\displaystyle\sum_1 \frac{x^n}{\sqrt[n]{n}}$

(d) $\displaystyle\sum_0 \frac{(-1)^n x^{2n}}{4^n(n!)^2}$ (e) $\displaystyle\sum_1 \left(\frac{n+2}{n}\right)^n x^n$ (f) $\displaystyle\sum_2 \frac{x^n}{(\ln n)^n}$

(g) $\displaystyle\sum_1 \frac{n^n x^n}{n!}$ (h) $\displaystyle\sum_1 \frac{n!\,x^n}{1\cdot 3\cdot 5\cdots(2n-1)}$

2. In the first half of the proof of Theorem 8.1, prove (2) in one step, without making two cases.

(Begin by writing $\sum |a_n x^n| = \sum |a_n c^n||x/c|^n$. This is faster than what is in the text, but seems trickier if you don't understand the motivation from scaling.)

3. In Example 8.1, we saw two series for which the result of Theorem 8.1 was proved automatically; i.e., the ratio test produced the radius of convergence R as well as the information that the series converged for $|x| < R$ and diverged for $|x| > R$. In fact, this will always be the case for any power series to which the ratio test can be applied.

(a) Given a power series, what is the condition on its coefficients that means the ratio test can be applied?

(b) Assuming this condition, prove Theorem 8.1 for such a series, by the ratio test.

8.2

1. For each series in Exercise 8.1/1 above, determine whether it converges at the endpoints $\pm R$ of the interval of convergence, if it is easy to do so.

2. The series $1 + 1 - 1 - 1 + 1 + 1 - 1 - 1 + \ldots$ diverges; what sum would Abel summation assign to it?

(Hint: sum the power series by breaking it into the series of even powers and the series of odd powers. What gives you the right to do this?)

8.3

1. From the power series (see Section 4.2)

$$x - \frac{x^2}{2} + \frac{x^3}{3} - \ldots + (-1)^{n-1}\frac{x^n}{n} + \ldots = \ln(1+x) ,$$

obtain the power series whose sum is $\ln\left(\dfrac{1+x}{1-x}\right)$.

2. What sum would Abel summation assign to $1 - 2 + 3 - 4 + 5 - \ldots$?

(See Definition 8.2; assume that power series can be differentiated term-by-term—cf. discussion at the beginning of Section 8.3).

8.4

1. Let $\sum_0^\infty a_n$ be a series and s_n its sequence of partial sums. Suppose $\sum a_n x^n$ converges for $|x| < 1$; let $f(x)$ be its sum. Then

(*)
$$\sum_0^\infty s_n x^n = \frac{f(x)}{1-x}, \quad \text{for } |x| < 1 .$$

(a) Illustrate the truth of (*) for the series: (i) $\sum 1$;   (ii) $\sum 1/2^n$.

(b) Prove (*), and show the hypothesis is satisfied if $\sum a_n$ converges.

Problems

8-1 Determine, with proof, the radius of convergence of $\sum (\sin n) x^n$.

(This is an example of a power series to which the ratio test does not apply; cf. Exercise 8.1/3. Note there are two things you have to show: one is harder than the other. The end of Chapter 5 may be of help.)

Answers to Questions

8.1

1. Radius of convergence: (a) 9   (b) 0   (c) 1   (d) ∞

2. (a) No, by the Tail-convergence Theorem.

(b) No, since $\sum |a_n x^n| = \sum |\pm a_n x^n|$.

3. If $c \in A$, then $-c \in A$ also, since $\sum |a_n (-c)^n| = \sum |a_n c^n|$.

8.2

1. (a) 1; div at 1, conv at -1 (b) 2; div at 2, -2 (c) 1; div at 1,-1

2. $1 - 2x + (2x)^2 - (2x)^3 + \ldots = 1/(1 + 2x)$; $f(1) = 1/3$, the Abel sum.

8.3

1. (a) $\cosh x = \sum_0 x^{2k}/(2k)!$ (b) $\sinh x = \sum_0 x^{2k+1}/(2k + 1)!$

8.4

1. Both give $1 + 2x + 3x^2 + \ldots + (n + 1)x^n + \ldots$.

9

Functions of One Variable

9.1 Functions.

This chapter is devoted to a brief review of some of the definitions and elementary properties associated with a real-valued function of one real variable. We call it just *function*, for short; it assigns to each real a in its domain a corresponding real number b. The formal definition avoids the vagueness of words like "assign" by in effect identifying a function with its graph. It runs:

Definition 9.1   A **function** f is a set G_f of ordered pairs (a, b) of numbers, such that no two ordered pairs have the same first entry:

(1) $$(a, b) \text{ and } (a, b') \in G_f \quad \Rightarrow \quad b = b' \ .$$

If $(a, b) \in G_f$, we say that f is **defined** at a, and write $b = f(a)$.

The **domain** of f is the set of numbers for which f is defined:

(2) $$D_f = \{a : (a, b) \in G_f \text{ for some } b\} \ .$$

When the ordered pairs (a, b) are visualized as points in the xy-plane, the set G_f is called the **graph** of the function; it is essentially the same as the function. Two functions are *equal*, or *the same*, if the sets G_f are the same, i.e., if they have the same graph. By (1), a subset G of the plane is the graph of a function if and only if each vertical line $x = a$ contains at most one point of the graph.

Side by side with this geometric view of function, there is an an analytic viewpoint which thinks of a function as a rule

$$a \mapsto f(a), \quad a \in D_f \ ,$$

assigning to each a in a set D_f a number designated by $f(a)$. The rule is usually given by an expression in an independent variable, like

$$\sqrt{4x + 1}, \quad |e^u \tan^{-1}(1 - u)|, \quad \text{erf}(\text{erf}\, t), \quad \int_0^x \sin(J_0(t))^2 \, dt \ .$$

From this point of view, (1) says that a must determine $f(a)$ uniquely, i.e., the function must be *single-valued*: for example, $\sqrt{4x + 1}$ signifies by the usual convention the positive square root.

For the first hundred and fifty years of analysis, there was controversy over which view afforded the proper definition of function. Though the analytic viewpoint represents the common workaday notion, using it to define function isn't a good idea. Here are some of the objections:

⋆ it is vague: just what is an "expression in x" or a "rule"?

⋆ it is inaccurate: the expressions $\cos^2 x$ and $1 - \sin^2 x$ are certainly different, and lead to different rules, yet presumably one would want to say they define the same function;

⋆ it is not general enough: many graphs arising in practice—as the solution to a differential equation, for instance—can be calculated numerically as precisely as you wish, but there is no currently available expression in x to describe them, i.e., no analytic representation in terms of known functions.

A third viewpoint somewhat different from the geometric and analytic approaches looks upon the function as a **mapping** (or **map**):

$$f : D_f \to \mathbb{R} \,,$$

which associates with each point $a \in D_f$ the corresponding point $b = f(a)$ on the number line. Though we shall occasionally use maps in the first part of this text—in this Chapter, 12, and 13, for instance—the mapping viewpoint is most useful when dealing with functions of several variables, particularly vector-valued functions. (The words "map" and "mapping" are synonyms for "function".)

Three different viewpoints are necessary, because they suggest different sorts of questions to ask about functions. From the *geometric* point of view, one might ask if f is increasing or decreasing, is convex, or has maxima and minima. The *analytic* view leads to the operations of algebra and calculus, with the resulting equations and inequalities. Thinking of f as a *mapping* leads one to ask whether it is injective (i.e., takes distinct a's to distinct b's), whether it has an inverse, and what it does to different types of sets. Does it take intervals into intervals? Are there points which are mapped to themselves?

A word about functional notation. The analytic viewpoint suggests the notation $f(x)$, or even introducing a dependent variable y and writing $y = f(x)$ or $y = y(x)$; this is standard practice in science and engineering, where there are always variables, and the function is thought of primarily as expressing a relation between them. The geometric and mapping viewpoints, on the other hand, which often do not use variables, suggest using just f as the notation.

We will be eclectic, choosing what is best adapted to the work at hand. Using just f makes certain operations awkward, and gives naming problems: "sin" and "exp" are all right, but by what variableless name shall we designate the function x ?

Every function comes with the domain D_f where it is defined. If the function is given by an expression in x, its domain is always understood (unless otherwise specified) to be the *natural* domain: all values of x for which the expression makes sense. If the domain is for some reason taken to be smaller than this, we get the *restricted* function (with its restricted domain), which is technically not the same as the original function, since the two have different graphs. Thus:

$$\sin x; \qquad \sin x, \quad 0 \le x \le \pi; \qquad \sin x, \quad -\pi/2 \le x \le \pi/2;$$

are three different functions: only the first is periodic, only the second is non-negative, and only the third is increasing. In the same way, x^2/x and x are different functions, since the natural domain of the first does not include 0.

Finally, note that the *sequences* $\{a_n\}$ we have been studying are really a very special type of function $a(n)$ whose domain is the set of natural numbers $\{0, 1, 2, 3, \ldots\}$; this is the "correct" definition of sequence, and the one used by computers. Many of the properties of functions (increasing, bounded, etc.) are analogues of properties of sequences, and we will use this to guide our work.

9.2   Algebraic operations on functions.

To build more elaborate functions from simpler ones, operations are used. Leaving until later the operations of differentiation and integration, which require the use of limits, there are first of all the basic algebraic operations: in the notation using variables, they are

$$f(x) + g(x), \qquad f(x)g(x), \qquad f(x)/g(x), \qquad cf(x),$$

defined by adding, multiplying, or dividing the two function-values, or multiplying the function-value by a fixed number. The natural domain for each of these is the set of x-values for which both $f(x)$ and $g(x)$ are defined (and for $f(x)/g(x)$, for which in addition $g(x) \neq 0$).

The operations above were all derived from the arithmetic operations on real numbers. By contrast, *composition* is an operation which has no analogue just for numbers. It is what you get by pressing successively two function keys on a calculator. If you are working with variables, it arises naturally in making a substitution, and that is probably the best way to think of it:

$$w = f(y), \quad y = g(x) \quad \Rightarrow \quad w = f(g(x)) \ .$$

The formal definition of composition is given in terms of mapping.

Definition 9.2   Given two functions $g : D_g \to \mathbb{R}$ and $f : D_f \to \mathbb{R}$, we define their **composition** $f \circ g : D_{f \circ g} \to \mathbb{R}$ by

$$(3) \qquad\qquad\qquad (f \circ g)(a) = f(g(a)), \quad a \in D_{f \circ g} \ ;$$

$$D_{f \circ g} = \{a \in \mathbb{R} : g \text{ is defined at } a \text{ and } f \text{ is defined at } g(a)\} \ .$$

Note that $f \circ g$ is read from right to left: first the mapping g is performed, then the mapping f; it is this convention that makes (3) true.

Example 9.2   Rewrite and find the domain: (a) $\sin x \circ \sqrt{x}$;   (b) $\sqrt{x} \circ \sin x$.

Solution.   (a) $\sin x \circ \sqrt{x} = \sin(\sqrt{x})$;   domain $= \{x : x \geq 0\}$.

(b) $\sqrt{x} \circ \sin x = \sqrt{\sin x}$;   domain $= \{[2k\pi, (2k+1)\pi]\}$, $k = 0, \pm 1, \pm 2, \ldots$; i.e., the set of intervals on which $\sin x$ is non-negative.

In the preceding descriptions of the operations, we have appealed to the analytic and mapping viewpoints on functions. Generally speaking, the geometric view gives little insight here.. There are however a few important exceptions to this. If $f(x)$ is changing relatively slowly, as compared with a rapid pure oscillation

like $\sin kx$, then the effects of adding and multiplying $f(x)$ and $\sin kx$ is sketched in the two figures shown above.

Two compositions interpreted geometrically by using the graph G_f of $f(x)$:

translation   if $a > 0$,

the graph of $f(x + a)$ is the graph G_f moved to the *left* a units;

the graph of $f(x - a)$ is the graph G_f moved to the *right* a units;

change of scale   if $a > 1$,

the graph of $f(x/a)$ is the graph G_f expanded horizontally by the factor a;

the graph of $f(ax)$ is the graph G_f compressed horizontally by $1/a$.

Questions 9.2

1. Define: $\operatorname{sgn} x = \begin{cases} 1, & \text{for } x > 0; \\ -1, & \text{for } x < 0; \end{cases}$ ("signum x": the sign of x).
Relate G_f to the graph of   (a) $f(x)\operatorname{sgn} x$;   (b) $\operatorname{sgn}(f(x))$.

2. If a and b are numbers, then $ab = 0 \Rightarrow a = 0$ or $b = 0$.
Is the analogous statement for functions true? That is, if f and g are defined for all x and $fg = 0$ (i.e., is the constant function 0), is either f or g the constant function 0? Proof or counterexample.

3. For which of these pairs of functions (f, g) is $f \circ g = g \circ f$? If not true for some pair, can you restrict the domain so that it becomes true?

   (a) (x^m, x^n), $m, n \in N$ (b) (e^x, x^2) (c) $(x^2, \sqrt{x})$

4. For each class of functions, tell whether or not it is preserved by translation and scale change, i.e., if $f(x)$ is in the class, then $f(x + a)$ and $f(ax)$ are in the class, for all $a \neq 0$ (A, B, k are arbitrary constants).

   (a) polynomials   (b) Ae^{kx} (c) $A\sin kx + B\cos kx$

9.3   Some properties of functions.

A mathematical analyst investigating a new function, say one that has arisen as a solution to a differential equation, is like a metallurgist investigating a new alloy: what are its properties, how does it behave?

We recall here a few of the properties $f(x)$ may have, concentrating on ones suggested by the geometric viewpoint. More subtle properties depending on limits or the Completeness Property (such as continuity, differentiability, or the existence of critical points) will be introduced formally in later chapters, though we will continue to use calculus freely in examples.

Definition 9.3A   Let $f(x)$ be a function with domain D. We say f is

increasing	if	$f(a) \leq f(b)$	for all pairs $a < b$ in D;
strictly increasing	if	$f(a) < f(b)$	for all pairs $a < b$ in D;
decreasing	if	$f(a) \geq f(b)$	for all pairs $a < b$ in D;
strictly decreasing	if	$f(a) > f(b)$	for all pairs $a < b$ in D;
monotone		if f is either increasing in D or decreasing in D;	
strictly monotone		if f is strictly increasing or strictly decreasing in D.	

Examples 9.3A   On their natural domains,

(a) e^x, x^3, and $\ln x$ are strictly increasing;

(b) e^{-x} is strictly decreasing;

(c) the function $\operatorname{sgn} x$ is increasing (see Question 9.2/1);

(d) $1/x$ is not decreasing (for example, $f(-1) < f(1)$).

Since inequalities reverse when multiplied through by -1, we can say

(4) $f(x)$ is (strictly) increasing $\Leftrightarrow$ $-f(x)$ is (strictly) decreasing.

This will be useful when we want to show that a theorem about increasing functions is also valid for decreasing functions.

Definition 9.3B

$f(x)$ is **even** if $f(-x) = f(x)$ for all $x \in D_f$;
$f(x)$ is **odd** if $f(-x) = -f(x)$ for all $x \in D_f$.

For both definitions the domain D_f of the function must be symmetric about the point 0 (i.e., $x \in D_f \Leftrightarrow -x \in D_f$), otherwise the equality makes no sense.

Geometrically, an even function is one whose graph is symmetric about the y-axis; an odd function is one whose graph is symmetric about the origin.

Examples 9.3B   Verify the following as easy exercises in using the definition.

(a) A polynomial with only odd powers of x is an odd function; one with only even powers is an even function; the function 0 is both even and odd.

(b) $f \cdot g$ and f/g are: $\begin{cases} \text{even, if } f \text{ and } g \text{ are both even or both odd;} \\ \text{odd, if one function is even and the other odd.} \end{cases}$

(c) The function $\cos x$ is even; $\sin x$ is odd; $\tan x$ is odd.

Proposition 9.3   *Suppose the domain of f is symmetric around 0. Then f has a unique representation as the sum of an even and an odd function:*

$$f(x) = E(x) + O(x), \qquad E(x) \text{ even, } O(x) \text{ odd.}$$

We leave the proof for the Exercises. As one example, if f is a polynomial, then E consists of its even powers, and O of its odd powers.

Definition 9.3C   We say $f(x)$ is **periodic** if there is a $c > 0$ such that
$$f(x + c) = f(x) \qquad \text{for all } x \in D_f.$$
The number c is called a **period** of f; the smallest such c (if it exists) is called the *minimal period* of f, or simply the *period* of f.

A periodic function with period c is one whose graph exactly coincides with itself when it is translated horizontally a distance $c > 0$ in either direction. The trigonometric functions are the simplest and most basic periodic functions. For example, $\sin x$ and $\cos x$ have period 2π, while $\tan x$ has period π.

If c is a period, so is $2c, 3c$, and so on. It can be shown that in this way the minimal period generates all the other periods; hence its importance. A constant function is periodic, but it has no minimal period. (Note that 0 is not considered a period, for if it were, every function would be periodic.)

In the subject *Fourier analysis* it is shown that any continuous periodic function can be expressed as the sum of an infinite series of sines and cosines. These *Fourier series* are more subtle to work with than power series, and important techniques in analysis were developed specifically to deal with Fourier series.

Here are two examples involving the interaction of properties in this section.

Example 9.3C   If a function f is even and monotone, it is constant.

Nothing further being said, we can assume we are supposed to prove it. Students have been known to respond to such questions on an assignment or exam with "That's nice", "So what?", or just "???".

Solution.   As noted after Definition 9.3B, if f is even, its domain D is symmetric about 0. Let $a, b \in D$, and suppose $0 \le a < b$; then $-b, -a \in D$, and $-b < -a$. So if, say, f is increasing, we have
$$f(a) \le f(b) \qquad \text{and} \qquad f(-b) \le f(-a).$$
But since f is even, the latter inequality implies
$$f(b) \le f(a).$$
Comparing, we see $f(a) = f(b)$. Since a and b were arbitrary, f is constant on the right half of D, and thus on all of D, since it is an even function. $\square$

If f is decreasing, then $-f$ is increasing by (4), therefore constant by the preceding; thus f is also constant. $\square\square$

Example 9.3D   Show that $\sin x$ is not a polynomial function.

You can't say, "because it doesn't have the form of a polynomial"; neither does $\sin^2 x + \cos^2 x$, yet it is the constant polynomial 1. You must instead find some property of $\sin x$ that a polynomial function cannot have, or vice-versa.

Solution.   A polynomial function has at most a finite number of zeros equal to its degree, whereas $\sin x$ has infinitely many zeros. $\square$

Questions 9.3

1. Tell whether the following are increasing, decreasing, or neither (over their domain).

(a) $|x|$ (b) x^n, $n \in \mathbb{N}$ (c) $1/x^2$, $x < 0$ (d) $\cos x$, $0 \le x \le \pi$

(e) $1/(x^2 + 1)$ (f) $[x]$ = greatest integer $\le x$

2. (a) Prove: (i) f odd, g odd $\Rightarrow fg$ even;   (ii) f odd, g even $\Rightarrow fg$ odd.

(b) Tell whether each is odd, even, or neither (use Example 9.3B):

$$x^3 \sin x, \quad (\cos x)/x, \quad e^x - e^{-x}, \quad \ln x, \quad \sin x^5, \quad \cos(1/x^3).$$

3. Is the product of increasing functions an increasing function? Either prove it, or give a counterexample and prove some amended version.

4. Suppose f and g have domain $(-\infty, \infty)$, i.e., all of $\mathbb{R}$. Will $f \circ g$ be periodic if f is periodic? if g is periodic?

9.4   Inverse functions.

Some important functions like $\sin^{-1} x$ or $\ln x$ entered analysis as the *inverse* of other functions (here, $\sin x$ and e^x, respectively.) We therefore review briefly how one defines and calculates with inverse functions. All three viewpoints about functions will be represented:

⋆ the mapping viewpoint, to define inverse functions;
⋆ the geometric viewpoint, to understand them intuitively;
⋆ the analytic viewpoint, to calculate with them.

The mapping viewpoint.   It is better not to use variables here.

Let f be a function defined on the interval $[a_1, a_2]$, and assume that on this interval f has these two properties:

Inv-1 f is strictly increasing;

Inv-2 f takes on every value between $b_1 = f(a_1)$ and $b_2 = f(a_2)$, i.e., if $b \in [b_1, b_2]$, there is an $a \in [a_1, a_2]$ such that $f(a) = b$.

Then f is a mapping of intervals:

$$f : [a_1, a_2] \to [b_1, b_2] \ ;$$

for each $b \in [b_1, b_2]$ there is an $a \in [a_1, a_2]$ such that $f(a) = b$, and there is only one such a, since the function is strictly increasing and so cannot repeat any of its values.

Since each b corresponds in this way to a unique a, it allows us to define the "backwards", or *inverse*, mapping

$$f^{-1} : [b_1, b_2] \to [a_1, a_2]$$

by the rule

(5) $$f^{-1}(b) = a \ \Leftrightarrow \ f(a) = b \ .$$

Inv-1 guarantees that the map f is *injective* $(a \neq a' \Rightarrow f(a) \neq f(a'))$, while Inv-2 says that f is *surjective* (given any b, there is an a such that $b = f(a)$). The two together say f is *bijective*, the property equivalent to the existence of an inverse map (cf. Appendix A.0 for this terminology.)

The geometric viewpoint. We consider the graphs of f and f^{-1}. Then the defining property (5) of the inverse map translates into

(6) $$(b, a) \in G_{f^{-1}} \quad \Leftrightarrow \quad (a, b) \in G_f \ .$$

How are (b, a) and (a, b) geometrically related? If we flip the xy-plane about the diagonal line $y = x$, the two axes are interchanged, and the two points (a, b) and (b, a) are carried into each other, as the figure on the left below shows. Thus from (6) we get the intuitive geometric picture of f^{-1} :

(7)   flipping the plane around $y = x$ carries G_f into $G_{f^{-1}}$.

Some prefer to think of (a, b) and (b, a) as reflections of each other in the diagonal line $y = x$, in which case the graph $G_{f^{-1}}$ is the reflection of G_f in the diagonal line.

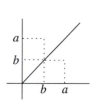 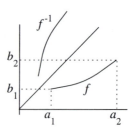

The analytic viewpoint. Expressed in terms of variables, the defining property (5) of inverse functions becomes

(8) $$y = f^{-1}(x) \quad \Leftrightarrow \quad x = f(y) \ .$$

This last shows how to get the expression for $f^{-1}(x)$:

(9)   get $f^{-1}(x)$ by solving $x = f(y)$ for y in terms of x .

Composing f and f^{-1} gives us also the useful equations

(10a) $$f(f^{-1}(x)) = x, \quad \text{for } b_1 \leq x \leq b_2 \ ;$$

(10b) $$f^{-1}(f(x)) = x, \quad \text{for } a_1 \leq x \leq a_2 \ .$$

All of the preceding is also valid, making the appropriate changes, for decreasing functions: if f is strictly decreasing on $[a_1, a_2]$ and satisfies the condition Inv-2, it has an inverse function on $[b_2, b_1]$ (note the subscript reversal)..

Example 9.4   Find f^{-1} if $f(x) = x^2 + 1$.

Solution.   To satisfy Inv-1 and Inv-2, we restrict the domain to the set $x \geq 0$, on which $f(x)$ is strictly increasing. Interchange the two variables and proceed as in (9); the restriction $x \geq 0$ turns into $y \geq 0$, and we get (cf. next figure)

$$x = y^2 + 1, \quad y \geq 0 \quad \Leftrightarrow \quad y = \sqrt{x - 1}, \quad x \geq 1 \ .$$

We use the positive square root since $y \geq 0$. The domain of $f^{-1}(x)$ is the range of $f(x)$: the set $x \geq 1$. Thus finally,

$$f^{-1}(x) = \sqrt{x - 1}, \quad x \geq 1 .$$

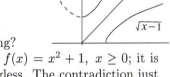

Remark. Since $f(x) = x^2 + 1$, we calculate

(11) $f^{-1}(f(-1)) = f^{-1}(2) = 1,$

which seems to contradict (10b) above. What's wrong?

The function f whose inverse was calculated is $f(x) = x^2 + 1$, $x \geq 0$; it is not defined at -1, so the calculation (11) is meaningless. The contradiction just underscores the fact that $x^2 + 1$, with its natural domain $\mathbb{R}$, does not have an inverse function.

Questions 9.4

1. What is the inverse function to each of the following? Restrict the domain of $f(x)$ if necessary; specify the domain of $f(x)$ and its inverse.

(a) $\sqrt{1 + x}$ (b) $\cos(3x + 1)$ (c) 2^{3x} (use $\ln x$)   (d) $x^2 - 4x$, $x \leq 2$

9.5   The elementary functions.

To get the class of functions generally called *elementary*, we start with the following four classes of functions:

(a) the *rational* functions: those writable in the form $p(x)/q(x)$, where $p(x)$ and $q(x)$ are polynomials;

(b) the basic *trigonometric* functions: $\cos x$, $\sin x$, $\tan x$, their three recipro-cals, and the six inverses $\cos^{-1} x, \sin^{-1} x, \ldots$;

(c) the *exponential* function e^x and its inverse, $\ln x$;

(d) the *algebraic* functions: those functions $y = y(x)$ which satisfy an equa-tion of the form

(12) $y^n + a_1(x)y^{n-1} + \ldots + a_n(x) = 0$

whose coefficients $a_k(x)$ are rational functions. (For example, any expression involving some combination of n-th roots, powers of x, and arithmetic operations is an algebraic function, but there are many other algebraic functions.)

The *elementary* functions are then all functions that you can get from the four classes above by addition, multiplication, division, and composition of functions. Thus it includes combinations such as

$$\sin^3(\sqrt{x - 2})10^{x^2} \quad \text{and} \quad \ln(\tan^{-1}(e^{\sqrt{x}} - x^3)\sec 22x .$$

Calling only these functions elementary is just a convention and not a terribly useful one at that. Scientists and engineers whose work involves other standard functions (like the Bessel functions $J_p(x)$) find them just as familiar-looking. With computers it is possible to work routinely with non-elementary functions which were previously inaccessible—functions which occur for example as the solutions to more complicated differential equations.

And on the other side, some elementary functions don't seem so elementary. Abel and Galois independently proved around 1825-1830 that if $n > 4$ in the equation (12), then in general the algebraic function $y(x)$ cannot be expressed by a formula involving n-th roots, the arithmetic operations, and composition. This ended a three-century search for analogs of the quadratic formula when $n \geq 5$ (analogs for $n = 3$ and 4 were known), and dealt yet another blow to the idea that a function should be given by an explicit expression in x.

Questions 9.5

1. Prove the following are elementary functions.

    (a) $|x|$ (b) $\operatorname{sgn} x$ (cf. Question 9.2/1)   (c) $\log_a(x)$.

Exercises

9.2

1. Let $u(x)$ be the *unit step* function: $u(x) = \begin{cases} 1, & \text{if } x \geq 0; \\ 0, & \text{if } x < 0 . \end{cases}$

Using the algebraic operations, translation and change of scale, and other composition of functions, express in terms of $u(x)$ the functions

(a) $f(x) = \begin{cases} 2, & 1 \leq x < 3; \\ 0, & \text{elsewhere.} \end{cases}$ (b) $f(x) = \begin{cases} 1, & 2n \leq x \leq 2n+1, \ n \in \mathbb{Z}; \\ 0, & \text{elsewhere.} \end{cases}$

2. Find a function $h(x)$ with this property: for any $f(x)$, the composite function $h(f(x))$ has a graph which agrees with the graph of $f(x)$ where $f(x) \geq 0$, but where $f(x) < 0$, it wipes away that part of the graph of f and replaces it with the corresponding piece of the x-axis.

3. Describe all functions $f(x)$ which are defined for $x \in \mathbb{R}$ and which satisfy $f(x) = 1/f(x)$. (There are more than two.)

4. The physical phenomenon of "beats" can be described mathematically as follows. By using the trigonometric addition formula

$$\sin a + \sin b = 2 \sin \frac{a+b}{2} \cos \frac{a-b}{2} ,$$

show that if you add together two sine waves with nearly equal large frequencies, the resulting function has as its graph a wave which oscillates rapidly, with a varying amplitude given by a wave which oscillates slowly (so that you hear a single pitch which gets regularly softer and louder, giving the "beats": wah, wah, wah, wah ... ; the slower the beats, the more nearly equal the frequencies).

9.3

1. Suppose $f(x)$ is defined for all x (it would be enough to assume it has a domain D symmetric about 0).

(a) Prove that $E(x) = \dfrac{f(x) + f(-x)}{2}$ is an even function.

(b) Show $f(x)$ can be expressed as the sum of $E(x)$ and an odd function $O(x)$.

(c) Show that the representation in (b) is unique; that is, if the function f has a representation $f(x) = E_1(x) + O_1(x)$, then $E = E_1$ and $O = O_1$.

(d) How does this representation look if $f(x)$ is a polynomial?

(e) How does it look if $f(x) = e^x$?   (If you have met these functions before, now you know where they come from.)

2. If f and g are defined for all x and are odd or even (four possibilities altogether), what can be said about the composition $f \circ g$? Prove it.

3. Suppose f and g are defined for all x and are decreasing functions. What if anything can you conclude about the composition $f \circ g$? (Try an example.) Prove it.

4. Give an example of a function which is periodic, non-constant, and yet has no minimal period.

5. Show: a periodic increasing function is constant. (What about $\tan x$?)

9.4

1. Suppose $f(x)$ coincides with its inverse over a suitable interval.

(a) How does the graph of $f(x)$ look?

(b) Verify that the following are examples of such functions, over a suitable interval: (i) $1/x$;   (ii) $\sqrt{1 - x^2}$.

(c) Suppose that $f(x) = \dfrac{ax + b}{cx + d}$. What conditions on a, b, c, d are necessary and sufficient in order that $f(x)$ coincide with its inverse function?

Problems

9-1 Let $I = [a_1, a_2]$ be an interval of positive length. Prove the only function $f(x)$ on I satisfying Inv-1 and Inv-2 on I (Section 9.4) and coinciding with its inverse function on I is the function $f(x) = x$, $x \in I$.

9-2 Let $f(x)$ satisfy Inv-1 and Inv-2 (see 9.4) for $x \geq 0$, and let $g(x)$ be its inverse function. Show that the solutions of the equation $f(x) = g(x)$ are exactly the same as the solutions of $f(x) = x$. Show the analogous statement for $x > 0$ and a strictly *decreasing* function is false.

9-3 Suppose $f(x)$ is defined for all x; then $f \circ \cos x$ is periodic and 2π is a period. Conversely, if $g(x)$ is defined for all x and is periodic, with 2π as a period, can one find a function f such that $g(x) = f \circ \cos x$?

    (a) Show by counterexample that the answer is no.

    (b) Strengthen the hypotheses on $g(x)$ and prove the amended statement.

Answers to Questions

9.2

    1. (a) The graph is reflected in the x-axis, for $x < 0$; $f(0)$ is deleted.

    (b) The portions of the graph above and below the x-axis are replaced respectively by horizontal line segments taken from the lines $y = 1$ and $y = -1$; the points where the graph crosses the x-axis are deleted.

f(x)   *g(x)*

    2. No; see the sketch at the right.

    3. (a) Yes; both are equal to x^{mn}.   (b) No; $e^{x^2} \neq (e^x)^2 = e^{2x}$.

    (c) No, since $(\sqrt{x})^2 = x$, $x \geq 0$, while $\sqrt{x^2} = |x|$ for all x. It becomes true if $f(x)$ is replaced by $f(x) = x^2$, $x \geq 0$.

    4. True for all of them (for (c) by the trigonometric addition formulas for $\sin(a + b)$ and $\cos(a + b)$).

9.3

    1. (a) neither (b) increasing, if n is odd; neither, if n is even; (c) increasing (d) decreasing (e) neither (f) increasing ((b), (c), (d) strictly ...).

    2. (a) (i) $f(-x)g(-x) = (-f(x))(-g(x)) = f(x)g(x)$ (ii) is similar

    (b) even, odd, odd, neither, odd, even

    3. No: $x \cdot x = x^2$, for $x < 0$, is an example.

    If $f(x)$, $g(x)$ are non-negative it is true, since by the multiplication law for inequalities (note that $a < b$),

$$f(a) \leq f(b),\ g(a) \leq g(b) \Rightarrow f(a)g(a) \leq f(b)g(b) .$$

    4. No: $\sin x^2$ is not periodic. Yes, since $f(g(x + P)) = f(g(x))$.

9.4

    1. Follow in each case the procedure described in Example 9.4. Only the domain of $\cos(3x + 1)$ has to be restricted, to $[-1/3, (\pi - 1)/3]$. The inverse functions then are

    (a) $x^2 - 1$, $x \geq 0$;   (b) $(\cos^{-1} x - 1)/3$, $[-1, 1]$;   (c) $\ln x/3 \ln 2$, $x > 0$;
    (d) $2 - \sqrt{4 + x}$. $x \geq -4$.

9.5

    1. (a) $= \sqrt{x^2}$ (b) $= |x|/x$ (c) $= \ln x/\ln a$

10

Local and Global Behavior

10.1 Intervals. Estimating functions.

If you think of them as functions, all sequences have essentially the same domain—the non-negative integers. By contrast, functions $f(x)$ can have a variety of domains: though these are almost always made up of intervals, there are different kinds of intervals, and they strongly influence the behavior of functions defined on them. We need therefore some terminology for describing intervals. We classify them according to, their

> length: **finite** or **infinite**
> endpoints: **closed** if the endpoints are included; **open** if they are not

A few examples should make the usage and notation clear.

$$
\begin{aligned}
(a, b) &= \{x : a < x < b\} & \text{finite open interval} \\
[a, b] &= \{x : a \le x \le b\} & \text{finite closed interval} \\
(a, \infty) &= \{x : x > a\} & \text{infinite open interval} \\
(-\infty, a] &= \{x : x \le a\} & \text{infinite closed interval}
\end{aligned}
$$

An interval like $[a, b)$ could be called *half-open*; the interval $(-\infty, \infty)$, which has no endpoints, is considered to be both open and closed. *

> The fuss over endpoints may seem picky, but for want of a nail the battle was lost: analysis proofs have imploded because an endpoint was needed but missing.

A finite interval can be specified by giving its endpoints, but for many purposes it is more convenient to describe it symmetrically by giving its center a and radius δ. The *δ-neighborhood of a* is the open interval

$$(a - \delta, a + \delta).$$

We can say that x belongs to this interval in any of four equivalent ways; which one we use will depend upon the context. They are:

$$x \in (a - \delta, a + \delta), \qquad a - \delta < x < a + \delta, \qquad |x - a| < \delta, \qquad x \underset{\delta}{\approx} a \ .$$

The last is a non-standard but expressive notation we will use if δ is small, i.e., if x is being thought of as approximately equal to a.

---

* We do not consider (a, a), which is empty, to be an interval; in books which do, it too is both open and closed.

Note that the open interval (a, b) is in some books denoted by $]a, b[$.

In what follows, we will extend functional notation to sets, writing
$$f(I) = \{f(x) : x \in I\} \, .$$
We call $f(I)$ the *range* of $f(x)$ over I, or the *image* of I under the mapping f. For example if $f(x) = \sin x$, then
$$f((-\infty, \infty)) = [-1, 1], \qquad f([0, \pi]) = [0, 1], \qquad f((0, \pi)) = (0, 1].$$

Using the notation $f(I)$, we can extend to functions the terminology used for sets and sequences; it will be needed in estimating the size of functions.

Definition 10.1A   Given an interval I on which $f(x)$ is defined, we say

$\qquad b$ is an **upper bound** for $f(x)$ on I $\Leftrightarrow$ b is an upper bound for $f(I)$

$\qquad\qquad\qquad\qquad\qquad\qquad\qquad\qquad \Leftrightarrow$ $f(x) \le b$ for $x \in I$;

$\qquad f(x)$ is **bounded above** on I $\Leftrightarrow$ f has an upper bound on I .

The concepts *lower bound* and *bounded below* are defined similarly. We say $f(x)$ is **bounded** on I if it is bounded above and below, i.e., one can find numbers b_1 and b_2 such that
$$b_1 \le f(x) \le b_2 \quad \text{for} \ \ x \in I \, .$$

If no I is specified, we understand it to be the natural domain D_f of $f(x)$ (which may be a collection of intervals, as in the functions $\sec x$ or $1/x$, for instance).

Examples 10.1A.   Which of these functions is bounded above or below?

$\qquad$ (a) $3 \cos x$ $\qquad$ (b) e^{-x} on $[0, \infty)$ $\qquad$ (c) $1 - x^2$ $\qquad$ (d) $\tan x$

Solution.

$\qquad$ (a) $-3 \le 3 \cos x \le 3$ for all x;   bounded;

$\qquad$ (b) $0 < e^{-x} \le 1$ for $0 \le x < \infty$;   bounded;

$\qquad$ (c) $1 - x^2 \le 1$ for all x;   bounded above;

$\qquad$ (d) $\tan x$ is not bounded above or below.

Definition 10.1B   Suppose $f(x)$ is defined on an interval I. We define the

$\qquad$ **supremum** of $f(x)$ on I $=$ sup $f(I)$ (notation: $\sup_I f(x)$);

$\qquad$ **maximum** of $f(x)$ on I $=$ max $f(I)$ (notation: $\max_I f(x)$);

the definitions and notation for the *infimum* and *minimum* are analogous.

The definitions of $\sup f(I)$, $\max f(I)$, etc., given in Section 6.5 show that

(1)
$$f(x) \text{ has a maximum on } I \ \Leftrightarrow \ \sup_I f(x) = f(\overline{m}), \text{ for some } \overline{m} \in I;$$
$$f(x) \text{ has a minimum on } I \ \Leftrightarrow \ \inf_I f(x) = f(\underline{m}), \text{ for some } \underline{m} \in I.$$

Example 10.1B   Find the sup, inf, max, and min of $f(x)$ over I and J:
$$f(x) = \sin x, \text{ and } I = (-\infty, \infty), \ J = (0, \pi/2).$$

Solution.   Over I : $\sup f(x) = \max f(x) = 1$; $\inf f(x) = \min f(x) = -1$.
Over J : $\sup f(x) = 1$, $\inf f(x) = 0$; $\max f(x)$ and $\min f(x)$ do not exist.

The Completeness Property (Section 6.5) for the set $f(I)$ translates into the following theorem (note that $f(I)$ is not empty, since an interval I is non-empty and $f(x)$ is understood to be defined on I).

Theorem 10.1   Completeness Property for functions.

Suppose $f(x)$ is defined on an interval I. If $f(x)$ is bounded above on I, then $\sup_{I} f(x)$ exists;   if $f(x)$ is bounded below, then $\inf_{I} f(x)$ exists.

Estimating functions: inequalities and absolute values.

If you stop to think about it, finding upper and lower bounds for $f(x)$ is really just estimating $f(x)$ by comparing it with constant functions. More generally, we estimate functions by using inequalities and absolute values to compare them with other functions:
$$f(x) \le g(x), \quad x \in I; \qquad |f(x)| \le h(x), \quad x \in I.$$
The usual rules for inequalities and absolute value apply, since $f(x)$ is a real number for any given value of x in the domain. Thus for example we have, wherever $f(x)$ and $g(x)$ are both defined,
$$|f(x)g(x)| = |f(x)||g(x)|, \qquad |f(x) + g(x)| \le |f(x)| + |g(x)|.$$

Because of the dangers lurking in the use of inequalities with negative numbers, use $|\ |$ whenever possible, to guarantee that quantities will be non-negative. For example, by Section 2.4 (2a), we have

(2) $f(x)$ is bounded $\Leftrightarrow$ $|f(x)| \le K$,   for some constant K.

This is usually safer than expressing boundedness by: $b_1 \le f(x) \le b_2$. The following example—like the one in Section 5.5—illustrates the difference.

Example 10.1C $f(x)$ and $g(x)$ bounded on I $\Rightarrow$ $f(x)g(x)$ bounded on I.

Solution.   On the interval I, we have, using (2),
$$|f(x)| \le K, \quad |g(x)| \le L \quad \Rightarrow \quad |f(x)g(x)| = |f(x)||g(x)| \le KL. \qquad \square$$

Many students try to avoid $|\ |$, writing the argument incorrectly:
$$b_1 \ \le \ f(x) \ \le b_2 \ ;$$
$$c_1 \ \le \ g(x) \ \le c_2 \ ;$$
$$\Rightarrow \qquad b_1 c_1 \ \le \ f(x)g(x) \ \le \ b_2 c_2 \ . \qquad \square$$

What's wrong with the argument above? (Question 10.1/4)

Here is an example where $|\ |$ is not needed. It involves intuitive ideas about integrals, as well as properties from Chapter 9. It concerns the *error function*

$$\operatorname{erf} x \;=\; \int_0^x e^{-t^2/2}\,dt \;,$$

which is used in statistics; it is not an elementary function, and the integral has to be calculated by numerical methods.

Example 10.1D Show $\operatorname{erf} x = \displaystyle\int_0^x e^{-t^2/2}\,dt$
is bounded above on the interval $[0,\infty)$.

The idea is to estimate the integral by estimating the integrand. It is easiest to do this when $t > 1$, so we divide the interval in two at the point $t = 1$.

Solution. The integrand is positive, so $\operatorname{erf} x$ is increasing on $[0,\infty)$; therefore it suffices to show $\operatorname{erf} x$ is bounded above for $x \geq 1$. Since

$$\operatorname{erf} x \;=\; \int_0^1 e^{-t^2/2}\,dt + \int_1^x e^{-t^2/2}\,dt \quad \text{for } x \geq 1,$$

it suffices to show this last integral is bounded above for $x \geq 1$. But here we have an easy estimate for the integrand:

$$\begin{aligned}
t^2 \;&\geq\; t & &\text{for } t \geq 1, \\
\Rightarrow\quad e^{t^2/2} \;&\geq\; e^{t/2} & &\text{for } t \geq 1,\ \text{since } e^x \text{ is increasing;} \\
\Rightarrow\quad e^{-t^2/2} \;&\leq\; e^{-t/2} & &\text{for } t \geq 1,\ \text{by the reciprocal law.}
\end{aligned}$$

Since both functions are positive, the areas under their graphs obey the same inequality; thus for $x \geq 1$, the above implies that

$$\int_1^x e^{-t^2/2}\,dt \;\leq\; \int_1^x e^{-t/2}\,dt \;=\; -2e^{-t/2}\Big|_1^x \;;$$

$$\leq\; 2e^{-1/2}, \qquad \text{for } x \geq 1\ .$$

This shows that the integral on the left is bounded above as a function of x, which completes the argument, in view of our work at the start. □

Questions 10.1

1. For each, tell if it is bounded (on its domain); if so, give bounds.

   (a) $x \sin x$ \qquad (b) $\dfrac{x^2+1}{x^2+4}$ \qquad (c) $\dfrac{x-2}{x+1}$

2. If $f(x) = x^2 - x$, what is $f(I)$ for each of the following I?

   (a) $(-\infty,\infty)$ \qquad (b) $(0,1)$ \qquad (c) $[1,2]$

3. Give the values of sup, max, inf, and min $f(x)$ on I, if they exist:

   $$f(x) = \frac{x}{1+x}; \quad I = [0,\infty),\ \ (0,\infty),\ \ (-1,\infty),\ \ (-\infty,\infty).$$

4. What's wrong with the argument in Example 10.1C?

10.2   Approximating functions.

An important problem in computation is to approximate a function over an interval, replacing a complicated expression (or a numerical solution) by a simpler function which is sufficiently close to the original one.

To write an approximation, we use any of the equivalent notations (all have the same meaning, but $\underset{\epsilon}{\approx}$ is not standard notation):

(3) $\qquad |f(x) - g(x)| < \epsilon , \qquad f(x) \underset{\epsilon}{\approx} g(x) , \qquad$ for $x \in I$;

or in the inequality form,

(4) $\qquad\qquad g(x) - \epsilon \; < \; f(x) \; < \; g(x) + \epsilon, \quad$ for $x \in I$;

or in the error form,

(5) $\qquad\qquad f(x) = g(x) + e(x) , \quad$ where $|e(x)| < \epsilon$ for $x \in I$.

The geometric picture in each case is that we mark off a band of vertical width 2ϵ whose center line is the graph of $g(x)$, and then the graph of $f(x)$ must lie entirely within this band, for values of x in the interval I. This geometric interpretation comes from the inequalities (4).

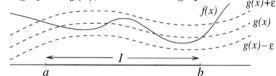

Approximating functions over intervals which are small is often done using the first few terms of a power series. If the signs alternate and the terms decrease in size, then the proof of Cauchy's test (Section 7.6) shows that the error will be less than the size of the next term. Here is an example, using the power series whose sum is $\sin x$; if you don't know this series, accept it for now, and a proof that it sums to $\sin x$ will be given in Chapter 17.

Example 10.2A   Find a δ-neighborhood of 0 over which $\sin x \underset{\epsilon}{\approx} x$, $\epsilon = .001$.

Solution.   Since $\sin x$ and x are odd functions, we can do the work assuming $x \geq 0$; the estimations will then be valid for $x < 0$ also.

The purpose of this preliminary step is to guarantee that $\sin x$ and x will be nonnegative, which avoids difficulties with inequalities.

The power series for $\sin x$ has terms which alternate in sign; it is

$$\sin x \; = \; x - \frac{x^3}{3!} + \frac{x^5}{5!} - \ldots + (-1)^n \frac{x^{2n+1}}{(2n+1)!} + \ldots \; ;$$

for any fixed value of x in the interval $0 \leq x \leq 1$, its terms decrease in size and tend to 0, as $n \to \infty$. So Cauchy's test applies, and we get by Section 7.6 (15),

$$|\sin x - x| \leq \frac{x^3}{3!}, \quad \text{for } 0 < x < 1 \; ;$$

$$< .001, \quad \text{if } x^3 < .006, \text{ or } x < .18 \; .$$

Thus we can take $\delta = .18$. $\square$

In what follows we use an inequality for integrals:

If $f(x) < g(x)$ for $x \in I$, and the integrals exist, then

(6) $$\int_a^b f(x)\,dx \;\; < \;\; \int_a^b g(x)\,dx, \qquad \text{for } a, b \in I, \quad a < b .$$

It remains true even if $f(x) = g(x)$ at a finite number of points on I. It is also true if $<$ is changed to $\leq$ in both the hypothesis and conclusion; we will prove this version in Chapter 19.

An approximation for $f(x)$ over $[a, b]$ gives us an approximation for its definite integral. For assume $a < b$; then on $[a, b]$,

(7) $$f(x) \underset{\epsilon}{\approx} g(x) \;\; \Rightarrow \;\; \int_a^b f(x)\,dx \underset{\epsilon(b-a)}{\approx} \int_a^b g(x)\,dx .$$

To see this, we write $f \underset{\epsilon}{\approx} g$ in the inequality form (4):

$$g(x) - \epsilon \;\; < \;\; f(x) \;\; < \;\; g(x) + \epsilon, \qquad x \in [a, b].$$

If we now integrate, and apply (6) to each inequality, we get

$$\int_a^b g(x)\,dx - \epsilon(b - a) \;\; < \;\; \int_a^b f(x)\,dx \;\; < \;\; \int_a^b g(x)\,dx + \epsilon(b - a),$$

which is the right side of (7) written in the form using inequalities.   □

Example 10.2B   Estimate the error in $\cos x \underset{\epsilon}{\approx} 1 - \dfrac{x^2}{2}$, for $|x| < .1$.

Solution.   Use the power series for $\cos x$ and imitate the previous example, or use the results of that example combined with (7). The latter gives, for $0 < x < .1$,

$$\int_0^x \sin t\,dt \underset{.001x}{\approx} \int_0^x t\,dt .$$

and evaluating the integrals gives, using the upper estimate $x < .1$,

$$1 - \cos x \underset{.0001}{\approx} \frac{x^2}{2} , \qquad \text{for } 0 < x < .1 .$$

Since both sides of the approximation are even functions, it is true also when x is in the interval $-.1 < x < 0$, and therefore for $|x| < .1$.   □

Questions 10.2

1. Find the interval on which this second-order approximation is valid:
$$(1 - x)(1 + x^2) \underset{\epsilon}{\approx} 1 - x + x^2, \qquad \epsilon = .001 .$$

2. If $f(x) \underset{\epsilon}{\approx} 1 + x$ and $g(x) \underset{\epsilon}{\approx} 1 - x$ for $|x| \leq 1$, estimate
$$\int_{-1}^1 \big(f(x) + g(x)\big)\,dx .$$

10.3 Local behavior.

Studying a function in a δ-neighborhood of some point x_0 is called studying its "local behavior at x_0". For example, you might be using a power series approximation which is only valid if you stay close to x_0. As other examples, the continuity or differentiability of $f(x)$ at x_0 are aspects of its local behavior there; so is its limit as $x \to x_0$, as we shall see.

Though the study takes place in some δ-neighborhood of x_0, we may not wish to specify δ exactly, perhaps because it is of no importance. In that case, we omit it from the notation, using one of the phrases,

$$for\ x \approx x_0, \qquad for\ x\ near\ x_0.$$

to mean

$$for\ x\ in\ some\ \delta\text{-neighborhood of } x_0.$$

These phrases are used just as we used "for $n \gg 1$" in our study of sequences; they mean nothing in themselves: they must be part of a mathematical sentence.

Examples 10.3A T or F? (a) $x^4 < x^2$ for $x \approx 0$ (b) $x^3 < x$ for $x \approx 0$

Solution. (a) is true; for example, $x^4 < x^2$ for $x \underset{1}{\approx} 0$ (i.e., for $|x| < 1$).

(b) is false, since $x^3 > x$ if $-1 < x < 0$; it is true however for $x \approx 0$, $x \geq 0$, since one can take $\delta = 1$, as in part (a).

Example 10.3B If $f(x)$ and $g(x)$ are bounded for $x \approx x_0$, so is $f(x) + g(x)$.

Proof.
$$
\begin{aligned}
|f(x)| \ &< \ K \quad \text{for } x \approx x_0, \quad \text{say for } |x - x_0| < \delta'; \\
|g(x)| \ &< \ L \quad \text{for } x \approx x_0, \quad \text{say for } |x - x_0| < \delta''; \quad \text{adding,} \\
|f(x) + g(x)| \ &\leq \ |f(x)| + |g(x)|, \quad \text{by the triangle inequality;} \\
&\leq \ K + L, \qquad \text{for } |x - x_0| < \min\ (\delta', \delta'');
\end{aligned}
$$
this shows $f(x) + g(x)$ is bounded for $x \approx x_0$. $\square$

Behavior at infinity.

Studying the local behavior of $f(x)$ at x_0 is like putting its graph under a microscope at the point $(x_0, f(x_0))$. However, sometimes one wants to use a telescope instead, studying the behavior of $f(x)$ on some infinite interval like (a, ∞) or $(-\infty, a)$. In this case we use a terminology like that for sequences:

$$
\begin{aligned}
for\ x \gg 1, \quad for\ large\ x \ &= \ \text{for } x \text{ in some interval } (a, \infty); \\
for\ x \ll -1, \quad for\ negatively\ large\ x \ &= \ \text{for } x \text{ in some interval } (-\infty, a); \\
for\ |x| \gg 1, \quad for\ large\ |x| \ &= \ \text{for } |x| > \text{ some number } a;
\end{aligned}
$$

where in each case, an appropriate value of a exists, but is unspecified.

We can also say that we are studying the "local behavior of $f(x)$ at ∞" (or at $-\infty$). These are not points on the x-axis, but the terminology is suggestive.

Such studies of functions "at infinity" have many applications in science and engineering, where initial disturbances die away with time—the function behavior may be complicated to start with, but becomes more regular and simple for large values of the time t.

Examples 10.3C   Let $f(x)$ be a polynomial with positive leading coefficient:
$$f(x) = a_0 x^n + a_1 x^{n-1} + \ldots + a_n, \qquad a_0 > 0 .$$
The following give examples of the terminology; the proofs are much easier if we use the limit theorems of the next chapter, so we postpone them.

(a) $f(x) > 0$ at ∞ (i.e., for $x \gg 1$);

(b) if n is even, then $f(x) > 0$ at $-\infty$ (i.e., for $x \ll -1$);

    if n is odd, then $f(x) < 0$ at $-\infty$;

(c) $1/f(x)$ is bounded at $\pm\infty$ (i.e., for $|x| \gg 1$).

Local properties at a point.

Finally, we note that some of the properties of functions make sense when restricted to a neighborhood of a point x_0, in which case the word "locally" can be attached to them. For example,

$f(x)$ is **locally increasing** at x_0 means $f(x)$ is increasing for $x \approx x_0$;

$f(x)$ is **locally bounded** at x_0 means $f(x)$ is bounded for $x \approx x_0$;

$f(x)$ is **locally positive** at x_0 means $f(x)$ is positive for $x \approx x_0$;

In each case, there is some unspecified δ-neighborhood of x_0 over which the property holds. Here are some examples of this usage; proofs are left for the Exercises.

Examples 10.3D

(a) A polynomial with positive constant term is locally positive at 0.

(b) $\sin x$ is locally increasing at every $x_0 \in (-\pi/2, \pi/2)$, but not at $\pm\pi/2$.

(c) The function defined by $f(x) = \begin{cases} 1/x, & x \neq 0; \\ 0, & x = 0, \end{cases}$ is locally bounded at any $x_0 \neq 0$, but not at 0.

Questions 10.3

1. Tell whether each of the following is true or false (no proofs required):

(a) $\sin x \leq x$ for $x \approx 0$;   (b) $e^{-x} < .01$ for $x \gg 1$;

(c) if $f(x)$ is a polynomial, then $1/f(x)$ is defined at ∞ and $-\infty$;

(d) given $\epsilon > 0$, $\sin x \underset{\epsilon}{\approx} x$ for $x \approx 0$;

(e) x^2 is locally bounded at any x_0, but not locally bounded at ∞;

(f) x^2 is locally monotone at any x_0, and also at ∞ and $-\infty$.

10.4   Local and global properties of functions.

To explain the title of this section, we begin with some examples.

We can formulate a notion of boundedness which is intermediate between boundedness on an interval I and local boundedness at a point x_0.

Definition 10.4A   We say $f(x)$ is **locally bounded** on the open interval I if it is locally bounded at every point of I :

$$\text{for all } x_0 \in I, \quad f(x) \text{ is bounded for } x \approx x_0 .$$

Example 10.4A   Prove $1/x$ is locally bounded on $(0, \infty)$.

Solution. Show: for any $a > 0$, $f(x)$ is locally bounded for $x \approx a$.
To see this, take $\delta = a/2$; by the reciprocal law for inequalities:

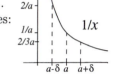

$$\frac{a}{2} < x < \frac{3a}{2} \quad \Rightarrow \quad \frac{2}{3a} < \frac{1}{x} < \frac{2}{a} .$$

Thus $f(x)$ is bounded for $x \underset{\delta}{\approx} a$, where $\delta = a/2$.

Note carefully the difference between "bounded" and "locally bounded" on an interval I. The function $1/x$ is locally bounded on $(0, \infty)$, as Example 10.4A shows. But it is not bounded on $(0, \infty)$, since it grows arbitrarily large as x approaches 0 from the right.

We can formulate something analogous for the "increasing" property.

Definition 10.4B   We say $f(x)$ is **locally increasing** on the open interval I if it is locally increasing at every point x_0 of I.

Example 10.4B $f(x) = \tan x$ is locally increasing on every interval of its domain, $(-\pi/2, \pi/2)$, $(\pi/2, 3\pi/2), \ldots$.

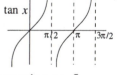

It is not increasing, however: for example, $\pi/4 < \pi$, but $f(\pi/4) > f(\pi)$.

Both of these definitions—locally bounded on I, locally increasing on I— assume that I is an open interval. To extend the definitions to a closed interval, they need to be modified if x_0 is one of the endpoints, since $f(x)$ may not be defined outside of I.

Suppose therefore $I = [a, b]$. In the preceding definitions, if $x_0 = a$ or $x_0 = b$, replace "for $x \approx a$" or "for $x \approx b$" respectively by:

(8a)   "for $x \approx a, \ x \geq a$"   (notation: *for $x \approx a^+$*);

(8b)   "for $x \approx b, \ x \leq b$"   (notation: *for $x \approx b^-$*).

Example 10.4C   Show that the function $\sqrt{x}$ is locally bounded on $[0, \infty)$.

Solution.   If $a > 0$, take $\delta = a$, say; then $0 \leq \sqrt{x} \leq \sqrt{2a}$ for $x \underset{\delta}{\approx} a$.

If $a = 0$, we use the modified definition. Take $\delta = 1$, say; then

$$0 \leq \sqrt{x} \leq 1, \quad \text{for } x \underset{\delta}{\approx} 0^+ .$$

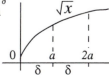

Note that the function is not bounded on $[0, \infty)$.

When we say $f(x)$ is locally increasing, or locally bounded, without specifying I, we mean the domain of f is a collection of non-overlapping intervals having positive length, and f is locally whatever on each of these intervals. (Recall Example 10.4B: $\tan x$ is locally increasing on each of the infinitely many open intervals that comprise its domain.)

Local vs. global.

Properties such as "locally increasing on I" and "locally bounded on I" are called *local properties* of a function. A property is **local** if to verify that it holds on an interval, it is enough to check that it holds in a neighborhood of each point on this interval. By contrast, a property like being bounded on I, or periodic on $(-\infty, \infty)$, cannot be checked point-by-point, neighborhood-by-neighborhood; you must look at the function on the interval I as a whole, to see if the property holds. Such properties are called **global** properties.

In general, global properties imply local properties, but not conversely. For example, consider the relation between boundedness and local boundedness on an interval I. It is clear that

(9) $f(x)$ bounded on I $\Rightarrow$ $f(x)$ locally bounded on I ,

since a global upper bound b for $|f(x)|$ will also be an upper bound on every neighborhood inside I, i.e., a local upper bound at each point $x_0 \in I$. The converse to (9) however is false, as the example of $1/x$ on $(0, \infty)$ shows: locally bounded on I does not imply bounded on I .

On the other hand, the converse to (9) turns out to be true, provided the interval I is finite and closed, or as one says, *compact*. We state it formally. It is a typical illustration of the way the nature of I enters into theorems of analysis.

Theorem 10.4   Let $f(x)$ be defined on an interval $I = [a, b]$. Then

$f(x)$ locally bounded on I $\Rightarrow$ $f(x)$ bounded on I .

We shall wait until Chapter 13 to prove this theorem, since it belongs naturally in the circle of ideas presented there. The theorem is of the type:

local property $\Rightarrow$ global property ;

though it is a fairly modest first example of this type of theorem, we can say that in general some of the most important (and non-trivial) theorems of analysis are of this local-to-global type. The proofs almost always invoke the Completeness Property in some form; watch for it.

Pointwise properties.

The local and global properties we have been talking about should be carefully distinguished from a simpler type of property of $f(x)$ that we shall call a **pointwise** property.(*) This is a property of $f(x)$ on an interval I which can be verified point-by-point in I, without having to look at a neighborhood of each point.

(*) With apologies to English; "punctual" would be a better choice and a real word to boot, but it has other connotations.

For example *positivity* is a pointwise property: to verify that $f(x)$ is positive on I, we check that at each $a \in I$, $f(a) > 0$, and for this it is not necessary to look at a neighborhood of a in I.

Example 10.4D Classify the two properties:

(a) $f(x)$ is bounded by K on I;   (b) $f(x)$ is bounded on I.

Solution. The first is a pointwise property, since to verify it, we check that $|f(a)| \leq K$ at each point $a \in I$. The second is global: no such pointwise check is possible, since we do not know in advance what the bound is.

A final controversial remark.

One of the main reasons for distinguishing between these three types of properties is to help in assessing the depth of a theorem. As we said, the deepest theorems are those of the type: local property $\Rightarrow$ global property. From this viewpoint, there is some justification for also classifying as *global*

$\star$ any property asserting that some point exists (since the proof will probably have to invoke the Completeness Principle);

$\star$ or any property whose verification requires you to choose one or more points.

The Questions incorporate these two suggestions in the parenthesized Answers.

Questions 10.4

1. (a) Define the notion: $f(x)$ is *locally constant* on an open I.

(b) Is the *unit step function:* $u(x) = \begin{cases} 1, & x \geq 0; \\ 0, & x < 0; \end{cases}$ locally constant?

Prove it, or explain why not.

(c) Find an f which is locally constant (on its domain), but not constant.

2. Classify each property of $f(x)$ as pointwise, local, global, or nonexistent.

(a) even function

(b) has a supremum on $[a, b]$

(c) locally odd function

(d) $y = f(x)$ satisfies the equation $y^5 + 2xy - 3x = 0$ on I

(e) is a linear function on I

(f) has a maximum on $[a, b]$

(g) $f(x) \neq 3$ for $x \in I$

(h) has a maximum on I at the point a

(i) $\max_I f(x) = 5$

(j) $f(x)$ has a zero on I

Exercises

10.1

1. If $f(x) = \cos \pi x$, what is $f(I)$ for each of the following I?

(a) $(-\infty, \infty)$ (b) $(0, 1)$ (c) $(1/2,\ 3/2)$

2. Examine the boundedness, sup, inf, max, and min (on D_f) of:

(a) $1/(1 + x^2)$ (b) $\tan^{-1} x$

3. For each of the following, tell if it is bounded (on its domain).

(a) $\dfrac{x^2 - 1}{x^2 - 4}$ (b) $\sin x \tan^{-1} x$ (c) $\ln(2 + \cos x)$

4. Prove that $\sum_1^k a_n \cos nx$ is bounded on $(-\infty, \infty)$.

5. Give an example of a function $f(x)$ for which $f([-1, 1]) = (-\infty, \infty)$.

6. For each of the following $f(x)$ and each interval I, give the values of sup, inf, max, and min $f(x)$, if they exist.

(a) $f(x) = \dfrac{1}{2 + \sin x}$; $I = [0, \pi],\ (-\pi/2, \pi/2),\ (-\infty, \infty)$;

(b) $f(x) = xe^{-x}$; $I = [0, \infty),\ (0, \infty),\ (-\infty, \infty)$.

7. (a) Give an $f(x)$ such that $1/f(x)$ is defined for all x and is

(i) unbounded   (ii) bounded, but $f(x)$ is decreasing.

(b) Suppose $f(x)$ is defined for all x ; under what hypotheses on $f(x)$ will $1/f(x)$ be bounded for all x?

8. Estimate the value of $\displaystyle\int_0^1 \dfrac{x^4}{1 + x^6}\, dx$ by estimating the integrand, and prove the value of the integral is $\leq 1/5$.

9. Suppose $f(x)$ and $g(x)$ are defined for all x. Prove if true; give a counterexample if false:

(a) $f(x)$ bounded $\Rightarrow f(g(x))$ bounded;
(b) $f(x)$ bounded $\Rightarrow g(f(x))$ bounded.

10.2

1. If you approximate $\ln(1 + x)$ by using the first three terms of the series

$$\ln(1 + x) = x - \frac{x^2}{2} + \frac{x^3}{3} - \frac{x^4}{4} + \dots,$$ estimate the error when $0 \leq x \leq .1$.

(Use Cauchy's test; cf. Section 7.6 (15).)

2. Estimate the error in the approximation $\cos x \underset{\epsilon}{\approx} 1 - \dfrac{x^2}{2}$, $|x| < .1$ by using the series $\cos x = 1 - x^2/2! + x^4/4! - x^6/6! + \dots$. (Use Cauchy's test.)

10.3

1. Prove the statements in Example 10.3D:

(a) A polynomial with positive constant term is locally positive at 0.

(b) $\sin x$ is locally increasing at every $x_0 \in (-\pi/2, \pi/2)$, but not at $\pm\pi/2$.

(c) The function defined by $f(x) = 1/x$, $x \neq 0$; $f(0) = 0$ is locally bounded at any $x_0 \neq 0$, but not at 0.

2. Suppose that $f(x)$ is defined for all x and periodic. Prove that if $f(x)$ is bounded for $x \gg 1$, then it is bounded for all x.

3. For what values of the positive real number k are the functions $f(x)$ bounded for $x \approx 0^+$?

$$\text{(a)} \quad f(x) = \int_x^1 (1/t^k)\, dt \qquad \text{(b)} \quad f(x) = \int_x^1 (e^t/t^k)\, dt$$

4. Let $f(x) = \sec(1/x)$, $x \neq 0$; $f(0) = 0$. Is $f(x)$ defined for $x \approx 0$? Prove it, or explain why not.

10.4

1. Classify the following properties as pointwise, local, or global.

(a) $f(x)$ is defined on I

(b) the supremum of $f(x)$ on I is 2

(c) $f(x) < 1/x$ on $(0, 1)$

(d) $f(x)$ attains a minimum on I

(e) $f(x)$ is constant on I

(f) $f(x)$ is periodic with period 2

(g) $f(x)$ is periodic

2. Classify as pointwise, local at some $a \in I$, local on I, or global (assume the series converge in the relevant interval or neighborhood).

(a) $f(x)$ can be represented by a series $\sum a_n x^n$ in a neighborhood of 0;

(b) $f(x)$ can be represented by a series $\sum a_n x^n$ in $(-1, 1)$;

(c) $f(x)$ can be represented by $\sum_1^\infty x^n/n^2$ in $(-1, 1)$;

(d) for each $x_0 \in (-1, 1)$, $f(x)$ can be represented by a series of the form $\sum a_n(x - x_0)^n$ convergent in some neighborhood of x_0.

Problems

10-1 Let $f(x) = \int_0^x \dfrac{\cos t}{1 + t^2} \, dt$; it is not an elementary function.

(a) Is it odd, even, or neither? (Indicate reason.)

(b) Is it increasing, decreasing, or neither? (Indicate reason.)

(c) Using its geometric interpretation as area, show intuitively that it has a maximum, tell where it occurs, and estimate its maximum value roughly.

10-2 Prove that a function which is locally constant on $[0, 1)$ is actually constant on $[0, 1)$ (cf. Question 10.4/1). (Suggestion: one possibility would be to consider sup S, where $S = \{a < 1 : f(x) \text{ is constant on } [0, a)\}$.)

10-3 Prove that a function which is locally increasing on an interval I is increasing on I. (Suggestion: use an indirect argument and bisection.)

Answers to Questions

10.1

1. (a) unbounded   (b) bounded; $1/4 \le f(x) \le 1$ (c) unbounded

2. (a) $[-1/4, \infty)$ (b) $[-1/4, 0)$ (c) $[0, 2]$

3. (a) sup $= 1$, no max; inf, min $= 0$ (b) as in (a), but no min

(c) sup $= 1$, no max, min, inf   (d) none

4. The argument looks plausible, but as written it requires that one of the functions and its bounds be non-negative.

10.2

1. $|f(x) - g(x)| = |x|^3 < .001$ if $|x| < .1$

2. $f(x) + g(x) \underset{2\epsilon}{\approx} 2$, therefore by (7) we get $\int_{-1}^{1} (f + g) \, dx \underset{4\epsilon}{\approx} 4$

10.3

1. (a) F (if $x < 0$)   (b) T   (c) T (since $f(x)$ has only finitely many zeros)

(d) T   (e) T   (f) F (not locally monotone at 0)

10.4

1. (a) For any $a \in I$, $f(x) = k$ for $x \approx a$.

(b) No; it is not locally constant at 0, since it is 0 on the left side of any δ-neighborhood of 0, and 1 on the right side.

(c) The function sgn x (see Question 9.2/1) is an example; or change the definition of $u(x)$ by leaving $u(0)$ undefined.

2. (a) pointwise   (b) global   (c) makes no sense   (d) pointwise

(e) pointwise (or global)   (f) global   (g) pointwise

(h) pointwise: ($f(x) \le f(a)$ for all $x \in I$)   (i) pointwise (or global)

(j) pointwise (or global)

11

Continuity and Limits

11.1 Continuous functions.

Continuity is one of the really important ideas in analysis; this chapter and the next two are devoted to it, but it is also fundamental in everything to come later. We begin with some introductory remarks.

Suppose x and y are two variables, related by an equation $y = f(x)$. We say that y *varies continuously with* x if, roughly, small changes in x produce only small changes in y.

For instance, do the roots of $x^5 + ax + b = 0$ vary continuously with the coefficients a and b? That is, if we vary a and b a little, do the roots change by only a small amount? There is no explicit algebraic formula for the roots that we can use to answer this question.

Or suppose we have an integral depending on a parameter s, such as

$$y = \int_0^\pi \frac{\sin st}{t} \, dt \ ;$$

for each value of s, there is a corresponding value of y. Does y vary continuously with s? (There is no elementary expression in s for the value of the integral.)

This idea of continuous variation is essential in science and engineering. Suppose for example that a and b in the fifth degree equation above are experimentally determined constants, and that x is some physical quantity whose value we are trying to calculate. If the quantity x does not vary continuously with a and b, then small errors in measuring a and b can produce big variations in the calculated value of x. This means the equation cannot be used for calculating the value of x: the roots of the equation are well-determined mathematically, but not scientifically.

We now formulate mathematically this idea of continuity. To avoid having to introduce a dependent variable, we work directly with the function $f(x)$. The definition says more or less that $f(x)$ should be arbitrarily close to $f(x_0)$, provided x stays sufficiently close to x_0 .

Definition 11.1A We say $f(x)$ is **continuous** at x_0 if it is defined for $x \approx x_0$, and

(1) $$\text{given any } \epsilon > 0, \quad f(x) \underset{\epsilon}{\approx} f(x_0) \quad \text{for } x \approx x_0 \ .$$

We say $f(x)$ is **continuous on the open interval** I if it is continuous at every point of I.

Example 11.1A Show x^2 is continuous on $I = (-a, a), \ a > 0$.

Solution.   Fix any $x_0 \in I$. Then, given $\epsilon > 0$, and any $x \in I$,

$$|x^2 - x_0^2| \ = \ |x - x_0||x + x_0|;$$
$$\leq \ |x - x_0|(|x| + |x_0|);$$
$$< \ |x - x_0| \cdot 2a \ < \ \epsilon, \quad \text{if } x \underset{\epsilon/2a}{\approx} x_0 . \qquad \square$$

We need to extend the definition of continuity to closed intervals I, in order to take care of functions like $\sqrt{1 - x^2}$, whose domain is $[-1, 1]$. The problem is how to define continuity at the endpoints. Since we can't ask anything of the function at points where it isn't defined, we use the same definition, but restrict ourselves to right or left half-neighborhoods at the endpoints.

Recall from 10.4 (8a) the terminology *for $x \approx a^+$*, meaning: for $x \approx a, \ x \geq a$.

Definition 11.1B   Assuming $f(x)$ is defined for the relevant x-values, we say

$f(x)$ is **right-continuous** at x_0 if, given $\epsilon > 0$, $f(x) \underset{\epsilon}{\approx} f(x_0)$ for $x \approx x_0^+$;

$f(x)$ is **left-continuous** at x_0 if, given $\epsilon > 0$, $f(x) \underset{\epsilon}{\approx} f(x_0)$ for $x \approx x_0^-$;

(2) $f(x)$ is **continuous on** $[a, b]$ if $f(x)$ is $\begin{cases} \text{continuous on } (a, b), \\ \text{right-continuous at } a, \\ \text{left-continuous at } b. \end{cases}$

Even if $f(x)$ is defined on a bigger interval than $[a, b]$, for it to be continuous on $[a, b]$ we only ask it to be one-sided continuous at the endpoints. To see why, consider the function $h(x)$, defined to be 1 for $x \in [0, 1]$, and 0 elsewhere. It is not continuous at 0 or 1, yet we want to say it is continuous on $[0, 1]$. The solution is to require only one-sided continuity at the endpoints.

Definition 11.1C We say $f(x)$ is **continuous** if its domain is an interval I of positive or infinite length, and it is continuous on I.

Why don't we just say $f(x)$ is continuous if it is continuous on its domain? Then we would have to say $1/x$ is continuous, which seems unreasonable.

Continuity at x_0 is an aspect of the local behavior of f at x_0, since we verify it by looking at $f(x)$ in a neighborhood of x_0. Continuity on I is a local property of $f(x)$ since the definitions say we verify it by checking that $f(x)$ is continuous at each point of I.

Example 11.1B   Show $\sin x$ is continuous.

Proof. Represent x as the arclength AP, so the point P is $(\cos x, \sin x)$. Since the perpendicular PR is the shortest distance from P to the horizontal, we see that

$$|PR| \ \leq \ |PP_0|, \quad \text{i.e.,}$$
(3) $|\sin x - \sin x_0| \ \leq \ |x - x_0| .$

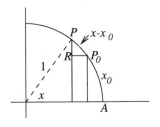

By Definition 11.A, we see that $\sin x$ is continuous at x_0, since (3) shows that

$$\text{given } \epsilon > 0, \quad \sin x \underset{\epsilon}{\approx} \sin x_0 \text{ for } x \underset{\epsilon}{\approx} x_0.$$

Though the picture is drawn for x_0 in the first quadrant, the reasoning is valid regardless of the position of x_0. Since x_0 was arbitrary, this shows $\sin x$ is continuous for all x, i.e., that $\sin x$ is continuous.   □

Example 11.1C   Let $f(x) = \displaystyle\int_0^\pi \frac{\sin xt}{t}\, dt$. Show $f(x)$ is continuous.

Remarks.   The value of the definite integral depends on the value of the parameter x, so the integral is a function of x. Integrals like this one which depend on a parameter—the Fourier and Laplace transforms are good examples—are important in the applications of analysis to science and engineering. Showing the integral "depends continuously on x" is what we are asked to do.

Proof.   Let x_0 be any fixed x-value. We want to show (1) holds. We have

$$|f(x) - f(x_0)| = \left| \int_0^\pi \frac{\sin xt}{t}\, dt - \int_0^\pi \frac{\sin x_0 t}{t}\, dt \right|$$

$$= \left| \int_0^\pi \frac{\sin xt - \sin x_0 t}{t}\, dt \right|$$

$$\leq \int_0^\pi \frac{|\sin xt - \sin x_0 t|}{t}\, dt \ , \quad \text{by (4) below;}$$

now use (3) above and 10.2 (6):

$$\leq \int_0^\pi \frac{|(x - x_0)t|}{t}\, dt \ = \ \pi|x - x_0| \ .$$

Thus,

$$\text{given } \epsilon > 0, \quad |f(x) - f(x_0)| \ \leq \ \pi\epsilon, \quad \text{for } x \underset{\epsilon}{\approx} x_0.$$

This completes the proof, by the K-ϵ principle (for functions—cf. 11.3)   □

The third line of the argument used the important inequality for estimating the size of integrals that we will prove in Chapter 19:

(4) $\left| \int_a^b f(x)\, dx \right| \ \leq \ \int_a^b |f(x)|\, dx \ .$

Discontinuities.

To understand the notion of continuity, it helps to look at a few examples of functions which are *not* continuous at some point x_0.

We call a point x_0 where f is not continuous a **point of discontinuity** for the function $f(x)$ provided that it is also *isolated*, that is, $f(x)$ is continuous at all points near x_0 but different from it (in symbols: for $x \underset{\neq}{\approx} x_0$).

DISCONTINUITY TYPES

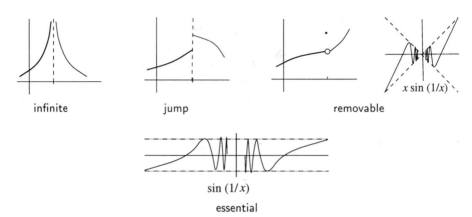

infinite   jump   removable

$x \sin (1/x)$

$\sin (1/x)$

essential

As shown above, there are several types of discontinuities, classified according to the geometric behavior of $f(x)$ at the point.

Neither $\sin(1/x)$ nor $x \sin(1/x)$ is defined when $x = 0$, but they are continuous elsewhere, so 0 is a point of discontinuity for both functions. However, there is an important difference between them, as the picture suggests. If we complete the definition of $f(x) = x \sin(1/x)$ by putting $f(0) = 0$, then it becomes continuous at 0. This is easy to show, and we will do it formally later, but the geometric reason is that the graph is confined between the two lines $y = \pm x$.

By contrast, there is no way we can define $g(x) = \sin(1/x)$ at $x = 0$ so as to make it continuous at 0. For $g(x)$ takes on the values 0 and 1 infinitely often in any δ-neighborhood of 0, and there is no number $g(0)$ which is simultaneously close to 0 and 1.

We say that $x \sin(1/x)$ has a **removable** discontinuity at 0, since it is possible to define (or change) its value at 0 so that the altered function becomes continuous at 0. It is not possible to do this with $\sin(1/x)$, which is said to have an **essential** discontinuity at 0. More about this in 11.4.

> We said earlier that functions had to be continuous to have scientific meaning. But discontinuous functions are actually of great importance in the applications, since processes where change takes place very rapidly (explosions, for example, or the throw of a switch in an electric circuit) are often modeled as discontinuities.
>
> Of course, we're talking here about the macro-world. In the world of quantum mechanics, discontinuous functions and divergent series are everywhere.

The mathematical description of the different types of discontinuity pictured above is most easily given using the idea of *limit* for a function; we turn to this now in 11.2. Then we will give some limit theorems—analogues of the ones we proved for sequences—and using the help these technical tools give us, return to the subject of continuity in 11.4 and 11.5, as well as the following two chapters.

Questions 11.1

1. (a) By Example 11.1A, x^2 is continuous on $(-a, a)$ for any $a > 0$. Show this implies x^2 is continuous on $(-\infty, \infty)$. (Write it out!)

Would the proof in Example 11.1A have worked directly for $I = (-\infty, \infty)$?

(b) Would the implication in (a) be true if "continuous" were replaced everywhere by "bounded"? What about the two concepts makes the difference?

2. Prove $1/x$ is continuous at $x = 2$.

3. Does "absolute" continuity imply continuity? That is, if $|f(x)|$ is continuous on I, will $f(x)$ be continuous on I?

11.2   Limits of functions.

To be continuous at x_0, the function $f(x)$ must first of all be defined at x_0. To have a limit as $x \to x_0$, it need not be. That's the essential difference between the two definitions, which otherwise look very similar. To make up for the lack of $f(x_0)$, we have to supply a number L.

A simple example illustrates the difference: $f(x) = x^2/x$. Here $f(0)$ does not exist, so $f(x)$ cannot be continuous at 0; but since $f(x) = x$, $x \neq 0$, we can say

$$\lim_{x \to 0} \frac{x^2}{x} = \lim_{x \to 0} x = 0.$$

Definition 11.2A   The limit of a function.

Let $f(x)$ be defined for $x \approx x_0$, but not necessarily for $x = x_0$ (we abbreviate this by: for $x \underset{\neq}{\approx} x_0$).

We say $f(x)$ has the **limit** L as $x \to x_0$ if there is a number L such that

(5)   given $\epsilon > 0$, $f(x) \underset{\epsilon}{\approx} L$ for $x \underset{\neq}{\approx} x_0$.

If this is so, we write

$$\lim_{x \to x_0} f(x) = L, \qquad \text{or} \qquad f(x) \to L \text{ as } x \to x_0.$$

Referring to the picture, the definition says $f(x)$ must stay entirely inside the two horizontal lines of level $L + \epsilon$ and $L - \epsilon$, when x is sufficiently near, but not equal to, the point x_0.

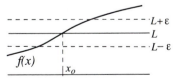

Notice that the approximation in (5) must be valid on both sides of x_0; that is, $f(x)$ must approach the value L no matter from which side x approaches x_0. If we restrict x to lie only on one side of x_0, we get *one-sided limits*.

Definition 11.2B   Assume $f(x)$ defined for $x \underset{\neq}{\approx} x_0^+$ or $x \underset{\neq}{\approx} x_0^-$, respectively.

(6)   **right-hand limit** $\lim\limits_{x \to x_0^+} f(x) = L$:   given $\epsilon > 0$, $f(x) \underset{\epsilon}{\approx} L$ for $x \underset{\neq}{\approx} x_0^+$

(7)   **left-hand limit** $\lim\limits_{x \to x_0^-} f(x) = L$:   given $\epsilon > 0$, $f(x) \underset{\epsilon}{\approx} L$ for $x \underset{\neq}{\approx} x_0^-$

Theorem 11.2   *The limit on the left below exists if and only if both limits on the right exist and are equal; if this is so, all three limits are equal:*

(8) $\lim\limits_{x \to x_0} f(x) = L$ $\Leftrightarrow$ $\lim\limits_{x \to x_0^+} f(x) = L$ and $\lim\limits_{x \to x_0^-} f(x) = L$.

   One source of confusion is the use of "right" and "left" in limits and continuity: these words refer to the direction of approach to x_0, not to the position of x_0 as an endpoint of the interval. At the left-hand endpoint a of $[a, b]$. we want the right-hand limit $x \to a^+$, for instance.

Examples 11.2A   Show directly from the definitions that

   (a) $\lim\limits_{x \to 0} x \sin 1/x = 0$;   (b) $\lim\limits_{x \to 1^-} \sqrt{1 - x^2} = 0$.

Solution.   For both we use the $|f(x)| < \epsilon$ form of the limit statement.

   (a)   Given $\epsilon > 0$,
$$|x \sin 1/x| \;=\; |x|\,|\sin 1/x|$$
$$\le |x|, \quad < \epsilon \text{ for } |x| < \epsilon, \; x \ne 0. \qquad \square$$

   (b)   Note that the function is not defined for $x > 1$. Given $\epsilon > 0$,
$$\sqrt{1 - x^2} = \sqrt{1 + x}\,\sqrt{1 - x}$$
$$< \sqrt{2}\,\sqrt{1 - x}, \qquad \text{for } x < 1;$$
$$< \epsilon \qquad \text{if } 1 - x < \epsilon^2/2. \qquad \square$$

Example 11.2B   Let $f(x) = \dfrac{|x^2 - 4|}{x + 2}$;   find $\lim\limits_{x \to -2} f(x)$.

Solution. This is an example where $f(x)$ is undefined at x_0. We have
$$f(x) = \frac{|x + 2||x - 2|}{x + 2} = \begin{cases} |x - 2|, & \text{if } x + 2 > 0, \text{ i.e., } x > -2; \\ -|x - 2|, & \text{if } x + 2 < 0, \text{ i.e., } x < -2 . \end{cases}$$
Therefore,
$$\lim\limits_{x \to -2^+} f(x) = 4, \qquad \lim\limits_{x \to -2^-} f(x) = -4.$$

Using (8), we see that $\lim\limits_{x \to -2}$ does not exist; the function actually has a jump discontinuity at $x = -2$. $\square$

We get another type of left- and right-hand limit by letting respectively either $x \to \infty$ or $x \to -\infty$; the first of these resembles the limit of a sequence.

Definition 11.2C   Limits at infinity.   We define

$$(9) \qquad \lim_{x \to \infty} f(x) = L : \quad \text{given } \epsilon > 0, \quad f(x) \underset{\epsilon}{\approx} L \text{ for } x \gg 1.$$

The corresponding definition for $-\infty$ substitutes $x \ll 1$ in (9).

Examples 11.2C   Show directly from the definition that

(a) $\displaystyle \lim_{x \to \infty} \frac{1}{1 + x^2} = 0;$ \qquad\qquad (b) $\displaystyle \lim_{x \to \infty} \frac{2x}{x+1} = 2$.

Solution.   In both cases, given $\epsilon > 0$, we have

(a) $\displaystyle \frac{1}{1+x^2} < \epsilon$ if $1 + x^2 > \dfrac{1}{\epsilon}$, for example if $x > \dfrac{1}{\sqrt{\epsilon}}$;

(b) $\displaystyle \left| \frac{2x}{x+1} - 2 \right| = \left| \frac{-2}{x+1} \right| < \epsilon,$ if $x + 1 > \dfrac{2}{\epsilon}$, for example if $x > \dfrac{2}{\epsilon}$.

To summarize, the definitions of the three kinds of limits are given below in symbols on the right, with the equivalent phrases on the left.

given $\epsilon > 0$,	$f(x) \underset{\epsilon}{\approx} L$	for $x \underset{\neq}{\approx} a :$	$\displaystyle \lim_{x \to a} f(x) = L$;
given $\epsilon > 0$,	$f(x) \underset{\epsilon}{\approx} L$	for $x \underset{\neq}{\approx} a^+ :$	$\displaystyle \lim_{x \to a^+} f(x) = L$;
given $\epsilon > 0$,	$f(x) \underset{\epsilon}{\approx} L$	for $x \underset{\neq}{\approx} a^- :$	$\displaystyle \lim_{x \to a^-} f(x) = L$;
given $\epsilon > 0$,	$f(x) \underset{\epsilon}{\approx} L$	for $x \gg 1 :$	$\displaystyle \lim_{x \to \infty} f(x) = L$;

The remaining definitions and theorems in this chapter will be valid for all of these types of limits; the statements and proofs for each type of limit differ only in the selection of the " $x \to \dots$" and the corresponding "for $x \dots$" phrases. To indicate that all these cases are being covered, we will write "for $x \underset{\neq}{\approx} a$, etc."

As an example, we extend the definition of limit to infinite limits.

Definition 11.2D   Infinite limits.   Let $f(x)$ be defined for $x \underset{\neq}{\approx} x_0$, etc.

$$(10) \qquad \lim_{x \to x_0} f(x) = \infty : \quad \text{given any } b > 0, \quad f(x) > b \text{ for } x \underset{\neq}{\approx} x_0, \text{ etc.}$$

The definition of $\displaystyle \lim_{x \to x_0} f(x) = -\infty$ substitutes $f(x) < -b$.

Examples 11.2D

(a) $\quad \lim\limits_{x \to 0+} \dfrac{1}{x} = \infty, \quad \lim\limits_{x \to 0-} \dfrac{1}{x} = -\infty, \quad \lim\limits_{x \to 0} \dfrac{1}{x^2} = \infty$.

(b) Show that $\lim\limits_{x \to \infty} x^2(k + \cos x) = \infty \;\Leftrightarrow\; k > 1.$

Solution. (b) Assume first that $k > 1.$

Since $\quad k + \cos x \ge k - 1$ for all $x,$ we have that, given $b > 0,$

$$x^2(k + \cos x) \;\ge\; x^2(k-1) \;>\; b \quad \text{for } x > \sqrt{\dfrac{b}{k-1}}\,.$$

If, however, $k \le 1,$ then $x^2(k + \cos x) \le 0$ when $x = \pi, 3\pi, 5\pi, \ldots ,$ thus it is not true that $x^2(k + \cos x) > b > 0,$ for $x \gg 1.$

Questions 11.2

1. Prove: $\lim\limits_{x \to a} f(x) = 0,\; g(x)$ locally bounded at $a \;\Rightarrow\; \lim\limits_{x \to a} f(x)g(x) = 0.$

2. Prove: $\lim\limits_{x \to a} f(x) = \infty,\; g(x)$ locally bounded below at a

$$\Rightarrow\; \lim\limits_{x \to a} f(x) + g(x) = \infty.$$

3. T or F? $\quad \lim\limits_{x \to a} f(x) = 0 \;\Rightarrow\; \lim\limits_{x \to a} 1/f(x) =$ either ∞ or $-\infty$

4. $\lim\limits_{x \to a} f(x) = \infty \;\Rightarrow\; \lim\limits_{x \to a} f(x)g(x) = \infty;$ what would be a reasonable hypothesis on $g(x)$ to guarantee this implication? (The weaker the hypothesis, the stronger the statement; the hypothesis should not involve $f(x)$.)

11.3 Limit theorems for functions.

We follow now along the general lines of what we did in Chapters 4 and 5 for sequences, discussing first how limit statements can be rephrased in terms of errors, then giving the analogous limit theorems. There are no surprises. For that reason, most of the proofs are omitted, since they are essentially the same as the proofs for sequences.

All of these results will be valid for limits as $x \to x_0,$ $x \to x_0^+,$ $x \to x_0^-,$ and $x \to \infty;$ as noted before, we indicate this by adding an "etc."to the relevant phrase: "for $x \underset{\neq}{\approx} x_0,$ etc.", "as $x \to x_0,$ etc."

Principle 11.3A   Error form for limit. Write $f(x) = L + e(x).$ Then

(11) $\qquad\qquad f(x) \to L \;\Leftrightarrow\; e(x) \to 0,\qquad$ as $x \to x_0,$ etc.

Just as in Chapter 4, the proof is immediate, but the point of view that the error form introduces can be significant. We will use it shortly.

Principle 11.3B   The K-ϵ principle for limits of functions.

If you can prove, for some K not depending on x or ϵ, that

(12)   given $\epsilon > 0$, $f(x) \underset{K\epsilon}{\approx} L$,   for $x \underset{\neq}{\approx} x_0$, etc.,

then $f(x) \to L$ as $x \to x_0$.

Theorem 11.3A   Algebraic limit theorems.

If a, b are constants, and $f(x) \to L$, $g(x) \to M$ as $x \to x_0$, etc.,

(13)   **Linearity theorem** $af(x) + bg(x) \to aL + bM$ as $x \to x_0$;

(14)   **Product theorem** $f(x) \cdot g(x) \to L \cdot M$ as $x \to x_0$;

(15)   **Quotient theorem** $f(x)/g(x) \to L/M$ as $x \to x_0$;

(for the Quotient Theorem, assume $g(x) \neq 0$ for $x \underset{\neq}{\approx} x_0$, and $M \neq 0$.)

Proofs.   These go exactly like the proofs in Chapter 5 of the corresponding theorems for sequences. To prove (14) for example, use the error form (11): writing e and e' for the two errors (we omit the variable x), we have

(16) $f \cdot g \;=\; (L+e)(M+e') = L \cdot M + (Me + Le' + ee')$.

We have to show the error term in parentheses is small. By hypothesis, $e \to 0$ and $e' \to 0$ as $x \to x_0$. Therefore,

given $\epsilon > 0$ (and < 1, say), $|e| < \epsilon$ and $|e'| < \epsilon$ for $x \underset{\neq}{\approx} x_0$;

so that for the error term in (16) we get the estimation

$$|Me + Le' + ee'| < (|M| + |L| + 1)\epsilon \quad \text{for } x \underset{\neq}{\approx} x_0 \; ;$$

this proves the error term has limit 0, according to the definition of limit and the K-ϵ principle (here $K = |M| + |L| + 1$); this proves (14). $\square$

Just as for sequences, there are analogues of the preceding for infinite limits, which we state so as to have them to refer to.

Theorem 11.3A∞ Infinite limit theorems.

In the statements below, the limits are taken as $x \to x_0$, etc., while the properties are assumed to hold for $x \underset{\neq}{\approx} x_0$, etc.

(17) $f(x) \to \infty$, $\begin{cases} g(x) \to \infty, \text{ or} \\ g(x) \text{ bounded below} \end{cases}$ $\Rightarrow$ $f(x) + g(x) \to \infty$.

(18) $f(x) \to \infty$, $\begin{cases} g(x) \to L > 0, \text{ or} \\ g(x) > k > 0 \text{ for some } k \end{cases}$ $\Rightarrow$ $f(x)g(x) \to \infty$.

(19) $f(x) \to \infty$ $\Rightarrow$ $\dfrac{1}{f(x)} \to 0$;   if $f(x) > 0$, the converse is true.

Proofs.   (17) is Question 11.2/2; the others are left as exercises.

Theorem 11.3B   Squeeze theorem.

Suppose $f(x) \leq g(x) \leq h(x)$ for $x \underset{\neq}{\approx} x_0$, etc. Then:

$$f(x) \to L \text{ and } h(x) \to L \text{ as } x \to x_0 \quad \Rightarrow \quad g(x) \to L \text{ as } x \to x_0 .$$

Theorem 11.3B∞   Squeeze theorem for infinite limits.

Suppose $f(x) \geq g(x)$ for $x \underset{\neq}{\approx} x_0$, etc. Then

$$\lim_{x \to x_0} g(x) = \infty \quad \Rightarrow \quad \lim_{x \to x_0} f(x) = \infty.$$

Proofs.

The proof of the second theorem is immediate from the definitions.

The proof of the first is like the proof for sequences (Theorem 5.2). Briefly, given $\epsilon > 0$, we get the following chain of inequalities, in which the first and last inequalities come from our hypotheses $f(x) \underset{\epsilon}{\approx} L$ and $h(x) \underset{\epsilon}{\approx} L$:

$$L - \epsilon \;\leq\; f(x) \;\leq\; g(x) \;\leq\; h(x) \;\leq\; L + \epsilon, \quad \text{for } x \underset{\neq}{\approx} x_0.$$

These inequalities show $g(x) \underset{\epsilon}{\approx} L$ for $x \underset{\neq}{\approx} x_0$, i.e., $g(x) \to L$ as $x \to x_0$. □

Example 11.3A   Show that $x^{1/n} \to 1$ as $x \to 1$.

Solution. Use the Squeeze Theorem, making two cases, $x > 1$ and $x < 1$:

1	$<$	$x^{1/n}$	$<$	x	if $x > 1$;		x	$<$	$x^{1/n}$	$<$	1	if $x < 1$.
$\downarrow$		$\Downarrow$		$\downarrow$			$\downarrow$		$\Downarrow$		$\downarrow$	
1		1		1			1		1		1	

The first case shows that the right-hand limit is 1, the second that the left-hand limit is 1; therefore the limit is 1 by (8). □

Example 11.3B   Let $f(x) = \displaystyle\int_1^x \frac{\sqrt{1+t}}{t}\, dt$. Show $f(x) \to \infty$ as $x \to \infty$.

Solution. It is not necessary to evaluate the integral explicitly; by the Infinite Squeeze Theorem, only a lower estimate for $f(x)$ is needed. We have

$$\frac{\sqrt{1+t}}{t} \;\geq\; \frac{\sqrt{t}}{t} \;=\; t^{-1/2}, \qquad t > 0 .$$

Integrating both sides from 1 to x, the inequality is preserved (10.2, (6)); we get

$$f(x) \;\geq\; 2\sqrt{x} - 2,$$

so $f(x) \to \infty$ as $x \to \infty$ by Theorem 11.3B∞ above. □

Finally, we have the analogues for limits of functions of the two Location Theorems we proved earlier in Chapter 5 for sequences. In both of them, we assume the limit exists, which implies that the function has to be defined for $x \underset{\neq}{\approx} x_0$, etc.

Theorem 11.3C   Limit location (for functions).   *If the limits exist,*

(20) $$ f(x) \leq M \quad \text{for } x \underset{\neq}{\approx} x_0 \quad \Rightarrow \quad \lim_{x \to x_0} f(x) \leq M \; ; $$

(21) $$ f(x) \leq g(x) \quad \text{for } x \underset{\neq}{\approx} x_0 \quad \Rightarrow \quad \lim_{x \to x_0} f(x) \leq \lim_{x \to x_0} g(x) \, . $$

Theorem 11.3D   Function location theorem.   *If the limit exists,*

(22) $$ \lim_{x \to x_0} f(x) < M \quad \Rightarrow \quad f(x) < M \quad \text{for } x \underset{\neq}{\approx} x_0 \, . $$

Proofs. Same as the corresponding proofs for sequences (5.3A and 5.3B).   □

Example 11.3C   Let $f(x) = \displaystyle\int_0^x \frac{dt}{\sqrt{1 - t^4}}$.   Estimate $\displaystyle\lim_{x \to 1-} f(x)$ from above.

Solution. We estimate the integrand by using $\sqrt{1 - t^2}$. Since $0 \leq t < 1$ we have

$$ t^4 \leq t^2 \quad \Rightarrow \quad \sqrt{1 - t^4} \geq \sqrt{1 - t^2} $$

$$ \Rightarrow \quad 1/\sqrt{1 - t^4} \leq 1/\sqrt{1 - t^2} \; ; $$

$$ \Rightarrow \quad \int_0^x \frac{dt}{\sqrt{1 - t^4}} \leq \int_0^x \frac{dt}{\sqrt{1 - t^2}} \; ; $$

$$ = \sin^{-1} x \leq \frac{\pi}{2} , \quad \text{for } 0 < x < 1. $$

Thus $\displaystyle\lim_{x \to 1-} f(x) < \frac{\pi}{2}$,   by (20) above.   □

Example 11.3D   Let $f(x) = \dfrac{x^3 + 9}{1 - x^2 - x^3}$. Show $f(x) < -.9$, for $x \gg 1$.

Solution. By the Algebraic Limits Theorem 11.3A, we have

$$ \lim_{x \to \infty} f(x) = \lim_{x \to \infty} \frac{1 + 9/x^3}{1/x^3 - 1/x - 1} = -1. $$

Therefore $f(x) < -.9$ for $x \gg 1$, by the Function Location Theorem 11.3D.   □

Questions 11.3

1. If $\displaystyle\lim_{x \to \infty} f(x) = 0$, then $\displaystyle\lim_{x \to \infty} f(x) \sin x = 0$. Criticize and repair the following "proof":

$$ -1 \leq \sin x \leq 1 \; \Rightarrow \; -f(x) \leq f(x) \sin x \leq f(x); $$

now apply the Squeeze Theorem 11.3B.

2. Let $f(x) = a_0 x^n + a_1 x^{n-1} + \ldots + a_n$. Prove, using theorems:

(a) $a_n > 0 \; \Rightarrow \; f(x) > 0$ for $x \approx 0$;   (b) $a_0 > 0 \; \Rightarrow \; f(x) > 0$ for $x \gg 1$.

3. Prove the Limit Location Theorem (20). (Let $L = \lim f(x)$.)

11.4   Limits and continuous functions.

We return now to the study of continuous functions, making use of these ideas about limits. First comes a simple reformulation of continuity.

Theorem 11.4A   Limit form of continuity. *Let $f(x)$ be defined for $x \approx x_0$.*

$$f(x) \text{ is continuous at } x_0 \quad \Leftrightarrow \quad \lim_{x \to x_0} f(x) = f(x_0) .$$

Proof.   Comparing the statement above with Definition 11.1A of continuity and Definition 11.2A of limit, evidently what we must show is: given $\epsilon > 0$,

$$f(x) \underset{\epsilon}{\approx} f(x_0) \quad \text{for } x \approx x_0 \quad \Leftrightarrow \quad f(x) \underset{\epsilon}{\approx} f(x_0) \quad \text{for } x \underset{\neq}{\approx} x_0 .$$

The direction $\Rightarrow$ is trivial; so is $\Leftarrow$, since $f(x) = f(x_0)$ if $x = x_0$. $\qquad \square$

The reason for reformulating continuity in terms of limits is that it lets us make use of the limit theorems. Here are some examples.

Theorem 11.4B   Positivity theorem for continuous functions.

$$f(x) \text{ continuous at } x_0, \quad f(x_0) > 0 \quad \Rightarrow \quad f(x) > 0 \ \text{ for } x \approx x_0.$$

First proof.   Using 11.4A, the hypothesis says $\lim_{x \to x_0} f(x) > 0$. But according to the Function Location Theorem 11.3D,

$$\lim_{x \to x_0} f(x) > 0 \quad \Rightarrow \quad f(x) > 0 \ \text{ for } x \underset{\neq}{\approx} x_0 ;$$

this holds for $x = x_0$ as well, since by hypothesis, $f(x_0) > 0$. $\qquad \square$

Second proof.   Since we did not prove the Function Location Theorem, let's prove 11.4B from scratch.

Choose an ϵ so that $0 < \epsilon < f(x_0)$. Since f is continuous at x_0,

$$f(x) \underset{\epsilon}{\approx} f(x_0), \quad \text{for } x \approx x_0,$$

which when combined with the previous inequality, says

$$0 \ < \ f(x_0) - \epsilon \ < \ f(x) \ < \ f(x_0) + \epsilon, \qquad \text{for } x \approx x_0 .$$

The two left-hand inequalities prove the theorem. $\qquad \square$

As another example of the use of limits to establish results about continuity, Theorem 11.3A for limits of sums, products, and quotients implies almost immediately the corresponding results for continuous functions:

Theorem 11.4C   Algebraic operations on continuous functions.

If f and g are continuous at x_0, and a, b are constants, then the following functions are also continuous at x_0:

$$af + bg, \qquad f \cdot g, \qquad f/g \quad (\text{if } g(x_0) \neq 0).$$

Proof. Using the Limit Form of Continuity 11.4A, the three statements follow immediately from the three corresponding statements in the Algebraic Limits Theorem 11.3A.

For the quotient statement, we must also verify that $g(x) \neq 0$ for $x \approx x_0$. This follows from the Positivity Theorem 11.4B:

$$g(x_0) \neq 0 \;\Rightarrow\; g(x_0) > 0 \text{ or } g(x_0) < 0;$$
$$\Rightarrow\; g(x) > 0 \text{ or } g(x) < 0 \quad \text{for } x \approx x_0, \quad \text{by 11.4B};$$
$$\Rightarrow\; g(x) \neq 0 \quad \text{for } x \approx x_0. \qquad \square$$

Theorem 11.4C extends immediately to intervals, because continuity is a local property. *On an interval I, the sum, product, and quotient (where defined) of continuous functions is again continuous.*

Try to remember this. Even at the end of a semester of steady work with continuous functions, there will be analysis students who when asked, say, why

$$f(x) \;=\; \frac{x^3 + x - 1}{1 + x^2}$$

is continuous, will assume the fetal position and try to prove directly that $f(x) \underset{\epsilon}{\approx} f(x_0)$ for $x \approx x_0$. Theorems are there to save you from working; grown-ups cite theorems.

Examples 11.4A

(a) Any polynomial $a_o x^n + \ldots + a_n$ is continuous for all x; namely, since the functions 1 and x are continuous for all x, so are $x^2, x^3, \ldots$ by the Product Theorem, and so therefore are the polynomials, by the Linearity Theorem.

(b) All rational functions—those functions writable as quotients $p(x)/q(x)$ of two polynomials—are continuous, except at the finite number of points where the denominator is 0; this follows from (a) above and the quotient statement in Theorem 11.4C.

As another example of how the limit form of continuity is helpful, we use it to describe mathematically the types of discontinuity we talked about in Section 11.1. Earlier we illustrated these different types by pictures; how can we describe them analytically, using limits?

1. Removable discontinuity $\lim\limits_{x \to x_0} f(x) = L$, but $L \neq f(x_0)$.

Usually what happens is that $f(x)$ is undefined at x_0. In any event, the discontinuity can be removed by defining (or redefining) $f(x_0)$ to be L. For example,

$$f(x) = \frac{x^2 - 1}{x - 1} \qquad \text{has a removable discontinuity at } x = 1;$$
$$\text{remove it by defining } f(1) = 2;$$

$$f(x) = \frac{\sin x}{x} \qquad \text{has a removable discontinuity at } x = 0;$$
$$\text{remove it by defining } f(0) = 1 \text{ (see Exercise 11.3/1).}$$

2. Jump discontinuity $\lim\limits_{x \to x_0^+} f(x) \neq \lim\limits_{x \to x_0^-} f(x)$, but both limits exist.

(a) $\operatorname{sgn} x = \begin{cases} 1 & \text{for } x > 0 \\ -1 & \text{for } x < 0 \end{cases}$ has a jump discontinuity at 0, since
$$\lim_{x \to 0+} f(x) = 1, \quad \lim_{x \to 0-} f(x) = -1.$$

(b) $f(x) = \dfrac{|x^2 - 1|}{x - 1} = \dfrac{|x - 1||x + 1|}{x - 1}$ has a jump discontinuity at 1, since
$$\lim_{x \to 1+} f(x) = \lim_{x \to 1+} |x + 1| = 2; \quad \lim_{x \to 1-} f(x) = \lim_{x \to 1-} -|x + 1| = -2.$$

3. Infinite discontinuity The right- or left-hand limit is ∞ or $-\infty$.

(a) $1/x^2$ at 0; here $\lim\limits_{x \to 0} 1/x^2 = \infty$.

(b) $1/x$ at 0, $\tan x$ at $\pi/2$; in both of these the two one-sided limits are ∞ and $-\infty$ (so that $\lim f(x) \neq \infty$, despite what many seem to think.)

4. Essential discontinuity Any discontinuity not of the preceding three types; for example: $\sin(1/x)$ at 0 (cf. 11.1, p. 154).

We have shown the standard arithmetic operations preserve continuity; we show finally this is also true of the operation which composes functions.

Theorem 11.4D Composition theorem. *Let* $x = g(t)$, $x_0 = g(t_0)$;
$$\left. \begin{array}{l} g(t) \text{ continuous at } t_0 \\ f(x) \text{ continuous at } x_0 \end{array} \right\} \quad \Rightarrow \quad f(g(t)) \text{ continuous at } t_0 \,.$$

Proof.   Given $\epsilon > 0$, there is a $\delta > 0$ such that
$$f(x) \underset{\epsilon}{\approx} f(x_0) \qquad \text{for } x \underset{\delta}{\approx} x_0, \qquad \text{(continuity of } f \text{ at } x_0\text{)}. \text{ Also,}$$
$$g(t) \underset{\delta}{\approx} g(t_0) \qquad \text{for } t \approx t_0, \qquad \text{(continuity of } g \text{ at } t_0\text{)};$$
$$x \underset{\delta}{\approx} x_0 \qquad \text{for } t \approx t_0, \qquad \text{(since } g(t) = x\text{)}.$$

We combine the third line with the first line, substituting $g(t)$ for x; we get:
$$\text{given } \epsilon > 0, \quad f(g(t)) \underset{\epsilon}{\approx} f(g(t_0)) \quad \text{for } t \approx t_0. \qquad \square$$

Notice that the above is one of the times where we need the δ explicitly. The theorem implies its analog about continuity on an interval:

Theorem 11.4D$'$ *Let* $x = g(t)$, *and suppose* $g(a) \leq g(t) \leq g(b)$ *for* $a \leq t \leq b$.
$$\left. \begin{array}{l} g(t) \text{ continuous on } [a, b] \\ f(x) \text{ continuous on } [g(a), g(b)] \end{array} \right\} \quad \Rightarrow \quad f(g(t)) \text{ continuous on } [a, b] \,.$$

The composition of continuous functions is therefore continuous, wherever it makes sense (i.e., on any interval over which it is defined).

Examples 11.4B

(a) $\sin x$ is continuous, by Examples 11.1B;

$\cos x = \sin(x + \pi/2)$ is continuous by Theorem 11.4D$'$;

$\tan x = \sin x / \cos x$, $\sec x$, and $\csc x$ are continuous wherever they are defined, since they are quotients of continuous functions.

(b) Other composite functions like $\sin(x^2 + 1)$, $\cos^3(1/x)$ are also continuous by the Composition Theorem 11.4D$'$.

Questions 11.4

1. Is the Positivity Theorem true if $<$ is replaced throughout by $\leq$?

2. Assuming $f(x)$ is continuous at $x = a$, formulate a sort of converse to Theorem 11.4B, and prove it. (The actual converse is trivial.))

3. Prove that if $f(x)$ is continuous, then $|f(x)|$ is continuous. Is the converse true? Give a proof or a counterexample.

4. (a)   Four continuous functions $y = f(x)$ satisfy $y^2 = x^2$. What are they?

(b)   Infinitely many continuous functions $y = f(x)$ satisfy $y^2 = 1 + \sin x$; describe some of them.

5. Suppose $f(x)$ has a jump discontinuity at $x = a$. Write down a formula for the number measuring the size and direction of the jump.

11.5   Continuity and sequences.

In our work with sequences, we have taken for granted statements like
$$\sin 1/n \to 0 \quad \text{and} \quad e^{1/n} \to 1, \quad \text{as } n \to \infty .$$
One of the exercises called for a direct proof that if $a_n \to L$, then $\sqrt{a_n} \to \sqrt{L}$.

In each of these, there was a convergent sequence $x_n \to a$ and a function $f(x)$; what we took for granted or were asked to prove in each case was that
$$f(x_n) \to f(a).$$
What makes this true is the *continuity* of $f(x)$ at a, as we shall now see.

Theorem 11.5   Sequential continuity theorem.
$$x_n \to a, \qquad f(x) \text{ continuous at } a \quad \Rightarrow \quad f(x_n) \to f(a) .$$

Proof.   By hypothesis, given $\epsilon > 0$, there is a $\delta > 0$ such that

$f(x) \underset{\epsilon}{\approx} f(a)$ for $x \underset{\delta}{\approx} a$,   since $f(x)$ is continuous at a.   Also,

$x_n \underset{\delta}{\approx} a$ for $n \gg 1$,   since $x_n \to a$.   Thus

$f(x_n) \underset{\epsilon}{\approx} f(a)$ for $n \gg 1$;   this shows $f(x_n) \to f(a)$. $\square$

Remarks.   Once again, we need the δ explicitly. Note the similarity of the proof to the proof of Theorem 11.4D. Formally, we can think of $f(x_n)$ as the composition of the functions $f(x)$ and $x_n = x(n)$, so that the two theorems are almost the same.

The theorem and its proof also apply to one-sided continuity, by making minor modifications: if $f(x)$ is right-hand continuous at a, the theorem would say

$$x_n \geq a, \quad x_n \to a, \quad f(x) \text{ continuous at } a^+ \Rightarrow f(x_n) \to f(a).$$

The theorem can be used to show a function $f(x)$ is not continuous at a, by giving a sequence x_n such that $x_n \to a$, but $\lim f(x_n) \neq f(a)$. This approach to discontinuities avoids a lot of negative statements. Here is an example.

Example 11.5   Show that $f(x) = \cos 1/x$ has an essential discontinuity at 0.

Solution.   It suffices to show $\lim\limits_{x \to 0^+} f(x)$ does not exist (nor is it ∞): then by the results in Section 11.4, $f(x)$ cannot be continuous at a, nor can it have a removable, jump, or infinite discontinuity at a.

The limit cannot be ∞, since $|\cos 1/x| \leq 1$ for all $x \neq 0$.

Reasoning indirectly, suppose $\lim\limits_{x \to 0^+} f(x) = L$. Define $f(0) = L$; then the function becomes right-hand continuous at 0. Now consider the two sequences

$$x_n = \frac{1}{2n\pi}, \quad (f(x_n) = 1 \text{ for all } n); \quad x_n' = \frac{1}{(2n+1)\pi}, \quad (f(x_n') = -1 \text{ for all } n).$$

Since $x_n \to 0^+$ and $x_n' \to 0^+$, Theorem 11.5 says (using right-hand continuity),

$$f(x_n) \to f(0); \qquad\qquad f(x_n') \to f(0);$$

comparing with the above, we see that $f(0) = 1$ and $f(0) = -1$, and we have our contradiction. Thus the limit does not exist, and the discontinuity is essential. $\square$

If $f(x)$ is not defined at a, the following version of the Sequential Continuity Theorem, phrased in terms of limits, can be used. In it, we must assume $x_n \neq a$ for all n since otherwise $f(x_n)$ might not be defined.

Theorem 11.5A   **Limit Form of Sequential Continuity.**

    If $x_n \to a$, $x_n \neq a$, and $\lim\limits_{x \to a} f(x) = L$,   then $\lim\limits_{n \to \infty} f(x_n) = L$.

The proof is more or less immediate, using the Limit Form of Continuity; we leave it for the Exercises. The theorem has a sort of converse:

Theorem 11.5B   *Let $f(x)$ be defined for $x \underset{\neq}{\approx} a$, and suppose that for all $\{x_n\}$ such that $x_n \to a$, $x_n \neq a$, we have $\lim\limits_{n \to \infty} f(x_n) = L$. Then $\lim\limits_{x \to a} f(x) = L$.*

We don't need this theorem and shall skip the proof, since it is indirect and involves negating the definition of limit; it is left as a Problem. Theorem 11.5B is not very useful in proving $\lim f(x)$ exists, since one has to examine all sequences converging to a. It is however of theoretical interest, since it gives a way of using the limit theorems for sequences (Sections 5.1-5.3) to prove the limit theorems for functions (Section 11.3). See Exercises 11.5 for an example.

Questions 11.5

1. (a) Use the Sequential Continuity Theorem to show that the function
defined by $f(x) = \begin{cases} 1, & x = 1/n; \\ 0, & \text{elsewhere} \end{cases}$ is discontinuous at $x = 0$.

(b) Show that it cannot be made continuous by altering $f(0)$.

2. In the statement of the Sequential Continuity Theorem, why isn't it necessary to assume that the $f(x_n)$ are defined?

Exercises

11.1

1. Prove from the definition that $\dfrac{x}{1+x}$ is continuous at $x = 1$.

2. Prove that $\cos x$ is continuous directly from its trigonometric definition (i.e., as in Example 11.1B).

3. Give another proof that $\sin x$ is continuous based on the trigonometric identity in Exercise 9.2/4 (change b to $-b$). (Use: $|\sin u| \le |u|$ for all u.)

4. Using the exponential law, show that if e^x is continuous at 0, it is continuous for all x.

5. Prove: if $f(x)$ is continuous at x_0, it is locally bounded at x_0.

6. Prove as in Example 11.1C that $f(x) = \displaystyle\int_0^\pi \dfrac{\cos xt}{1+t}\,dt$ is continuous at $x = 0$. (You will need to estimate $1 - \cos a$; use the picture in Example 11.1B.)

11.2

1. Prove from the definition of limit that $\displaystyle\lim_{x \to 0} \dfrac{1-x}{x^2+1} = 1$.

2. Let $f(x)$ be even; prove that $\displaystyle\lim_{x \to 0^+} f(x) = L \Rightarrow \lim_{x \to 0} f(x) = L$.

3. Prove: if $f(x) \to \infty$ as $x \to a$, and $g(x) \ge b > 0$ for $x \approx a$, then $f(x)g(x) \to \infty$ as $x \to a$.

11.3

1. (a) Prove that $\displaystyle\lim_{x \to 0^+} \dfrac{\sin x}{x} = 1$ by first establishing the inequalities

$$\cos x \le \frac{\sin x}{x} \le \frac{1}{\cos x} \quad \text{for } x \approx 0^+ \underset{\neq}{} .$$

(Hint: compare the area of the circular sector with
the area of the two triangles.)

(b) Deduce that $\displaystyle\lim_{x \to 0} \dfrac{\sin x}{x} = 1$ by using one of the previous exercises.

2. Deduce from the limit in the preceding exercise that $\displaystyle\lim_{x \to 0} \dfrac{1 - \cos x}{x} = 0$.

(Use: $\sin^2 x = 1 - \cos^2 x$.)

3. Prove by using the Squeeze Theorem, or otherwise:

(a) $\displaystyle \lim_{x \to 0^+} \int_0^\pi \frac{\sin t}{1 + xt} \, dt = 2$ (b) $\displaystyle \int_0^1 \frac{t^2}{1 + t^4 x} \, dt$ is continuous at 0 .

(Warning: you can't do these by moving lim to the right of the integral sign! At this point we don't have any theorems telling us this is legitimate. One of the goals of the book is to see when this is legal.)

4. Prove the Function Location Theorem 11.3D.

5. Prove (19): $f(x) \to \infty \Rightarrow 1/f(x) \to 0$; the converse requires additional hypotheses: give them.

6. Let $f(x) = a_0 x^n + a_1 x^{n-1} + \ldots + a_n$, where $n \geq 1$. Using limit theorems as much as possible, prove that

(a) $\displaystyle \lim_{x \to \infty} f(x) = \infty$, if $a_0 > 0$; $\displaystyle \lim_{x \to \infty} f(x) = -\infty$, if $a_o < 0$.

(b) if $a_0 > 0$, then $f(x) > 0$ for $x \gg 1$. (Use part (a).)

11.4

1. Prove: $f(x) = \sqrt{x} \cos(1/x)$ is continuous at 0 if we define $f(0) = 0$.

2. If $f(x)$ is continuous on $[a, b]$ and strictly increasing on (a, b), prove it is strictly increasing on $[a, b]$.

3. If $f(x)$ has discontinuities, in general so will $f(g(x))$. For example, suppose $f(x) = \text{sgn } x$ (see Question 9.2/1)) and $g(x)$ is a polynomial. Where are the discontinuities of sgn $(g(x))$, and what kind are they? (Begin by taking very simple examples for $g(x)$ until you think you see what is going on.)

4. Prove that $\max(f, g)$ and $\min(f, g)$ are continuous, if f and g are (cf. Exercise 2.4/1, Question 11.4/3).

11.5

1. (a) Prove that if $f(x)$ is continuous, and $f(x) = 0$ when x is a rational number, then $f(x) = 0$ for all x.

(b) Prove that if $f(x)$ and $g(x)$ are continuous, and $f(x) \leq g(x)$ when x is a rational number, then $f(x) \leq g(x)$ for all x. Show that $\leq$ cannot be replaced by $<$ throughout.

2. Prove that $\displaystyle \lim_{x \to \infty} \sin x$ does not exist, by using the Limit Form of the Sequential Continuity.

3. State and prove the analogue of Theorem 11.5A (the Limit Form of Sequential Continuity) when $L = \infty$.

4. Let $f(x) = \begin{cases} 0, & \text{if } x \text{ is rational}; \\ 1, & \text{if } x \text{ is irrational}. \end{cases}$ Prove $f(x)$ is not continuous at any point a. Theorem 2.5, or the ideas in its proof, will be helpful.)

5. (a) Prove the Limit Form of Sequential Continuity 11.5A from scratch.

(b) Deduce it from the Sequential Continuity Theorem 11.5.

6. By using Theorems 11.5A and 11.5B, deduce each of these limit theorems for functions from the corresponding limit theorem for sequences:

(a) the Product Theorem 11.3A (14) from the Product Theorem 5.1 (2);

(b) the Squeeze Theorem 11.3B from the Squeeze Theorem 5.2 .

Problems

11-1 Suppose $f(x)$ is continuous for all x and $f(a+b) = f(a) + f(b)$ for all a and b. Prove that $f(x) = Cx$, where $C = f(1)$, as follows:

(a) prove, in order, that it is true when $x = n$, $1/n$, and m/n, where m, n are integers, $n \neq 0$;

(b) use the continuity of f to show it is true for all x.

11-2 Call a function "multiplicatively periodic" if there is a positive number $c \neq 1$ such that $f(cx) = f(x)$ for all $x \in \mathbb{R}$. Prove that if a multiplicatively periodic function is continuous at 0, then it is a constant function.

11-3 Define the "ruler function" $f(x)$ as follows:

$$f(x) = \begin{cases} 1/2^n, & \text{if } x = b/2^n \text{ for some odd integer } b; \\ 0, & \text{otherwise.} \end{cases}$$

(a) Prove that $f(x)$ is discontinuous at the points $b/2^n$, (b odd).

(b) Prove $f(x)$ is continuous at all other points.

11-4 Prove Theorem 11.5B by contraposition; in other words, show that if $\lim_{x \to a} f(x) \neq L$, then it is possible to find a sequence x_n such that $x_n \to a$, yet $\lim_{n \to \infty} f(x_n) \neq L$.

(Hint: if you wish to try this problem, it is essential that you work with a clear version of the hypothesis $\lim_{x \to a} f(x) \neq L$. To get this, you will need to study Appendix B first; don't attempt the problem without doing this.)

Answers to Questions

11.1

1. (a) Given $c \in (-\infty, \infty)$, choose any $a > 0$ such that $|c| < a$.

Since x^2 is continuous on $(-a, a)$, it is continuous at $c \in (-a, a)$. This shows x^2 is continuous for all x.

No, the proof requires an upper estimate for $|x + x_0|$, which does not exist if the interval $I = (-\infty, \infty)$.

(b) No; x^2 is bounded on $(-a, a)$, but not on $(-\infty, \infty)$. The difference is that continuity is a local property, so if it holds on every $(-a, a)$, it holds on $(-\infty, \infty)$. Boundedness however is a global property.

2. Given $0 < \epsilon < 1$, $\left| \dfrac{1}{x} - \dfrac{1}{2} \right| = \dfrac{|x - 2|}{2|x|} < \dfrac{|x - 2|}{2} < \dfrac{\epsilon}{2}$, if $x \underset{\epsilon}{\approx} 2$;

note that $x \underset{\epsilon}{\approx} 2 \Rightarrow x > 1$, which justifies the next-to-last inequality above.

3. No; the function $f(x) = 1, \ x \geq 0; \ = -1, \ x < 0$ is a counterexample.

11.2

1. We know that $|g(x)| < B$ for $x \approx a$. Since $\lim\limits_{x \to a} f(x) = 0$, we can say: given $\epsilon > 0$, $|f(x)| < \epsilon/B$ for $x \underset{\neq}{\approx} a$; therefore

$|f(x)g(x)| = |f(x)||g(x)| < (\epsilon/B) \cdot B = \epsilon$, for $x \underset{\neq}{\approx} a$; i.e., $\lim\limits_{x \to a} f(x)g(x) = 0$.

2. Given $b > 0$, the hypotheses give the first two inequalities below:

$$g(x) > B \qquad \text{for } x \approx a;$$
$$f(x) > b - B \qquad \text{for } x \underset{\neq}{\approx} a; \quad \text{adding,}$$
$$f(x) + g(x) > b \quad \text{for } x \underset{\neq}{\approx} a, \quad \text{as was to be proved.}$$

3. False; take $f(x) = (-1)^n/n$, for $x \in [1/(n+1), \ 1/n)$.

4. $g(x) \geq k$ for some fixed $k > 0$ (weaker than $g(x) \to \infty$ or $g(x) \to L > 0$)

11.3

1. What if $f(x) < 0$? Use instead $|f(x)| < \epsilon \Rightarrow |f(x) \sin x| < \epsilon$.

2. (a) $\lim\limits_{x \to 0} f(x) = a_n$, by 11.3A; finish by 11.3D.

(b) In outline: write $f(x) = x^n g(x)$; then $\lim\limits_{x \to \infty} g(x) = a_0$, by 11.3A, so $g(x) > 0$ for $x \gg 1$, by 11.3D.

3. Given $\epsilon > 0$, the hypotheses give: $L - \epsilon < f(x) \leq M$ for $x \underset{\neq}{\approx} x_0$; therefore $L - \epsilon < M$ for all $\epsilon > 0$, whence $L \leq M$ (use contraposition).

11.4

1. No; $f(x) = x$ at 0 is a counterexample.

2. If $f(x)$ is continuous and $f(x) \geq 0$ for $x \underset{\neq}{\approx} a$, then $f(a) \geq 0$.

(The first $\geq$ can be made $>$, but you get a weaker theorem.) The proof is immediate from Theorem 11.4A and the Limit Location Theorem.

3. $|x|$ is continuous (this is obvious if $x \neq 0$, since continuity is a local property and both x and $-x$ are continuous; for $x = 0$, it follows immediately from Definition 11.1A). Therefore $|f(x)|$ is continuous by Theorem 11.4D$'$. The converse is false: take $f(x)$ as in Answer 11.1/3 above.

4. (a) $y = |x|$ is one. (b) Any combination of positive arches from $1 + \sin x$ and negative arches from $-(1 + \sin x)$ will be continuous.

5. $\displaystyle\lim_{x \to a+} f(x) - \lim_{x \to a-} f(x)$

11.5

1. (a) $\displaystyle\lim_{n \to \infty} 1/n = 0$ but $\displaystyle\lim_{n \to \infty} f(1/n) = 1 \neq f(0)$; therefore it cannot be continuous at 0 by Theorem 11.5.

(b) Use the sequences $1/n$ and say π/n. Both $\to 0$, and $f(1/n) = 1$, while $f(\pi/n) = 0$. Thus if $f(x)$ could be made continuous at 0 by choosing $f(0)$ correctly, we would have the contradiction

$$\lim_{n \to \infty} f(1/n) = 1 = f(0) \; ; \qquad \lim_{n \to \infty} f(\pi/n) = 0 = f(0) \; .$$

2. Since $f(x)$ is continuous at a, it is defined in some δ-neighborhood of a, and the x_n will lie in this δ-neighborhood for $n \gg 1$, by the Sequence Location Theorem.

Thus $f(x_n))$ will be defined at least for $n \gg 1$, so the statement involving the limit makes sense.

12

The Intermediate Value Theorem

12.1 The existence of zeros.

Continuity is a local property of functions. But in this chapter and the next, we will prove it implies certain global properties, basic to analysis and its applications. We begin with one such application: root-finding.

The oldest use of algebra is to solve one-variable equations: $f(x) = 0$. Solving means finding the *zeros* of $f(x)$: those real c for which $f(c) = 0$. In increasing detail, we can ask about

(i) *existence*: are there any zeros?
(ii) *number*: are there finitely many? how many, or about how many?
(iii) *approximate location*: find small intervals containing only one zero;
(iv) *calculation*: determine the zero "exactly", or to a given accuracy.

The main theorem of this section, Bolzano's Theorem, is applicable to all of these questions. It is about continuous functions which change sign.

Definition 12.1   We say $f(x)$ *changes sign* on the closed finite interval $[a, b]$ if it is defined on this interval and has opposite signs at a and b:

$$f(a) < 0, \quad f(b) > 0 \quad \text{or} \quad f(a) > 0, \quad f(b) < 0 .$$

(The efficiency experts write the condition: $f(a)f(b) < 0$.)

If $f(x)$ changes sign on $[a, b]$, one might think it should have a zero on $[a, b]$, but the picture gives a counterexample by using a discontinuous function. Bolzano's Theorem says that if $f(x)$ doesn't cheat this way by jumping, it will have a zero on the interval.

Theorem 12.1   Bolzano's Theorem.   *Let $f(x)$ be continuous on $[a, b]$.*

$$f(x) \text{ changes sign on } [a, b] \quad \Rightarrow \quad f(x) \text{ has a zero on } [a, b] .$$

The conclusion asserts that a certain real number exists, so we can expect that the proof will require the Completeness Property. We will use it in the form of the Nested Interval Theorem 6.1.

To construct the nested intervals, we will use the same bisection process used in 6.3 to prove the Bolzano-Weierstrass Theorem. This bisection method is not just something needed for proofs: using a calculator or computer, it is actually an effective, easily managed, popular method for finding the zeros.

Proof. We consider the case where $f(x)$ changes from $-$ to $+$, proving

(1) $f(a) < 0,$ $f(b) > 0$ $\Rightarrow$ $f(c) = 0$ for some c in $[a, b]$.

Bisection will give a sequence $[a_i, b_i]$ of nested intervals such that $f(x)$ changes from $-$ to $+$ on each; c will be the unique point inside all of them.

The starting interval $[a_0, b_0]$ is just $[a, b]$ itself. To get the next interval in the sequence, divide $[a_0, b_0]$ in two by its midpoint x_0; then choose as $[a_1, b_1]$ the half-interval on which $f(x)$ goes from $-$ to $+$:

$$\text{if } f(x_0) > 0, \quad \text{let } [a_1, b_1] = [a, x_0] ;$$

$$\text{if } f(x_0) < 0, \quad \text{let } [a_1, b_1] = [x_0, b] .$$

In either case, we have $f(a_1) < 0$. $f(b_1) > 0$. This gives a new interval $[a_1.b_1]$ of half the length, on which $f(x)$ still changes sign from $-$ to $+$.

(If at the midpoint we find that $f(x_0) = 0$, the above doesn't apply, but in that case we can stop and pack up: we've found a zero.)

We continue this process with $[a_1, b_1]$, bisecting it and choosing as $[a_2, b_2]$ the half on which $f(x)$ goes from $-$ to $+$. If at any stage the midpoint is a zero of $f(x)$, we are done; if not, we get an infinite sequence of nested intervals

$$[a, b] \supset [a_1, b_1] \supset [a_2, b_2] \supset \ldots \supset [a_n, b_n] \supset \ldots$$

such that

(2) $f(a_n) < 0,$ $f(b_n) > 0,$ and $b_n - a_n \to 0$.

By the Nested Interval Theorem 6.1, there is a unique c inside all these intervals, and

$$\lim a_n = c , \quad \lim b_n = c.$$

To finish, we show that $f(c) = 0$. Since $f(x)$ is continuous on $[a, b]$, the Sequential Continuity Theorem 11.5 implies that

$$\lim f(a_n) = f(c) , \quad \lim f(b_n) = f(c) .$$

According to (2), we have $f(a_n) < 0$ and $f(b_n) > 0$ for all n; it follows by the Limit Location Theorem for sequences 5.3A that

$$\lim f(a_n) \leq 0 , \quad \lim f(b_n) \geq 0 , \quad \text{i.e.,}$$

$$f(c) \leq 0 , \quad\quad f(c) \geq 0 ,$$

which implies that $f(c) = 0$, proving (1). $\square$

If $f(x)$ changes sign from $+$ to $-$, the argument need not be repeated; just apply (1) to $-f(x)$: it changes from $-$ to $+$, and therefore has a zero c on $[a, b]$. But a zero of $-f(x)$ is also a zero of $f(x)$. $\square\square$

Corollary 12.1   Intermediate value theorem.

Assume $f(x)$ is continuous on $[a,b]$, and $f(a) \le f(b)$. Then for $k \in \mathbb{R}$,

(3) $f(a) \le k \le f(b)$ $\Rightarrow$ $k = f(c)$ for some $c \in [a,b]$.

If $f(a) \ge f(b)$, the hypothesis is $f(a) \ge k \ge f(b)$. Both cases are covered by the usual way the theorem is quoted (expressively, if somewhat informally):

If $f(x)$ is continuous on $[a,b]$, it takes on all values between $f(a)$ and $f(b)$, as x varies over $[a,b]$.

Proof. If $k = f(a)$ or $k = f(b)$ we are done. If not, we reduce to the Bolzano Theorem by lowering the graph of $f(x)$ by k units; in other words, we consider the function $f(x) - k$: it is continuous and changes sign on the interval, since

$$f(a) < k < f(b)   \Rightarrow   f(a) - k < 0   \text{ and }   f(b) - k > 0 .$$

Therefore, by Bolzano's Theorem, $f(x) - k$ has a zero c on $[a,b]$, i.e.,

$$f(c) = k   \text{ for some }   c \in [a,b] .$$

(If $f(a) \ge f(b)$, we reason similarly, or apply (3) to $-f(x)$.)   □

Since Bolzano's Theorem is essentially the special case of the Intermediate Value Theorem when $k = 0$, the two theorems are equivalent: either is an immediate corollary of the other, and they just give two slightly different ways of looking at the same fact. We will need both viewpoints.

Questions 12.1

1. (a) Show that $f(x) = x^3 + 3x^2 - 1$ has at least one zero on each of the intervals $[0,1], [-1,0], [-3,-2]$. Deduce from this that it has exactly one zero on each of the intervals.

(b) If you start with the interval $[-3,1]$ and apply the bisection process to find a zero of $f(x)$, which of the above zeros will you find?

What if you start with $[-3,3]$?

2. Show $e^x - 3x$ has at least two positive zeros.

3. If $f(x)$ is continuous and changes sign on $[a,b]$, it can have more than one zero on this interval, but the picture suggests the number of zeros should be *odd*. Is this true?

4. If a computer carries out the bisection process starting with $[a,b]$, how many steps will it need to locate the zero to within ϵ?

5. Show $\sin x = a$ has a solution between $-\pi/2$ and $\pi/2$ for every a on the interval $[-1,1]$.

12.2   Applications of Bolzano's theorem.

We illustrate now with some more examples of how Bolzano's Theorem can be used to prove the existence of zeros, locate them approximately, and count them.

Example 12.2A   Existence of a zero.

(4)   *A polynomial of odd degree has at least one (real) zero.*

Proof.   Let $f(x) = a_0 x^n + a_1 x^{n-1} + \ldots + a_n$; assume first that $a_0 > 0$. We are going to apply Bolzano's Theorem to $f(x)$; looking forward to this, we claim

(5) $f(x) < 0$ for $x \ll -1$, $f(x) > 0$ for $x \gg 1$.

This is so because the $a_0 x^n$ term dominates in $f(x)$ when $|x| \gg 1$; to make this clear, we first isolate x^n by factoring it out of the polynomial:

$$f(x) = x^n \left(a_0 + a_1 \frac{1}{x} + \ldots + a_n \frac{1}{x^n} \right) = x^n g(x) \ .$$

Since $1/x \to 0$ as $|x| \to \infty$, the Algebraic Operations Theorem 11.3A shows

$$g(x) \to a_0, \quad \text{as} \quad |x| \to \infty \ .$$

Since we assumed $a_0 > 0$, by the Function Location Theorem 11.3D we get

$$g(x) > 0 \quad \text{for} \quad |x| \gg 1 \ ;$$

and therefore (since n is odd),

$$x^n g(x) < 0 \quad \text{for} \quad x \ll -1, \qquad x^n g(x) > 0 \quad \text{for} \quad x \gg 1 \ ,$$

which is exactly (5).

We have thus shown that $f(x)$ changes sign as x goes from negatively large to positively large values. Since a polynomial is continuous on the entire x-axis, Bolzano's Theorem tells us it has a zero. ☐

We assumed $a_0 > 0$; if $a_0 < 0$, the above applies to the polynomial $-f(x)$ and shows it has a zero, which will then be a zero of $f(x)$ as well. ☐☐

The argument fails if n is even, and the corresponding result would not be true, as the zeroless $1 + x^2$ shows—or the brain-dead polynomial 1.

Remarks.   In computer searches for zeros of a polynomial, one looks for intervals on which $f(x)$ changes sign, and then uses the bisection process or Newton's method to find a zero inside. (Newton's method is faster; the bisection method is more reliable and doesn't require calculation of derivatives.)

A difficult point is the *isolation* of the zeros: one would like to have each interval on which $f(x)$ changes sign so small there is only one zero of $f(x)$ inside it; otherwise one will miss some zeros. Numerical techniques for isolating the zeros have been devised, but there is always room for improving them.

Here is an example using Bolzano to locate a zero approximately.

Example 12.2B   Approximate location of a zero   Let
$$f(x) = x^3 + hx - 1 , \qquad h \approx 0^+ .$$
Think of $f(x)$ as a small perturbation of $x^3 - 1$. Since $x^3 - 1$ has a unique zero at 1, the polynomial $f(x)$ should have a zero close to 1; call it $z(h)$ since its value depends on h. Give the approximate value of $z(h)$, and prove it is a right-continuous function at 0.

Solution. Write $z = 1 + \epsilon$, where $\epsilon \approx 0$. Since $f(z) = 0$, we get on substituting $z = 1 + \epsilon$ into the above expression for $f(x)$,
$$(1 + \epsilon)^3 + h(1 + \epsilon) - 1 = 0 .$$
Multiply out the polynomial, ignoring all terms in h and ϵ of higher order (since h is assumed small and we hope ϵ is also small), and you get
$$1 + 3\epsilon + h - 1 \approx 0;$$
$$\epsilon \approx -\frac{h}{3} , \qquad z \approx 1 - \frac{h}{3} .$$

To prove $z(h)$ is right-continuous at 0, i.e., that $z(h) \to 1$ as $h \to 0^+$, it suffices by the Squeeze Theorem 11.3B to prove that z lies in $[1 - h, 1]$. But this follows from Bolzano's Theorem, since we can calculate that $f(x)$ changes sign on the interval:
$$f(1 - h) = -2h + 2h^2 - h^3 ,$$
$$< 0, \quad \text{for} \quad h \approx 0^+; \qquad f(1) = h > 0 . \qquad \square$$

In trying to establish the existence of a zero, it may be better to recast the equation in the form $f(x) = g(x)$, since the graphs of $f(x)$ and $g(x)$ may be easier to plot or visualize than the graph of $f(x) - g(x)$. Then one can use the following third form of Bolzano's Theorem.

Intersection Principle 12.2

(a)   *The roots of $f(x) = g(x)$ are the x-coordinates of the points where the graphs of $f(x)$ and $g(x)$ intersect.*

(b)   *If $f(x)$ and $g(x)$ are continuous on $[a, b]$, and on this interval their graphs change their "above-below" position, i.e., the graph that is below at a becomes the one above at b:*
$$f(a) < g(a) \quad \text{and} \quad f(b) > g(b), \qquad \text{or} \qquad f(a) > g(a) \quad \text{and} \quad f(b) < g(b) ,$$
then the two graphs intersect over some point $c \in [a, b]$.

Proof.   Part (a) is self-evident from the picture.

For part (b), either of the conditions shows $f(x) - g(x)$ changes sign on $[a, b]$. So by Bolzano's Theorem, the function has a zero $c \in [a, b]$, meaning that $f(c) = g(c)$. Let $d = f(c) = g(c)$; then the point (c, d) lies in both graphs. $\square$

Example 12.2C   Counting zeros

Approximately how many zeros has $x - A \sin x$, if A is large?

Solution. The zeros are the same as the roots of the equation

(7) $\sin x = x/A$;

we choose this form for the equation, since x/A is easy to plot or visualize for different values of the parameter A. By the Intersection Principle, we have to estimate how many intersections the graphs of $\sin x$ and x/A have.

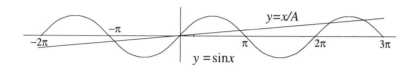

Since $|\sin x| \le 1$, the solutions to (7) all satisfy $|x/A| \le 1$, and they therefore lie in the interval $[-A, A]$, because

$$|x/A| \le 1 \quad \Rightarrow \quad |x| \le A \quad \Rightarrow \quad x \in [-A, A].$$

Over each interval $[2n\pi, (2n + 2)\pi]$ lying inside $[-A, A]$ there will be two intersections, as the picture shows (this isn't quite right at the ends of $[-A, A]$, and 0 is in two intervals, but we won't fuss). Since there are about $2A/2\pi$ of these intervals of length 2π inside $[-A, A]$, each with two intersections, there are in all approximately $2A/\pi$ zeros.   □

Questions 12.2

1. Two flag-wavers standing side-by-side in a parade move their flagpoles simultaneously over the course of a second from one horizontal position to the opposite horizontal position; one does it counterclockwise, the other clockwise. Show that at some moment in time the two flagpoles will be parallel.

2. How many zeros has $x - a \ln x + b$ if $a < 0$? if $a > 0$?

3. Let $f(x) = x^2 - x + h$, $h \approx 0$.
   (a) Without using the quadratic formula, find the approximate location of the greater zero of $f(x)$, in terms of h.
   (b) Is your estimate in part (a) too high or too low? (Use Bolzano.)
   (c) Check your answer by using the quadratic formula, estimating the square root by using the binomial series:

$$(1 + A)^{1/2} = 1 + (1/2)A - (1/8)A^2 + \dots \ .$$

4. A polynomial of degree n can have up to n zeros; the polynomials x^n only have one zero, however, and $x^{2n} + 1$ has none. Give a less trivial example by finding polynomials of odd degree n, containing about $n/2$ terms, and having only one zero.

12.3   Graphical continuity.

Often one hears a continuous function described as one whose "graph is unbroken"—roughly speaking, one whose graph can be traced out by a moving pen which never leaves the paper. This is hard to make precise, and if one tried to use it as the definition of continuity, it would not be at all clear how to prove even that a polynomial function was continuous.

But if we restrict ourselves to functions which are strictly monotone on an interval $[a, b]$, a reasonable interpretation of "unbroken graph" is that the function have what we will call the

Intermediate Value Property   A function $f(x)$ has the IVP on $[a, b]$ if it is defined on $[a, b]$ and it takes on all values between $f(a)$ and $f(b)$ as x varies between a and b.

If a strictly monotone $f(x)$ has the IVP, then by the above, its graph never skips over any y-values, and it never skips over any x-values either, since f is defined on $[a, b]$; thus it is reasonable to say its graph is "unbroken".

The Intermediate Value Theorem (Cor. 12.1) says a continuous function has the IVP; we now prove the converse of this, for strictly monotone functions.

Theorem 12.3   Continuity theorem for monotone functions.
If the function $f(x)$ is strictly monotone and has the Intermediate Value Property on $[a, b]$, then it is continuous on $[a, b]$.

Proof. To be definite, suppose $f(x)$ is strictly increasing on $[a, b]$. Let x_0 be a point in (a, b), i.e., assume x_0 is not an endpoint. We show f is continuous at x_0.

Given a small $\epsilon > 0$, draw in the two horizontal lines as shown at the heights $f(x_0) \pm \epsilon$.

Locate the points x_1 and x_2 shown, where

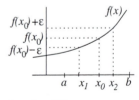

(8) $f(x_1) = f(x_0) - \epsilon, \quad f(x_2) = f(x_0) + \epsilon.$

These points exist since we are assuming that $f(x)$ has the intermediate value property on $[a, b]$; they are unique since $f(x)$ is strictly increasing.

Since $x_1 < x < x_2$, we have $f(x_1) < f(x) < f(x_2)$, so that by (8),

(9) $f(x_0) - \epsilon \ \leq \ f(x) \ \leq \ f(x_0) + \epsilon, \qquad \text{for } x \in [x_1, x_2].$

Therefore

$$f(x) \underset{\epsilon}{\approx} f(x_0) \quad \text{for} \ \ x \approx x_0 \ . \qquad\qquad \square$$

The argument assumes that ϵ is small enough so that the horizontal lines intersect the graph of $f(x)$ over the interval $[a, b]$. At an endpoint, the argument must be modified a little. If for example $x_0 = a$, replace (9) by

$$f(a) \ \leq \ f(x) \ \leq \ f(a) + \epsilon, \qquad a \leq x \leq x_2 \ .$$

This implies that $f(x) \to f(a)$ as $x \to a^+$, which is the definition of right-continuity at the endpoint a.

Questions 12.3

1. The proof assumed $f(x)$ was strictly increasing; deduce the theorem is also true if $f(x)$ is strictly decreasing.

2. Give an example of a function $f(x)$ which is defined on $[a, b]$, takes on all the intermediate values between $f(a)$ and $f(b)$ as x ranges over the interval $[a, b]$ (but does not do so if x ranges over any smaller closed interval), and which is nevertheless not continuous on $[a, b]$.

3. Prove from scratch that with the hypotheses of Theorem 12.3, $f(x)$ will be left-continuous at b.

12.4   Inverse functions.

We talked about inverse functions in Chapter 9; there we described how they looked from the three function viewpoints: graph, expression in x, and mapping. At that time, we simply assumed what we needed to make sure inverse functions existed. With the above Continuity Theorem and the Intermediate Value Theorem, we are now in a position to do better.

Theorem 12.4   Inverse function theorem.   *If $y = f(x)$ is continuous and strictly increasing on $[a, b]$, it has an inverse function $x = g(y)$ on $[f(a), f(b)]$ which is continuous and strictly increasing.*

The theorem is also true for strictly decreasing functions; in that case $[f(a), f(b)]$ should be replaced by $[f(b), f(a)]$.

Proof.   There are three things to prove.

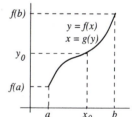

(A)   The inverse function is *defined* on $[f(a), f(b)]$.

For, if y_o is a point in this interval, by the Intermediate Value Theorem (Cor. 12.1) there is a point x_0 in $[a, b]$ such that $f(x_0) = y_0$; it is unique because $f(x)$ is strictly increasing. Therefore one can define g at y_0 by

$$g(y_0) \;=\; x_0 \;.$$

(B)   The function $g(y)$ is *strictly increasing*: $y_0 < y_1 \;\Rightarrow\; g(y_0) < g(y_1)$.

To prove this, set $x_0 = g(y_0)$ and $x_1 = g(y_1)$. We have

$$x_0 \;\geq\; x_1 \;\Rightarrow\; y_0 \;\geq\; y_1, \quad \text{since } f \text{ is strictly increasing .}$$

But this statement is the contrapositive of the one we wanted to prove.

(C)   The function $g(y)$ is *continuous*.

Namely, it is strictly increasing on $[f(a), f(b)]$ and takes on every value between a and b as y ranges over this interval, since if $a \leq x_0 \leq b$, we have $g(y_0) = x_0$, where $y_0 = f(x_0)$. Therefore, by Theorem 12.3, it is continuous. $\square$.

For example, we conclude from Theorem 12.4 that for every positive integer n, the function $x^{1/n}$ exists and is continuous on the interval $[0, \infty)$, since it is the inverse to the strictly increasing continuous function x^n on this interval. This shows for instance that *positive numbers have n-th roots*, an elegant application of the notion of continuous function.

Questions 12.4

1. Section 9.4 makes assumptions under which the inverse function to $f(x)$ on $[a, b]$ exists; Section 12.4 makes other assumptions. Which set of assumptions is stronger (i.e., more restrictive on the function)?

Exercises

12.1

1. Prove that a continuous function whose values are always rational numbers is a constant function.

2. (a) Find intervals of unit length on which $f(x) = 2x^4 - 8x^3 + 24x - 17$ has its zeros.

(b) For each of the following starting intervals, tell which of the zeros of $f(x)$ will be found by the bisection method associated with the proof of Bolzano's Theorem. (Label the zeros $x_1 < x_2 < x_3 < x_4$.)

    (i) $[-4, 2]$ (ii) $[-2, 4]$ (iii) $[0, 4]$

3. Give an example of a function with two zeros, one of which cannot be found by the bisection method associated with Bolzano's Theorem. Can Newton's method find this zero?

4. (a) Show that if $0 \le a \le (1/e)$, then $x e^{-x} = a$ has a non-negative solution. (Use the Intermediate Value Theorem; assume anything that you need from calculus.)

(b) Show it has two solutions if $0 < a < 1/e$.

5. Show that for values of $y \ge 1$, the equation $y^2 \cos x - e^x = 0$ has a solution for x lying in the interval $[0, \pi/2)$.

(Hint: use the Intermediate Value Theorem.)

12.2

1. Take a topographic map of the White Mountains and draw a circle on it somewhere. Show that there are always two diametrally opposite points on the circle (i.e., at the ends of the same diameter) whose corresponding points on the mountain are at the same height above sea level.

2. (a) Approximately where are the roots of $x^4 - \epsilon x^2 - 1 = 0$ if ϵ is a small positive number? (Do not use the quadratic formula.)

(b) Let r be the highest root; r depends on ϵ, that is, $r = r(\epsilon)$. Show that $\lim_{\epsilon \to 0} r(\epsilon)$ exists and determine its value.

(Follow the similar example in the text for both parts. The type of reasoning in part (a) is an example of what are called *perturbation methods*: varying the parameters in a problem slightly, and studying how this affects the solutions.)

3. Show that $f(x) = x - \tan x$ has an infinity of positive solutions $x_1, x_2, \ldots$. Find a simple function $f(n)$ such that $x_n - f(n) \to 0$ as $n \to \infty$. (Do not write it as $x_n \to f(n)$, since this use of $\to$ has not been defined.)

4. For which value(s) of the parameter A will

$$f(x) = x - A \sin x$$

have exactly five zeros? Express your answer in terms of the numbers x_i of the previous exercise.

12.3

1. Without looking, write out a proof from scratch of the Continuity Theorem for monotone functions 12.3, for the case where $f(x)$ is decreasing; take the interval I to be open, to avoid having to consider continuity at the endpoints.

12.4

1. In your own words, give the outline of an argument which shows that each real number a has one and only one fifth root $\sqrt[5]{a}$, by using the theorems of this chapter (but not Theorem 12.4).

2. Over what range of x-values can each of these equations be solved for x in terms of y? What can you say about the resulting function $x = f(y)$? (You'll need to use calculus.)

(a) $x^5 + x - 2 = y$ (b) $e^x - y = x^2 y$.

Problems

12-1 Give a proof of Bolzano's Theorem, using the set $S = \{x : f(x) < 0\}$; assume $f(a) < 0$, $f(b) > 0$.

(Begin by using S to locate the zero c. The Positivity Theorem for continuous functions (11.4B) is helpful.)

12-2 Suppose a function $f(x)$ is continuous on a closed interval $I = [a, b]$, and that $f(I) = [f(a), f(b)]$. Suppose further that as x varies over I, $f(x)$ never repeats a value. Prove $f(x)$ is strictly increasing.

12-3 Finding chords By a *chord* of a function, we mean a horizontal line segment whose two ends both lie on the graph of the function.

Suppose $f(x)$ is a continuous function on $[-a, a]$, whose graph is higher at 0 than at either a or $-a$, i.e., $f(0) > f(-a)$, $f(0) > f(a)$.

Prove that $f(x)$ has a chord of length a.

12-4 Prove that a polynomial of degree n cannot have more than $n - 1$ chords of length k (where k is a given positive number), unless it is a constant polynomial (i.e., $n = 0$). (See the previous problem for the definition of "chord".)

12-5 Inscribing equilateral triangles Let C be a smooth, convex, closed curve, i.e., one without endpoints, and such that a line segment joining any two points on C lies inside C. (An ellipse or an oval are examples. "Smooth" means it has a tangent line at each point.)

Let P be any point on C. Show convincingly that you can always find two other points Q and R on C such that PQR is an equilateral triangle. (Try some sketches.)

12-6 Circumscribing squares Let C be a continuous closed curve, i.e., one without endpoints. Show convincingly that it is always possible to circumscribe a square around C, that is, find a square all four of whose sides touch C (so that the square cannot be shrunk without rotation to a smaller rectangle enclosing C). (Try some sketches.)

12-7 Danish ham sandwich A Danish open-faced ham sandwich consists of a thin slice of rye bread and a thin slice of ham. Both have irregular shapes, but assume they have uniform thickness. Prove it is possible with a single vertical knife-cut to divide any ham sandwich in two so that each person gets half the ham and half the bread.

Answers to Questions

12.1

1. (a) $f(-3) = -1, f(-2) = 3; f(-1) = 1, f(0) = -1, f(1) = 3$; therefore $f(x)$ changes sign on each of the three intervals, so by Bolzano's Theorem it has a least one zero on each interval.

Since it is a polynomial of degree 3, it has at most 3 zeros altogether; therefore it has exactly 3 zeros, and thus exactly one on each interval.

(b) For $[-3, 1]$, the midpoint is -1, and $f(x)$ changes sign on $[-3, -1]$; this will be the next interval therefore, and we will find the lowest zero.

For $[-3, 3]$ the midpoint is 0, and $f(x)$ changes sign on $[0, 3]$; this time we find the greatest zero.

2. $f(0) = 1$, $f(1) = e - 3 < 0$, $f(2) = e^2 - 6 > 0$; thus $f(x)$ changes sign on $[0, 1]$ and $[1, 2]$, so it has at least two zeros, by Bolzano's Theorem.

3. No, the graph of the function might be tangent to the x-axis. (Call this a multiple zero if you like, but there's really only one zero there.)

4. After n steps, the subinterval has length $(b - a)/2^n$; this will be $< \epsilon$ if $n > [\log_2 (b - a)]/\epsilon$.

5. Since $\sin(-\pi/2) = -1$ and $\sin(\pi/2) = 1$, the Intermediate Value Theorem shows that on $[-\pi/2, \pi/2]$, $\sin x$ takes on every value between -1 and 1.

12.2

1. Let $\theta_i(t)$ be the angle the flagpole makes with the horizontal, as a function of time, for the i-th flag-waver ($i = 1, 2$). Then for both functions, $0 \le t \le 1$; as t goes from 0 to 1, one function goes from 0 to π, the other from π to 0. Since we can assume they are continuous functions, by the Intersection Principle their graphs must cross at some point (t_0, θ_o); at this moment in time, the two flagpoles will make the same angle θ_o, i.e., be parallel.

2. Write it in the form $\ln x = x/a + b/a$. Using the Intersection Principle, it is clear from the graphs that if $a < 0$, the line has negative slope and there is one solution; while if $a > 0$, it has positive slope, and there can be 0, 1, or 2 solutions, depending on the value of a and b.

3. (a) If $h = 0$, the zeros are 0 and 1; if $h \approx 0, h \neq 0$, call the greater zero $z = 1 + \epsilon$. Expanding out $f(z) = 0$, and dropping the terms in ϵ and h of higher degree, we get $\epsilon + h = 0$, so $z \approx 1 - h$.

(b) Calculating, we have $f(1 - h) = h^2$, while say $f(1/2) < 0$; thus the true zero is inside $[1/2, 1 - h]$, which shows that $1 - h$ is too high.

(c) The exact root is $[1 + \sqrt{1 - 4h}]/2 \approx 1 - h - h^2$, which checks. (Use the binomial series: $(1 + x)^k = 1 + kx + (k(k - 1)/2)x^2 + \ldots$, for any real k and $|x| < 1$, to make the approximation.)

4. Any polynomial with positive coefficients and only odd-degree terms is strictly increasing for all x (this is obvious for $x > 0$, and it follows for $x < 0$ since the polynomial is an odd function); thus it has only one zero.

12.3

1. $-f(x)$ will be strictly increasing and will have the Intermediate Value Property on $[a, b]$; therefore $-f(x)$ is continuous, so $f(x)$ is also.

2. For example, on $[0, 1]$, let $f(x) = \begin{cases} 2x, & 0 \leq x < 1/2; \\ x, & 1/2 \leq x \leq 1. \end{cases}$

3. Given a small $\epsilon > 0$, find x_1 such that $f(x_1) = f(b) - \epsilon$; this is possible since $f(x)$ has the Intermediate Value Property. Then

$$f(b) - \epsilon \; < \; f(x) \; < \; f(b), \qquad x_1 < x < b,$$

since $f(x)$ is strictly increasing. This shows $f(x) \underset{\epsilon}{\approx} f(b)$, if $x \approx b^-$.

12.4

1. Neither is stronger: both sets are equivalent. Both assume $f(x)$ is defined and strictly monotone; the other assumptions are then equivalent:

$f(x)$ has the Intermediate Value Property $\Leftrightarrow$ $f(x)$ is continuous.

The direction $\Rightarrow$ is the Continuity Theorem for monotone functions 12.3; the direction $\Leftarrow$ is the Intermediate Value Theorem, Cor. 12.1.

13

Continuous Functions on Compact Intervals

13.1 Compact intervals.

This chapter continues our study of continuity with some theorems about continuous functions; these theorems are among the most important in analysis. Though not long, it is fairly dense, and you should study it carefully.

The three main theorems in it are all "local-to-global" theorems, in the language of Chapter 10; that is, they say that on some interval I,

$$f(x) \text{ has a local property} \quad \Rightarrow \quad f(x) \text{ has a global property} .$$

In each case, the local property is *continuity on I*. The proofs are very similar; all use the Completeness Property, in the form of the Bolzano-Weierstrass Theorem 6.3, as well as the Sequential Ccontinuity Theorem 11.5. If you are hazy about either, review them now.

What distinguishes these theorems and makes it possible to prove results of the local-to-global type is that the interval I is special: in each of the theorems, I must be a finite closed interval $[a, b]$. From now on, we will call such an interval a **compact** interval. What makes such intervals special is that among all possible types of interval, they alone have the property called *sequential compactness*. Here is the definition; it may seem odd at first, but it is exactly what the proofs require, and you will quickly get used to it.

Definition 13.1 A set $S \subseteq \mathbb{R}$ is said to be **sequentially compact** if every sequence of points in S has a subsequence converging to a point in S.

Theorem 13.1 Sequential compactness theorem.

A compact interval $[a, b]$ is sequentially compact .

Proof. Let $\{x_n\}$ be any sequence in $[a, b]$; it is bounded since the interval is finite. By Bolzano-Weierstrass 6.3, it has a convergent subsequence $\{x_{n_i}\}$; put

$$\lim_{i \to \infty} x_{n_i} = c .$$

Since every $x_n \in [a, b]$,

$$a \leq x_{n_i} \leq b \qquad \text{for all } i.$$

Therefore by the Limit Location Theorem for sequences 5.3A,

$$a \leq \lim_{i \to \infty} x_{n_i} \leq b ; \quad \text{i.e., } c \in [a, b]. \qquad \square$$

Questions 13.1

1. Recall the different types of intervals:

$[a, b]$ finite closed; *compact* $[a, \infty), (-\infty, a]$ semi-infinite closed
(a, b) finite open $(a, \infty), (-\infty, a)$ semi-infinite open
$(a, b], [a, b)$ finite half-open $(-\infty, \infty)$ infinite (open and closed).

Show that the compact intervals are the only intervals with the sequential compactness property, i.e., that none of the other intervals has it.

13.2 Bounded continuous functions.

Here is the first of the three theorems referred to in the introduction; the global property of $f(x)$ on I is *boundedness*.

Theorem 13.2   Boundedness theorem.

If $f(x)$ is continuous on a compact interval I, then $f(x)$ is bounded on I.

The conclusion would be false if "compact" were omitted: the function x on the interval $[0, \infty)$ and the function $1/x$ on $(0, 1]$ are both continuous but unbounded. It would also be false if "continuous" were omitted; see Question 13.2/4.

Proof.   We prove first that $f(x)$ is bounded above, by proving the contrapositive (cf. Appendix A.2): if $f(x)$ is not bounded above on the compact interval I, then it is not continuous at some point of I.

Suppose $f(x)$ is not bounded above on I. We construct a convergent sequence of points in I on which $f(x) \to \infty$.

To do this, for each positive integer n, we choose a point x_n in I such that

(1) $$f(x_n) > n .$$

This is possible, for if there were an n for which no such x_n existed, $f(x)$ would be bounded above by the number n, contrary to our assumption.

The sequence $\{x_n\}$ will not in general converge, but since I is compact, the Sequential Compactness Theorem 13.1 shows that from $\{x_n\}$ we can choose a subsequence $\{x_{n_i}\}$ converging to a point c of I:

$$\lim_{i \to \infty} x_{n_i} = c , \quad \text{where } c \in I .$$

We note first of all that since the subsequence is infinite,

(2) $$\lim_{i \to \infty} n_i = \infty .$$

According to (1), we have

$$f(x_{n_i}) > n_i ,$$

from which it follows from (2) (and Theorem 5.2∞, to be very picky), that

$$\lim_{i \to \infty} f(x_{n_i}) = \infty .$$

But this shows that $f(x)$ cannot be continuous at $c \in I$, for if it were, the Sequential Continuity Theorem 11.5 would tell us that

$$\lim_{i \to \infty} f(x_{n_i}) = f(c) ,$$

a contradiction, since $c \in I$ implies that $f(c)$ is defined and finite. This completes the proof (by contraposition) that $f(x)$ is bounded above. ☐

To show that $f(x)$ is also bounded below, we note that $-f(x)$ is continuous, therefore by what we have just proved, it is bounded above on I:

$$-f(x) \; < \; K \qquad \text{for all} \;\; x \in I .$$

Changing signs reverses the inequality:

$$f(x) \; > \; -K \qquad \text{for all} \;\; x \in I ,$$

which shows $f(x)$ is bounded below on I . ☐☐

Questions 13.2

1. Prove that a function $f(x)$ continuous for all x is bounded on every finite interval, whether it is compact or not.

2. Describe all intervals on which $1/x$ is bounded.

3. Let $f(x)$ be continuous; prove that it is bounded on any finite union (cf. Appendix A.0) of compact intervals, but not necessarily on an infinite union.

4. Give an example of a discontinuous function on $[0, 1]$ which is not bounded.

13.3 Extremal points of continuous functions.

Our second theorem is (along with its generalizations to higher-dimensional space) considered by many to be the "fundamental theorem of analysis".

Theorem 13.3 Maximum theorem.

Let $f(x)$ be continuous on the compact interval I. Then $f(x)$ has a maximum and minimum on I, that is, there exist points $\bar{x}, \underline{x} \in I$ such that

$$f(\bar{x}) \; = \; \sup_{x \in I} f(x) , \qquad f(\underline{x}) \; = \; \inf_{x \in I} f(x) .$$

Once again, it is important to see that the two hypotheses are necessary. If the interval I is not compact, an example would be the continuous function x which has no maximum or minimum on $(0, 1)$, and no maximum on $[0, \infty)$.

If the function $f(x)$ is allowed to be discontinuous, then the $f(x)$ below is defined on the compact interval $[-1, 1]$ but has no maximum or minimum on this interval:

$$f(x) = 1/x, \; x \neq 0; \quad f(0) = 0 .$$

Proof. Since $f(x)$ is continuous on a compact interval, the Boundedness Theorem 13.2 shows that $f(x)$ is bounded above on I. Therefore the number

$$M = \sup_{x \in I} f(x)$$

exists, by the Completeness Property for sets 6.5; then by definition of the supremum,

(3) $f(x) \leq M$ for all $x \in I$.

We now construct a sequence of points in I which will turn out to converge to the maximum point $\bar{x}$ (which we don't know exists, so far).

To do this, for each integer $n > 0$, we choose a point $x_n \in I$ such that

(4) $f(x_n) \geq M - 1/n$.

This is possible, since if there were an n for which no such x_n existed, then $M - 1/n$ would be an upper bound for $f(x)$ on I, contradicting the fact that M is the *least* upper bound.

By the Sequential Compactness Theorem 13.1, the sequence $\{x_n\}$ has a subsequence $\{x_{n_i}\}$ converging to a point of I:

$$x_{n_i} \to \bar{x} , \qquad \bar{x} \in I .$$

We now have the point $\bar{x}$ we were looking for; we still have to show it is the maximum point for $f(x)$ on I. To see this, we apply the Squeeze Theorem for sequences 5.2 to the inequalities in (3) and (4) above:

$$
\begin{array}{ccccc}
M - 1/n_i & \leq & f(x_{n_i}) & \leq & M \\
\downarrow & & \Downarrow & & \downarrow \qquad \text{as } i \to \infty . \\
M & & M & & M
\end{array}
$$

This shows

(5) $f(x_{n_i}) \to M,$ as $i \to \infty$.

On the other hand, since $f(x)$ is continuous at $\bar{x}$, the Sequential Continuity Theorem 11.5 tells us that

$$f(x_{n_i}) \to f(\bar{x}) , \qquad \text{as } i \to \infty .$$

Since the limit of a sequence is unique, it follows from (5) and the line above that $f(\bar{x}) = M$, which proves that $\bar{x}$ is a maximum point for $f(x)$ on I .   □

To see that $f(x)$ also attains its minimum on I, we apply the above to $-f(x)$, since a maximum point for $-f(x)$ is a minimum point for $f(x)$.   □□

Questions 13.3

1. Show the Maximum Theorem is stronger than the Boundedness Theorem (cf. Appendix A.1 for "stronger").

2. The proof of the Maximum Theorem uses the Completeness Property twice—where?

3. Give an example of a continuous function on $(0, 1]$ with no maximum or minimum on this interval, but which does not have the limit ∞ or $-\infty$ as $x \to 0^+$.

13.4   The mapping viewpoint.

If we look at a continuous function as a map $f : D_f \to \mathbb{R}$, where D_f is the domain of the function, we get an elegant way to formulate as a single theorem the main results of this chapter and the last one. Recall from Section 10.1:

Definition 13.4 Let f be defined on the set $S \subseteq D_f$; we call the set

$$f(S) = \{f(x) : x \in S\}$$

the **image of S** under the map f .

Examples 13.4 If $f(x) = x^2$, then $f(\mathbb{R}) = [0, \infty)$, $f\big((-1,1)\big) = [0,1)$.

If $u(x)$ is the unit step function (cf. Exercise 9.2/1), then $u\big([-1,1]\big) = \{0,1\}$.

This last example, of an image consisting of two points, shows that if f is discontinuous, the image of an interval under the map f need not even be an interval. However, if f is continuous, the next theorem shows that the situation is better.

Theorem 13.4   Continuous mapping theorem.

If $f(x)$ is defined and continuous on the compact interval I, then $f(I)$ is a compact interval; i.e.,

the continuous image of a compact interval is a compact interval.

Proof.   By the Maximum Theorem 13.3, there are points $\underline{x}, \bar{x} \in I$ such that

(6) $\qquad f(\underline{x}) = m = \inf_{x \in I} f(x), \qquad f(\bar{x}) = M = \sup_{x \in I} f(x) .$

To prove the theorem, we show (7) holds, by showing $\subseteq$ and $\supseteq$ hold:

(7) $\qquad\qquad\qquad f(I) = [m, M] .$

$f(I) \subseteq [m, M]:$ $x_0 \in I \Rightarrow m \le f(x_0) \le M \Rightarrow f(x_0) \in [m, M] .$

$f(I) \supseteq [m, M]:$ to prove this, we must show that

(8) $\qquad y_0 \in [m, M] \quad \Rightarrow \quad y_0 = f(x_0)$ for some $x_0 \in I$.

The hypothesis of (8) is by (6) the same as

$\qquad\qquad f(\underline{x}) \le y_0 \le f(\bar{x}) ;$ (note: $\underline{x}, \bar{x} \in I$).

Since $f(x)$ is continuous on I, the Intermediate Value Theorem (Cor 12.1) shows there is a point x_0 between $\underline{x}$ and $\bar{x}$ (and thus in I) such that $y_0 = f(x_0)$.

This proves (8), which completes the proof of (7) and of the theorem. $\square$

The proof of the preceding theorem used the Maximum Theorem and the Intermediate Value Theorem. Conversely, both of these follow easily from the Continuous Mapping Theorem (see Exercise 13.4/1). So this statement about maps is equivalent to our two deepest results so far. It is a concise and elegant way of stating both of them together. Its generalization to higher-dimensional spaces is an important theorem of what is called *point-set topology*.

Questions 13.4

1. Let $f(x) = \cos x$. Give examples of intervals I such that

    (a) I is finite open, yet $f(I)$ is compact;

    (b) I is semi-infinite closed, yet $f(I)$ is compact.

Why aren't these counterexamples to the Continuous Mapping Theorem ?

2. Find functions $f(x), g(x)$ such that

    (a) $f\big([-1, 1]\big) = (-1, 1)$;   (b) $g\big([0, \pi/2]\big) = [0, \infty)$.

13.5   Uniform continuity.

Uniform continuity is stronger than continuity; it is a global property, formulated only for a function on an interval: "uniform continuity at a point" makes no sense. We will need uniform continuity when we study integration; we are taking it up now because the main theorem about it is a lot like the others in this chapter, and the same techniques are used in its proof.

Definition 13.5   We say $f(x)$ is **uniformly continuous on the interval** I if:

given $\epsilon > 0$,   there is a $\delta > 0$ (depending only on ϵ) such that

$$(9) \qquad\qquad f(x') \underset{\epsilon}{\approx} f(x'') \quad \text{if} \quad x' \underset{\delta}{\approx} x'', \quad x', x'' \in I \ .$$

Example 13.5A   Show x^2 is not uniformly continuous on $[0, \infty)$.

Solution.   The argument is informal, but we hope convincing.

Given $\epsilon > 0$, we have (if x' or $x'' \neq 0$),

$$|x'^2 - x''^2| < \epsilon \quad \Leftrightarrow \quad |x' - x''||x' + x''| < \epsilon \ ;$$

$$\Leftrightarrow \quad |x' - x''| < \frac{\epsilon}{|x' + x''|} = \delta \ .$$

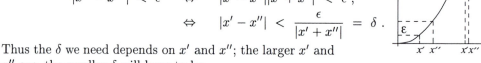

Thus the δ we need depends on x' and x''; the larger x' and x'' are, the smaller δ will have to be.

The picture shows this: the further out you go, the smaller δ will have to be to make $|x'^2 - x''^2| < \epsilon$: no one δ will work over the whole interval $[0, \infty)$.   □

To make it easier to compare uniform continuity with ordinary continuity, we change x', x'' to x, a and state the two definitions side by side:

Uniform continuity on I

Given $\epsilon > 0$, there is a $\delta > 0$ (depending only on ϵ) such that

$$f(x) \underset{\epsilon}{\approx} f(a) \quad \text{for} \quad x \underset{\delta}{\approx} a, \quad x, a \in I \ .$$

Ordinary continuity on I

Given $\epsilon > 0$, there is a $\delta > 0$ (depending on ϵ and a) such that

$$f(x) \underset{\epsilon}{\approx} f(a) \quad \text{for} \quad x \underset{\delta}{\approx} a, \quad x, a \in I \ .$$

Thus the difference between the two definitions is that:

⋆ for ordinary continuity on I, the δ is allowed to depend on a, that is, on where you are in the interval, whereas

⋆ for uniform continuity, the δ is to be chosen *uniformly* with respect to a, that is, one δ must do for every point $a \in I$.

If $f(x)$ is uniformly continuous on I, it is continuous on I, so that uniform continuity is a stronger assumption on $f(x)$ than ordinary continuity. But the third main theorem of this chapter says that the distinction between the two types of continuity disappears if the interval I is compact.

Theorem 13.5 Uniform continuity theorem. *If I is a compact interval,*

$$f(x) \text{ continuous on } I \quad \Rightarrow \quad f(x) \text{ uniformly continuous on } I \ .$$

Proof. We use contraposition. If $f(x)$ is not uniformly continuous on I, then there is an $\epsilon_0 > 0$ for which *no* δ can be found which will make the implication below true:

$$x' \underset{\delta}{\approx} x'' \quad \Rightarrow \quad f(x') \underset{\epsilon_0}{\approx} f(x'') \ .$$

In particular, none of these δ will make it true: $\delta = 1/2, 1/3, \ldots, 1/n, \ldots$. Since none of these δ will work, that must mean that for each positive integer n, there is a pair of points x'_n, x''_n in I such that

(10) $$|x'_n - x''_n| \ < \ 1/n, \quad \text{but}$$

(11) $$|f(x'_n) - f(x''_n)| \ \geq \ \epsilon_0 \ .$$

Since I is compact, the Compactness Theorem 13.1 says the sequence $\{x'_n\}$ has a subsequence $\{x'_{n_i}\}$ converging to a point $c \in I$:

(12) $$\lim_{i \to \infty} x'_{n_i} \ = \ c \ , \qquad c \in I. \qquad \text{Also,}$$

(13) $$\lim_{i \to \infty} (x'_{n_i} - x''_{n_i}) \ = \ 0 \qquad \text{by (10)} \ .$$

Subtracting the sequence (13) from (12) and using the Linearity Theorem 5.1A for limits of sequences, we deduce that also

(14) $$\lim_{i \to \infty} x''_{n_i} \ = \ c \ .$$

We now show $f(x)$ is not continuous at c. If it were, the Sequential Continuity Theorem 11.5, with (12) and (14), would imply that as $i \to \infty$,

$$f(x'_{n_i}) \ - \ f(x''_{n_i}) \ \to \ f(c) - f(c) \ = \ 0,$$

and therefore, by definition of the limit of a sequence,

$$|f(x'_{n_i}) - f(x''_{n_i})| \ < \ \epsilon_0 \ , \qquad \text{for } i \gg 1 \ .$$

But this contradicts how the points x'_n, x''_n were selected: see (11). Thus $f(x)$ is not continuous at c, and the proof by contraposition is complete. □

Conclusion. The three main theorems of this chapter may be summarized:

If $f(x)$ is continuous on a compact interval, then on this interval

  (a) $f(x)$ is bounded,
  (b) $f(x)$ has a maximum and minimum,
  (c) $f(x)$ is uniformly continuous.

There are really only two theorems, since (b) implies (a); but since (a) has to be proved first, we'll leave it at three. Each has the form:

  local property of $f(x)$ on I $\Rightarrow$ *global property of $f(x)$ on I.*

They are therefore powerful results, and it is right that their proofs should require some effort. Each proof required the Completeness Property, in the Bolzano-Weierstrass Theorem form. We will need these three theorems at critical moments in our study of analysis.

Questions 13.5

1. Prove directly from the definition (i.e., without quoting the Uniform Continuity Theorem) that x^2 is uniformly continuous on $[-a, a]$.

Exercises

13.1

1. (a) Show that if S is the union (cf. A.0) of a finite number of compact intervals, then S is sequentially compact. (The intervals need not overlap.)

   (b) Show by counterexample that this is no longer true if S is the union of an infinite number of compact intervals.

2. (a) Find a function $y = f(x)$ whose domain is $[0, 1]$ and whose range of values is the non-negative y-axis: $0 \le y < \infty$.

   (b) Find a continuous function with the same properties as in (a).

13.2

1. (a) Let $x(t)$ and $y(t)$ be continuous on $[a, b]$. As t varies over this interval the point $(x(t), y(t))$ traces out a curve in the xy-plane. Prove this curve is contained within some large square centered at the origin.

   (b) Show by example that (a) may not be true if the interval used on the t-axis is of the form (a, b).

13.3

1. Suppose that $f(x)$ is continuous and > 0 on $I = \mathbb{R}$, and $\lim\limits_{x \to \pm\infty} f(x) = 0$.

   (a) Prove $f(x)$ has no minimum on I.

   (b) Prove $f(x)$ has a maximum on I (note that I is not compact)..

   (c) Prove (b) under weaker hypotheses than positivity on all of I.

2. Using the ideas of this section (i.e., without using derivatives), prove that a polynomial of even degree either has a maximum or a minimum on $(-\infty, \infty)$. Give a simple criterion for deciding which it has.

3. (a) With the same hypotheses as Exercise 13.2/1 above, prove there is a point on the curve which is closest to the origin, in the sense that there is no point on the curve which is closer.

(b) Give a counterexample to (a) if the interval is open and finite.

13.4

1. Let $f(x)$ be a function defined for all x that maps compact intervals to compact intervals: *I compact interval* $\Rightarrow$ $f(I)$ *compact interval*. Do *not* assume $f(x)$ is continuous.

(a) Prove that on any compact interval I the function attains its maximum and minimum (as in the conclusion of the Maximum Theorem).

(b) Prove that $f(x)$ has the *Intermediate Value Property*: for any compact interval $[a, b]$, $f(x)$ takes on all values between $f(a)$ and $f(b)$ as x varies over the interval.

This exercise shows the Maximum Theorem and Intermediate Value Theorem follow from the Continuous Mapping Theorem, and therefore taken together they are equivalent to the Continuous Mapping Theorem.

2. Give an example of a function defined for all x which is not continuous, yet which satisfies the italicized property on line 2 of the previous exercise.

13.5

1. Prove directly from the definition of uniform continuity that $\sin x$ is uniformly continuous on $(-\infty, \infty)$. (The unit circle picture helps to make the estimations needed; see Example 11.1B.))

2. Prove that a function which is continuous and periodic on R is uniformly continuous on R. (This gives an alternative proof for Exercise 13.5/1.)

3. Prove that if $f(x)$ is uniformly continuous on I and J, where these are two intervals which overlap, but are not assumed to be compact, then $f(x)$ is uniformly continuous on the interval $I \cup J$.

4. Is $\sqrt{x}$ uniformly continuous on $[0, \infty)$? Justify your answer. (Note that its slope at 0 is infinite. There are several approaches to this exercise; one of them uses the preceding exercise.)

5. Assuming the laws of logarithms, and that $\ln x$ is continuous at 1, prove $\ln x$ is uniformly continuous on $[1, \infty)$, a non-compact interval.

6. Let $f(x)$ be a function defined on an interval I (not assumed compact), and assume that its secants have bounded slope, i.e., for any two distinct points on the graph of $f(x)$ over I, the slope λ of the line joining them is bounded: $|\lambda| < K$, where K is some fixed number not depending on the two points selected.

(a) Prove $f(x)$ is uniformly continuous on I.

(b) Does $\sqrt{x}$ on $[0, 1]$ satisfy the above hypotheses on $f(x)$? Is it uniformly continuous on the interval?

Problems

13-1 Suppose $f(x)$ is continuous on some open interval I, and c is a maximum point for $f(x)$ inside this interval.

Common sense suggests $f(x)$ should be increasing immediately to the left of c and decreasing immediately to the right of c. Is this true? Either prove it, or give a counterexample. (Note that a constant function is considered to be both increasing and decreasing.)

13-2 Deduce the Maximum Theorem from the Boundedness Theorem by an indirect argument which is different from the one in Section 13.3. Namely, assume $f(x)$ has no maximum on I;

let $M = \sup_I f(x)$, apply the Boundedness Thereom to $\dfrac{1}{M - f(x)}$.

13-3 Prove directly from the definition of uniform continuity that a polynomial $f(x)$ is uniformly continuous on a compact interval.

(Hint: $f(x') - f(x'') = (x' - x'') \cdot g(x', x'')$, where g is a polynomial in two variables; prove this, and then use it. Note that at this point we do not have any definitions or theorems about functions of two variables.)

13-4 Let $f(x)$ be continuous on the compact interval I, and suppose that $f(x)$ has an infinity of maximum points (i.e. points where it takes on its maximum value on I), an infinity of minimum points on I, and that between any two maximum points lies at least one minimum point.

Prove $f(x)$ is constant on I. Why isn't $\sin(1/x)$ a counterexample?

13-5 Let $f(x)$ have the continuous mapping property (as in Exer.13.4/1):

I is a compact interval $\Rightarrow f(I)$ *is a compact interval.* Is f necessarily continuous for all x? Prove it or give a counterexample.

13-6 (a) Prove that if $f(x)$ is uniformly continuous on I, and a_n is a Cauchy sequence in I, then $f(a_n)$ is a Cauchy sequence.

(b) Use part (a) to prove that $1/x$ is not uniformly continuous on $(0, 1]$.

13-7 Prove that if $f(x)$ is uniformly continuous on $I = (a, b)$, then it is bounded above on I. (Use an indirect argument: follow the proof of the Boundedness Theorem, but use part (a) of the preceding problem as a substitute for the non-compactness of I.)

Answers to Questions

13.1

1. The infinite intervals $[a, \infty)$ contain a tail of the sequence $\{n\}$, which has no convergent subsequence. Similarly for $(-\infty, a]$.

A finite interval of the form $(a, b]$ or (a, b) contains a tail of the sequence $\{a + 1/n\}$, which converges to the point $a \notin I$; therefore any subsequence also converges to $a \notin I$, by the Subsequence Theorem 5.4. Similarly, intervals open at the right-hand endpoint, like $[a, b)$, contain a tail of $\{b - 1/n\}$.

13.2

1. If the interval is $(a, b), (a, b]$, or $[a, b)$, then $f(x)$ is bounded on the compact interval $[a, b]$, therefore on any of the three smaller intervals also.

2. Any interval, finite or infinite, not containing 0 and not having 0 as its right or left endpoint (as do $(-1, 0)$ or $(0, 1]$).

3. If $|f(x)| \le B_k$ on the interval I_k, for $k = 1, 2, \ldots, n$, then we can bound f by saying: $|f(x)| \le \max(B_1, \ldots, B_n)$ on $I_1 \cup I_2 \cup \ldots \cup I_n$.

The argument doesn't work for an infinite set of intervals, since max may not exist; counterexample: x is not bounded on $[0, 1] \cup [1, 2] \cup \ldots$.

4. The function $f(x) = 1/x$ on $(0, 1], \quad f(0) = 0$.

13.3

1. The Maximum Theorem implies the Boundedness Theorem, since if $f(x)$ has a maximum point $\bar{x}$ and a minimum point $\underline{x}$ on I, it is bounded on I:

$$f(\underline{x}) \ \le \ f(x) \ \le \ f(\bar{x}) \quad \text{for all} \ \ x \in I.$$

2. It uses the Completeness Property to find $M = \sup f(x)$, and it uses the Compactness Theorem 13.1 (which in turn uses the Bolzano-Weierstrass Theorem) to find a subsequence converging to $\bar{x}$. (Two numbers M and $\bar{x}$ are found, both by the Completeness Property.)

3. $(1/x) \sin(1/x)$ is an example; it oscillates in wider and wider swings as $x \to 0^+$.

13.4

1. (a) $I = (-2\pi, 2\pi)$;   (b) $I = [0, \infty)$;   in both cases $f(I) = [-1, 1]$.
The theorem says nothing about $f(I)$ if I is not compact.

2. (a) $f(x) = 2x, \quad -\frac{1}{2} < x < \frac{1}{2}; \quad f(x) = 0$ elsewhere in $[-1, 1]$
(b) $g(x) = \tan x, \ x \ne \pi/2; \quad g(\pi/2) = 0$.

Note that the functions cannot be continuous, since that would violate the Continuous Mapping Theorem: I is compact, but $f(I)$ would not be.

13.5

1. For $-a \le x \le a$, we have
$$|x'^2 - x''^2| = |x' - x''||x' + x''| \le |x' - x''| \cdot 2a$$
$$< \epsilon \quad \text{if } |x' - x''| < \epsilon/2a.$$

14

Differentiation: Local Properties

14.1   The Derivative.

We have finally reached the subject in analysis you probably are most comfortable with, since it is the starting point for most calculus courses. For that reason, we can go rather quickly. The derivative is a local notion; this chapter deals with local properties, while the next will discuss how the derivative influences the global behavior of $f(x)$.

We begin in this first section with the definitions; in the next we use the limit theorems to discuss the differentiation formulas, while the last sections discuss the relation of the derivative to other local properties.

Definition 14.1A   The derivative at a point.
Let $f(x)$ be defined for $x \approx a$. We write

$$(1) \qquad \lim_{x \to a} \frac{f(x) - f(a)}{x - a} = f'(a)$$

if the limit exists; its value $f'(a)$ is called the **derivative** of $f(x)$ at a, and we say f is **differentiable** at a, or *has a derivative* at a.

> The function on the left side of (1) is called the *difference quotient*; note that its limit, if it exists, can never be evaluated by taking the limits of the numerator and denominator separately, since both of these limits are 0.

Two alternative ways of writing (1) are common and useful in different contexts. We may convert (1) into a limit at 0 by changing the variable from x to $x - a$ (which we rename Δx), so that (1) takes the form

$$\lim_{\Delta x \to 0} \frac{f(a + \Delta x) - f(a)}{\Delta x} = f'(a) ; \qquad \Delta x = x - a.$$

Or if we view the function primarily as a relation between variables, as engineers and scientists usually do, we use new variables $y = f(x)$, $\Delta y = y - f(a)$, and then (1) is written

$$\lim_{\Delta x \to 0} \frac{\Delta y}{\Delta x} \;=\; \left. \frac{dy}{dx} \right|_{x=a} ; \qquad\qquad \begin{aligned} y &= f(x), \\ \Delta y &= y - f(a). \end{aligned}$$

Interpretations of the derivative and the difference quotient correspond to the different ways of thinking about $f(x)$. The two most common are:

f(x) is defined by its graph:

$$f'(a) = \begin{array}{c} \text{slope of the tangent} \\ \text{line to the graph of} \\ f(x) \text{ at } (a, f(a)); \end{array} \qquad \frac{f(x) - f(a)}{x - a} = \begin{array}{c} \text{slope of} \\ \text{the secant;} \end{array}$$

f(x) is a relation between variables:

$$\left. \frac{dy}{dx} \right|_a = \begin{array}{c} \text{rate of change of } y \\ \text{with respect to } x \\ \text{when } x = a \,; \end{array} \qquad \frac{\Delta y}{\Delta x} = \begin{array}{c} \text{average rate of change} \\ \text{over } [a, a + \Delta x]. \end{array}$$

When it first appears, the derivative is often defined, or at least described, as the slope of the tangent line or as the instantaneous rate of change. These are only ways of interpreting the derivative, however. The correct procedure is to take the derivative, defined by (1), as the primary notion; the slope of the tangent and rate of change are then defined by the above—that is, slope and instantaneous rate of change are derivatives, not the other way around.

The derivative $f'(a)$ is a number. From this, we define the derivative as a function, denoted by $f'(x)$, as follows.

Definition 14.1B   The derivative function.

We say $f(x)$ is **differentiable** on the open interval I if it is differentiable at every point of I; when that is so, its **derivative** on I is defined to be the function $f'(x)$ given by the rule: $x_0 \mapsto f'(x_0), \quad x_0 \in I.$

If the interval I is closed, say $I = [a, b]$, we handle differentiability at the endpoints by using right-hand and left-hand derivatives, according to the following definition.

Definition 14.1C   Assuming the limits exist, we define

$$f'(x_0^+) = \lim_{x \to x_0^+} \frac{f(x) - f(x_0)}{x - x_0} \qquad \begin{array}{c} \textbf{right-hand} \\ \textbf{derivative} \end{array} \qquad \begin{array}{c} \text{assume } f(x) \text{ is} \\ \text{defined for } x \approx x_0^+; \end{array}$$

$$f'(x_0^-) = \lim_{x \to x_0^-} \frac{f(x) - f(x_0)}{x - x_0} \qquad \begin{array}{c} \textbf{left-hand} \\ \textbf{derivative} \end{array} \qquad \begin{array}{c} \text{assume } f(x) \text{ is} \\ \text{defined for } x \approx x_0^-; \end{array}$$

and we say that $f(x)$ is differentiable on the right or left, respectively. (Note: at the left endpoint you take the right-hand derivative!)

According to the general result about the equality of right- and left-hand limits (cf. 11.2 (8)), if the function $f(x)$ is defined for $x \approx a$,

(2) $\qquad\qquad\qquad f'(x_0) \text{ exists } \quad \Leftrightarrow \quad f'(x_0^+) = f'(x_0^-) \,.$

If no interval is specified, the domain of $f'(x)$ is understood to be all points where $f(x)$ is differentiable; if the domain of $f(x)$ is a closed interval, differentiability at the endpoints means on the right or the left.

Example 14.1A   Let $f(x) = |x|$. Find its derivative on

   (a) $I = (-\infty, \infty)$;   (b) $I = [0, \infty)$;   (c) $I = (-\infty, 0]$.

Solution. $f'(x) = \begin{cases} 1, & x > 0 \\ -1, & x < 0 \end{cases} = \operatorname{sgn} x.$ At 0, we have

$$f'(0^-) = \lim_{x \to 0^-} \frac{|x| - 0}{x - 0} = \lim_{x \to 0^-} \frac{-x}{x} = -1; \quad \text{similarly,} \quad f'(0^+) = 1 .$$

Thus by (2), $f'(0)$ does not exist and $f(x)$ is not differentiable on $(-\infty, \infty)$, nor on any interval I containing 0 in its interior.

However, $f(x)$ *is* differentiable on $[0, \infty)$, i.e., if the domain of $f(x)$ is restricted to $[0, \infty)$, since in that case we only use the right-hand derivative at 0. When we specify an interval, it restricts the domain of the function as well as its derivative. Thus we can say

$$f'(x) = 1 \quad \text{on} \quad [0, \infty), \qquad f'(x) = -1 \quad \text{on} \quad (-\infty, 0] . \qquad \square$$

Example 14.1B   Let $f(x) = \sqrt{1 - x^2}$, $g(x) = (\sqrt{1 - x^2})^3$.   Discuss f', g'.

Solution. Both functions are defined only on $[-1, 1]$. The usual rules for differentiation give

$$f'(x) = \frac{-x}{\sqrt{1 - x^2}} , \qquad g'(x) = -3x\sqrt{1 - x^2} .$$

The first function is only differentiable on the open interval $(-1, 1)$; the second function however is differentiable on $[-1, 1]$. $\qquad \square$

In order to avoid endlessly repeating an obvious hypothesis, from now on when we speak of $f'(a)$, $f'(a^+)$, or $f'(b^-)$, we will automatically be assuming that $f(x)$ is defined respectively for $x \approx a$, $x \approx a^+$, or $x \approx b^-$.

Different notations for the derivative are used, depending on the context.

$\star$ If no independent variable is explicitly named, we use f' or Df . This is variable-free, but it is hard to write the value at a specific point.

$\star$ If an independent variable x is given, we use any of

$$f'(x), \qquad Df(x), \qquad D_x f(x), \qquad \frac{df}{dx}, \qquad \frac{d}{dx} f(x) ,$$

though only the first accommodates points easily: $f'(a), f'(x_0)$. The fourth can be used for this, however, by writing

$$\frac{df}{dx}\bigg|_{x=a} \quad \text{or} \quad \frac{df}{dx}\bigg|_a , \quad \text{and} \quad \frac{df}{dx}\bigg|_{x_0} \quad \text{or} \quad \left(\frac{df}{dx}\right)_0 .$$

$\star$ If a dependent variable is also given, say $y = f(x)$, we can use

$$y'(x), \qquad y', \qquad \frac{dy}{dx}, \qquad \dot{y} \text{ if } x \text{ represents time.}$$

Notations for the value of $f(x)$ and $f'(x)$ at a point x_0 would then be

   function: $y(x_0), \quad y_0$;   derivative: $y'(x_0), \quad y'_0, \quad \left(\dfrac{dy}{dx}\right)_0 .$

Theorem 14.1 $f(x)$ *differentiable at* a $\Rightarrow$ $f(x)$ *continuous at* a;
$\quad\quad\quad\quad\quad\quad$ $f(x)$ *differentiable on* I $\Rightarrow$ $f(x)$ *continuous on* I.

Proof.   Using the differentiability assumption and the Product Theorem 11.3A,

$$\lim_{x \to a} \big(f(x) - f(a)\big) = \lim_{x \to a} \frac{f(x) - f(a)}{x - a} \cdot \lim_{x \to a} (x - a) = f'(a) \cdot 0 = 0,$$

which shows that $\lim_{x \to a} f(x) = f(a)$, proving the first statement.

The same proof, using $x \to a^+$ or $x \to a^-$, shows that if $f(x)$ is differentiable on the right or left at a, it is continuous on the right or left at a.

Now the second statement follows immediately, since f is differentiable on I or continuous on I if it is, respectively, differentiable or continuous at every point of I (with the appropriate right or left notion at the endpoints of I.) $\square$

Differentiability on I is a local property of $f(x)$, verified by checking that f is differentiable at each point of I. So we can say that the preceding theorem is of the type: local $\Rightarrow$ local, and is easy. It shows that differentiability is a stronger property than continuity (since it implies continuity); it is a more restrictive property than continuity.

In short, it is harder for a function to be differentiable than to be continuous. The function $|x|$ is continuous but not differentiable, which shows the converse to Theorem 14.1 is not true.

On the other hand, the folk-feeling is that if $f(x)$ is continuous, there will be at worst only a finite number of points where it fails to be differentiable. While this is certainly true for the ordinary functions one deals with in analysis, it is not so in general; there exist in fact continuous functions which are nowhere differentiable, first constructed (or discovered) in the mid-1800's by Weierstrass. For a long time just curiosities, they are now in the public eye because their graphs are typical "fractal" curves: under the microscope they wiggle infinitely often, no matter how high one turns up the magnification.

In Edmund Landau's calculus book, written about 100 years ago, immediately after the definition of derivative are several very long pages devoted to constructing a continuous function which is nowhere differentiable. The book was written for students who fought duels at 6:00 AM, then went to their lectures on metaphysics at 7:00.

Questions 14.1

1. Let $F(x) = \dfrac{f(x) - f(a)}{x - a}$; suppose $f(x)$ continuous for $x = a$. Show

   (a) $F(x)$ has a removable discontinuity at a $\Leftrightarrow$ $f'(a)$ exists;

   (b) $F(x)$ has a jump discontinuity at a $\Leftrightarrow$ $f'(a^+) \neq f'(a^-)$ (but both derivatives exist).

2. Let $f(x) = x|x|$. Determine $f'(0^+)$ and $f'(0^-)$; does $f'(0)$ exist?

14.2   Differentiation formulas.

Two kinds of formulas are used in calculating derivatives.

There are specific formulas for the derivatives of some basic functions:

$$D \sin x = \cos x, \qquad De^x = e^x, \qquad Dx^n = nx^{n-1} \ \ (n \text{ a positive integer}),$$

and so on. These are proved by going back to the definition of derivative.

Then there are the general rules which allow us to extend differentiation to more complicated functions by telling how differentiation is related to other operations on functions: sum, product, quotient, composition, inverse. These are familiar from calculus, and stated in the next three theorems; all are proved by using the corresponding limit theorems.

Theorem 14.2A   Algebraic differentiation rules

Let I be an interval on which the functions u and v are differentiable (and for (5), on which $v \neq 0$). Then the functions below are also differentiable on I, and their derivatives are given by the indicated formula.

(3) **Linearity rule** $(au + bv)' = au' + bv', \qquad (a, b \text{ constants});$

(4) **Product rule** $(uv)' = u'v + uv' \ ;$

(5) **Quotient rule** $\left(\dfrac{u}{v}\right)' = \dfrac{vu' - uv'}{v^2} \ , \qquad (v \neq 0).$

Proof.   The linearity rule is easy to prove, using the definition of derivative and the Linearity Theorem for limits (11.3A).

The quotient rule for $1/v$ is also not difficult, using the definition of derivative and the Quotient Theorem for limits 11.3A (cf. Question 14.1/2). For the general case u/v, apply the product rule to $u(1/v)$.

To prove the product rule (4), at each $a \in I$, we have schematically,

$$\frac{u(x)v(x) - u(a)v(a)}{x - a} \quad = \quad \frac{u(x) - u(a)}{x - a} v(a) \quad + \quad u(x) \frac{v(x) - v(a)}{x - a} , \qquad \text{for } x \approx a;$$

$$\Downarrow \qquad\qquad\qquad \downarrow \qquad\qquad\qquad \downarrow$$

$$(uv)'(a) \qquad = \qquad u'(a)v(a) \qquad + \qquad u(a)v'(a), \qquad \text{as } x \to a.$$

The two limits on the right exist, since u and v are differentiable at a, and by the continuity of $u(x)$ at a (Theorem 14.1). Therefore by the Linearity Theorem for limits (11.3A), the limit on the left exists, i.e., $u(x)v(x)$ has a derivative at a, and this derivative has the value given on the right.   □

Where did we get the top line from? The numerator on the left is the change in uv as we move from a to x; the idea is to make this change in two steps, changing first u, then v; schematically:

$$u(a)v(a) \quad \longrightarrow \quad u(x)v(a) \quad \longrightarrow \quad u(x)v(x) \ .$$

The change in uv is the sum of the changes at each step:

$$u(x)v(x) - u(a)v(a) = [u(x) - u(a)]v(a) + u(x)[v(x) - v(a)].$$

and this gives the top line, after dividing through by $x - a$.

Theorem 14.2B   Chain rule

If f and g are differentiable, then over any interval where $f(g(x))$ is defined, it is differentiable, and

(7) $$f(g(x))' = f'(g(x))g'(x) \ .$$

In terms of variables, if $y = f(x)$ and $x = g(t)$, then

(8) $$\frac{dy}{dt} = \frac{dy}{dx}\frac{dx}{dt} \ .$$

Proof. It is most convenient to use the variables. Set

$$y_0 = f(x_0), \quad x_0 = g(t_0), \quad \Delta y = y - y_0, \quad \Delta x = x - x_0, \quad \Delta t = t - t_0.$$

Note that since $x = g(t)$ is continuous at t_0, we have

(9) $$\Delta x \to 0 \quad \text{as} \quad \Delta t \to 0 \ .$$

To clarify the argument, we make the simplifying assumption:

(10) $$\Delta x \neq 0 \quad \text{for} \quad \Delta t \underset{\neq}{\approx} 0 \ ;$$

the general case is left for the Problems. In summary form the proof then is:

$$\begin{array}{ccccc}
\dfrac{\Delta y}{\Delta t} & = & \dfrac{\Delta y}{\Delta x} & \cdot & \dfrac{\Delta x}{\Delta t}, \quad \text{for } \Delta t \underset{\neq}{\approx} 0; \\
\Downarrow & & \downarrow & & \downarrow \\
\dfrac{dy}{dt} & = & \dfrac{dy}{dx} & \cdot & \dfrac{dx}{dt}, \quad \text{as } \Delta t \to 0 \ .
\end{array}$$

In more detail: the first line makes sense because of our assumption (10). The two limits on the right exist, by (9) and the fact that f and g are assumed differentiable. Therefore, by the Product Theorem for limits 11.3A, the limit on the left exists and is the product of the limits on the right. $\quad\square$

Theorem 14.2C   Differentiation of inverse functions.

Let $y = f(x)$ and $x = g(y)$ be inverse functions defined respectively on intervals I and $J = f(I)$. If f is differentiable on I and $f'(x) \neq 0$ on I, then g is differentiable on J and

(11) $$\frac{dx}{dy} = \frac{1}{dy/dx} \ .$$

More precisely, if we set $y_0 = f(x_0)$, then whenever $f'(x_0) \neq 0$, g is differentiable at y_0, and

$$g'(y_0) = \frac{1}{f'(x_0)} \ .$$

Proof. Easy, using the variable notation. See the Exercises. $\quad\square$

Questions 14.2

1. Prove the sum rule for differentiation of $u(x) + v(x)$, at $x = a$.

2. Prove the rule for differentiating $1/v(x)$, at a point $x = a$, directly from the definition of derivative.

3. Criticize the following "proof" of Theorem 14.2C:

Since f and g are inverse functions, $g(f(x)) = x$; applying the chain rule gives $g'(f(x))f'(x) = 1$, from which $g'(y) = 1/f'(x)$.

4. Let $f(x) = \sin x$, $-\pi/2 \le x \le \pi/2$. Use (11) to find the derivative of its inverse function, $g(x) = \sin^{-1} x$, $-1 \le x \le 1$.

You may assume as known: $D \sin x = \cos x$, for all x.

(It is easiest to introduce a dependent variable, writing $y = \sin^{-1} x$. Remember that your answer must be expressed in terms of x.)

14.3   Derivatives and Local Properties.

We have already proved that differentiability implies continuity. Here are the elementary results relating derivatives with other local properties. See Chapter 10 for the definitions of these, if you've forgotten them.

Theorem 14.3A   *Suppose $f(x)$ is differentiable on an open interval I. Then*

$$f(x) \quad \text{locally increasing on } I \quad \Rightarrow \quad f'(x) \ge 0 \text{ on } I \text{ .}$$
$$f(x) \quad \text{locally decreasing on } I \quad \Rightarrow \quad f'(x) \le 0 \text{ on } I \text{ .}$$

Proof.   We verify the first statement for any point $a \in I$.

$$f(x) \text{ locally increasing at } a \quad \Rightarrow \quad f(x) \ge f(a), \qquad \text{for } x \approx a^+,$$

$$\Rightarrow \quad \frac{f(x) - f(a)}{x - a} \ge 0 \qquad \text{for } x \underset{>}{\approx} a,$$

$$\Rightarrow \quad \lim_{x \to a^+} \frac{f(x) - f(a)}{x - a} \ge 0 \text{ ,}$$

by the Limit Location Theorem for functions 11.3C. The last line says $f'(a^+) \ge 0$; since by hypothesis $f'(a)$ exists, we have (cf. 14.1(2))

$$f'(a) = f'(a^+) \ge 0. \qquad \square$$

If $f(x)$ is locally decreasing, then $-f(x)$ is locally increasing, so by the first part, $-f'(x) \ge 0$, i.e., $f'(x) \le 0$, on I.   □□

The theorem is also true for a closed interval I, interpreting the derivative at an endpoint as the left- or right-hand derivative, and "locally increasing at an endpoint" as only applying to that side of the endpoint lying inside I.

If $f(x)$ is locally *strictly* increasing on I, it doesn't follow that $f'(x) > 0$ on I, as the example $f(x) = x^3$ shows. In the proof, even if the difference quotient were positive, the Limit Location Theorem would only show that its limit was ≥ 0.

Definition 14.3A   Let $f(x)$ be defined on an open interval I; then

$c \in I$ is a **local maximum point** if $f(c) \geq f(x)$ for $x \approx c$;

$c \in I$ is a **local minimum point** if $f(c) \leq f(x)$ for $x \approx c$.

Sometimes these points are called *relative* maximum and minimum points. In everyday mathematical speech, the word "point" is often omitted, but this can lead to confusion, since "maximum" by itself should mean the greatest y-value, not the x-value where it is attained.

Remarks.   A local maximum point for $f(x)$ on I may also be a maximum point for $f(x)$ on I, though in general this need not be the case. In the accompanying picture, $f(x)$ has on the interval $[a, b]$:

a local maximum point at d,   a maximum point at a,

a local minimum point at c,   a minimum point at b.

The words "local extremum" cover both local maximum and minimum. Note that Definition 14.3A implies that the point must be interior to I, i.e., not an endpoint, to qualify as a local extremum. The reason is that, if $f(x)$ is differentiable, we want its derivative to be 0 at a local maximum point. Thus in the picture, a is not considered to be a local maximum point, even though there is a one-sided neighborhood in which it is king of the hill.

Theorem 14.3B   *Suppose $f(x)$ is differentiable on the open interval I.*

$a \in I$ *is a local extremum point* $\Rightarrow$ $f'(a) = 0$.

Proof.   Suppose for example that a is a local maximum point. Then

$$\frac{f(x) - f(a)}{x - a} \leq 0 \quad \text{for } x \approx a^+; \qquad \frac{f(x) - f(a)}{x - a} \geq 0 \quad \text{for } x \approx a^- .$$

Take limits as $x \to a^+$ and $x \to a^-$ respectively. This gives

$$f'(a^+) \leq 0 \qquad \text{and} \qquad f'(a^-) \geq 0 ,$$

by the Limit Location Theorem 11.3C. Since $f'(a)$ exists, both one-sided derivatives must $= f'(a)$, by (2) above. Therefore $f'(a) = 0$.   □

If f has a local minimum at a, one can apply the above to $-f(x)$, which has a local maximum at a.   □□

Definition 14.3B   A point where $f'(x) = 0$ is called a **critical point** for $f(x)$.

Of course we want to use Theorem 14.3B to find local extremum points. The difficulty is that its converse is not true: a critical point need not be a local extremum point, as is shown by the function x^3 at $x = 0$.

There is a well-known second derivative test to see whether a critical point is in fact a local extremum point; we will discuss it in Chapter 16. However, it is grossly overused, it is often clumsy to apply in practice, and it sometimes fails (as it does even for the simple function x^4, whose minimum is clearly at $x = 0$).

Here is a simpler way you can often use to decide if a critical point is actually an extremum point. It follows immediately from Theorem 14.3B.

Isolation Principle 14.3C.

If we can find an I on which $f(x)$ is differentiable, such that

(i) I contains exactly one critical point x_0 of $f(x)$, and

(ii)   the Maximum Theorem (13.3) predicts that $f(x)$ has one or more local extremum points on I,

then $f(x)$ has only one local extremum point on I, and it is at x_0.

> "There's no place anywhere near this place that's anything like this place, so this must be the place."
>
> —sign on the wall at the old Durgin-Park eatery in the Quincy Market

The following example shows a typical use of the principle. Note how the Maximum Theorem from Chapter 13 can be used even though the domain $(-\infty, \infty)$ of $f(x)$ is not compact.

Example 14.3   Let $f(x) = xe^{-x}$. Find and classify its extremum points.

Solution.   The function is differentiable on $I = (-\infty, \infty)$. We begin by finding its critical points. Calculating and combining terms, we get
$$f'(x) = e^{-x}(1 - x);$$
therefore
$$f'(x) = 0 \quad \Leftrightarrow \quad x = 1.$$

So there is just one critical point, at $x = 1$. But is it an extremum point?

According to the Isolation Principle, it will be a local extremum, and the unique one, if we can use the Maximum Theorem 13.3 to show that $f(x)$ has a local extremum somewhere on $I = (-\infty, \infty)$. But we can use it: a preliminary rough sketch of the graph to see where to calculate, and then calculation shows
$$f(0) = 0, \qquad f(1) = 1/e, \qquad f(2) < 1/e \ ;$$
the Maximum Theorem shows then that $f(x)$ has a maximum point on $[0, 2]$, which must be a local one, since it is not either of the endpoints.

Thus the function has a unique maximum point, at $x = 1$. It has no minimum, since we have shown there is only one extremum point; that it has no minimum is also clear from the fact that $xe^{-x} \to -\infty$ as $x \to -\infty$.   □

Questions 14.3

1. True or false? If false, give a counterexample. I is an interval.

(a)   If $f(x)$ is continuous on I, a local maximum point is a critical point.

(b) $f(x)$ is differentiable and strictly decreasing on $I \ \Rightarrow \ f'(x) < 0$ on I.

(c) I compact, $f(x)$ differentiable on $I \ \Rightarrow \ f(x)$ has a local extremum on I.

(d) I compact, $f(x)$ continuous on I $\Rightarrow$ $f(x)$ is differentiable at all but a finite number of points.

(e) $f(x)$ increasing on I $\Rightarrow$ $f(x)$ has no local extrema on I.

(f) $f(x)$ increasing for $x \approx a^-$, decreasing for $x \approx a^+$ $\Rightarrow$ a is a local maximum point for $f(x)$.

2. Without looking at the text, prove from scratch that if $f(x)$ is differentiable and locally decreasing on the open interval I, then $f'(x) \leq 0$ on I.

3. Without looking at the text, prove from scratch that if $f(x)$ is differentiable on the open interval I and has a local minimum at the point c in this interval, then $f'(c) = 0$.

Exercises

14.1

1. (a) Let $f(x) = \sin x$. Derive the formula for $f'(a)$ at an arbitrary point a, by using the standard trigonometric identities and trigonometric limits (cf. Exercises 11.3/1,2).

(Like many proofs of derivative formulas, this is best done by making the change of variable $x = a + \Delta x$, and working with Δx instead of x.)

(b) Show the two limits needed in part (a) can be interpreted as derivatives of certain functions at 0.

2. Let $f(x) = e^x$. Assuming that $f'(0) = 1$, prove that $f'(x) = e^x$. (Use the remark in Exercise 1 above.)

3. Let $f(x)$ be an even function which is differentiable at 0. Prove from the definitions that $f'(0) = 0$. (Consider $f'(0^+), f'(0^-)$.)

4. (a) Suppose $f(0) = f'(0) = 0$. Find $\lim\limits_{x \to 0} \dfrac{f(x)}{x}$, with proof.

(b) Suppose $|f(x)| \leq x^2$ for $x \approx 0$. Prove that $f'(0) = 0$.

5. Let $f(x) = x \sin(1/x)$, $x \neq 0$; $f(0) = 0$. Prove that $f(x)$ is continuous at 0, but not differentiable at 0.

6. (a) Prove: if $f'(a)$ exists, then $f(x) = L(x) + e(x)$, where

$$L(x) = f(a) + f'(a)(x - a) \quad \text{and} \quad \lim_{x \to a} \frac{e(x)}{x - a} = 0 .$$

(b) $L(x)$ is called the *linearization* of $f(x)$ at a; explain part (a) geometrically by interpreting $f(x), L(x)$, and $e(x)$ graphically.

7. Prove a sort of converse to Exercise 6a:

If $f(x)$ is defined for $x \approx a$, and there is a number k such that

$$f(x) = f(a) + k(x - a) + e(x), \qquad \text{where} \quad \lim_{x \to a} \frac{e(x)}{x - a} = 0,$$

then $f(x)$ is differentiable at a, and $k = f'(a)$.

This shows that $L(x)$ is that linear function through $(a, f(a))$ which gives the best approximation to $f(x)$ for $x \approx a$, in the sense that it is the only linear function for which the error term $e(x)$ is negligible compared with $x - a$.

14.2

1. Let $f(x) = \tan x$, $-\pi/2 < x < \pi/2$, and $g(x) = \tan^{-1} x$, its inverse function. Derive the formula for $g'(x)$, either from the rule (11) for differentiating inverse functions, or better, as in Question 14.2/4.

2. Let $u(x)$ be differentiable and non-negative. Prove $Du^k = ku^{k-1}u'$ if

    (a) $k = n$, a positive integer (use induction, cf. Appendix A.4);
    (b) $k = 1/n$, assuming $u(x)$ positive (use (11));
    (c) $k = m/n$, a rational number, assuming $u(x) > 0$ (use (a) and (b)).

3. Prove the rule (11) for differentiating inverse functions, using variable notation. (Start with $\Delta x/\Delta y = 1/(\Delta y/\Delta x)$; the proof is almost immediate, but you must be a little careful, since denominators cannot be 0.)

4. If $f(x)$ is differentiable, prove that

    (a) $f(x)$ even $\Rightarrow$ $f'(x)$ odd;   (b) $f(x)$ odd $\Rightarrow$ $f'(x)$ even.

5. If $f(x)$ is differentiable and periodic, then $f'(x)$ is periodic. Proof or counterexample.

14.3

1. The following refer to an interval I on which $f'(x)$ exists.

    (A) $f(x)$ locally increasing $\Rightarrow$ $f'(x) \geq 0$
    (B) $f(x)$ locally increasing $\Rightarrow$ $f'(x) > 0$
    (C) $f(x)$ locally strictly increasing $\Rightarrow$ $f'(x) \geq 0$
    (D) $f(x)$ locally strictly increasing $\Rightarrow$ $f'(x) > 0$.

Make a "lattice diagram" (as it is called) showing these four statements, such that if two are connected by a line, the higher one is a stronger statement than the lower one (cf. Appendix A.1 for "stronger" and "weaker"; at right a typical diagram is drawn.)

Label on your diagram which statements are true, and which false.

2. Find the extremum points of the following functions on the indicated interval. Give reasoning as in Example 14.3 to show you have them all.

    (a) $\dfrac{x^2}{1 + x^4}$ $(-\infty, \infty)$ (b) $\dfrac{\ln x}{x}$, $[1, \infty)$

3. Let $f(x)$ be a polynomial of degree n.

(a) What is the maximum number of critical points $f(x)$ can have? For each n, give an example, with proof, of a polynomial having the maximum number for that n.

(b) Answer the same question with "maximum" replaced by "minimum".

Problems

14-1 Suppose $f(x)$ is differentiable at a. Define a new function by

$$F(\Delta x) = \frac{f(a + \Delta x) - f(a - \Delta x)}{2\Delta x}, \qquad x = a + \Delta x.$$

(a) Find $\lim_{\Delta x \to 0} F(\Delta x)$, and interpret F geometrically.

(b) Can the limit of part (a) exist even if f is not differentiable at a? If so, give an example; if not, prove it.

(c) Suppose the right- and left-hand derivatives of $f(x)$ exist at a, but they are not equal. What is $\lim_{\Delta x \to 0} F(\Delta x)$? Prove it.

14-2 Suppose $f(x)$ is differentiable for all x, and for all a, b, we have
$$f(a + b) = f(a) + f(b).$$
Prove that $f'(x) = f'(0)$ for all x. What kind of function is $f(x)$?

14-3 Suppose $f(x)$ is differentiable for all x, and for all a, b,
$$f(a + b) = f(a) + f(b) + 2ab .$$
Deduce that $f'(x) = f'(0) + 2x$. Give two examples of such a function.

14-4 Suppose $f(x)$ is differentiable for all x. State hypotheses under which $|f(x)|$ will be differentiable for all x, and prove it. (Try examples first and draw pictures.)

14-5 Prove the function defined by $f(x) = x^2 \sin(1/x)$ if $x \neq 0$; $f(0) = 0$ is differentiable for all x, but $f'(x)$ is not a continuous function (cf. Exercise 14.1/4).

14-6 (a) Draw the graph of a function which is continuous on $[-1, 1]$, and differentiable at 0, but not differentiable in any δ-neighborhood of 0. (Hint: cf. preceding problem.)

(b) Give an example of a function defined for all x which is differentiable at 0, but not even continuous at any other point.

14-7 Prove the Chain Rule 14.3B for the general case (i.e., without the assumption $\Delta x \underset{\neq}{\ne} 0$ for $\Delta t \approx 0$), as follows.

(a) Show the hypothesis of differentiability can be written in the form of the two equations (cf. Exercise 14.1/6)

$$\Delta y = f'(x_0)\Delta x + e_1, \qquad \Delta x = g'(t_0)\Delta t + e_2 ;$$

where

$$\lim_{\Delta x \to 0} \frac{e_1}{\Delta x} = 0, \qquad \lim_{\Delta t \to 0} \frac{e_2}{\Delta t} = 0.$$

(b) Substitute the second equation into the first, to get

$$\Delta y = f'(x_0)g'(t_0)\Delta t + e,$$

where the error term e is expressed in terms of the error terms e_1 and e_2.

(c) Then use the expression for e to show that

$$\lim_{\Delta t \to 0} \frac{e}{\Delta t} = \lim_{\Delta t \to 0} \epsilon \cdot \frac{\Delta x}{\Delta t}, \qquad \text{where} \quad \epsilon = \begin{cases} e_1/\Delta x, & \Delta x \ne 0, \\ 0, & \Delta x = 0; \end{cases}$$

then show this limit is 0 and finish the argument using (b).

(The simple proof of the chain rule we gave before used Δx in the denominator of a fraction, so we had to assume it was non-zero. By using here the linearizations and introducing the new function ϵ, we avoid ever having to write Δx in a denominator.)

All the same, G. H. Hardy's *Pure Mathematics*, the book that introduced rigor into calculus, went through several editions before someone pointed out that its proof of the Chain Rule (the one we gave earlier) assumed $\Delta x \underset{\neq}{\ne} 0$ for $\Delta t \approx 0$.

Answers to Questions

14.1

1. (a) $F(x)$ has a removable discontinuity at a

$$\Leftrightarrow \quad \lim_{x \to a} F(x) \text{ exists} \quad \Leftrightarrow \quad f'(a) \text{ exists}.$$

(b) $F(x)$ has a jump discontinuity

$$\Leftrightarrow \quad \lim_{x \to a^+} F(x) \quad \text{and} \quad \lim_{x \to a^-} F(x) \text{ exist but are not equal}$$
$$\Leftrightarrow \quad f'(a^+) \quad \text{and} \quad f'(a^-) \text{ exist, but } f'(a) \text{ does not (14.1(2)).}$$

2. $f(x) = x^2,\ x \ge 0; \quad = -x^2,\ x \le 0.$
Both derivatives are 0, so $f'(0)$ exists.

14.2

1. $\dfrac{[u(x) + v(x)] - [u(a) + v(a)]}{x - a} = \dfrac{u(x) - u(a)}{x - a} + \dfrac{v(x) - v(a)}{x - a};$

now take limits as $x \to a$ and apply Theorem 11.3A; since the two limits on the right exist, so does the one on the left, and it equals $u'(a) + v'(a)$.

2. $\dfrac{1/v(x) - 1/v(a)}{x - a} = \dfrac{v(a) - v(x)}{(x - a)v(x)v(a)} \longrightarrow -v'(a)\dfrac{1}{v(a)v(a)}$, as $x \to a$,

by the Algebraic Limit Theorems 11.3A, and because $v(x) \to v(a)$ (since $v(x)$ is continuous at a by Theorem 14.1).

3. The argument assumes $g(x)$ is differentiable, but is otherwise correct; it thus proves a weaker statement than Theorem 14.2C.

4. Write $y = \sin^{-1} x$, then $x = \sin y$, so

$$dx/dy = \cos y, \quad -\pi/2 \le y \le \pi/2 \ ;$$

therefore by Theorem 14.2C,

$$dy/dx = 1/\cos y = 1/\sqrt{1 - \sin^2 y} = 1/\sqrt{1 - x^2},$$

where we use the positive square root since $\cos y \ge 0$ on the interval.

14.3

1. (a) false; $f(x) = -|x|$ (0 is a local maximum, but not a critical point)

(b) false; $f(x) = -x^3$ ($f'(0) = 0$, but the function is strictly decreasing)

(c) false; $f(x) = x$ on $[0, 1]$

(d) false; on $[-1, 1]$, $f(x)$ looks like $x \sin(1/x)$, but instead of curves is made up entirely of line segments—it has a sharp corner (like $|x|$ at 0) at each of its local extremum points, and there are an infinity of them on the interval. It is continuous everywhere (continuous at 0 by the same reasoning as for $x \sin(1/x)$.)

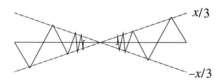

(e) false; $f(x) = c$ (increasing, and every point is a local extremum)

(f) true

2. Same proof as in the text, with the inequalities reversed.

3. Same proof as in the text, with the inequalities reversed.

15

Differentiation: Global Properties

15.1   The mean-value theorem.

The last chapter used basic facts about limits to relate the derivative $f'(x)$ to local properties of $f(x)$. We turn now to global properties. We will need the deeper results about continuity discussed in Chapter 13, those depending on completeness and compactness.

Here is the key theorem from which everything follows.

Theorem 15.1   Mean-value theorem.

Let $f(x)$ be continuous on $[a, b]$ and differentiable on (a, b). Then there is at least one point $c \in (a, b)$ such that

(1) $$f(b) - f(a) = f'(c)(b - a) .$$

Remarks.   The theorem has an easy geometric interpretation if you divide both sides by $b - a$, writing it in the following form (which is the one we shall actually use to prove the theorem):

(1a) $$\frac{f(b) - f(a)}{b - a} = f'(c) .$$

In this form, the theorem says there is at least one point c inside the interval at which the slope of the tangent equals the slope of the secant PQ.

Note that since differentiability implies continuity (Theorem 14.1), the hypotheses can also be stated: $f(x)$ is differentiable on (a, b) and continuous at the endpoints (as always, only one-sided continuity is required at the endpoints).

Examples 15.1 Discuss the applicability of the Mean-value Theorem to

(a) $f(x) = |x|$ on $[-1, 2]$;   (b) $f(x) = \sqrt{x}$ on $[0, a]$.

Solution. (a) Since $|x|$ is not differentiable at the point 0 on the interval, the hypothesis is not satisfied, so the theorem does not apply. In fact, the conclusion does not hold, since the left side of (1a) has the value $1/3$, while $f'(x)$ can only be ± 1; that is, no such point c exists.

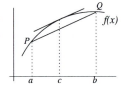

Notice that non-differentiability at a single point is enough to make the theorem inapplicable; the picture shows how. "For want of a nail..."

(b) The function $f(x) = \sqrt{x}$ is differentiable on $(0, a]$, but not at 0 where it has infinite slope. However, since it is right-continuous at 0, it satisfies the hypotheses. So the conclusion also holds, and we can verify this by explicit calculation: since $f'(x) = 1/(2\sqrt{x})$, we have

$$\frac{f(a) - f(0)}{a - 0} = \frac{1}{\sqrt{a}}; \qquad f'(c) = \frac{1}{\sqrt{a}}, \quad \text{if } c = \frac{a}{4}. \qquad \square$$

We prove the Mean-value Theorem now. The special case where the function $f(x)$ is 0 at both endpoints is usually called *Rolle's Theorem*; this is stated and proved first, then used to prove the general case.

Lemma 15.1   Rolle's theorem.

Let $f(x)$ be continuous on $[a, b]$ and differentiable on (a, b), and suppose that $f(a) = f(b) = 0$. Then there is at least one point $c \in (a, b)$ such that $f'(c) = 0$.

Proof.   By the Maximum Theorem 13.3, $f(x)$ has a maximum and a minimum point on $[a, b]$.

If one of these points—call it c—is not an endpoint, then it is actually a local extremal point for $f(x)$, so that $f'(c) = 0$, by Theorem 14.3B, and $c \in (a, b)$.

If however, it is the endpoints a and b that are the maximum and minimum points of $f(x)$ on the interval, then since $f(a) = f(b) = 0$, the function must be 0 on the whole interval, and then $f'(c) = 0$ for all $c \in (a, b)$. $\qquad \square$

Proof of the Mean-value Theorem.   Set

$$g(x) = f(x) - L(x),$$

where $L(x)$ is the function whose graph is the secant line PQ shown, i.e.,

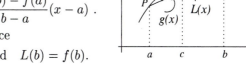

(2) $\qquad\qquad L(x) = f(a) + \dfrac{f(b) - f(a)}{b - a}(x - a) .$

Rolle's Theorem applies to $g(x)$, since

$$L(a) = f(a) \quad \text{and} \quad L(b) = f(b).$$

Therefore there is a point $c \in (a, b)$ such that $g'(c) = 0$, that is

$$L'(c) = f'(c) \qquad \text{for some } c \in (a, b) .$$

But this is the same as (1a) above, since $L'(c)$ is the quantity on the left side of equation (1a); you can either see this by differentiating (2), or by noting that for all x, $L'(x)$ is the slope of the secant PQ. $\qquad \square\square$

Questions 15.1

1. Discuss the applicability of the Mean-value Theorem to

(i) $\sqrt{1 - x^2}$ on $[-1, 1]$ (ii) $y = x^{1/3}$ on $[-1, 1]$.

(Do the hypotheses hold? Does the conclusion? If it does, locate c.)

2. There can be several c that satisfy (1a). If $f(x)$ is a polynomial of degree $n > 1$, what is the maximum number possible of such points c?

3. Why does the statement of the Mean-value Theorem imply that the Completeness Property will be used in its proof, and where in the proof is it used?

15.2 Applications of the mean-value theorem.

Since the Mean-value Theorem is a local-to-global theorem (requiring the Maximum Theorem in its proof), one might expect it can in turn be used to prove other local-to-global theorems. Here are typical examples; they are like converses to Theorem 14.3A ("locally increasing $\Rightarrow f'(x) \geq 0$"), but not exactly.

Theorem 15.2 *Let $f(x)$ be differentiable on the interval I. Then on I,*

$$
\begin{aligned}
f'(x) > 0 &\quad \Rightarrow \quad f(x) \text{ is strictly increasing ;} \\
f'(x) < 0 &\quad \Rightarrow \quad f(x) \text{ is strictly decreasing ;}
\end{aligned}
\tag{3}
$$

$$
\begin{aligned}
f'(x) \geq 0 &\quad \Rightarrow \quad f(x) \text{ is increasing ;} \\
f'(x) \leq 0 &\quad \Rightarrow \quad f(x) \text{ is decreasing ;}
\end{aligned}
\tag{4}
$$

$$
f'(x) = 0 \quad \Rightarrow \quad f(x) \text{ is constant .}
\tag{5}
$$

Proof. Suppose $x_1 < x_2$, where $x_1, x_2 \in I$.

By the Mean-value Theorem 15.1, there is a point $c \in I$ such that
$$ f(x_2) - f(x_1) = f'(c)(x_2 - x_1) . $$
To prove, say, the first statement in (4)—the others go similarly—we have
$$
\begin{aligned}
f'(x) \geq 0 \text{ on } I \quad &\Rightarrow \quad f'(c) \geq 0 \\
&\Rightarrow \quad f(x_2) \geq f(x_1) \\
&\Rightarrow \quad f(x) \text{ is increasing .} \qquad \square
\end{aligned}
$$

The sort of information provided by the preceding theorem:

$$\text{properties of } f'(x) \ \Rightarrow \ \text{properties of } f(x),$$

is generally the subject of *integration*; the Mean-value Theorem can therefore be thought of as a baby step in the direction of integration. But it is also a basic step, since the theorems connecting differentiation and integration depend on the Mean-value Theorem, in particular on the "obvious" result (5).

The Mean-value Theorem is primarily a theoretical tool for estimating the difference between two values of an abstractly given function $f(x)$. But as always, one gets used to it by practicing with specific functions.

Example 15.2. Using the Mean-value Theorem, show that

  (a) $\sin x < x$, for all $x > 0$;

  (b) given $\epsilon > 0$, $\ln x_2 - \ln x_1 < \epsilon(x_2 - x_1)$, if $x_2 > x_1 \gg 1$.

Solution.   (a) If $x > 1$, it is true since $\sin x \leq 1 < x$. If $0 < x \leq 1$, we have by the Mean-value Theorem (1),

$$\sin x - \sin 0 = (\cos c)(x - 0), \quad \text{for some } c \text{ such that } 0 < c < x \leq 1.$$

Since $0 < \cos c < 1$ in this interval, we conclude that $\sin x < x$.   □

(b) $\ln x_2 - \ln x_1 = (1/c)(x_2 - x_1)$, for some c such that $x_1 < c < x_2$. The result now follows since $1/c < 1/x_1 < \epsilon$ if $x_1 > 1/\epsilon$.   □

The Mean-value Theorem (1) remains true if $b < a$, since both sides simply change sign. Therefore, if we replace b by x, we can write the theorem in the following useful form; this is the statement that will be generalized in the next two chapters, where we will use higher derivatives.

Approximation form of the mean-value theorem.

Let $f(x)$ be differentiable on an interval I, and let $a \in I$. Then, for each $x \in I$, there is at least one c between a and x such that

$$(6) \qquad\qquad f(x) = f(a) + f'(c)(x - a) .$$

To get insight into what (6) says, we compare it with a similar-looking but more elementary statement. Namely, from the definition of derivative, one can deduce (cf. Exercise 14.1/6; the two-line proof is given in Theorem 16.1 at the beginning of the next chapter):

Linearization approximation.   For $x \approx a$, we have

$$(7) \qquad\qquad f(x) \approx f(a) + f'(a)(x - a),$$

with an error that is negligible compared with $x - a$.

Both (6) and (7) estimate $f(x)$ based on information about $f'(x)$. The linearization approximation (7) only requires knowledge of $f'(x)$ at a, but it does not give very precise information. The mean-value approximation (6) does give a precise statement, provided we have an estimate for $f'(x)$ over the whole interval (since we do not know where c is exactly).

For example, contrast (6) and (7) for the function $\sin x$, for $x > 0$.

$\star$ Mean-value Theorem: $\sin x < x$, for $x > 0$ (cf. Example 15.2a);

$\star$ Linearization Approximation: $\sin x \approx x$, for $x \approx 0$.

The first is a precise statement, the second is not; but the two statements are giving different kinds of information.

Questions 15.2

1. Prove: $f'(x) < 0$ on $I \Rightarrow f(x)$ is strictly decreasing on I.

2. Prove that if $f'(x) = 0$ on an interval I, then $f(x)$ is constant on I. Is this true if I is replaced by D_f, the domain of f?

3. Prove that $e^x > 1 + x$, if $x > 0$. (Use elementary facts about e^x, not the infinite series representation.)

4. Prove that given any number $a > 1$, $e^x < 1 + ax$, for $x \underset{\neq}{\approx} 0^+$.

5. In (6), why must one assume f is differentiable on the whole interval, when only its derivative at c enters into the formula?

15.3   Extension of the mean-value theorem.

An elegant application of the Mean-value Theorem is its use in proving l'Hospital's rule for evaluating indeterminate forms of type $0/0$ or ∞/∞. For this we need the theorem in a slightly strengthened form which uses two functions simultaneously. In the statement, we call the independent variable t, as it makes the proof a bit more natural.

Theorem 15.3   The Cauchy mean-value theorem.

Suppose that $f(t)$ and $g(t)$ have continuous first derivatives on $[a,b]$, and $g'(t) \neq 0$ on (a,b). Then there is a point $c \in (a,b)$ such that

$$(8) \qquad \frac{f(b) - f(a)}{g(b) - g(a)} = \frac{f'(c)}{g'(c)}, \qquad a < c < b .$$

Note that this does not follow from the ordinary Mean-value Theorem 15.1, since if we apply that theorem separately to the numerator and denominator of the left side of (8), we would get on the right side the expression $f'(c_1)/g'(c_2)$, with different c_i. The whole point of Cauchy's form is that even though there are two functions, a single number c can be found.

Proof.   We give the essence of the argument first, then fill in a few details. Set

$$(9) \qquad\qquad x = g(t), \qquad y = f(t) .$$

As t varies from a to b, the point (x, y) traces out a smooth curve in the xy-plane, as shown. The hypotheses imply (see below) that this curve is the graph of a differentiable function $y(x)$ on the interval $(x(a), x(b))$, and by the chain rule,

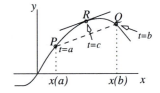

$$(10) \qquad\qquad \frac{dy}{dx} = \frac{dy/dt}{dx/dt} = \frac{f'(t)}{g'(t)} .$$

By the secant interpretation (1a) of the Mean-value Theorem 15.1, the slope of the secant PQ is equal to the slope of the curve at some intermediate point R. Suppose $t = c$ at the point R. Then (8) is proved, since

$$\underset{\text{slope of } PQ}{\frac{f(b) - f(a)}{g(b) - g(a)}} \quad = \quad \underset{\text{slope at } R}{\left.\frac{dy}{dx}\right|_{t=c}} \quad = \quad \underset{\text{using (10)}}{\frac{f'(c)}{g'(c)}} . \qquad\qquad \square$$

To complete the argument, we have to see that y is a differentiable function of x between P and Q. For this, note that g' is continuous and non-zero, so by Bolzano's Theorem 12.1 it cannot change sign; say $g' > 0$ on (a, b). Then $g(t)$ is differentiable and strictly increasing on $[a, b]$ (see Exercise 15.2/3).

This shows first of all that the left denominator in (8) is non-zero. It also shows that g has a differentiable inverse function, by Theorems 12.4 and 14.2C:

$$t = h(x) \ , \qquad x(a) < x < x(b) \ ,$$

which in turn shows the composite function $y = f(h(x))$, $x(a) < x < x(b)$, is differentiable, and the chain rule (10) holds. $\qquad \qquad \square\square$

The hypotheses are stronger than needed; it would suffice for the above argument to have the derivatives exist in (a, b), with g' continuous and nonzero in (a, b), and f and g continuous at a and b .

For a simpler proof of the the Cauchy Mean-value Theorem, using even weaker hypotheses, but less geometrically appealing, see Exercise 15.3/2.

Questions 15.3

1. If $f(t) = \cos t$ and $g(t) = \sin t$, what point c satisfies (8) for the interval $[0, \pi/2]$? Answer two ways:

    (a) by using (8);

    (b) by going back to the proof and interpreting c geometrically.

2. Illustrate the remark just before the proof starts, by calculating the points c_1 and c_2 that would be obtained by applying the Mean-value Theorem separately to the functions $f(t)$ and $g(t)$ of the previous question.

3. In (8), what hypothesis guarantees that $g(b) - g(a) \neq 0$? Show that it can be weakened by dropping a continuity requirement.

15.4 L'Hospital's rule for indeterminate forms.

In comparing functions, it is often important to be able to evaluate limits of quotients: $\lim_{x \to a} f(x)/g(x)$. Usually f and g are continuous; in that case, by the Quotient Theorem for limits and the Limit Form of Continuity 11.4, we get

$$\lim_{x \to a} \frac{f(x)}{g(x)} = \frac{\lim f(x)}{\lim g(x)} = \frac{f(a)}{g(a)} \ .$$

If however both $f(a) = 0$ and $g(a) = 0$, this method produces what is called an **indeterminate form of type** $0/0$, and we must look for another way. Sometimes $f(x)/g(x)$ can be massaged algebraically so that its limit becomes clear; most students turn first (even when they shouldn't) to l'Hospital's rule.

We first consider an elementary case of it, then the full rule.

Theorem 15.4A   L'Hospital's rule (elementary case).

If $f(a) = g(a) = 0$, and the right side below is defined, then

(11)
$$\lim_{x \to a} \frac{f(x)}{g(x)} = \frac{f'(a)}{g'(a)} \ .$$

Proof.   This is easy application of the definition of derivative, and is left as Exercise 15.4/1.   □

Examples 15.4A Evaluate each of the following indeterminate forms:

(a) $\displaystyle\lim_{x \to 0} \frac{x^3 - 2x}{x^3 + x}$ (b) $\displaystyle\lim_{x \to 0} \frac{\sin x}{x}$ (c) $\displaystyle\lim_{x \to 0} \frac{1 - \cos x}{\sin x}$ (d) $\displaystyle\lim_{x \to 0} \frac{1 - \cos x}{x \sin x}$

Solution. (a) One can use (11) but it's silly to do so; after cancelling an x, the Algebraic Limit Theorems immediately give -2 as the limit.

(b) Using (11) gives the limit $(\cos 0)/1 = 1$, but it's a cheat (Question 4).

(c) Using (11), the limit $= \dfrac{\sin 0}{\cos 0} = 0$.

(d) Using (11), $\displaystyle\lim_{x \to 0} \frac{1 - \cos x}{x \sin x} = \left. \frac{\sin x}{\sin x + x \cos x} \right|_{x=0} = \frac{0}{0}$;

here the elementary case of l'Hospital's rule fails, since $f'(a)/g'(a)$ is not defined. Since it is of the type $0/0$, one hopes to be able to reapply l'Hospital's rule, continuing the differentiation; to justify this, however, we need the full version of the theorem, which we give now.

Theorem 15.4B   L'Hospital's rule for 0/0 .

Suppose that $f(x)$ and $g(x)$ have continuous first derivatives for $x \approx a$, and $f(a) = g(a) = 0$, but $g'(x) \neq 0$ for $x \underset{\neq}{\approx} a$. Then, if the limit on the right exists,

(12)
$$\lim_{x \to a} \frac{f(x)}{g(x)} = \lim_{x \to a} \frac{f'(x)}{g'(x)}.$$

Proof.   By the Cauchy Mean-value Theorem 15.3, for $x \approx a$ we have, for some c between a and x,

$$\frac{f(x)}{g(x)} = \frac{f'(c)}{g'(c)} \ .$$

Therefore

$$\lim_{x \to a} \frac{f(x)}{g(x)} = \lim_{x \to a} \frac{f'(c)}{g'(c)} = \lim_{x \to a} \frac{f'(x)}{g'(x)} \ ,$$

since c lies between a and x.   □

Examples 15.4B   Evaluate: (a) $\displaystyle\lim_{x\to0}\frac{1-\cos x}{x\sin x}$ (b) $\displaystyle\lim_{x\to0}\frac{e^x-1-x}{x^2}$

Solution. (a) Continuing the work in Example 15.4A(d) above,

$$\lim_{x\to0}\frac{1-\cos x}{x\sin x}=\lim_{x\to0}\frac{\sin x}{\sin x+x\cos x}=\lim_{x\to0}\frac{\cos x}{2\cos x-x\sin x}=\frac{1}{2}.$$

Here Theorem 15.4B is applied twice. Using the falling-domino reasoning: the first limit exists *if* the second one does, by Theorem 15.4B; the second limit exists *if* the third one does, by Theorem 15.4B; but the third limit exists by the Quotient Theorem for limits, and is 1/2. So all three limits exist and are 1/2. □

   (b) $\displaystyle\lim_{x\to0}\frac{e^x-1-x}{x^2}=\lim_{x\to0}\frac{e^x-1}{2x}=\lim_{x\to0}\frac{e^x}{2}=\frac{1}{2}.$ □

L'Hospital's rule works for one-sided limits as $x\to a^+$ or $x\to a^-$, making the relevant changes in the hypotheses. It also holds for limits taken as $x\to\infty$, since the change of variable $x=1/t$ reduces it to the case $t\to0^+$.

The rule is also valid for indeterminate forms of type ∞/∞ . One might expect to handle these—in theory or in practice—by rewriting the fraction so as to convert it to the 0/0 form:

(13) $\displaystyle\lim_{x\to a}\frac{f(x)}{g(x)}=\lim_{x\to a}\frac{1/g(x)}{1/f(x)}$ (a can be ∞) ;

alas, this doesn't lead to a proof and rarely works in practice. One has to prove the rule from scratch, using again the Cauchy Mean-value Theorem. We state this form of the rule for the most common case, $x\to\infty$; the other cases are stated by making the usual changes.

Theorem 15.4C   L'Hospital's rule for ∞/∞

   If $f(x)$ and $g(x)$ are differentiable for $x\gg1$, and $f(x)\to\infty$, $g(x)\to\infty$ as $x\to\infty$, then, if the limit on the right exists,

(14) $\displaystyle\lim_{x\to\infty}\frac{f(x)}{g(x)}=\lim_{x\to\infty}\frac{f'(x)}{g'(x)}.$

Proof.   Left for the Problems. Another proof will be given in Chapter 20.

Questions 15.4

   1. Evaluate: (a) $\displaystyle\lim_{x\to1}\frac{\ln x}{x-1}$ (b) $\displaystyle\lim_{x\to0}\frac{x^2}{1-\cos x}$ (c) $\displaystyle\lim_{x\to1}\frac{x^2-2x+1}{x^3-3x+1}$.

   2. Evaluate $\displaystyle\lim_{x\to\infty}\frac{\ln x}{x^r}$,   if $r>0$.

   3. Consider $\displaystyle\lim_{x\to\infty}\frac{e^x}{x}$; try evaluating it by using (13) to convert it to an indeterminate form of type 0/0, and then applying l'Hospital's rule for forms of type 0/0.

   4. Why is it cheating to use l'Hospital's rule on Example 15.4A(b)?

Exercises

15.1

1. Each of the following illustrates on the interval $[-1, 1]$ how the conclusion of the Mean-value Theorem can fail to hold if the hypothesis about $f(x)$ is not satisfied. For each, tell how the hypothesis fails, and show there is no point $c \in (-1, 1)$ satisfying the conclusion of the theorem.

$\qquad$ (a) $x^{2/3}$ $\qquad$ (b) $1/x$ $\qquad$ (c) $\tan \pi x$.

2. The conclusion of the Mean-value Theorem says there is at least one point c, but there may actually be more. Give an example of a function on $[0, 1]$ which is not linear, but which satisfies the hypothesis of the Mean-value Theorem, and for which there are infinitely many c.

3. Prove that if $f(x)$ is differentiable on $I = [a, b]$ and changes sign from $-$ to $+$ on I (cf. Section 12.1), then $f'(x) > 0$ at some point of I.

4. Let $f(x) = \dfrac{x^3 + x^2 + 2x}{2x^6 + x^2 + 1}$; show $f'(x) = 1$ somewhere on $[0, 1]$, without differentiating or solving any equations.

15.2

1. (a) Prove that if $f'(x)$ is bounded on a finite interval I, then $f(x)$ is bounded on I.

(Note that I is not assumed to be compact. Do not use integration; $f'(x)$ is not assumed to be integrable.)

Is the result true if "finite" is omitted? (You cannot answer this by showing that it is needed for your argument! cf. the end of Appendix A.3.)

(b) Show that the converse is false. (Give a counterexample; the function should be differentiable on I.)

2. Using standard facts about $\ln x$ (but not its series representation) and the Mean-value Theorem, show that

$\qquad$ (a) $\ln(1 + x) < x$, for $x > 0$;

$\qquad$ (b) for any given $a \in (0, 1)$, we have $\ln(1 + x) > ax$, for $x \underset{\neq}{\approx} 0^+$.

3. Prove that if g' is continuous and positive in (a, b), and g is continuous on $[a, b]$, then g is strictly increasing on $[a, b]$.

(This is of course known on (a, b) by Theorem 15.2; you are asked to extend it to the endpoints as well.)

15.3

1. Taking $f(t) = t^3$, $g(t) = t^2$, on the interval $[0, b]$:

(a) What value c satisfies the Cauchy Mean-value Theorem?

(b) What values c_1 and c_2 would you get applying the Mean-value Theorem to the functions $f(t)$ and $g(t)$ separately?

2. Here is another way of stating and proving the Cauchy Mean-value Theorem, which uses weaker hypotheses.

Let f and g be continuous on $[a, b]$ and differentiable on (a, b). Then there is a point $c \in (a, b)$ such that

(15) $$f'(c)[g(b) - g(a)] = g'(c)[f(b) - f(a)] .$$

(a) Deduce the two Mean-value Theorems in the text (Theorems 15.1 and 15.3) from this formulation.

(b) Prove (15) by applying the Mean-value Theorem to
$$F(t) = f(t)[g(b) - g(a)] - g(t)[f(b) - f(a)].$$

15.4

1. Prove the elementary case of l'Hospital's rule, Theorem 15.4A.

2. Evaluate $\lim\limits_{x \to 0} \dfrac{x - \tan x}{x - \sin x}$.

3. Prove l'Hospital's rule in the $0/0$ form for the case where a is replaced by ∞, by making the change of variable $x = 1/t$ to reduce it to the case $t \to 0^{+}$.

4. (a) Write out a careful statement of l'Hospital's rule in the ∞/∞ form for the case where $x \to a^{+}$.

(b) Use it to evaluate this important limit: $\lim\limits_{x \to 0^{+}} x \ln x$.

Problems

15-1   (a) If a polynomial has n distinct real zeros, prove that its derivative has at least $n - 1$ distinct real zeros.

(b) Deduce that a polynomial of degree $n > 0$ has at most n distinct real zeros. (Use induction; cf. Appendix A.4).

15-2   Suppose $f'(x) = ag(x)$ and $g'(x) = bf(x)$, where a and b are non-zero constants; show that between any two zeros of f is a zero of g, and vice-versa.

Illustrate with the function $f(x) = \sin kx$.

15-3   (a) Prove: given $\epsilon > 0$, $\ln x < (x^{\epsilon} - 1)/\epsilon$, for $x > 1$.

(b) Deduce: $\lim\limits_{x \to \infty} \dfrac{\ln x}{x^{\epsilon}} = 0$, for any $\epsilon > 0$ (use the Squeeze Theorem).

(c) Do part (b) by l'Hospital's rule.

15-4 Prove l'Hospital's rule for the case ∞/∞ (as $x \to \infty$) along the following lines.

Let $L = \lim\limits_{x \to \infty} \dfrac{f'(x)}{g'(x)}$; choose a so that $\dfrac{f'(x)}{g'(x)} \underset{\epsilon}{\approx} L$ for $x > a$.

Prove the two approximations below (valid for $x \gg 1$); deduce Theorem 15.4C:

$$\frac{f(x)}{g(x)} \underset{\epsilon}{\approx} \frac{f(x) - f(a)}{g(x) - g(a)} \underset{\epsilon}{\approx} L \ .$$

(Hint: for the first, write $f(x) - f(a) = f(x)[1 - f(a)/f(x)]$, and use limit theorems; for the second, use the Cauchy Mean-value Theorem.)

15-5 Suppose $f(x)$ is differentiable on an interval I (which may be open), and $f'(x)$ is bounded on I. Prove that $f(x)$ is uniformly continuous on I.

(Give a direct proof, based on the ideas of this chapter and definition of uniform continuity. Note that the Uniform Continuity Theorem 13.5 is not applicable, since the interval I is not assumed compact.)

Answers to Questions

15.1

1. (i) $f(x)$ is continuous at the endpoints ± 1, though not differentiable; the Mean-value Theorem applies: $f'(c) = 0$ for $c = 0$.

(ii) Since $f'(0)$ does not exist, the Mean-value Theorem does not apply; nonetheless, the conclusion holds, since $f'(c) = 1$ when $c = \pm(1/3)^{3/2}$.

2. The points c are the solutions of $f'(x) = k$ where k is the slope of the secant (as in (1a)). Since $f'(x) - k$ is a polynomial of degree $n - 1$, it has at most $n - 1$ zeros; therefore there can be up to $n - 1$ points c satisfying the conclusion of the Mean-value Theorem.

3. The theorem asserts that "there exists a number ...", which suggests the Completeness Property is being invoked somewhere. It occurs at the point in the proof of Rolle's Theorem where the Maximum Theorem is used to prove the existence of a local extremum point, at which the derivative will be zero.

(The Maximum Theorem is proved using the Compactness Theorem (13.1), which in turn uses Bolzano-Weierstrass Theorem (6.3), which uses the Nested Intervals Theorem (6.1), which uses the Completeness Property in its proof, and this is the house that Jack built.)

15.2

1. By the Mean-value Theorem, if $x_1 < x_2$ are two points in I, we have

(1) $\qquad\qquad f(x_2) = f(x_1) + f'(c)(x_2 - x_1), \qquad x_1 < c < x_2$.

Since $c \in I$, $f'(c) < 0$; therefore $f'(c)(x_2 - x_1) < 0$, which shows in turn that $f(x_2) < f(x_1)$, i.e., that $f(x)$ is strictly decreasing on I.

2. Again using the Mean-value Theorem in the form of equation (1), we see that $c \in I \implies f'(c) = 0$; therefore $f(x_2) = f(x_1)$ for all $x_1 < x_2$ in I.

This shows $f(x)$ is constant on I.

The above would not be true if I were replaced by D_f; a counterexample would be $f(x) = \text{sgn } x$ (cf. Question 9.2/1). Formally, the equation (1) is only valid on an interval on which $f(x)$ is differentiable; thus for sgn x, if $x_1 < 0 < x_2$, then x_1 and x_2 do not lie on such an interval and (1) cannot be applied.

3. Since $De^x = e^x$, the Mean-value Theorem gives

$$e^x = 1 + e^c x, \qquad \text{where } 0 < c < x \ .$$

Since $e^c > 1$ for $c > 0$, we get $e^x > 1 + x$, for $x > 0$.

4. Since e^x is continuous at 0 (it is differentiable there), $\lim\limits_{x \to 0^+} e^x = 1$. So,

given $a > 1$, $\quad e^x < a$ for $x \underset{\neq}{\approx} 0^+$, $\quad$ by the Func. Loc. Thm. 11.3D; thus

$$e^c < a \ \text{ for } x \underset{\neq}{\approx} 0^+, \ \ 0 < c < x;$$

so by the Mean-value Theorem (as in Question 15.2/3), we get

$$e^x < 1 + ax, \ \text{ for } x \underset{\neq}{\approx} 0^+.$$

5. For Theorem 15.2, we must assume something about $f'(x)$ over all of I, even though we only need to know it is true for $f'(c)$, because we have no idea in advance where c is.

15.3

1. (a) Calculating, $\dfrac{-1}{1} = \dfrac{-\sin c}{\cos c}$, from which $c = \pi/4$.

(b) Geometrically, the curve $x = \cos t$, $y = \sin t$ is the circular arc from $(1, 0)$ to $(0, 1)$; the secant has slope -1, which is the slope of the tangent at the midpoint $t = \pi/4$.

2. Calculating separately, $0 - 1 = -\sin c_1 \cdot (\pi/2); \quad 1 - 0 = \cos c_2 \cdot (\pi/2)$, so that $c_1 = \sin^{-1}(2/\pi)$, $c_2 = \cos^{-1}(2/\pi)$. (Note that $c_1 + c_2 = \pi/2$, but $c_1 \neq c_2$.)

3. It is the hypothesis $g'(t) \neq 0$ on $[a, b]$. This shows that $g(b) - g(a) \neq 0$: for otherwise, the approximation form (6) of the Mean-value Theorem would say that $g'(c) = 0$ for some $c \in (a, b)$, contrary to the hypothesis. (This reasoning is valid if $g'(t)$ merely exists on $[a, b]$; it does not have to be continuous.)

15.4

1. (a) 1 $\quad$ (b) 2 $\quad$ (use l'Hospital twice) $\quad$ (c) 0 $\quad$ (not 2/3 !!)

2. $\lim\limits_{x \to \infty} \dfrac{\ln x}{x^r} = \lim\limits_{x \to \infty} \dfrac{1}{rx^r} = 0$, if $r > 0$.

3. $\lim\limits_{x \to \infty} \dfrac{e^x}{x} = \lim\limits_{x \to \infty} \dfrac{1/x}{e^{-x}} = \lim\limits_{x \to \infty} \dfrac{-1/x^2}{-e^{-x}} = \lim\limits_{x \to \infty} \dfrac{e^x}{x^2}$

—looks like the l'Hospital Express is running in reverse.

4. The limit asked for is, by definition, $\sin'(0)$, and therefore it's sort of cheating, since this has to be known in order to apply l'Hospital's rule.

16

Linearization and Convexity

16.1 Linearization.

To study the local behavior of a function $f(x)$ near a point a, and calculate its values numerically, we usually approximate $f(x)$ by a polynomial function. The degree of the polynomial is called the **order** of the approximation; as the order increases, so does the accuracy of the numerical values, as well as the amount of information conveyed about $f(x)$.

We begin by considering the simplest case: first-order approximation, in which $f(x)$ is approximated near a by a polynomial which is *linear*. We will study the error term, and give some applications.

In the next chapter this will be extended to approximations by higher-order polynomials and power series. If you prefer to study integration now, you can proceed to Chapter 18, skipping these two chapters (or just Chapter 17) with no loss of continuity.

Theorem-Definition 16.1A   *If $f(x)$ is differentiable at a, then*

$$(1) \qquad f(x) = f(a) + f'(a)(x-a) + e(x), \qquad \text{where} \quad \lim_{x \to a} \frac{e(x)}{x-a} = 0 .$$

We call the polynomial $f(a) + f'(a)(x-a)$ the **linearization** of $f(x)$ at a.

Remarks.   The linearization is the first-order approximation to $f(x)$ at a. The formula for it shows its graph is a line of slope $f'(a)$ passing through the point $(a, f(a))$; so it is the tangent line at this point.

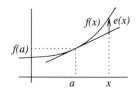

The term $e(x)$ in (1) is the error term; the limit statement in (1) says this error is negligible compared with $x - a$, when $x \approx a$.

Proof.   To prove (1), start with the definition of derivative, writing the limit in the error form:

$$\frac{f(x) - f(a)}{x - a} = f'(a) + e_1(x), \qquad \text{where} \quad \lim_{x \to a} e_1(x) = 0 .$$

To get (1), multiply the equation by $x - a$, and set $e(x) = e_1(x)(x - a)$.   □

We show next that the error term $e(x)$ in (1) can be expressed in terms of the second derivative $f''(x)$; the rough statement is

$$e(x) \approx \frac{f''(a)}{2}(x - a)^2 .$$

This makes sense geometrically, as the figure shows. For example, at a the slope $f'(x)$ is increasing rapidly, so its derivative $f''(a)$ is positive and large, and we see that the formula predicts that $e(x)$ is positive and large. Similarly, at b the slope $f'(x)$ is decreasing slowly, so $f''(a)$ is negative and small, and we see that also $e(x)$ is negative and small, if you stay near b.

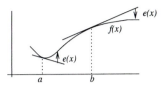

Theorem 16.1B   Linearization error theorem.

Suppose $f''(x)$ exists in an interval I containing a. Then for each $x \in I$, there is a point c lying between a and x such that

(2) $$f(x) = f(a) + f'(a)(x - a) + \frac{f''(c)}{2}(x - a)^2 .$$

Remarks.   The unknown point c plays here the same role as it does in the Mean-value Theorem (in the approximation form, 15.2 (6); we call the point c_1 to avoid confusion with the c in (2) above): for some c_1 between a and x,

(3) $$f(x) = f(a) + f'(c_1)(x - a) .$$

Compare the two theorems: the constant function $f(a)$ in (3) corresponds in (2) to the linear function $f(a) + f'(a)(x - a)$. The error terms in both equations involve the next derivative, evaluated at some point lying between a and x.

The proof we shall give of the Linearization Error Theorem parallels the proof of the Mean-value Theorem (in the secant form, 15.1A) very closely. To make the arguments look as similar as possible, in (2) we write b instead of x:

(2a) $$f(b) = f(a) + f'(a)(b - a) + \frac{f''(c)}{2}(b - a)^2 .$$

We prove a special case of (2a) first, as a lemma.

Lemma 16.1   Extended Rolle's theorem.

Suppose $f''(x)$ exists on $[a, b]$, and $f(a) = f'(a) = 0$, $f(b) = 0$. Then there exists a point $c \in (a, b)$, such that $f''(c) = 0$.

(The theorem is also valid for $[b, a]$, using $f(b) = 0, f'(a) = 0$.)

Proof.   Assume $a < b$. We apply Rolle's Theorem (Lemma 15.1) twice:

$$f(a) = f(b) = 0 \quad \Rightarrow \quad f'(c_1) = 0, \quad \text{for some } c_1 \in (a, b) ;$$
$$f'(a) = f'(c_1) = 0 \quad \Rightarrow \quad f''(c) = 0, \quad \text{for some } c \in (a, c_1) ;$$

We have $c \in (a, b)$ since $a < c < c_1 < b$.   (The proof is similar if $b < a$.)   □

The geometric significance of the point c is shown at the right. The hypothesis $f(a) = f'(a) = 0$ means that the graph of $f(x)$ is tangent to the x-axis at a, while $f(b) = 0$ says the graph intersects the x-axis at another point b. Under these conditions, there will be a point of inflection c; and at such a point we have $f''(c) = 0$.

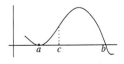

Proof of Theorem 16.1B. We want to find a c between a and x such that

(2) $$f(x) = f(a) + f'(a)(x - a) + \frac{f''(c)}{2}(x - a)^2 .$$

To do this, we introduce the quadratic polynomial

(4) $$P(x) = f(a) + f'(a)(x - a) + C(x - a)^2 \qquad (C \text{ constant}) .$$

The first two coefficients have been chosen because they are the ones which make $P(x)$ agree with $f(x)$ at a both in value and slope (check this):

(5) $$P(a) = f(a), \qquad P'(a) = f'(a) .$$

Then we choose C so that

(6) $$f(b) = P(b), \qquad \text{i.e.,}$$

(2a) $$f(b) = f(a) + f'(a)(b - a) + C(b - a)^2 .$$

Now consider the function

$$g(x) = f(x) - P(x) .$$

By (5) and (6), the function $g(x)$ satisfies the hypotheses of Lemma 16.1. Therefore there is a point c between a and b such that $g''(c) = 0$, i.e.,

(7) $$f''(c) = P''(c) .$$

Using (4) to calculate $P''(c)$, we get $P''(c) = 2C$; this and (7) tell us that

$$C = \frac{P''(c)}{2} = \frac{f''(c)}{2} .$$

Substituting $f''(c)/2$ for C into (6a) then gives us equation (2a). $\qquad\square$

The above proof follows in outline the proof of the Mean-value Theorem 15.1A; compare the two. The quadratic polynomial $P(x)$ here plays the role of the linear polynomial $L(x)$ in the earlier proof.

In that earlier proof, $L(x)$ was chosen so its graph was the secant line; i.e., it was chosen so that $f(a) = L(a)$, $f(b) = L(b)$. Since a linear polynomial is determined by two points, these were the only conditions that could be imposed on $L(x)$.

Here we have a quadratic polynomial; since it has three coefficients, we can make it satisfy three conditions. These are the ones in (5) and (6): agreeing with $f(x)$ in value at a and b, and also having the same slope as $f(x)$ at the point a.

Questions 16.1

1. Show that Lemma 16.1 applies to $f(x) = x^2(x - 1)$ on $[0, 1]$. What is the point c? Sketch the graph.

2. Use Theorem 16.1 to prove $\cos x > 1 - x^2/2$, for $x \underset{\neq}{\approx} 0$.

3. In order to understand better the role of the mysterious constant C in the proof of Theorem 16.1B, rewrite the proof of the Mean-value Theorem 15.1A so it imitates equations (4)-(6a) and continues in the same way.

16.2   Applications to convexity.

We give some applications of the Linearization Error Theorem to the study of a function $f(x)$. Here is a classic first application: the second derivative test which tells whether a critical point is a maximum or a minimum point.

A local maximum or minimum for $f(x)$ is said to be **strict** if the strict inequality $<$ or $>$ holds, rather than just $\leq$ or $\geq$. For example,

$f(x)$ has a *strict local maximum* at a if $f(x) < f(a)$ for $x \underset{\neq}{\approx} a$.

Theorem 16.2A   Second derivative test for extrema.

Suppose $f''(x)$ is continuous at $x = a$ and $f'(a) = 0$. Then

(8)
$$f''(a) > 0 \quad \Rightarrow \quad f(x) \text{ has a strict local minimum at } a;$$
$$f''(a) < 0 \quad \Rightarrow \quad f(x) \text{ has a strict local maximum at } a.$$

Proof.   Since $f''(x)$ is continuous at a, it exists for $x \approx a$, so we can apply the Linearization Error Theorem (2). Since $f'(a) = 0$, it says there is a point c between a and x such that

(9)
$$f(x) - f(a) = \frac{f''(c)}{2}(x - a)^2, \qquad \text{for } x \approx a .$$

Suppose that $f''(a) > 0$. Since $f''(x)$ is continuous at a, the Positivity Theorem 11.4B says that $f''(x) > 0$ for $x \approx a$, and therefore $f''(c) > 0$ also, since c lies between a and x.

This tells us that the right side of (9) is positive if $x \underset{\neq}{\approx} a$, so that

$$f(x) > f(a) \qquad \text{for } x \underset{\neq}{\approx} a ,$$

which shows that $x = a$ is a strict local minimum point for $f(x)$.   □

The proof that $f(x)$ has a strict local maximum if $f''(a) < 0$ is similar, or else one can apply the above to the function $g(x) = -f(x)$.   □□

The test fails if $f''(a) = 0$, as it does for $f(x) = x^4$, which nonetheless has a local minimum at 0. When this happens, there are tests involving still higher derivatives, but they are rarely used; Taylor's Theorem (17.2) or common sense works better.

Even the second derivative test is overused; students spend time calculating messy second derivatives to test points which can be easily seen to be maximum or minimum points just by looking at the function or making a crude sketch of its graph.

At a critical point the graph of $f(x)$ has a horizontal tangent line, and the second derivative test tells whether the graph lies above or below the tangent line. At other points the tangent line will not be horizontal, but it still makes sense to ask whether the graph of $f(x)$ lies entirely above or below the tangent line. This leads to the notion of convexity.

Definition 16.2A   Let $f(x)$ be differentiable on I. We call $f(x)$ **convex on I** if over all of I its graph lies above the tangent line at $a, f(a)$, for all $a \in I$, i.e.,

(10) $f(x) \geq f(a) + f'(a)(x - a),$ for all $a, x \in I$;

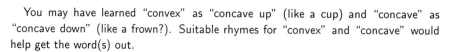

$f(x)$ is *strictly convex on I* if the inequality $>$ holds for $x \neq a$.

Similarly, $f(x)$ is **concave on I** (*strictly concave* if $<$ holds) if

(10a) $f(x) \leq f(a) + f'(a)(x - a),$ for all $a, x \in I$

You may have learned "convex" as "concave up" (like a cup) and "concave" as "concave down" (like a frown?). Suitable rhymes for "convex" and "concave" would help get the word(s) out.

Theorem 16.2B   Second derivative test for convexity.

Assume $f''(x)$ *exists on the open interval I. Then*

(11) $f''(x) \geq 0$ on I $\Rightarrow$ $f(x)$ convex on I .

Proof.   By the Linearization Error Theorem 16.1B, we have for any $a, x \in I$,

$$f(x) = f(a) + f'(a)(x - a) + \frac{f''(c)}{2}(x - a)^2 , \qquad \begin{cases} a < c < x , & \text{or} \\ x < c < a . \end{cases}$$

Since evidently $c \in I$, our hypothesis shows $f''(c)(x - a)^2 \geq 0$; this implies the inequality (10) holds, so that $f(x)$ is convex on I.   □

A similar argument shows

(11a) $f''(x) > 0$ on I $\Rightarrow$ $f(x)$ is strictly convex on I .

Because the definition of convexity only assumes $f'(x)$ exists, we give a second criterion which does not require the existence of $f''(x)$.

Theorem 16.2C   First derivative test for convexity.

If $f(x)$ *is differentiable on I, then*

$$f(x) \text{ is convex on } I \iff f'(x) \text{ is increasing on } I .$$

Proof.   We prove it first in the direction $\Rightarrow$.

Let $a < b$ be two points on I. Apply (10) first at a and then at b:

$$f(b) \geq f(a) + f'(a)(b - a); \qquad f(a) \geq f(b) + f'(b)(a - b) .$$

If we solve these inequalities for $f'(a)$ and $f'(b)$, we get respectively (watch the signs and the inequality directions carefully!):

$$f'(a) \leq \frac{f(b) - f(a)}{b - a} , \qquad f'(b) \geq \frac{f(b) - f(a)}{b - a} .$$

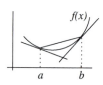

The inequalities show that $f'(a) \leq f'(b)$, so $f'(x)$ is increasing. □

The picture gives the geometry behind this argument: the slope of the secant line lies between the slope of the graph at a and at b .

To prove it in the direction $\Leftarrow$, we use contraposition; that is, we prove

$$f(x) \text{ not convex on } I \quad \Rightarrow \quad f(x) \text{ is not increasing on } I \ .$$

Namely, if $f(x)$ is not convex, by (10) this means that for some $a, b \in I$,

$$f(b) < f(a) + f'(a)(b - a) \ .$$

If we suppose that $a < b$, the inequality is preserved when we divide:

$$\frac{f(b) - f(a)}{b - a} < f'(a) \ .$$

By the Mean-value Theorem 15.1A in the secant form, the left side can be written $f'(c)$, for some c such that $a < c < b$. Therefore we have on I,

$$c > a, \quad \text{but} \quad f'(c) < f'(a) \ ,$$

which shows $f'(x)$ is not increasing on I. The proof is similar if $a > b$. $\square\square$

Since the notion of convexity is often used for continuous functions which are piecewise linear—the graph is made up of connected line segments—and therefore not differentiable everywhere, it is important to be able to to define convexity without even assuming differentiability. To distinguish it from the earlier definition we will refer to it as *geometric convexity*. The definition would be similar for *geometrically concave*, replacing "below" by "above" in (12).

Definition 16.2B   Geometric convexity.

Let $f(x)$ be defined on any type of interval I. For any subinterval $[a, b] \subset I$, we let $P : (a, f(a))$ and $Q : (b, f(b))$ be the two points of the graph lying over the endpoints of the interval. We say $f(x)$ is **geometrically convex** on I if

(12)   the graph of $f(x)$ lies on or below the chord PQ,   for all $[a, b] \subset I$.

An equivalent analytic formulation of this is:

$$(12a) \qquad \frac{f(x) - f(a)}{x - a} \ \leq \ \frac{f(b) - f(a)}{b - a} \ , \quad \text{for any triple } a < x < b \text{ in } I \ .$$

The equivalence of (12) and (12a) can be seen both geometrically and analytically. Referring to the picture, we see that for any $x \in (a, b)$,

(*) S lies on or below $R \iff$ slope $PS \leq$ slope PQ .

The left side of (*) is (12), while the right side is (12a). $\square$

Analytically, using the equation for PQ,

$$y - \text{coordinate of } R \ = \ f(a) + \frac{f(b) - f(a)}{b - a}(x - a) \ .$$

This shows that the analytical formulation of (12) is

$$f(x) \ \leq \ f(a) + \frac{f(b) - f(a)}{b - a}(x - a), \quad a < x < b;$$

this is the same as (12a), since dividing by $x - a$ preserves the inequality. $\square\square$

A geometrically convex function need not be differentiable, but it always turns out to be continuous, even though that is not assumed to start with. On the other hand, if a function is differentiable, then geometric convexity is equivalent to convexity, that is, the inequalities (10) and (12) mutually imply each other. We leave the proofs of both these facts for the Exercises and Problems.

Questions 16.2

1. Define what you think "$f(x)$ is locally convex at a" should mean. Does $f''(a) > 0$ imply $f(x)$ is locally convex at a? If so, prove it; if not, amend the statement and prove it.

2. Give two functions f and g which are convex on $(-\infty, \infty)$, such that:

(a) $f(x)$ is a polynomial with n terms;   (b) $g(x) \to 0$ as $x \to \infty$.

3. Assuming $f''(x)$ exists on I, prove that if $f(x)$ is convex on I, then $f''(x) \geq 0$ on I.

4. Prove (11a) from scratch, without looking at the proof of (11).

Exercises

16.1

1. Using Theorem 16.1B, prove these estimates, on the given interval.

(a) $e^x > 1 + x + x^2/2$, for $x \in [0, 1]$; find a quadratic upper estimate;

(b) $\ln(1 + x) > x - x^2$, for $x \in [0, 1]$; find a quadratic upper estimate.

2. Write out the proof of the Extended Rolle's Theorem (Lemma 6.1) for the case $b < a$.

3. In the neighborhood of what point $a \in [0, 2]$ will the linearization at a of $3x + 4x^3 - x^4$ give the least accurate approximation to the function; why?

4. Suppose $f''(x)$ exists on I, and $f(a) = f(b) = f(c) = 0$, where $a, b, c \in I$ and $a < b < c$. Prove $f''(x_0) = 0$ for some $x_0 \in (a, c)$.

16.2

1. Show by counterexample that the converse of the second derivative test is not true. Modify it to the strongest statement that does hold, and prove it.

2. Assume $f(x)$ is differentiable on I; prove that if $f(x)$ is geometrically convex, then it is convex. (The converse is Problem 16-2.)

3. Suppose the real line $(-\infty, \infty)$ is divided into intervals of maximal length on which $f(x)$ is convex or concave. What is the maximum number of such intervals if $p(x)$ is a polynomial of degree n and

(a) $f(x) = p(x)$?   (b) $f(x) = e^{-x}p(x)$?

4. (a) Let $f(x)$ be geometrically concave on I, and $a, b \in I$. Show the value of $f(x)$ at the average of a and b is $\geq$ the average of $f(a)$ and $f(b)$.

(b) Prove the arithmetic-geometric mean inequality: $\sqrt{ab} \leq (a+b)/2$, (cf. Problem 2-3), by showing $\ln x$ is geometrically concave (you can use the result in Problem 16-2), and deducing the inequality from this.

Problems

16-1 Prove that on an open I, a geometrically convex function $f(x)$ is continuous. (Show $\lim\limits_{\Delta x \to 0^-} \dfrac{\Delta y}{\Delta x}$ exists at each point of I; deduce $\lim\limits_{\Delta x \to 0^-} \Delta y = 0$.)

16-2 Prove that if $f(x)$ is differentiable on I and convex, then it is geometrically convex.

16-3   **The error in linear interpolation.**

To estimate $f(1.13)$, knowing $f(1.1)$ and $f(1.2)$, *linear interpolation* is used: you assume $f(x)$ is linear between 1.1 and 1.2, and calculate the value of $f(1.13)$ accordingly. The problem is to estimate the error in this procedure.

So suppose $f''(x)$ exists on $[a, b]$, and let $L(x)$ be the linear function agreeing with $f(x)$ at the endpoints. Then

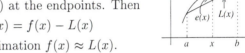

$$e(x) = f(x) - L(x)$$

measures the error in the approximation $f(x) \approx L(x)$.

What is the maximum value of $|e(x)|$ on $[a, b]$? Estimate it as follows.

(a) Let x_0 be a maximum point for $|e(x)|$; say x_0 lies in the right half of the interval $[a, b]$. We may assume that x_0 is a *local* maximum or minimum point for $e(x)$; justify this.

(b) Use Theorem 16.1B to express $e(b)$ in terms of $e(x_0), e'(x_0)$, and powers of $(b - x_0)$. Deduce from this that

$(*)$ $|e(x_0)| \leq \dfrac{|e''(c)|}{8}(b-a)^2,$ where $a < c < b$.

(c) Deduce from part (b) that if $|f''(x)| \leq M$ on $[a, b]$, we get the estimate

$$|e(x)| \leq \frac{M}{8}(b-a)^2, \qquad \text{for } a \leq x \leq b .$$

16-4 This uses the result of Problem 16-3; in it, avoid calculations by using the simple estimation $\sin a \leq a, \quad a > 0$.

(a) You are given values of $\sin x$ for $x = .05, .10, .15, \dots$, and use linear interpolation to calculate $\sin .12$ and $\sin .48$. In each case, tell how many decimal places the answer is correct to; explain geometrically why one answer is more accurate than the other..

(b) Over approximately what interval $[0, a]$ could you be sure that the linear interpolation process for $\sin x$ would be accurate to within .0001?

16-5 If the trapezoidal rule is used to estimate $\int_a^b f(x)\,dx$, taking Δx as the width of the trapezoids, show using Problem 16-3 that

$$|\text{error in trapezoidal rule}| \le \frac{M}{8}(\Delta x)^2 (b-a)\,,$$

where M is the maximum of $|f''(x)|$ on $[a, b]$.

(Since the formula for the error depends on $(\Delta x)^2$, the trapezoidal rule is called a *second-order* method; it means the error goes down by a factor of $1/4$ each time the width of the trapezoids is halved.

A better value for the denominator in the above error estimate would be 12, rather than 8, but the proof is harder and the gain not that much.)

Answers to Questions

16.1

1. $f'(x) = 3x^2 - 2x$, $f''(x) = 6x - 2$; $f(0) = f'(0) = 0 = f(1)$; $c = 1/3$.

2. $\cos x = 1 - \dfrac{\cos c}{2} x^2$, for c between 9 and x.

Since $0 < \cos c < 1$ in the interval $0 < |x| < \pi/2$, we conclude that

$$\cos x > 1 - x^2/2, \text{ for } 0 < |x| < \pi/2.$$

3. Let $L(x) = f(a) + C(x - a)$, where C is chosen so $f(b) = L(b)$, i.e.,

$(*)$ $f(b) = f(a) + C(b - a)$.

Then the function $g(x) = f(x) - L(x)$ satisfies the hypotheses of Rolle's Theorem on $[a, b]$, so there is a point c on the interval such that $g'(c) = 0$, i.e.,

$$f'(c) = L'(c) = C\,;$$

substituting $C = f'(c)$ into $(*)$ gives the Mean-value Theorem.

(This way of presenting the argument avoids having to make C explicit, and is better adapted to generalizing the theorem than the more geometric argument given in Section 15.1.)

16.2

1. "$f(x)$ locally convex at a" should mean: "$f(x)$ is convex for $x \approx a$".

As given, the hypotheses in the statement are too weak; they should be strengthened to:

If $f''(x)$ is continuous for $x \approx a$ and $f''(a) > 0$, then $f(x)$ is locally convex at a.

To prove it, the hypotheses imply $f''(x) > 0$ for $x \approx a$, by the Positivity Theorem for continuous functions. Hence Theorem 16.2B applies, and $f(x)$ is convex for $x \approx a$, i.e., locally convex at a.

2. (a) $f(x) = 1 + x^2 + x^4 + \ldots + x^{2n}$ (b) $g(x) = e^{-x}$

3. $f(x)$ convex $\Rightarrow$ $f'(x)$ increasing on I $\Rightarrow$ $f''(x) \ge 0$ (Theorem 15.2).

4. The only change: $f''(c)(x - a)^2 > 0$ if $x \ne a$, since $f''(c) > 0$.

17

Taylor Approximation

17.1 Taylor polynomials.

In Section 16.1 we obtained a formula for the error involved in approximating $f(x)$ near the point a by the first degree polynomial

$$(1) \qquad f(a) + f'(a)(x - a) \, ,$$

which at a agrees with $f(x)$ in value and slope.

In this chapter we get better approximations to $f(x)$ by using polynomials of higher degree that agree closely with $f(x)$ at a. We begin by describing these polynomials; they generalize the linear polynomial (1) above.

Definition 17.1. Two functions f and g have **n-th order agreement** at a if they are n times differentiable at a and

$$(2) \qquad f(a) = g(a), \ f'(a) = g'(a), \ \ldots, \ f^{(n)}(a) = g^{(n)}(a) \, .$$

For example, $\sin x$ and $x - x^3$ have second-order agreement at 0. Namely, both functions satisfy $f(0) = 0$, $f'(0) = 1$, $f''(0) = 0$; however at 0 they have different third derivatives (-1 and -6, respectively), so they do not have third-order agreement at 0.

As another example, the polynomial in (1) has first-order agreement at a with $f(x)$, and is the only such linear polynomial.

Theorem-Definition 17.1 *Suppose $f^{(n)}$ exists. Then the polynomial*

$$(3) \qquad T_n(x) = f(a) + f'(a)(x - a) + \ldots + \frac{f^{(n)}(a)}{n!}(x - a)^n \, ,$$

called the **n-th order Taylor polynomial** *for $f(x)$ at a, is the unique polynomial of degree n in powers of $x - a$ having n-th order agreement with $f(x)$ at a.*

Remarks. Note that all the Taylor polynomials have the same form, so that from $T_n(x)$ one can obtain any of the lower-order Taylor polynomials $T_k(x)$ just by truncating $T_n(x)$ at the term of degree k.

Since $f(x)$ and $T_n(x)$ have n-th order agreement at k, the Taylor polynomials $T_n(x)$ give a good polynomial approximation to $f(x)$ for $x \approx a$. In the next section we will study the error in this approximation, and then give applications.

Proof. Let $p(x)$ be a polynomial of degree n written in powers of $x - a$:

$$p(x) = c_0 + c_1(x - a) + \ldots + c_n(x - a)^n , \qquad c_i \in \mathbb{R} .$$

After k differentiations, $k \leq n$, the constants $c_0, \ldots, c_{k-1}$ disappear, giving

$$p^{(k)}(x) = k!c_k + \text{terms having } x - a \text{ as a factor;}$$

this shows $p^{(k)}(a) = k!c_k$. Thus $f(x)$ and $p(x)$ have n-th order agreement at a

$$\Leftrightarrow \quad f^{(k)}(a) = k!c_k, \qquad\qquad k = 0, 1, \ldots, n ;$$

$$\Leftrightarrow \quad c_k = \frac{f^{(k)}(a)}{k!} , \qquad\qquad k = 0, 1, \ldots, n .$$

These are precisely the coefficients of $T_n(x)$, which is therefore the only polynomial $\sum_0^n c_i(x - a)^i$ having n-th order agreement with $f(x)$ at a. $\qquad\square$

Taylor polynomials for the standard functions

From the formula (3) for the Taylor polynomials, we get the polynomials which approximate the standard elementary functions. We take $a = 0$. All of them, except perhaps the last, are sufficiently useful to be committed to memory.

(4) $$\frac{1}{1 - x} \approx 1 + x + x^2 + \ldots + x^n$$

(5) $$e^x \approx 1 + x + \frac{x^2}{2!} + \ldots + \frac{x^n}{n!}$$

(6) $$\sin x \approx x - \frac{x^3}{3!} + \frac{x^5}{5!} - \ldots + (-1)^n \frac{x^{2n+1}}{(2n + 1)!}$$

(7) $$\cos x \approx 1 - \frac{x^2}{2!} + \frac{x^4}{4!} - \ldots + (-1)^n \frac{x^{2n}}{(2n)!}$$

(8) $$(1 + x)^r \approx 1 + rx + \frac{r(r - 1)}{2!}x^2 + \ldots + \frac{r(r - 1)\cdots(r - n + 1)}{n!}x^n ,$$

$$\text{for any } r \in \mathbb{R}$$

(9) $$\ln(1 + x) \approx x - \frac{x^2}{2} + \frac{x^3}{3} - \ldots + (-1)^{n-1}\frac{x^n}{n}$$

Notice that the even-order derivatives of $\sin x$ at 0 are all equal to 0, so that the Taylor polynomials are equal in pairs:

$$T_1(x) = T_2(x), \quad T_3(x) = T_4(x), \quad \ldots .$$

A similar remark applies to the Taylor polynomials for $\cos x$ at 0:

$$T_0(x) = T_1(x), \quad T_2(x) = T_3(x), \quad \ldots .$$

In the above examples, it is easy to guess the law of formation for the coefficients after computing a few of them. But this is rarely the case. A more typical example would be $\tan x$: the polynomial below gives no clue what T_9 would be:

$$\tan x \approx x + \frac{x^3}{3} + \frac{2x^5}{3 \cdot 5} + \frac{17x^7}{5 \cdot 7 \cdot 9} = T_7(x) .$$

Questions 17.1

1. Calculate from (3) the polynomials T_n for   (a) $\cos x$ at 0;   (b) $\ln x$ at 1.

2. Why is the Taylor polynomial in powers of x given for $\ln(1 + x)$, rather than for the more natural choice $\ln x$?

17.2   Taylor's theorem with the Lagrange remainder.

The approximation $f(x) \approx T_n(x)$ for $x \approx a$ is made precise by the following n-th order version of the Linearization Error Theorem 16.1.

Theorem 17.2   Taylor's theorem.

Suppose $f(x)$ is $n + 1$ times differentiable in an open interval I containing a and x. Then for some c between a and x, we have

$$(10) \qquad f(x) \; = \; f(a) + f'(a)(x - a) + \ldots + \frac{f^{(n)}(a)}{n!}(x - a)^n + R_n(x) \; ;$$

$$(11) \qquad R_n(x) \; = \; \frac{f^{(n+1)}(c)}{(n + 1)!}(x - a)^{n+1} \; , \qquad \begin{cases} a < c < x, & \text{or} \\ x < c < a \, . \end{cases}$$

$R_n(x)$ is called the **remainder term** or *error term*. It has other forms, but the one above, called the *Lagrange* or *derivative* remainder, is the easiest to remember: It looks like the next term of the series, except that the derivative is at c instead of a.

The point c is not necessarily unique. Note that since c depends on x, $f^{(n+1)}(c)$ is *not* a constant, so the right side of (10) is *not* a polynomial of degree $n + 1$.

If $n = 0$, Taylor's Theorem is just the Mean-value Theorem in the approximation form (15.2 (6)). If $n = 1$, it is just the Linearization Error Theorem 16.1. We model the proof of Taylor's Theorem on the proofs of these two theorems.

As before, imagine that x in (10) has been replaced by b; we begin by proving as a lemma the special case of Taylor's Theorem in which the first n derivatives are 0 at a, and $f(b) = 0$ also.

Lemma 17.2   Extended Rolle's theorem.

If $f^{(n+1)}$ exists on $[a, b]$, and $f(a) = f'(a) = \ldots = f^{(n)}(a) = f(b) = 0$, then for some c between a and b, $f^{(n+1)}(c) = 0$.

Proof.   Assume $a < b$. Apply Rolle's Theorem (Lemma 15.1) $n + 1$ times:

$$f(a) = f(b) = 0 \qquad \Rightarrow \qquad f'(c_1) = 0, \qquad \text{for some } c_1, \; a < c_1 < b \; ;$$
$$f'(a) = f'(c_1) = 0 \qquad \Rightarrow \qquad f''(c_2) = 0, \qquad \text{for some } c_2, \; a < c_2 < c_1 \; ;$$
$$\vdots \qquad\qquad\qquad \vdots \qquad\qquad\qquad \vdots$$
$$f^{(n)}(a) = f^{(n)}(c_n) = 0 \qquad \Rightarrow \qquad f^{(n+1)}(c) = 0, \qquad \text{for some } c, \; a < c < c_n \; ;$$

and $c \in (a, b)$, since $a < c < c_n < \ldots < c_1 < b$. (The case $b < a$ is similar.). $\square$

Proof of Theorem 17.2.   We now use the lemma to prove (10)-(11), writing b for x. First we construct the analog of the quadratic polynomial $P(x)$ used in the proof of the Linearization Error Theorem 16.1.

According to (3), for any value of the constant C, the polynomial

$$(12) \qquad P(x) \; = \; f(a) + f'(a)(x-a) + \ldots + \frac{f^{(n)}(a)}{n!}(x-a)^n + C(x-a)^{n+1}$$

has n-th order agreement with $f(x)$ at a. To make it also agree in value with $f(x)$ at b, we choose C so that $f(b) = P(b)$, i.e.,

$$(13) \qquad f(b) \; = \; f(a) + f'(a)(b-a) + \ldots + \frac{f^{(n)}(a)}{n!}(b-a)^n + C(b-a)^{n+1} \;.$$

The desired value for C is found by solving (13) for C; this is possible since the coefficient of C is non-zero (note that $a \neq b$).

We now consider the difference function

$$(14) \qquad\qquad\qquad g(x) \; = \; f(x) - P(x) \;.$$

By the preceding paragraph, it satisfies the hypotheses of the Extended Rolle's Theorem, so for some c between a and x,

$$g^{(n+1)}(c) \; = \; 0 \;;$$

differentiating both (14) and (12) $n+1$ times, the above gives

$$f^{(n+1)}(c) \; = \; P^{(n+1)}(c) \; = \; (n+1)! \cdot C \;.$$

This shows that

$$C = \frac{f^{(n+1)}(c)}{(n+1)!} \;;$$

substituting this value of C into (13) and replacing b by x gives (10)-(11). $\square$

Questions 17.2

1. If $f(x) = x^3(1-x)$, show that $f'''(c) = 0$ for some $c \in (0,1)$. Then determine c explicitly.

2. Just from looking at (10) and (11), it is clear that, in general, c cannot be a constant (i.e., it depends on x). Why? For what $f(x)$ is it a constant?

3. (a) Use (10) and (11) to write x^3 as a polynomial in powers of $x + 1$.

(b) Do part (a) by setting $u = x + 1$ and writing x^3 as a polynomial in u, then switching back to x again.

4. Let $f(x)$ be twice differentiable on (a,b), and suppose $f(x)$ is zero at the three points a, b, and c, where $c \in (a,b)$. Prove $f''(k) = 0$ for some $k \in (a,b)$.

17.3 Estimating error in Taylor approximation.

The expression for the error (remainder) term in $f(x) \approx T_n(x)$ is

(11) $\qquad R_n(x) \;=\; \dfrac{f^{(n+1)}(c)}{(n+1)!}(x-a)^{(n+1)} , \qquad c$ between a and x.

The formula is exact, but we don't know where c is. So what we get is not the exact error, but rather a way of estimating the error, provided that we can estimate the $(n+1)$-st derivative occurring in (11).

For example, let $f(x) = e^x$. Then Taylor's formula at $a = 0$ is

(15) $\qquad e^x = 1 + x + \dfrac{x^2}{2!} + \ldots + \dfrac{x^n}{n!} + \dfrac{e^c x^{n+1}}{(n+1)!} , \qquad 0 < |c| < |x|.$

Note the concise way of writing "c between 0 and x", which covers both of the cases $0 < c < x$ and $x < c < 0$.

In using the formula, there are three things to play with: the size of the error, the size of the interval, and n — i.e., what order approximation to use; we can fix two of these, and ask how the third is affected. Different possibilities are illustrated by the Examples, Questions, and Exercises.

Example 17.3A. Is the third-order Taylor approximation to e^x at 0 good to two decimal places, for $|x| \le .5$?

Solution. In (15) we take $n = 3$. Since $e < 3$ and $|c| \le .5$,

$$e^c < \sqrt{3} < 1.75;$$

so

$$|R_3(x)| < \frac{(1.75)(.5)^4}{4!} < .005, \qquad \text{for } |x| \le .5 .$$

By convention, this is the meaning of "good to two decimal places"; answer: yes.

Example 17.3B Calculate e to two decimal places.

Solution. Taking $x = 1$ in (15) gives

(16) $\qquad e = 1 + 1 + \dfrac{1}{2!} + \ldots + \dfrac{1}{n!} + \dfrac{e^c}{(n+1)!} , \qquad 0 < c < 1 .$

If we suppose that it is known in advance that $e < 3$, say, then by experimenting with the error term, we see that $n = 5$ is good enough, since

$$R_5(1) = \frac{e^c}{6!} < \frac{3}{6!} = \frac{1}{240} < .005 .$$

So, taking $n = 5$ and calculating,

$$e \approx 1 + 1 + \frac{1}{2} + \frac{1}{6} + \frac{1}{24} + \frac{1}{120} = 2.716 .$$

Since the missing later terms are all positive, 2.716 is too small, so $e \approx 2.72$, to two decimal places. $\qquad\square$

If you feel it was cheating, or there was something circular about assuming at the beginning that $e < 3$ in order to estimate the remainder term, this simple estimate for e can also be deduced from (16); see Question 1.

In doing this sort of estimating, one should take advantage of missing terms in the Taylor polynomial. Thus in the approximation

$$\sin x \approx x - \frac{x^3}{3!} \; ,$$

the polynomial on the right should be viewed as $T_4(x)$, not $T_3(x)$, so that the error term is

$$R_4(x) \;=\; \frac{\cos c}{5!} \, x^5 \; ,$$

rather than the much larger (for $|x| \approx 0$)

$$R_3(x) \;=\; \frac{\sin c}{4!} \, x^4 \; .$$

Questions 17.3

1. Use (16) to derive the crude estimation $e < 3$, by taking $n = 2$, and noting that $0 < c < 1$.

2. Using $\cos x \approx 1 - \frac{x^2}{2}$, find an approximate value of $\cos .1$ and an estimate of the decimal place accuracy, viewing the polynomial as   (a) $T_2(x)$ (b) $T_3(x)$.

17.4   Taylor series

Another application of Taylor's formula is to represent a function $f(x)$ as the sum of a power series. Familiar examples are

(17) $$\frac{1}{1-x} \;=\; 1 + x + x^2 + \ldots + x^n + \ldots, \qquad |x| < 1 \; ;$$

(18) $$e^x \;=\; 1 + x + \frac{x^2}{2!} + \ldots + \frac{x^n}{n!} + \ldots, \qquad \text{all } x \in R.$$

To do this, we suppose $f(x)$ has derivatives of all orders at $a = 0$, and form the power series, called the **Taylor series at 0**, or the **MacLaurin series**:

(19) $$f(0) + f'(0)x + \frac{f''(0)}{2!}x^2 + \ldots + \frac{f^{(n)}(0)}{n!}x^n + \ldots \; .$$

Let R be the radius of convergence of this power series; we pose the

Question.   *Does the Taylor series (19) actually sum to $f(x)$, for $|x| < R$?*

To answer, we use the definition of convergence. Our question asks if

(20) $$f(x) \;=\; f(0) + f'(0)x + \frac{f''(0)}{2!}x^2 + \ldots + \frac{f^{(n)}(0)}{n!}x^n + \ldots, \qquad |x| < R \; ;$$

or as we may also write it,

(20a) $$f(x) = T_n(x) + \frac{f^{(n+1)}(c)}{(n+1)!} \, x^{n+1} \; ,$$

where $T_n(x)$ is both the n-th order Taylor polynomial for $f(x)$ at $x = 0$, as well as the n-th partial sum of the series (20).

By definition of convergence, the series converges to $f(x)$ if and only if

$$\lim_{n \to \infty} T_n(x) = f(x) \qquad \text{for } |x| < R;$$

i.e., if and only if

$$(21) \qquad \qquad \lim_{n \to \infty} \frac{f^{(n+1)}(c)}{(n+1)!} x^{n+1} = 0 \qquad \text{for } |x| < R .$$

If (20) holds (or (21), which is the same thing), we say the function $f(x)$ can be represented in the interval $(-R, R)$ by its Taylor series and say it is **analytic** at 0.

For most functions, it is difficult to prove the Taylor series converges to the function, or for that matter even to calculate the Taylor series in the first place. The difficulty lies in $f^{(n)}(x)$: the higher derivatives are just too hard to calculate and estimate.

Fortunately, there are other methods for expanding functions in power series, like the use of substitution and term-by-term differentiation or integration. We used these a bit in Chapter 8, and will return to them later, in Chapter 22. The general rule we will prove is that an analytic function can have only one series representation in powers of x, so that any acceptable method of finding it will lead to the Taylor series (20).

Despite the general difficulties with the direct approach to verifying series expansions by using Taylor's theorem, some of the simpler functions can be handled this way, using (21) to prove the convergence of the series to $f(x)$.

Example 17.4   Prove the series expansion (18) for e^x is valid.

Solution.   We have to verify (21) holds when $f(x) = e^x$, and $-\infty < x < \infty$, i.e., given x, show the remainder term is small for $n \gg 1..$

Given any x, choose N so that $|x| < N/2$. Then estimate $e^c < A$ as follows:

$\qquad$ if $\ 0 < c < x, \ $ then $\ 0 < e^c < e^x; \ $ let $A = e^x$;

$\qquad$ if $\ x < c < 0, \ $ then $\ e^x < e^c < 1; \ $ let $A = 1$.

In either case, for $n > N$ we have

$$\frac{|e^c x^n|}{n!} \ \le \ \frac{A|x|^n}{n!} \ \le \ \frac{A|x|^N}{N!} \cdot \frac{|x|}{N+1} \cdot \frac{|x|}{N+2} \cdot \ \ldots \ \cdot \frac{|x|}{n} \ ;$$

$$\le \ \frac{A|x|^N}{N!} \cdot \frac{1}{2} \cdot \frac{1}{2} \cdot \ \ldots \ \cdot \frac{1}{2} \ \le \ \frac{K}{2^{n-N}} \ ,$$

where K is a constant not depending on n. So by the Squeeze Theorem 5.2,

$$\lim_{n \to \infty} \frac{e^c x^n}{n!} = 0 \ ,$$

which shows the remainder term goes to 0 with n, and hence proves (18). $\qquad \square$

Series expansions can be made around other points than $a = 0$. Using the approximation by Taylor polynomials around the point a, we would write

$$(22) \qquad \qquad f(x) = \sum_{0}^{\infty} \frac{f^{(k)}(a)}{k!} (x - a)^k \ , \qquad |x - a| < R \ ,$$

if the radius of convergence for the series is R and the error term $\to 0$:

(23) $$\lim_{n \to \infty} \frac{f^{(n)}(c)}{n!}(x-a)^n = 0 , \qquad |x-a| < R .$$

In practice however one usually changes the variable by setting $x = u + a$, so as to be able to work around 0 instead of a. This is what we did in (9) for example: to study $\ln x$ near $x = 1$, we studied instead $\ln(1+x)$ near $x = 0$.

In doing numerical calculations with power series, the difficulty of calculating and estimating derivatives usually makes it impracticable to estimate errors by using the remainder term (11). A common procedure is to calculate up to the n-th term, and then use the next term in the series as an estimate of the error. This amounts to assuming that

$$f^{(n)}(c) \approx f^{(n)}(a) ;$$

since $f^{(n)}$ is continuous (it is even differentiable), this will be all right if $|x-a|$ is small, as it should be to give rapid convergence and therefore effective computation.

Questions 17.4

1. How would you calculate from their series: (a) $\sin 13$? (b) $e^{10.5}$?

Exercises

17.1

1. Calculate the formula for the n-th Taylor polynomial for $(1+x)^r$ at the point $a = 0$, r being any real number.

17.2

1. Give a more formal proof of the Extended Rolle's Theorem (Lemma 17.2) by using mathematical induction (Appendix A.4; the dots in the book's proof are as usual an attempt to conceal the induction.)

2. (a) Let $P(x)$ be a polynomial of degree n. Write it as a polynomial in powers of $x - a$ by substituting $x = u + a$, using algebra to write $P(u + a)$ as a polynomial in u, and then setting $u = x - a$ (you could instead use synthetic division, if you remember it):

$$P(x) = c_0 + c_1(x-a) + c_2(x-a)^2 + \ldots + c_n(x-a)^n.$$

Prove that the resulting polynomial on the right is the same as the Taylor polynomial $T_n(x)$ for the function $P(x)$ at the point a.

(b) Use the two procedures described in part (a) to write $x^3 - 2x + 2$ in powers of $x + 1$ by two different methods.

3. Suppose $g(t)$ is a function which has an $(n+1)$-th derivative $g^{(n+1)}$ on some open interval I. Suppose that $g(t)$ has $n + 2$ distinct roots in this interval:

$$g(a) = g(a_1) = \ldots = g(a_n) = 0, \qquad g(b) = 0; \qquad a < a_1 < \ldots < a_n < b .$$

Prove there is a point c in the interval (a, b) such that $g^{(n+1)}(c) = 0$.

4 (a) Prove that if $f'''(x)$ exists on $[a, b]$, and $f(x)$ and $f'(x)$ are both zero at a and b, then $f'''(c) = 0$ for some $c \in (a, b)$.

(b) Show part (a) applies to $(x - a)^2(x - b)^2$, and find c explicitly.

17.3

1. Determine explicitly the Taylor polynomial $T_2(x)$ for e^{-x} at 0. What is the maximum error made in using this as an approximation to the function over the interval $[0, .1]$?

2. You wish to use $1 - x^2/2$ as an approximation to $\cos x$, with an error not greater than .0001. Estimate over what interval this will be valid.

3. You want to estimate $\sin x$ to three decimal places over $|x| < .5$. How large should n be in order that the n-th order Taylor polynomial give you this accuracy over the given interval? Show your answer is correct.

4. Calculate $\cos .1$ to seven decimal places, and show your answer has the desired accuracy. (This takes surpisingly little calculation.))

5. Calculate $\ln 1.1$ to three decimal places by using a suitable Taylor polynomial, and show your answer has the desired accuracy.

17.4

1. Prove the Taylor series at 0 of the following functions converge to the function for all x, or for the indicated values of x:

(a) $\sin x$ (b) $\cos x$ (c) $\dfrac{1}{1 - x}$, $x \in (-1, 0]$ (d) $\ln(1 + x)$, $x \in [0, 1)$

(For the last two, explicit remainders were given in Section 4.2. Here, use the general form of the remainder, as given in (11).)

Problems

17-1 A polynomial $P(x)$ has a **k-fold zero** at the point a if it has $(x - a)^k$ as a factor, but not $(x - a)^{(k+1)}$, i.e.,

$$P(x) = (x - a)^k Q(x), \qquad \text{where } Q(x) \text{ is a polynomial, } Q(a) \neq 0 .$$

(a) Prove: the point a is a k-fold zero for the polynomial $P(x)$ $\Leftrightarrow$

$$P(a) = P'(a) = \ldots = P^{(k-1)}(a) = 0, \quad P^{(k)}(a) \neq 0 .$$

(b) For what value(s) of the constant b will $2x^3 - bx^2 + 1$ have a double zero at some point (i.e., a 2-fold zero)?

(c) Using part (a), describe the connection between Exercise 17.2/3 and the Extended Rolle's Theorem (Lemma 17.2).

Answers to Questions

17.1

1. (a) $f(x) = \cos x$; successive derivatives are $-\sin x, -\cos x, \sin x, \cos x$, after which the cycle repeats. So the successive values of the derivatives at 0 are (starting with the "zero-th" derivative: $1, 0, -1, 0$, and then the cycle is repeated. Dividing these numbers by $n!$ for $n = 0, 1, 2, \ldots$ then gives the coefficients of the Taylor polynomials for $\cos x$ at 0.

(b) $f(x) = \ln x$, $f' = x^{-1}$, $f'' = -x^{-2}$, $f''' = 2x^{-3}$, $f^{(4)} = -3!x^4$, and in general, $f^{(n)} = (-1)^{n-1}(n-1)!x^{-n}$. Thus $f^{(n)}(1)/n! = (-1)^{n-1}/n$, and

$$T_n(x) = (x-1) - \frac{(x-1)^2}{2} + \frac{(x-1)^3}{3} - \ldots + \frac{(-1)^{n-1}(x-1)^n}{n} + \ldots .$$

2. $\lim_{x \to 0^+} \ln x = -\infty$, so $\ln x$ can't be approximated by a polynomial at 0; formally, if $f(x) = \ln x$, then $f(0), f'(0), \ldots$ are all undefined.

17.2

1. $f(x)$ satisfies the hypotheses of the Extended Rolle's Theorem (with $n = 2$), on $[0, 1]$, so $f'''(c) = 0$ for some $c \in (0, 1)$. Explicitly, $f'''(x) = 6 - 24x$, so $c = 1/4$.

2. If c were a constant, then (10)-(11) taken together show that $f(x)$ must be a polynomial of degree $n + 1$.

3. (a) The values of the function and its successive derivatives at -1 are: $-1, 3, -6, 6$, and 0 thereafter. After dividing these by $n!$, we get the polynomial

$$f(x) = -1 + 3(x+1) - 3(x+1)^2 + (x+1)^3 .$$

(b) $x^3 = (u-1)^3 = u^3 - 3u^2 + 3u - 1 = (x+1)^3 - \ldots - 1$, as in part (a).

4. By Rolle's Theorem (Lemma 15.1),

$$f'(c_1) = 0 \text{ for some } a < c_1 < c \text{ and } f'(c_2) = 0 \text{ for some } c < c_2 < b;$$

by Rolle's Theorem again, $f''(k) = 0$ for some $k \in (c_1, c_2) \subset (a, b)$.

17.3

1. Taking $n = 2$ and $x = 1$, we get $e = 5/2 + e^c/6 < 5/2 + e/6$, since $c < 1$. From this we get $5e/6 < 5/2$, so that $e < 3$.

2. We have $\cos .1 = 1 - .01/2 = .995000\ldots$; how many decimal places are accurate?

(a) $|R_2(.1)| = \dfrac{|\sin c|}{3!}(.1)^3 < (.001)/6$; this says it is accurate to 3 places;

(b) $|R_3(.1)| = \dfrac{|\cos c|}{4!}(.1)^4 < (.0001)/24$; this says it is accurate to 5 places.

17.4

1. (a) Since $13 \approx 4\pi$, and $\sin 13 = \sin(13 - 4\pi)$, substitute the small value $x = 13 - 4\pi$ into the series for $\sin x$ to calculate $\sin 13$.

(b) Substitute 1 and .5 into the series for e^x to calcluate e and $e^{.5}$; then $e^{10.5} = e^{10}e^{.5}$.

18

Integrability

18.1 Introduction. Partitions.

The integral $\displaystyle\int_a^b f(x)\,dx$ is the single most important tool of analysis.

There are several versions of it, at least two of which are basic: the integral of ordinary calculus, called the *Riemann integral,* and one introduced around 1900 in a PhD thesis, the *Lebesgue integral.* Both integrals have the same value when applied to the standard functions of analysis, but the Lebesgue integral is more general, since you can use it to integrate some highly discontinuous functions. These occur in some applications (probability theory, for instance), but for a first exposure it's best to stay with the Riemann integral, as we will do.

For those who need it or the curious, Chapter 23 gives a little introduction to the Lebesgue integral.

Before proceeding with its definition, it might be well to consider how the Riemann integral is *not* defined, since many have strong but incorrect ideas about it.

Many calculus students offer as the definition of the integral

$\int_a^b f(x)dx \;=\; F(b) - F(a), \qquad$ where $\;F'(x) = f(x).$

This is not a good definition for the integral, since given $f(x)$, we do not know whether such an $F(x)$ exists; in fact it doesn't if $f(x)$ has discontinuities on the interval $[a,b]$.

Many working scientists and engineers (and the rest of the calculus students) define the integral to be the area over $[a,b]$ and under the graph of $f(x)$. This is better, but still not a good definition, even if the function is positive. The trouble is that "area" itself has not been defined—in fact, area is actually defined by means of the integral, rather than the other way around. Another objection is that in many applications of integration, it is unnatural to drag in the idea of area, just as one does not talk about velocity in most applications of differentiation, and does not define the derivative in terms of velocity.

The correct definition of integral takes a while, and proceeds in stages. As presented here, the first part of the work is to identify and study the class of functions for which the integral can be defined—we call them *Riemann-integrable,* or simply *integrable.* We will do this in the present chapter; the actual definition of the integral will be taken up in the next chapter.

Definition 18.1A   A **partition** $\mathcal{P}$ of a compact interval $[a, b]$ is a strictly increasing finite sequence of numbers starting with a and ending with b:

(1) $$\mathcal{P} : a = x_0 < x_1 < \ldots < x_n = b \; .$$

Throughout this chapter, we assume $a < b$.

A partition divides up $[a, b]$ into smaller *subintervals*

$$[x_0, x_1], \; [x_1, x_2], \; \ldots, \; [x_{n-1}, x_n] \; .$$

For the subintervals and their lengths we will use Δx notation:

(2) $$[\Delta x_i] = [x_{i-1}, x_i] \; , \qquad \Delta x_i = x_i - x_{i-1} \; , \qquad i = 1, 2, \ldots, n.$$

We need some way of saying that a partition $\mathcal{P}$ is *fine*, by which we mean that all its subintervals are small. The way to do this is to look at the largest of its subintervals:

Definition 18.1B   The **mesh** $|\mathcal{P}|$ of a partition $\mathcal{P}$ is defined by

(3) $$|\mathcal{P}| = \max_i \Delta x_i \; .$$

Thus $\mathcal{P}$ will be fine if its mesh $|\mathcal{P}|$ is small.

Definition 18.1C   An **n-partition** is one containing n subintervals.

The *standard n-partition* $\mathcal{P}^{(n)}$ is the one in which all the subintervals have the same length. If we denote this common length by Δx, then for the standard n-partition of $[a, b]$, we have

(4) $$\Delta x = \frac{b - a}{n} \; , \qquad |\mathcal{P}^{(n)}| = \frac{b - a}{n} \; .$$

18.2   Integrability.

Given a partition $\mathcal{P}$ of $[a, b]$, we draw in the associated inscribed and circumscribed rectangles, and consider their total areas. The function will be called integrable if these two areas get arbitrarily close as the partition gets finer and finer. The two areas then have a common limit, whose value will be called the Riemann integral, $\int_a^b f(x)dx$.

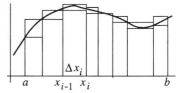

We now want to say these things analytically, without referring to areas.

Definition 18.2A   Let $f(x)$ be bounded on $[a, b]$, and let $\mathcal{P}$ be a partition of $[a, b]$. Using the notation in (2), let

(5) $$m_i = \inf_{[\Delta x_i]} f(x), \qquad M_i = \sup_{[\Delta x_i]} f(x), \qquad \Delta x_i = x_i - x_{i-1} \; .$$

We define the **lower sum** and **upper sum** for $f(x)$ over $\mathcal{P}$ to be respectively

(6) $$L_f(\mathcal{P}) = \sum_1^n m_i (\Delta x_i), \qquad U_f(\mathcal{P}) = \sum_1^n M_i (\Delta x_i) \; .$$

Geometrically, if $f(x) > 0$, the upper sum represents the total area of the circumscribed rectangles, the lower sum the total area of the inscribed rectangles, for the partition $\mathcal{P}$ and the function $f(x)$.

In the notation, the subscript f can be omitted if it is clear from the context.

Definition 18.2B   A function f is called **integrable** (or *Riemann-integrable*) on $[a, b]$ if it is defined and bounded on $[a, b]$, and it satisfies

(7)   given $\epsilon > 0,\quad L_f(\mathcal{P}) \underset{\epsilon}{\approx} U_f(\mathcal{P})$ for all $\mathcal{P}$ such that $|\mathcal{P}| \approx 0$.

Statement (7) looks like a limit definition; it is often written informally as

$$\lim_{|\mathcal{P}| \to 0} U_f(\mathcal{P}) - L_f(\mathcal{P}) = 0 .$$

This is misleading however: the "variable" in $U_f(\mathcal{P})$ is not the real number $|\mathcal{P}|$, but rather $\mathcal{P}$ itself.

Example 18.2   Prove $u(x) = \begin{cases} 1, & \text{if } x \geq 0, \\ 0, & \text{otherwise.} \end{cases}$ is integrable on any $[a, b]$.

Solution.   We prove (7). Let $\mathcal{P}$ be any partition of $[a, b]$. We have by (6) (the subscript u is omitted on U and L):

$$U(\mathcal{P}) - L(\mathcal{P}) = \sum_1^n (M_i - m_i)\Delta x_i .$$

If a subinterval $[\Delta x_i]$ does not contain 0, or has the form $[0, a]$, the corresponding term of the right-hand sum is 0, since either $M_i = m_i = 1$, or $M_i = m_i = 0$.

There will be at most one subinterval $[\Delta x_i]$ of $\mathcal{P}$ containing 0, but not of the form $[0, a]$; for this one, $M_i = 1$ and $m_i = 0$, so that the corresponding term of the right-hand sum will be Δx_i. Since $\Delta x_i \leq |\mathcal{P}|$ for all i (by the definition (3) of mesh), we conclude that

$$U(\mathcal{P}) - L(\mathcal{P}) \leq |\mathcal{P}| .$$

This proves that

$$U(\mathcal{P}) \underset{\epsilon}{\approx} L(\mathcal{P}) \qquad \text{if } |\mathcal{P}| < \epsilon ,$$

which establishes (7).   □

To prove integrability as we have defined it here, you have to show the upper and lower sums are close for all partitions $\mathcal{P}$ having sufficiently fine mesh. You cannot just show they are close for the standard n-partitions, for example. Once again, imagine a limit demon is giving you the worst partition it can think of—you have to prove it for that one.

Questions 18.2

1. The definition of integrability requires that $f(x)$ be bounded on the interval I. Show that if, say, $f(x)$ is not bounded above on $[a, b]$, then it cannot be integrable on $[a, b]$.

2. In the definition of the upper and lower sums, why are sup and inf used, rather than max and min?

3. Define $f(x)$ to be 1, if x is a rational number, 0 otherwise. Prove that $f(x)$ is not integrable on $[0, 1]$.

4. Prove that the function x is integrable on any interval $[a, b]$.
(Outline: for any $\mathcal{P}$, we have $U(\mathcal{P}) - L(\mathcal{P}) = \sum(\Delta x_i)^2 \leq |\mathcal{P}| \cdot \sum \Delta x_i, \ ... \ .$)

18.3   Integrability of monotone and continuous functions.

Most everyday functions are integrable—all bounded elementary functions, for instance. Lebesgue succeeded in characterizing the integrable functions in terms of their discontinuities, and the result is given in Chapter 23 where the Lebesgue integral is discussed. For now, to get further insight into how one proves integrability, in this section we will prove that two kinds of functions are always integrable: monotone functions and continuous functions.

Making use of a K-ϵ principle, in both cases we will prove $f(x)$ is integrable by establishing (7) in the following K-ϵ form:

(8)   given $\epsilon > 0$, $U_f(\mathcal{P}) \underset{K\epsilon}{\approx} L_f(\mathcal{P})$ for all $\mathcal{P}$ such that $|\mathcal{P}| \approx 0$.

Here K is a fixed constant depending only on $f(x)$ and not on the partition $\mathcal{P}$. (Once again, the proofs drop the subscripts on U_f and L_f.)

Theorem 18.3A   Integrability of monotone functions

If $f(x)$ is monotone on $[a, b]$, then $f(x)$ is integrable on $[a, b]$.

As the familiar picture on the right suggests, the idea is to show the difference between the upper and lower sum is small by interpreting it as the total area of the shaded rectangles. This is small because the shaded rectangles all fit into the narrow rectangle at the left, whose area is $K\epsilon$, where $K = f(b) - f(a)$.

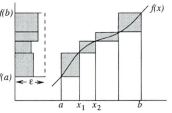

Proof. Suppose for definiteness that $f(x)$ is increasing on $[a, b]$.

Given $\epsilon > 0$, let $\mathcal{P}$ be any partition whose mesh $|\mathcal{P}| < \epsilon$. Since $f(x)$ is an increasing function on the interval $[\Delta x_i] = [x_{i-1}, x_i]$, we have

$$M_i = \sup_{[\Delta x_i]} f(x) = f(x_i), \qquad m_i = \inf_{[\Delta x_i]} f(x) = f(x_{i-1}) \ .$$

Therefore

$$U(\mathcal{P}) \ = \ f(x_1)\Delta x_1 + \ldots + f(x_n)\Delta x_n,$$

$$L(\mathcal{P}) \ = \ f(x_0)\Delta x_1 + \ldots + f(x_{n-1})\Delta x_n;$$

subtracting,

$$U(\mathcal{P}) - L(\mathcal{P}) \ = \ \sum_1^n (f(x_i) - f(x_{i-1})\Delta x_i,$$

$$\leq \ \sum_1^n (f(x_i) - f(x_{i-1}) \cdot \epsilon \ ,$$

since all the differences are positive ($f(x)$ is increasing), and since our hypothesis $|\mathcal{P}| < \epsilon$ implies that $\Delta x_i < \epsilon$ for all i.

In the last sum, successive terms cancel except for the first and last; noting that $x_n = b$ and $x_0 = a$, we get finally

$$U(\mathcal{P}) - L(\mathcal{P}) \leq (f(b) - f(a)) \cdot \epsilon \ ,$$

which proves (8), taking $K = f(b) - f(a)$.

(If $f(x)$ is decreasing, the same proof works, *mutatis mutandis*.)   □

Theorem 18.3B   Integrability of continuous functions.

If $f(x)$ is continuous on $[a, b]$, then it is integrable on $[a, b]$.

Proof.   Let $\mathcal{P}$ be any partition of $[a, b]$.

Since $f(x)$ is continuous, by the Maximum Theorem 13.3 it has a maximum point x_i' and a minimum point x_i'' on each subinterval $[\Delta x_i]$ of $\mathcal{P}$:

$$f(x_i') = \max_{[\Delta x_i)} f(x), \qquad f(x_i'') = \min_{[\Delta x_i)} f(x) \ .$$

Forming the upper and lower sums, and subtracting as in the previous proof,

(9) $$U(\mathcal{P}) - L(\mathcal{P}) = \sum_1^n (f(x_i') - f(x_i''))\Delta x_i \ .$$

We now want to estimate the right-hand side and show it is small.

Since f is continuous on $[a, b]$, it is uniformly continuous on $[a, b]$ by the Uniform Continuity Theorem 13.5. According to the definition of uniform continuity, this means that, given $\epsilon > 0$, there is a $\delta > 0$ such that

(10) $$|f(x') - f(x'')| < \epsilon \qquad \text{if} \quad |x' - x''| < \delta \ .$$

Suppose now that our partition $\mathcal{P}$ has mesh $|\mathcal{P}| < \delta$. Then since our maximum and minimum points x_i', x_i'' both lie in the same subinterval $[\Delta x_i]$,

$$|x_i' - x_i''| \leq \Delta x_i < \delta \ , \quad \text{for all } i \ .$$

Therefore by (10),

$$f(x_i') - f(x_i'') < \epsilon \quad \text{for all } i \ .$$

Using this to estimate the right side of (9), we see that if $|\mathcal{P}| < \delta$,

$$U(\mathcal{P}) - L(\mathcal{P}) \ < \ \sum_1^n \epsilon \cdot \Delta x_i \ = \ \epsilon \cdot \sum_1^n \Delta x_i \ = \ \epsilon(b - a) \ .$$

This proves (8), with $K = b - a$, showing that $f(x)$ is integrable.   □

Questions 18.3

1. In the proof of Theorem 18.3A, suppose that $\mathcal{P}$ is the standard n-partition of $[a, b]$. What is the value of $U(\mathcal{P}) - L(\mathcal{P})$?

2. Show that neither integrability theorem implies the other.

3. Suppose a function has this property: the difference between its sup and inf on any finite interval is always less than or equal to the length of the interval.

(a) Prove from the definition of integrability that it is integrable on any $[a, b]$.
(b) Prove another way that it is integrable.

18.4   Basic properties of integrable functions.

In order to be able to prove in the next chapter the standard facts about the definite integral that we have been using since Chapter 1, we need to know certain elementary facts about integrable functions. The proofs of these facts will provide good exercises in learning to work with integrability.

For the proofs, we recall from Section 6.5 some properties of the sup and inf of sets; here $A, B \subset \mathbb{R}$, $c \in \mathbb{R}$.

(11)   if $c > 0$, $\sup cA = c \cdot \sup A$, $\inf cA = c \cdot \inf A$;

(12) $\sup(-A) = -\inf A$, $\inf(-A) = -\sup A$;

(13) $\sup(A + B) \leq \sup A + \sup B$, $\inf(A + B) \geq \inf A + \inf B$.

Theorem 18.4A   Linearity property of integrability.   Let $c_1, c_2 \in \mathbb{R}$;

$f(x)$ and $g(x)$ integrable on $[a, b] \Rightarrow c_1 f(x) + c_2 g(x)$ integrable on $[a, b]$.

Proof.   It suffices to prove that $f(x) + g(x)$ and $cf(x)$ are integrable, for any $c \in \mathbb{R}$. We do this in three steps: (14), (15), and (16).

(14) $f(x)$ integrable $\Rightarrow$ $-f(x)$ integrable .

Namely, by hypothesis,

(14a)   given $\epsilon > 0$, $U_f(\mathcal{P}) \underset{\epsilon}{\approx} L_f(\mathcal{P})$ for $|\mathcal{P}| \approx 0$.

It follows that the same is true for the function $-f$, since from (12) and the definition of upper and lower sums, we have

$$\sup -f(x) = -\inf f(x) \Rightarrow U_{-f}(\mathcal{P}) = -L_f(\mathcal{P});$$
$$\inf -f(x) = -\sup f(x) \Rightarrow L_{-f}(\mathcal{P}) = -U_f(\mathcal{P});$$

so that (14a) for f immediately implies (14a) for $-f$; this proves (14).   □

(15) $f(x)$ integrable $\Rightarrow$ $cf(x)$ integrable .

If $c > 0$, this follows from (11) (see Question 1); if $c < 0$, then $(-c)f$ is integrable, therefore also cf, by (14).   □□

(16) $f(x), g(x)$ integrable $\Rightarrow$ $f(x) + g(x)$ integrable .

Namely, by (13), we have on any interval I,

$$\sup(f + g) \leq \sup f + \sup g , \qquad \inf(f + g) \geq \inf f + \inf g ,$$

and applying this to each subinterval of a partition $\mathcal{P}$ and summing, we get

$$U_{f+g}(\mathcal{P}) \leq U_f(\mathcal{P}) + U_f(\mathcal{P}) , \qquad L_{f+g}(\mathcal{P}) \geq L_f(\mathcal{P}) + L_g(\mathcal{P}) .$$

Multiply the second inequality by -1 (which reverses it), and add , getting

$$U_{f+g}(\mathcal{P}) - L_{f+g}(\mathcal{P}) \leq (U_f(\mathcal{P}) - L_f(\mathcal{P})) + (U_g(\mathcal{P}) - L_g(\mathcal{P})) .$$

From this estimate we deduce (16) immediately; namely, given $\epsilon > 0$, we have

$$U_f(\mathcal{P}) \underset{\epsilon}{\approx} L_f(\mathcal{P}) \quad \text{and} \quad U_g(\mathcal{P}) \underset{\epsilon}{\approx} \mathcal{L}_g(\mathcal{P}), \quad \text{when } |\mathcal{P}| \approx 0,$$

since f and g are assumed integrable. It follows from the estimate that

$$U_{f+g}(\mathcal{P}) \underset{2\epsilon}{\approx} L_{f+g}(\mathcal{P}) \qquad \text{when } |\mathcal{P}| \approx 0,$$

and this shows $f + g$ is integrable, by the K-ϵ principle. $\Box\Box\Box$

Theorem 18.4B   Absolute value property of integrability.

(17) $f(x)$ integrable on $[a, b]$ $\Rightarrow$ $|f(x)|$ is integrable on $[a, b]$.

Proof. We define the positive and negative functions

$$f^+(x) = \begin{cases} f(x), & \text{for } \{x : f(x) \geq 0\}; \\ 0, & \text{otherwise}; \end{cases} \quad f^-(x) = \begin{cases} f(x). & \text{for } \{x : f(x) \leq 0\}; \\ 0, & \text{otherwise}. \end{cases}$$

We first show that it suffices to prove

(18) $f(x)$ integrable on $[a, b]$ $\Rightarrow$ $f^+(x)$ is integrable on $[a, b]$.

For since $f^-(x) = -(-f(x))^+$, it follows from (14), (18), and then (14) again that also $f^-(x)$ is integrable; we deduce that $|f(x)|$ is integrable by applying the sum theorem (16) to

$$|f(x)| = f^+(x) - f^-(x) .$$

To prove (18), it is not difficult to show (see the Questions) that for any function $f(x)$ that is bounded on an interval I, we have

(19) $\sup_I f^+(x) - \inf_I f^+(x) \leq \sup_I f(x) - \inf_I f(x) ;$

in view of the definition of upper and lower sums, this implies immediately that

$$U_{f^+}(\mathcal{P}) - L_{f^+}(\mathcal{P}) \leq U_f(\mathcal{P}) - L_f(\mathcal{P}) ;$$

this proves (18), in view of the definition of integrability. $\Box$

Questions 18.4

1. Prove (15) for the case $c > 0$.

2. Prove (19); (make cases: $f(x) \geq 0$, $f(x) \leq 0$, neither of the preceding).

Exercises

18.1

1. If an n-partition $\mathcal{P}$ of $[a, b]$ has mesh $(b - a)/n$, is $\mathcal{P} = \mathcal{P}^{(n)}$, the standard n-partition? Give a proof or counterexample.

18.2

1. Prove directly from the definition of integrability that x^2 is integrable on any $[a, b] \subset [0, \infty)$.

2. Let $f(x) = \begin{cases} 0, & \text{if there exist integers } m, n > 0 \text{ such that } x = m/2^n; \\ 1, & \text{otherwise.} \end{cases}$

Prove $f(x)$ is not Riemann-integrable on $[0, 1]$.

3. Suppose $f(x)$ is integrable on $[a, b]$. Prove it is also integrable on any subinterval $[c, d] \subset [a, b]$.

18.3

1. Let n be a fixed integer > 0; consider the function

$$f_n(x) = \begin{cases} 0, & \text{if } x = 1/n, 2/n, \ldots, (n-1)/n \ ; \\ 1, & \text{otherwise .} \end{cases}$$

Prove directly from the definition of integrability that $f_n(x)$ is integrable on $[0, 1]$.

2. Prove directly from the definition of integrability: on an interval $[a, b]$, if $f(x)$ is differentiable and $f'(x)$ is bounded, then $f(x)$ is integrable.

(Of course, this follows from Theorem 18.3B, since a differentiable function is continuous; but prove it directly from the definition of integrability, without quoting 18.3B or any exercises on uniform continuity.)

3. Prove that the function $f(x) = \sin(1/x), x \neq 0; f(0) = 0$ is integrable on $[-1, 1]$. (You may use the theorems of this section.)

4. Let $f(x) = \begin{cases} 0, & \text{if } x = 1/n \text{ for some integer } n > 0, \\ 1, & \text{otherwise ;} \end{cases}$ prove $f(x)$ is integrable on $[0, 1]$.

18.4

1. Prove: if $f(x)$ is integrable on $[a, b]$, then $(f(x))^2$ is integrable on $[a, b]$.

(First assume $f \geq 0$; how are $\sup f^2$ and $\sup f$ related? Then deduce from this that it is true for any integrable f; use theorems.)

2. Prove that if f and g are non-negative and integrable, so is fg. (You will need to relate $\sup fg$ with $\sup f$ and $\sup g$. In putting it all together, the idea used in the proof of the product rule for differentiation could be helpful.)

Answers to Questions

18.2

1. Let $\mathcal{P}$ be any partition of $[a, b]$, and suppose for example that $f(x)$ is not bounded above. Then the upper sum $U_f(\mathcal{P})$ is undefined, since on at least one of the subintervals $[\Delta x_i]$, $\sup f(x)$ does not exist (for if it existed on all of them, the maximum of these finitely many $\sup f(x)$ would be an upper bound for $f(x)$ on $[a, b]$).

2. In general, functions do not have a maximum or minimum on a compact interval; only for continuous functions is this guaranteed (Maximum Theorem 13.3).

3. Every compact interval $[a, b]$ of positive length contains a rational and an irrational number (Theorem 2.5). Thus on every subinterval of any partition $\mathcal{P}$, we have

$$\sup_{[\Delta x_i]} f(x) = 1, \quad \inf_{[\Delta x_i]} f(x) = 0; \qquad \text{thus} \quad U_f(\mathcal{P}) = 1, \ L_f(\mathcal{P}) = 0 \ .$$

This shows f is not integrable on any interval $[a, b]$.

$$
\begin{aligned}
\text{4.} \quad U(\mathcal{P}) - L(\mathcal{P}) &= \sum x_i \Delta x_i - \sum x_{i-1} \Delta x_i = \sum (\Delta x_i)^2 \\
&\leq \sum \epsilon(\Delta x_i), \quad \text{if } |\mathcal{P}| \leq \epsilon \ (\text{since } \Delta x_i \leq \epsilon) \\
&= \epsilon \sum \Delta x_i = \epsilon(b - a), \quad \text{since } \sum \Delta x_i = b - a \ .
\end{aligned}
$$

Therefore

$$U(\mathcal{P}) \underset{(b-a)\epsilon}{\approx} L(\mathcal{P}), \quad \text{if } |\mathcal{P}| \leq \epsilon, \quad \text{which proves integrability.}$$

18.3

1. Since $\Delta x_i = (b - a)/n$, $\quad U(\mathcal{P}) - L(\mathcal{P}) = [f(b) - f(a)](b - a)/n$.

2. Continuous and monotone are independent properties (i.e., neither implies the other): the functions x^2 and $u(x)$ on $[-1, 1]$ illustrate this ($u(x)$ is the unit step function; cf. Example 18.2.): the first is continuous but not monotone, the second monotone but not continuous.

3. (a) Reasoning as in this section, if $|\mathcal{P}| < \epsilon$, then $\Delta x_i < \epsilon$, so that

$$U(\mathcal{P}) - L(\mathcal{P}) = \sum (\sup_{[\Delta x_i]} f - \inf_{[\Delta x_i]} f) \Delta x_i \ \leq \ \sum \Delta x_i \cdot \epsilon \ = \ (b-a)\epsilon \ ,$$

which proves that the definition of integrability (7) is satisfied.

(b) The hypothesis implies $f(x)$ is uniformly continuous. Namely, given $\epsilon > 0$, for any two points $x' < x''$ we have (letting $I = [x', x'']$),

$$|f(x'') - f(x')| \ \leq \ \sup_I f(x) - \inf_I f(x) \ \leq \ x'' - x' \quad \text{by hypothesis,}$$

$$\leq \ \epsilon, \quad \text{if } x' \underset{\epsilon}{\approx} x'' \ .$$

Since $f(x)$ is continuous, it is integrable, by Theorem 18.3B.

18.4

1. If $c > 0$, from (11) we see that

$$U_{cf}(\mathcal{P}) - L_{cf}(\mathcal{P}) \ = \ c\big(U_f(\mathcal{P}) - L_f(\mathcal{P})\big),$$

so that, given $\epsilon > 0$,

$$U_f(\mathcal{P}) \underset{\epsilon}{\approx} L_f(\mathcal{P}) \quad \text{for } |\mathcal{P}| \approx 0 \quad \Rightarrow \quad U_{cf}(\mathcal{P}) \underset{c\epsilon}{\approx} L_{cf}(\mathcal{P}) \quad \text{for } |\mathcal{P}| \approx 0 \ .$$

2. Referring to (19):

if $f(x) \geq 0$ on I, equality holds;

if $f(x) \leq 0$ on I, the left side of (19) is 0, and the right non-negative;

if $f(x)$ is both positive and negative on I, then (19) is still true, because

$$\sup f^+(x) = \sup f(x), \qquad \inf f^+(x) > \inf f(x) \ .$$

19

The Riemann Integral

19.1 Refinement of partitions.

In this chapter we finally define the Riemann integral of an integrable function $f(x)$, show it exists and is unique, and establish its most important properties. As a preliminary, we need one more idea involving partitions.

Definition 19.1   Refinement.

The partition $\mathcal{P}'$ is a **refinement** of the partition $\mathcal{P}$ if $\mathcal{P}'$ is formed by partitioning each subinterval of $\mathcal{P}$. (The notation we use is: $\mathcal{P}' \leq \mathcal{P}$.)

The figure illustrates $\mathcal{P}' \leq \mathcal{P}$:

$$\mathcal{P} : a < x_1 < x_2 < b,$$
$$\mathcal{P}' : a < y_1 < y_2 < \ldots < y_6 < b.$$

Note that, since every subinterval of $\mathcal{P}'$ lies inside a subinterval of P,

(1)   *refinement makes the mesh smaller:* $\mathcal{P}' \leq \mathcal{P} \;\Rightarrow\; |\mathcal{P}'| \leq |\mathcal{P}|$.

Example 19.1A   Successive bisection.

Repeated bisection of $[a, b]$ gives a sequence of standard n-partitions,

(2) $$\mathcal{P}^{(1)} \;\geq\; \mathcal{P}^{(2)} \;\geq\; \mathcal{P}^{(4)} \;\geq \ldots \geq\; \mathcal{P}^{(2^i)} \;\geq \ldots\;,$$

each of which is a refinement of the preceding, and whose mesh $\to 0$.

Example 19.1B   Common refinement.

Any two partitions $\mathcal{P}_1$ and $\mathcal{P}_2$ have a *least common refinement* $\mathcal{P}'$, which is obtained by combining the points of $\mathcal{P}_1$ and $\mathcal{P}_2$, deleting repetitions, and listing the combined set of points in natural order. The figure below illustrates:

$$\mathcal{P}_1 : a, x_1, x_2, b \;; \qquad \mathcal{P}_2 : a, y_1, y_2, y_3, b \;; \qquad \mathcal{P}' : a, y_1, x_1, x_2, y_3, b \;.$$

For the least common refinement, we see that

(3) $$\mathcal{P}' \leq \mathcal{P}_1 \qquad \text{and} \qquad \mathcal{P}' \leq \mathcal{P}_2 \;;$$

any partition $\mathcal{P}'$ satisfying (3) is called a *common refinement* of $\mathcal{P}_1$ and $\mathcal{P}_2$, regardless of how it is constructed.

Lemma 19.1   Upper and lower sum lemma.

Refinement lowers the upper sums and raises the lower sums:

(4) $$\mathcal{P}' \leq \mathcal{P} \quad \Rightarrow \quad U_f(\mathcal{P}') \leq U_f(\mathcal{P}), \qquad L_f(\mathcal{P}') \geq L_f(\mathcal{P}) .$$

Proof.   First, if I and J are two intervals on which $f(x)$ is bounded,

(5) $$I \subseteq J \quad \Rightarrow \quad \sup_I f(x) \leq \sup_J f(x), \qquad \inf_I f(x) \geq \inf_J f(x) .$$

The first inequality says that the tallest person among the first 100 in a concert ticket line is no bigger than the tallest person in the entire line; Question 3 asks for a more formal argument.

To prove the lemma for the upper sums, suppose $\mathcal{P}'$ partitions the i-th interval $[\Delta x_i]$ of $\mathcal{P}$ as shown into smaller intervals $I_1, \ldots, I_r$, of length $|I_k|$.

As before, we set

$$M_i = \sup_{[\Delta x_i]} f(x), \quad m_i = \inf_{[\Delta x_i]} f(x);$$

then, since the subintervals $I_k \subseteq [\Delta x_i]$, we have by (5)

$$\sup_{I_k} f(x) \quad \leq \quad M_i ,$$

$$\sum_{k=1}^{r} \sup_{I_k} f(x) \, |I_k| \quad \leq \quad \sum_{k=1}^{r} M_i \, |I_k| \quad = \quad M_i \, \Delta x_i .$$

In this inequality, the left-hand sum represents the contribution of the subintervals of $[\Delta x_i]$ to $U_f(\mathcal{P}')$. Summing both sides of the inequality over the subintervals $[\Delta x_i]$ of $\mathcal{P}$ gives then

$$U_f(\mathcal{P}') \quad \leq \quad U_f(\mathcal{P}) .$$

The proof of the inequality for the lower sums is similar.   □

Corollary 19.1   *For $f(x)$ on $[a, b]$, any lower sum is less than any upper sum:*

(6) $$\text{for any partitions } \mathcal{P}_1, \mathcal{P}_2 \text{ of } [a, b], \qquad L_f(\mathcal{P}_1) \leq U_f(\mathcal{P}_2) .$$

Proof.   The proof is a typical use of common refinement (Example 19.1B).

First of all, (6) is true if $\mathcal{P}_1 = \mathcal{P}_2$, according to Definition 18.2A of upper and lower sum, since $m_i \leq M_i$ for all i.

If $\mathcal{P}_1 \neq \mathcal{P}_2$, the two partitions have a common refinement $\mathcal{P}'$ (for example, their least common refinement). Then since $\mathcal{P}' \leq \mathcal{P}_1$ and $\mathcal{P}' \leq \mathcal{P}_2$, the preceding remark and (4) show

$$L(\mathcal{P}_1) \leq L(\mathcal{P}') \leq U(\mathcal{P}') \leq U(\mathcal{P}_2) .$$ □

Questions 19.1

1. (a) When does the standard m-partition $\mathcal{P}^{(m)}$ of $[a, b]$ refine $\mathcal{P}^{(n)}$?

(b) What standard k-partition has the largest mesh of all common refinements of $\mathcal{P}^{(m)}$ and $\mathcal{P}^{(n)}$?

2. What is "least" and "greatest" about the least common refinement?

3. Give a formal proof of (5); use the set $f(I)$ and $\sup f(I)$ (cf. Section 10.1.)

19.2 Definition of the Riemann integral.

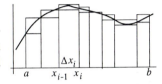

At last. We base the definition on the geometric observation that the area under the graph of $f(x)$ always lies between the total area of any set of inscribed rectangles and the total area of any set of circumscribed rectangles.

Theorem-Definition 19.2   The Riemann integral.

If $f(x)$ is integrable on $[a, b]$, there is a unique number $\mathcal{I}$ such that for any partitions $\mathcal{P}_1$ and $\mathcal{P}_2$ of $[a, b]$,

(7) $$L_f(\mathcal{P}_1) \leq \mathcal{I} \leq U_f(\mathcal{P}_2) ;$$

$\mathcal{I}$ is called the **Riemann integral** of $f(x)$ over $[a, b]$; it is denoted by

$$\mathcal{I} = \int_a^b f(x)\, dx, \qquad \text{or simply} \qquad \int_a^b f .$$

Proof. We first locate the number $\mathcal{I}$. Consider the sequence (2) of standard partitions $\mathcal{P}^{(1)}, \mathcal{P}^{(2)}, \ldots$ produced by successive bisections of $[a, b]$. The Upper Sum Lemma 19.1 and Corollary (6) tell us that (dropping the subscript f),

$$L(\mathcal{P}^{(1)}) \leq L(\mathcal{P}^{(2)}) \leq L(\mathcal{P}^{(4)}) \leq \cdots \leq U(\mathcal{P}^{(4)}) \leq U(\mathcal{P}^{(2)}) \leq U(\mathcal{P}^{(1)}) .$$

This shows the intervals $[L(\mathcal{P}^{(2^i)}), U(\mathcal{P}^{(2^i)})]$ form a sequence of nested intervals. Moreover, since $f(x)$ is integrable on $[a, b]$, and the mesh $|\mathcal{P}^{(2^i)}| \to 0$ as $i \to \infty$, the definition of integrability (18.2B, (7)) shows

$$\lim_{i \to \infty} U(\mathcal{P}^{(2^i)}) - L(\mathcal{P}^{(2^i)}) = 0 .$$

Therefore by the Nested Intervals Theorem 6.1, there is a unique number $\mathcal{I}$ which is inside all these intervals, and $\lim L(\mathcal{P}^{(2^i)}) = \mathcal{I}$, $\lim U(\mathcal{P}^{(2^i)}) = \mathcal{I}$.

Finally, we prove (7) holds for $\mathcal{I}$. Namely, we have for any partition $\mathcal{P}$,

$$L(\mathcal{P}) \leq U(\mathcal{P}^{(2^i)}) \qquad \text{for all } i, \quad \text{by (6)};$$

$$\Rightarrow \quad L(\mathcal{P}) \leq \lim U(\mathcal{P}^{(2^i)}) \quad \text{by the Limit Location Theorem 5.3A;}$$

$$\Rightarrow \quad L(\mathcal{P}) \leq \mathcal{I} .$$

The proof that $\mathcal{I} \leq U(\mathcal{P})$ for any $\mathcal{P}$ is similar, and left as a Question. $\square$

Corollary 19.2 *If* $f(x)$ *is integrable on* $[a, b]$, *then for any sequence* $\mathcal{P}_i$ *of partitions of* $[a, b]$ *such that* $|\mathcal{P}_i| \to 0$,

$$(8) \qquad \lim_{i \to \infty} U(\mathcal{P}_i) = \int_a^b f(x)dx, \qquad \lim_{i \to \infty} L(\mathcal{P}_i) = \int_a^b f(x)dx \ .$$

Proof. Let $\mathcal{I} = \int_a^b f(x)dx$. Since f is integrable, we have, given any $\epsilon > 0$,

$$L(\mathcal{P}_i) \underset{\epsilon}{\approx} U(\mathcal{P}_i) \qquad \text{for } i \gg 1 \ ; \quad \text{moreover,}$$

$$L(\mathcal{P}_i) \leq \mathcal{I} \leq U(\mathcal{P}_i) \qquad \text{for all } i \ , \quad \text{by (7); thus}$$

$$L(\mathcal{P}_i) \underset{\epsilon}{\approx} \mathcal{I} \quad \text{and} \quad U(\mathcal{P}_i) \underset{\epsilon}{\approx} \mathcal{I} \qquad \text{for } i \gg 1 \ .$$

This proves (8), according to the definition of limit of a sequence. $\square$

This corollary will be used frequently to evaluate integrals in a theoretical situation. It shows we have great freedom in the choice of what partitions we use, and whether we use upper or lower sums.

Example 19.2 Calculate $\int_0^1 x^2 dx$ directly from the definition.

This was first done by Archimedes, who calculated the area under a parabola by a method very close in spirit to the definition of the integral.

Solution. We know that x^2 is integrable on $[0, 1]$ by Theorem 18.3A or B, since it is both monotone and continuous.

We use the sequence $\mathcal{P}^{(n)}$ of standard n-partitions. The upper sums for this are illustrated; they are seen to be

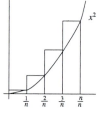

$$U(\mathcal{P}^{(n)}) = \frac{1}{n}\left(\left(\frac{1}{n}\right)^2 + \left(\frac{2}{n}\right)^2 + \ldots + \left(\frac{n}{n}\right)^2 \right) ,$$

so that by the Corollary (8) above,

$$(9) \qquad \int_0^1 x^2 dx = \lim_{n \to \infty} U(\mathcal{P}^{(n)}) = \lim_{n \to \infty} \frac{1^2 + \ldots + n^2}{n^3} \ .$$

To evaluate this limit, which has the form ∞/∞, we can use the exact formula for the sum of the first n squares (cf. Example A.4A in Appendix A.4); we leave this as a Question. The exact formula isn't really necessary however—an estimation will do, based on the old Greek formula

$$\text{volume of a pyramid} = \tfrac{1}{3}(\text{base})(\text{height}).$$

Estimate the sum $1^2 + 2^2 + \ldots + n^2$ by thinking of an Egyptian step pyramid, for which the successive layers are respectively squares of side $1, 2, \ldots, n$, and one unit high. Its volume is the sum of the first n squares; to estimate this volume we observe that the pyramid lies between the inner and outer smooth pyramids sketched, which gives us the inequalities

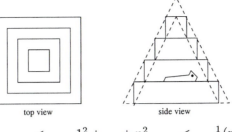

top view   side view

$$\tfrac{1}{3}n^2 \cdot n \qquad < \qquad 1^2 + \ldots + n^2 \qquad < \qquad \tfrac{1}{3}(n+1)^2(n+1)$$

inner pyramid volume   step pyramid volume   outer pyramid volume

Dividing these inequalities through by n^3 and using the Squeeze Theorem 5.2, we conclude that the limit on the right of (9) is $1/3$, which is therefore the value of the definite integral. $\qquad\qquad\qquad\qquad\qquad\qquad\qquad\qquad\qquad\qquad\qquad$ $\square$

The purpose of this example is to make the definitions more concrete by seeing how they work out for a particular function. For general functions, using (8) to evaluate the integral directly rarely works, since if we use for example the sequence of standard n-partitions, we will get

$$U(\mathcal{P}^{(n)}) \;=\; (M_1 + \ldots + M_n)\Delta x \,, \qquad \Delta x = \frac{b-a}{n} \,,$$

which as $n \to \infty$ is always an indeterminate form of type $\infty \cdot 0$.

To evaluate the limit therefore one must be able to estimate $\sum M_i$. The exercises give some cases where this can be done, but there is no general method. Calculus only got off the ground with the discovery that definite integrals could be evaluated by using antiderivatives, as in the next chapter.

For theoretical work with the integral, however, you need Definition 19.2 and the Corollary, as the rest of this chapter shows.

___

Questions 19.2

1. Evaluate the limit in (9) by using the exact formula for the sum of the first n squares (Example A.4A, Appendix A.4).

2. Carry out the evaluation of $\int_0^1 x^2 dx$ by using the lower sums.

3. Evaluate $\int_0^1 x\,dx$ by using the standard n-partitions and upper sums.

4. For Theorem 19.2, show that $\mathcal{I} \le U(\mathcal{P})$ for any partition $\mathcal{P}$.

___

19.3   Riemann sums.

According to Corollary 19.2, the definite integral $\int_a^b f(x)dx$ can be calculated as the limit of a sequence of upper or lower sums. But sometimes it is more convenient to calculate it as the limit of another type of sum, whose value lies between the upper and lower sums.

Definition 19.3A Let $f(x)$ be defined on $[a, b]$. A **Riemann sum** for $f(x)$ over the partition $\mathcal{P}$ is any sum of the form

(10)
$$S_f(\mathcal{P}) = \sum_1^n f(x_i')\Delta x_i, \qquad \text{where } x_i' \in [\Delta x_i].$$

There are infinitely many Riemann sums for $f(x)$ over a given partition $\mathcal{P}$, since there are infinitely many ways to choose the points x_i'. But every Riemann sum lies between the upper and lower sums for that $\mathcal{P}$:

(11)
$$L_f(\mathcal{P}) \leq S_f(\mathcal{P}) \leq U_f(\mathcal{P}),$$

since on the i-th interval of the partition, $m_i \leq f(x_i') \leq M_i$, and therefore

$$\sum_1^n m_i\Delta x_i \leq \sum_1^n f(x_i')\Delta x_i \leq \sum_1^n M_i\Delta x_i.$$

Theorem 19.3 The Riemann sum theorem.

Let $f(x)$ be integrable on $[a, b]$.

Let $\mathcal{P}_k$ be a sequence of partitions of $[a, b]$ such that $|\mathcal{P}_k| \to 0$.

For each k, let $S_f(\mathcal{P}_k)$ be a Riemann sum for $f(x)$ over $\mathcal{P}_k$. Then

(12)
$$\lim_{k \to \infty} S_f(\mathcal{P}_k) = \int_a^b f(x)\,dx.$$

Proof. We write the inequalities (11) for each $\mathcal{P}_k$, and let $k \to \infty$; we get

$$\begin{array}{ccccc}
L_f(\mathcal{P}_k) & \leq & S_f(\mathcal{P}_k) & \leq & U_f(\mathcal{P}_k) \\
\downarrow & & \Downarrow & & \downarrow \\
\int_a^b f(x)\,dx & & \int_a^b f(x)\,dx & & \int_a^b f(x)\,dx;
\end{array}$$

the two outside limits by (8), the middle limit by the Squeeze Theorem 5.2 . $\square$

Example 19.3 Evaluate $\displaystyle \lim_{n \to \infty} \frac{1}{n}\left(\sin\frac{\pi}{n} + \sin\frac{2\pi}{n} + \ldots + \sin\frac{n\pi}{n}\right)$.

Solution. When multiplied by π, the expression is a Riemann sum for $\sin x$ over the standard n-partition of the interval $[0, \pi]$, with x_i' always the right-hand endpoint of each subinterval $[\Delta x_i]$. (Note that $|\Delta x_i| = \pi/n$, and that the expression is not an upper or lower sum.)

Since $\sin x$ is integrable (it is continuous), by (12) the limit is

$$\frac{1}{\pi}\int_0^\pi \sin x\,dx = \frac{2}{\pi}. \qquad \square$$

Questions 19.3

1. Is the upper sum $U(\mathcal{P})$ for a function $f(x)$ over some partition $\mathcal{P}$ always a Riemann sum if   (a) $f(x)$ is monotone?   (b) $f(x)$ is continuous?
Give in each case a proof or a counterexample.

19.4   Basic properties of integrals.

These three properties of the integral are used all the time in calculations. Their proofs use the Riemann Sum Theorem 19.3.

Theorem 19.4A   Linearity theorem for integrals

If $f(x)$ and $g(x)$ are integrable on $[a, b]$, and c_1, c_2 are constants,

$$(14) \qquad \int_a^b [c_1 f(x) + c_2 g(x)] \, dx \;=\; c_1 \int_a^b f(x) dx + c_2 \int_a^b g(x) dx \; .$$

Proof.   By the linearity property of integrability (Theorem 18.4A), we know that the function $c_1 f(x) + c_2 g(x)$ is integrable. Therefore the Riemann Sum Theorem 19.3 can be used to evaluate the integral.

Take a sequence of partitions $\mathcal{P}_k$ such that $|\mathcal{P}_k| \to 0$, and for each, form a Riemann sum for $c_1 f(x) + c_2 g(x)$. Break up the Riemann sum for each k:

$$\sum [c_1 f(x_i') + c_2 g(x_i')] \Delta x_i \;=\; c_1 \sum f(x_i') \Delta x_i + c_2 \sum g(x_i') \Delta x_i \; .$$

Take the limit as $k \to \infty$; since all three functions are integrable, the Riemann Sum Theorem shows the three sums converge respectively to the three integrals:

$$(14) \qquad \int_a^b [c_1 f(x) + c_2 g(x)] \, dx \;=\; c_1 \int_a^b f(x) dx + c_2 \int_a^b g(x) dx \; ;$$

equality holds by the Linearity Theorem 5.1 for limits of sequences. $\square$

Theorem 19.4B   Comparison theorem for integrals.

If $f(x)$ and $g(x)$ are integrable on $[a, b]$,

$$(15) \qquad f(x) \leq g(x) \quad \text{on } [a, b] \quad \Rightarrow \quad \int_a^b f(x) dx \leq \int_a^b g(x) dx \; .$$

Proof.   Left as a Question. If $f < g$ and both are continuous, the $\leq$ can be changed to $<$ on the right-hand side; cf. Exercise 19.4/2b.

Theorem 19.4C   Absolute value theorem for integrals.

If $f(x)$ is integrable on $[a, b]$,

$$(16) \qquad \left| \int_a^b f(x) dx \right| \;\leq\; \int_a^b |f(x)| dx \; .$$

Proof.   By Theorem 18.4B, we know $|f(x)|$ is integrable. We have (cf. 2.4(2))

$$-|f(x)| \;\leq\; f(x) \;\leq\; |f(x)| \; .$$

By the Comparison Theorem (15) and the Linearity Theorem (14),

$$-\int_a^b |f(x)| dx \;\leq\; \int_a^b f(x) dx \;\leq\; \int_a^b |f(x)| dx \; ;$$

this is the same as (16), again by 2.4 (2). $\square$

Questions 19.4

1. Prove the Comparison Theorem 19.4B.

2. Assume f and g are integrable on $[a, b]$. Prove that over $[a, b]$, the integral of their average is the average of their integrals.

3. Prove that $\displaystyle\int_0^1 \frac{\cos x}{1 + x^2}\, dx \le \frac{\pi}{4}$.

4. Prove: if $f(x) \underset{\epsilon}{\approx} g(x)$ on $[a, b]$, then $\left| \displaystyle\int_a^b f(x)dx - \int_a^b g(x)dx \right| \le \epsilon\,(b-a)$.

19.5   The interval addition property.

We devote a separate section to this property, since the proofs have a slightly different character from the preceding ones, and are a little more technical. In the proofs, to make the notation expressive, we will use $\mathcal{P}'$ and $\mathcal{P}''$ for partitions which are related to $\mathcal{P}$, but are not refinements of it (as they have been in this chapter up to now).

Theorem 19.5   Interval addition for integrals.

Suppose $a < b < c$. Then if $f(x)$ is integrable on $[a, b]$ and $[b, c]$, it is integrable on $[a, c]$, and

(17) $$\int_a^c f(x)dx \;=\; \int_a^b f(x)dx + \int_b^c f(x)dx \;.$$

Proof. The proof divides naturally into two parts: the proof of integrability, and the proof of (17). We begin with the proof that $f(x)$ is integrable over $[a, c]$.

Fix any $\delta > 0$, and let $\mathcal{P}$ be any partition of $[a, c]$ with mesh $|\mathcal{P}| < \delta$.

As the picture shows, it gives partitions $\mathcal{P}'$ and $\mathcal{P}''$ of $[a, b]$ and $[b, c]$ respectively, each with mesh $< \delta$. The point b divides the subinterval I of the partition $\mathcal{P}$ in which it lies into two smaller subintervals I' and I'', belonging respectively to the partitions $\mathcal{P}'$ and $\mathcal{P}''$. (Assume $b \notin \mathcal{P}$; if $b \in \mathcal{P}$, the following argument simplifies to a few lines.)

Since $f(x)$ is integrable on $[a, b]$ and $[b, c]$, it is by definition bounded on these two intervals, and therefore on $[a, c]$ as well. So we can say $|f(x)| \le K$, and therefore also

(18) $$\sup |f(x)| \le K, \qquad x \in [a, c] \;.$$

We prove first that

(19) $$U(\mathcal{P}) \underset{3K\delta}{\approx} U(\mathcal{P}') + U(\mathcal{P}'') \;.$$

If we compare $U_f(\mathcal{P})$ with $U_f(\mathcal{P}')$ and $U_f(\mathcal{P}'')$—we drop the subscript f from now on—we see that all the terms in $U(\mathcal{P})$ occur in either $U(\mathcal{P}')$ or $U(\mathcal{P}'')$, except for the terms involving the intervals containing point b. Therefore, referring to the picture, (here $|I| = $ length of I),

$$\left| U(\mathcal{P}) - U(\mathcal{P}') - U(\mathcal{P}'') \right| = \left| \sup_I f \cdot |I| - \sup_{I'} f \cdot |I'| - \sup_{I''} f \cdot |I''| \right|$$

$$\leq 3K\delta ,$$

by the triangle inequality and (18), since the intervals I, I', I'' all have length $< \delta$. This proves (19):

(19) $$U(\mathcal{P}) \underset{3K\delta}{\approx} U(\mathcal{P}') + U(\mathcal{P}'') .$$

A similar argument using inf shows

(20) $$L(\mathcal{P}) \underset{3K\delta}{\approx} L(\mathcal{P}') + L(\mathcal{P}'') .$$

Since f is integrable over $[a,b]$ and $[b,c]$, we know that, given $\epsilon > 0$, there is a $\delta > 0$ such that

$$U(\mathcal{P}') \underset{\epsilon}{\approx} L(\mathcal{P}') \quad \text{and} \quad U(\mathcal{P}'') \underset{\epsilon}{\approx} L(\mathcal{P}'') \qquad \text{for } |\mathcal{P}'|, |\mathcal{P}''| < \delta .$$

Reducing δ if necessary, we may also suppose $\delta < \epsilon/3K$. Combining the above approximations with (19) and (20), since $3K\delta < \epsilon$ we see that

(21) $$U(\mathcal{P}) \underset{\epsilon}{\approx} U(\mathcal{P}') + U(\mathcal{P}'') \underset{2\epsilon}{\approx} L(\mathcal{P}') + L(\mathcal{P}'') \underset{\epsilon}{\approx} L(\mathcal{P}), \quad \text{for } |\mathcal{P}| < \delta .$$

Thus finally

$$U(\mathcal{P}) \underset{4\epsilon}{\approx} L(\mathcal{P}) \qquad \text{for } |\mathcal{P}| < \delta ,$$

which shows that f is integrable over $[a, c]$. $\square$

From this, the proof of the interval addition property (17) follows easily..

Choose a sequence of partitions $\mathcal{P}'_k$ of $[a, b]$ such that $|\mathcal{P}'_k| \to 0$, and a similar sequence $\mathcal{P}''_k$ for $[b, c]$. By putting the points of $\mathcal{P}'_k$ and $\mathcal{P}''_k$ together, we get a sequence $\mathcal{P}_k = \mathcal{P}'_k \bigcup \mathcal{P}''_k$ of partitions of $[a, c]$, such that $|\mathcal{P}_k| \to 0$. Choose Riemann sums $S_f(\mathcal{P}'_k)$ and $S_f(\mathcal{P}''_k)$ for the partitions on $[a, b]$ and $b, c]$. Put together, they give Riemann sums $S_f(\mathcal{P}_k)$ for the partitions of $[a, c]$. The rest of the argument is summarized by the following diagram:

(22) $$S_f(\mathcal{P}_k) = S_f(\mathcal{P}'_k) + S_f(\mathcal{P}''_k)$$
$$\downarrow \qquad\qquad \downarrow \qquad\qquad \downarrow$$
(23) $$\int_a^c f(x)dx = \int_a^b f(x)dx + \int_b^c f(x)dx.$$

The vertical arrows represent the limits as $k \to \infty$; the limits are valid by the Riemann Sum Theorem 19.3, since by hypothesis $f(x)$ is integrable on $[a, b]$ and $[b, c]$, and we have proved above that it is integrable on $[a, c]$. It follows that the bottom line is valid, by the Linearity Theorem 5.1 for limits of sequences, and this proves (17). $\square$

Until now we have defined and used $\int_a^b f(x)dx$ only when $a < b$. In order to make the Interval Addition Theorem true for any a, b, c, we extend the definition of the integral by allowing $a \geq b$, as follows.

Definition 19.5   Backwards integration.   We define

(24) $\displaystyle\int_a^a f(x)dx = 0$ for all a ; $\displaystyle\int_a^b f(x)dx = -\int_b^a f(x)dx$,   if $a > b$.

Corollary 19.5   General interval addition theorem.

The equality (17) is true for all positions of a, b, c, if $f(x)$ is integrable on the two intervals with endpoints a, b and b, c.

Proof.   Use (17) and the above definitions; there are several cases, all easy, and left as exercises.   □

Questions 19.5

1. In the proof of integrability in Theorem 19.5, how does the proof simplify for those partitions $\mathcal{P}$ which contain b ? (Going through this will help you understand the proof.)

19.6   Piecewise continuous and monotone functions *(optional)*

Many functions in common use in science and engineering have discontinuities and are not monotone, but they still can be integrated. Most of them will be taken care of by the following definition. (Others may require Lebesgue integration.)

Definition 19.6   A function $f(x)$ is **piecewise continuous** on $[a, b]$ if there is a partition $\mathcal{P} : a < x_1 < x_2 < \ldots < x_{n-1} < b$ such that $f(x)$ is continuous on each open subinterval (x_{i-1}, x_i).
        (We sometimes say $f(x)$ is piecewise continuous with respect to $\mathcal{P}$.)

The definition of **piecewise monotone** is the same, replacing everywhere "continuous" by "monotone".

As the examples below show, for greater generality it is convenient not to require $f(x)$ to be defined at the points of the partition, including the two endpoints.

Example 19.6   Which of the following functions are piecewise continuous or monotone on the indicated interval? (The graph of $w(x)$ is a square wave.)

(a) $\tan x$ on $[0, 2\pi]$ (b) $\cos 1/x$ on $[0, 1]$

(c) $w(x)$ on $[0, n]$, where $w(x) = \begin{cases} 1, & x \in (0,1), (2,3), \ldots, \\ 0, & x \in (1,2), (3,4), \ldots; \end{cases}$ $n \in \mathbb{N}$.

Solution.

(a) $\tan x$ is both piecewise continuous and piecewise monotone with respect to the partition $< 0, \pi/2, 3\pi/2, 2\pi >$, since it is continuous and increasing on the open subintervals. (It is not defined at $\pi/2$ or $3\pi/2$.)

(b) $\cos 1/x$ is continuous on $(0, 1]$, therefore piecewise continuous. It is not piecewise monotone: a partition $\mathcal{P}$ is finite, but the intervals $[1/n\pi, 1/(n+1)\pi]$ on which $\cos 1/x$ is monotone form an "infinite partition" of $(0, 1]$.

(c) The square wave function is both piecewise continuous and monotone with respect to the partition $\mathcal{P} : 0 < 1 < 2 < \ldots < n$.   □

We want to integrate a piecewise continuous or monotone function. To get over the difficulty that it may not be defined at the points of the partition, we will need the following lemma, which we will apply to each subinterval of the partition; it says roughly that, for integration, "endpoints don't matter".

Lemma 19.6 *If $f(x)$ is bounded on (a, b) and integrable on any closed subinterval $I \subset (a, b)$, then for any choice of $f(a)$ and $f(b)$, the function $f(x)$ will be integrable on $[a, b]$, and the value of the integral will be the same.*

Proof. $f(x)$ is bounded on $[a, b]$, since $f(a)$ and $f(b)$ are finite; let

$$(25) \qquad\qquad K = \sup_{[a,b]} |f(x)| \ .$$

We are given $\epsilon > 0$; we may assume $\epsilon \ll b - a$.

Since $f(x)$ is integrable on $[a + \epsilon, b - \epsilon]$, there is a $0 < \delta < \epsilon$ such that for any partition $\mathcal{P}'$ of this interval with mesh $< \delta$, we have

$$(26) \qquad\qquad U_f(\mathcal{P}') - L_f(\mathcal{P}') < \epsilon \ .$$

Now, any partition $\mathcal{P}$ of $[a, b]$ having mesh $< \delta$ induces a partition $\mathcal{P}'$ of $[a + \epsilon, b - \epsilon]$ having mesh $< \delta$. Comparing the difference between the upper and lower sums for these two partitions we get by (25)

$$(27) \qquad\qquad U_f(\mathcal{P}) - L_f(\mathcal{P}) \ < \ U_f(\mathcal{P}') - L_f(\mathcal{P}') + 8K\epsilon,$$

since $\sup f(x) - \inf f(x) \leq 2K$, the two end intervals have total length 2ϵ, and a safety factor of 2 is required since the subintervals containing $a + \epsilon$ and $b - \epsilon$ appear twice in the sums (cf. Question 2). Thus we get by (26) and (27)

$$U_f(\mathcal{P}) - L_f(\mathcal{P}) \ < \ (1 + 8K)\,\epsilon \ ,$$

where K depends on the choice of $f(a)$ and $f(b)$, but not on ϵ. By the K-ϵ principle, this proves that $f(x)$ is integrable on $[c, d]$.   □

Since $f(x)$ is integrable, the Riemann Sum Theorem 19.3 shows that

$$(28) \qquad\qquad \lim_{k \to \infty} S_f(\mathcal{P}_k) \ = \ \int_a^b f(x)dx$$

for any Riemann sums over a sequence of partitions $\mathcal{P}_k$ such that $|\mathcal{P}_k| \to 0$. Choose the Riemann sums so they never use the points a and b. Then the values $f(a)$ and $f(b)$ never enter into the sums, and therefore by (28) the integral does not depend on these values.   □□

Theorem 19.6   Integration of piecewise continuous functions.

If $f(x)$ is bounded and piecewise continuous or monotone on $[a, b]$, with respect to the partition $\mathcal{P} : a = x_0, x_1, \ldots, x_{n-1}, x_n = b$, then for any assigned values $f(x_i)$, $i = 0, \ldots, n$, $f(x)$ is integrable on $[a, b]$, and the integral, which does not depend on the choice of $f(x_i)$, is given by

$$(29) \qquad \int_a^b f(x)\, dx = \int_a^{x_1} f(x)\, dx + \ldots + \int_{x_{n-1}}^b f(x)\, dx \; .$$

Proof. $f(x)$ is integrable over each subinterval by Lemma 19.6, therefore it is integrable over $[a, b]$ by the Interval Addition Theorem 19.5 . The value of the integral is given by the same theorem; that it does not depend on the choice of the $f(x_i)$ follows from Lemma 19.6 again. $\square$

If a function $f(x)$ has points of discontinuity which are not infinite, so that it remains bounded, and its integral is to be calculated, usually one does not bother specifying values for $f(x)$ at the points of discontinuity, since they will not affect the value of the integral.

Questions 19.6

1. For each of the following tell whether on $[0, 1]$ it is piecewise continuous; piecewise monotonic; integrable.

(a) $f(x) = 1/2^i$, for $1/2^{i+1} < x \leq 1/2^i$; $i = 0, 1, 2, \ldots$

(b) $D \sin(1/x)$

2. Justify the $8K\epsilon$ term in (27) in more detail.

3. Could the same argument be made to end the proof of Lemma 19.6 if we used upper sums instead of Riemann sums in (28)?

4. Evaluate $\int_0^n w(x)\, dx$, where $n \in \mathbb{N}$ and $w(x)$ is the square wave function of Example 19.6(c).

Exercises

19.2

1. Evaluate $\int_0^1 e^x dx$ directly, by using (8) applied to the upper sums taken over the standard n-partition.

The upper sum is a geometric sum, so it can be summed by the standard formula 4.2 (4). The limit can be related to De^x at 0, or evaluated by l'Hospital's rule.

2. Evaluate $\int_0^{\pi/2} \sin x dx$ directly, by using the standard n-partitions, lower sums, and (8). You can use the interesting formula

$$\sum_{k=0}^{n-1} \sin(a + kb) = \frac{\sin(a + (n-1)b/2) \sin(nb/2)}{\sin(b/2)} .$$

3. (Fermat's method)   Evaluate $\int_1^a x^k dx$, where k is a positive integer, by using upper sums and (8); however, instead of the standard n-partition, put $r = a^{1/n}$ and use the n-partition $1 < r < r^2 < \ldots < r^{n-1} < a$.

4. Let $f_n(x) = \begin{cases} 0, & \text{if } x = 1/n, 2/n, \ldots, (n-1)/n, \\ 1, & \text{otherwise;} \end{cases}$, $\quad n \in \mathbb{N}.$

According to Exercise 18.3/1, this is integrable. Assuming this, evaluate $\int_0^1 f_n(x)\, dx$, using only the results in this section.

5. If you apply Corollary 19.2 to the function $f(x) = 1/x$, over the interval $[1, 2]$, using lower sums and the standard n-partition, it gives the limiting value of a certain sum, as $n \to \infty$. What is the sum, and what is its limit?

19.3

1. Suppose a function $f(x)$ is integrable on $[a, b]$ and $f(x) = 0$ whenever x is rational. (You are not told anything about the value of $f(x)$ when x is irrational.) Prove that $\int_a^b f(x)dx = 0$.

2 Evaluate $\lim_{n \to \infty} \sum_{k=0}^{2n} \frac{k}{n^2 + k^2}$.   Cite the theorems used. (Hint: divide top and bottom by n^2.)

3. A common way of estimating a definite integral numerically is to use the **trapezoidal rule**, which approximates the integral by a sum representing the total area of trapezoids lying between the inscribed and the circumscribed rectangles; $\mathcal{P}$ is taken to be the standard n-partition of $[a, b]$. The formula is

$$\int_a^b f(x)dx = \sum_1^n \frac{f(x_{i-1}) + f(x_i)}{2} \Delta x_i.$$

Is the trapezoidal sum a Riemann sum for $f(x)$ if

(a) $f(x)$ is monotone?   (b) $f(x)$ is continuous?

In each case, give a proof or counterexample.

19.4

1. Prove the **Translation Theorem** for integrals:

If $f(x)$ is integrable on $[a, b]$, then $f(x - c)$ is integrable on $[a + c, b + c]$, and

$$\int_{a+c}^{b+c} f(x - c)dx = \int_a^b f(x)dx .$$

(You don't yet have officially any theorems about changing variables in an integral. Prove it by going back to the definitions.)

2. (a) Prove the **Positivity Theorem** for integrals:

If on an interval $[a, b]$, $f(x)$ is continuous and non-negative, and $f(c) > 0$ for some point $c \in [a, b]$, then $\int_a^b f(x)dx > 0$. (Use Theorem 19.4B, carefully.)

(b) Deduce that if f and g are continuous, $f \leq g$ on $[a, b]$, and also $f(c) < g(c)$ for some $c \in [a, b]$, then $\int_a^b f(x)dx < \int_a^b g(x)dx$.

(c) Show that the conclusion in part (a) does not hold if "continuous" is changed to "integrable".

3. Show that $\displaystyle\int_0^{\pi/2} \frac{\sin x}{x(5 + x)} dx < \frac{\pi}{10}.$

4. Prove that on an interval $[a, b]$, if f and g are integrable, then $\max(f, g)$ is integrable and

$$\max\left(\int_a^b f(x)dx, \int_a^b g(x)dx \right) \leq \int_a^b \max(f(x), g(x)) \, dx .$$

(Use Exercise 2.4/1; cite theorems used.)

19.5

1. Suppose that for all $a > 0$, $\int_{-a}^a f(t)dt = 0$. Prove that $\int_0^x f(t)dt$ is an even function.

2. Prove the General Interval Addition Theorem for the case $b < a < c$.

3. Assume the General Interval Addition Theorem, i.e., that (17) holds for any points a, b, c. Show that (24) follows from (17).

19.6

1. Let $p(x)$ be a polynomial, and $f(x) = \text{sgn}\,(p(x))$ (cf. Question 9.2/1 for sgn x).

(a) Show that $f(x)$ is piecewise continuous on any $[a, b]$.

(b) Show that $\displaystyle\int_a^b f(x)\,dx = P - N$, where P and N are the total lengths of the intervals on which $f(x) > 0$ and $f(x) < 0$ respectively.

2. (a) How would you define "piecewise monotone" and "piecewise continuous" on an infinite interval? Both properties should hold for the *greatest integer* function:

$$[x] = \text{the greatest integer } n \text{ such that } n \leq x .$$

(b) Evaluate $\displaystyle\int_m^n [x]\,dx$, where m, n are integers, $m < n$.

3. Let $f(x)$ be continuous on $[a, b]$. According to Theorem 19.6, if the value of $f(x)$ is changed at a finite sequence of points, it remains integrable and the value of its integral is unchanged.

Show by example that the above does not necessarily hold true if $f(x)$ is changed at an infinite sequence of points: take the interval as $[0, 1]$ and the sequence to be $1, \frac{1}{2}, \frac{1}{4}, \frac{3}{4}, \frac{1}{8}, \frac{3}{8}, \frac{5}{8}, \frac{7}{8}, \cdots$.

Problems

19-1 (a) Assume $f(x)$ integrable on I and $a, x \in I$. Prove $F(x) = \int_a^x f(t)dt$ is continuous on I. (Use several theorems of this chapter.)

(b) Suppose I is compact and $f(x)$ is integrable on I. Define

$$G(x, y) = \int_x^y f(t)dt.$$

Prove there exist points x_0 and y_0 such that $G(x_0, y_0) = \max_{x,y \in I} G(x, y)$.

(Don't try to use guessed-at theorems about functions of two variables.)

19-2 Alternative definitions of integrability and the integral. Just as we defined differentiability and the derivative at the same time, it is possible to define integrability and the integral together. This problem and the next offer two such alternative definitions. They are equivalent to the definitions in the text (18.2 and 19.2).

Sequential Definition. *Let $f(x)$ be bounded on $[a,b]$, and $\mathcal{I}$ a number. If for every sequence $\mathcal{P}_k$ of partitions of $[a,b]$ such that $|\mathcal{P}_k| \to 0$, and for every Riemann sum $S_f(\mathcal{P}_k)$ for $f(x)$ over the partition $\mathcal{P}_k$ we have*

$$\lim_{k \to \infty} S_f(\mathcal{P}_k) = \mathcal{I},$$

then we say $f(x)$ is integrable on $[a,b]$, and write $\mathcal{I} = \int_a^b f(x)dx$.

(a) Prove that if $f(x)$ is integrable on $[a,b]$, and $\mathcal{I}$ is its integral, in the sense of Definitions 18.2 and 19.2, then it is integrable and $\mathcal{I}$ is its integral also in the sense of the Sequential Definition above. (This is easy, using theorems.)

(b) Prove the converse: if $f(x)$ on $[a,b]$ is integrable with integral $\mathcal{I}$ in the sense of the above definition, then it is integrable and its integral is $\mathcal{I}$ in the sense of definitions 18.2 and 19.2. (Not quite so easy.)

19-3 Here is a third definition.

Limit-style Definition. *Let $f(x)$ be bounded on $[a,b]$, and $\mathcal{I}$ a number. Suppose that, given $\epsilon > 0$, we have*

$$S_f(\mathcal{P}) \underset{\epsilon}{\approx} \mathcal{I}$$

for every partition $\mathcal{P}$ of $[a,b]$ such that $|\mathcal{P}| \approx 0$ and for every Riemann sum $S_f(\mathcal{P})$ over such a partition; then we say $f(x)$ is integrable over $[a,b]$ and write $\mathcal{I} = \int_a^b f(x)dx$.

(a) Prove that if $f(x)$ is integrable on $[a,b]$, and $\mathcal{I}$ is its integral, in the sense of Definitions 18.2 and 19.2, then it is integrable and $\mathcal{I}$ is its integral also in the sense of the Limit-style Definition above. (The converse is also true.)

(b) Prove the Limit-style Definition implies the Sequential Definition.

Answers to Questions

19.1

1. (a) if m is a multiple of n (b) $k =$ least common multiple of m, n.

2. It has the least number of points and the largest mesh of any common refinement.

3. We have

$$I \subseteq J \;\Rightarrow\; f(I) \subseteq f(J), \qquad \text{(cf. p. 138 for } f(I)\text{)};$$
$$\Rightarrow\; \sup f(I) \le \sup f(J), \qquad \inf f(I) \ge \inf f(J);$$
$$\Rightarrow\; \sup_I f(x) \le \sup_J f(x), \qquad \inf_I f(x) \ge \inf_J f(x);$$

The second line follows since $\sup f(J)$ is an upper bound for $f(I)$; the last line uses Definition 10.1B: $\sup_I f(x) = \sup f(I)$, $\inf_I f(x) = \inf f(I)$.

19.2

1. Using the formula for the sum of the first n squares (App. A.4, Ex. A.4A),

$$\lim_{n\to\infty} \frac{1^2 + 2^2 + \ldots + n^2}{n^3} = \lim_{n\to\infty} \frac{n(n+1)(2n+1)}{6n^3} = \lim_{n\to\infty} \frac{\left(1+\frac{1}{n}\right)\left(2+\frac{1}{n}\right)}{6} = \frac{1}{3}.$$

2. $L(\mathcal{P}^{(n)}) = (0^2 + 1^2 + \ldots + (n-1)^2)/n^3 = U(\mathcal{P}^{(n)}) - 1/n \to 1/3$.

3. $U(\mathcal{P}^{(n)}) = \dfrac{1+2+\ldots+n}{n^2} = \dfrac{n(n+1)}{2\,n^2} \to \dfrac{1}{2}$.

4.

$$L(\mathcal{P}_i) \;\le\; U(\mathcal{P}') \qquad \text{for all } i, \quad \text{by (6)};$$
$$\Rightarrow \quad \lim L(\mathcal{P}_i) \;\le\; U(\mathcal{P}') \quad \text{by the Limit Location Theorem 5.3A};$$
$$\Rightarrow \quad \mathcal{I} \;\le\; U(\mathcal{P}') \,.$$

19.3

1. (a) Yes; since $M_i = f(x_i)$ if $f(x)$ is increasing, $= f(x_{i-1})$ if decreasing.

(b) Yes, since $M_i = f(\bar{x}_i)$, where $\bar{x}_i$ is a maximum point for $f(x)$ on the interval $[\Delta x_i]$; such a point exists by the Maximum Theorem 13.3.

19.4

1. Let $\mathcal{P}_k$ be a sequence of partitions of $[a, b]$ such that $|\mathcal{P}_k| \to 0$. Let $S_f(\mathcal{P}_k)$ and $S_g(\mathcal{P}_k)$ be Riemann sums for the two functions that use the same chosen points x_i' in each subinterval of the partitions. Then

$$f(x_i') \le g(x_i') \;\Rightarrow\; S_f(\mathcal{P}_k) \;\le\; S_g(\mathcal{P}_k)$$
$$\Rightarrow\; \lim S_f(\mathcal{P}_k) \;\le\; \lim S_g(\mathcal{P}_k),$$

by a variant of the Limit Location Theorem 5.3 (15);

$$\Rightarrow \int_a^b f(x)\, dx \;\le\; \int_a^b g(x)\, dx,$$

by the Riemann Sum Theorem 19.3.

2. $\int_a^b \left(\dfrac{f(x) + g(x)}{2} \right) dx = \dfrac{1}{2} \left(\int_a^b f(x)\, dx + \int_a^b g(x)\, dx \right)$, by the Linearity Theorem for integrals 19.4A.

3. Using the Comparison Theorem for integrals 19.4B,

$$\cos x \leq 1 \;\Rightarrow\; \int_0^1 \frac{\cos x}{1 + x^2}\, dx \;\leq\; \int_0^1 \frac{dx}{1 + x^2} = \tan^{-1} 1 = \pi/4.$$

4. By the Linearity Theorem 19.4A and the Absolute Value Theorem 19.4C,

$$\left| \int_a^b f(x)\,dx - \int_a^b g(x)\, dx \right| = \left| \int_a^b f(x) - g(x)\, dx \right| \leq \int_a^b |f(x) - g(x)|\, dx \leq \epsilon(b - a).$$

19.5

1. If $b \in \mathcal{P}$, then the subintervals of $\mathcal{P}'$ and $\mathcal{P}''$ are the same as those of $\mathcal{P}$, so equality holds in (19) and (20). Thus the rest of the argument simplifies to:

Since f is integrable over $[a, b]$ and $[b, c]$, given $\epsilon > 0$, there is a $\delta > 0$ s.t.

$$U(\mathcal{P}') \underset{\epsilon}{\approx} L(\mathcal{P}') \quad \text{and} \quad U(\mathcal{P}'') \underset{\epsilon}{\approx} L(\mathcal{P}'') \qquad \text{for } |\mathcal{P}'|, |\mathcal{P}''| < \delta \, .$$

$$U(\mathcal{P}) = U(\mathcal{P}') + U(\mathcal{P}'') \underset{2\epsilon}{\approx} L(\mathcal{P}') + L(\mathcal{P}'') = L(\mathcal{P}), \quad \text{for } |\mathcal{P}| < \delta \, .$$

Thus finally, $U(\mathcal{P}) \underset{2\epsilon}{\approx} L(\mathcal{P}) \qquad$ if $|\mathcal{P}| < \delta$ and $b \in \mathcal{P}$.

19.6

1. (a) Monotonic increasing on $[0, 1]$ so piecewise monotonic on $[0, 1]$; integrable; not piecewise continuous, since there are an infinity of discontinuities.

(b) Not piecewise monotonic; piecewise continuous, but not integrable since not bounded.

2. We consider the endpoint a, which contributes $4K\epsilon$; the endpoint b is treated similarly and contributes another $4K\epsilon$.

Let $I_1, \ldots, I_m$ be the subintervals of $\mathcal{P}$ overlapping $[a, a + \epsilon]$. Since we have $|\mathcal{P}| = \delta < \epsilon$, the $I_k \subset [a, a + 2\epsilon]$, say.

It follows by using the triangle inequality that the part of $U_f(\mathcal{P}) - L_f(\mathcal{P})$ involving the I_k is in absolute value

$$\leq \sum_1^m |\sup_{I_k} f(x) - \inf_{I_k} f(x)| \, |I_k| \;\leq\; 2K \sum_1^m |I_k| \;\leq\; 2K \cdot 2\epsilon.$$

The above uses the estimate $|\sup_{I_k} f(x) - \inf_{I_k} f(x)| \leq 2K$, which follows from the triangle inequality and (25), or intuitively because (25) says that over the interval, $K \geq$ the maximum distance between the graph of $f(x)$ and the x-axis.

3. Even though it is true that $\lim U_f(\mathcal{P}_k) = \int_a^b f(x)dx$, you can't use the upper sums in the text's proof of the lemma, since the value of $U_f(\mathcal{P}_k)$ depends on the values of $f(a)$ and $f(b)$.

4. By Theorem 19.6, $\int_0^n w(x)\, dx = k$, if $n = 2k$ or $n = 2k - 1$.

20

Derivatives and Integrals

20.1   First fundamental theorem of calculus.

Theorem 20.1   *Assume that on a finite closed interval* $[a, b]$ *the function* $F(x)$ *is differentiable and its derivative* $f(x) = F'(x)$ *is integrable. Then*

$$(1) \qquad \int_a^b f(x)dx \;=\; F(b) - F(a).$$

Proof.   Assume that $a < b$ (the case $a > b$ is left as a Question); we first construct, given any partition

$$\mathcal{P}: \; a = x_0 < x_1 < \ldots < x_n = b,$$

a special Riemann sum for it which has the exact value $F(b) - F(a)$.

Namely, $F(x)$ is differentiable on each subinterval $[\Delta x_i]$, so the Mean-value Theorem 15.1 tells us there is a point c_i in $[\Delta x_i]$ such that

$$(2) \qquad F(x_i) - F(x_{i-1}) \;=\; F'(c_i)\Delta x_i \;.$$

Using this c_i in each subinterval, we form the corresponding Riemann sum for the partition; according to (2), its value is — remember that $F'(x) = f(x)$ —

$$\sum_1^n f(c_i)\Delta x_i \;=\; \sum_1^n \big(F(x_i) - F(x_{i-1})\big).$$

But the sum on the right is a "telescoping sum", i.e., the terms cancel in pairs, leaving only the two end terms, $F(x_n) - F(x_0)$; thus we see that

$$(3) \qquad \sum_1^n f(c_i)\Delta x_i \;=\; F(b) - F(a) \;.$$

To prove (1) now, we use the Riemann Sum Theorem 19.3 to calculate the definite integral; $f(x)$ satisfies the hypothesis of that theorem since it is integrable on $[a, b]$. Use the sequence of standard n-partitions $\mathcal{P}^{(n)}$ and the special Riemann sums constructed above. Then

$$\int_a^b f(x)\,dx \;=\; \lim_{n\to\infty} \sum_1^n f(c_i)\Delta x_i \;, \qquad \text{by the Riemann sum theorem;}$$

$$= \; \lim_{n\to\infty} F(b) - F(a), \qquad \text{by (3);}$$

$$= \; F(b) - F(a), \qquad \text{which proves (1).} \qquad \square$$

Questions 20.1

1. In two of the three lines which conclude the proof, one takes the limit as $n \to \infty$. In the first line, why not use the customary notation for this, omitting the lim and writing instead $\sum_1^\infty$? In the second line, how can one take $\lim_{n \to \infty}$ of a quantity $F(b) - F(a)$ which doesn't depend on n?

2. Finish the proof of the First Fundamental Theorem by showing that if (1) holds for $a < b$, it also holds for $a \geq b$.

20.2   Existence and uniqueness of antiderivatives.

To use Theorem 20.1, we generally start with $f(x)$ and then have to find an antiderivative $F(x)$ somehow; much of what is called "integration" in calculus is actually just "antidifferentiation", i.e., a collection of methods for finding antiderivatives.

To state Theorem 20.1, however, we must start with $F(x)$, not with $f(x)$, since we do not know whether a given $f(x)$ actually has an antiderivative.

Indeed, any calculus student will tell you there are many functions with no antiderivative: $\sin x^2$, for example. The folk expressions for this are well-known phrases like, "you can't integrate $\sin x^2$," "you can't find $\int \sin x^2 \, dx$—it isn't in the tables". But what's meant by this is only that there is no elementary function $F(x)$ whose derivative is $\sin x^2$; is there perhaps a non-elementary function $F(x)$?

Theorem 20.2A   Second fundamental theorem of calculus.

Let $f(x)$ be continuous on an interval I, let $a \in I$ and for $x \in I$ set

(4) $$F(x) \; = \; \int_a^x f(t)dt, \qquad \text{for all } x \in I \; .$$

Then $F(x)$ is differentiable on I, and $F'(x) = f(x)$ for all $x \in I$:

Every Continuous Function has an Antiderivative.

Referring to the picture on the next page, $F(x)$ represents the area under the curve between the vertical lines at a and x; the theorem says that as x moves, the rate of change of area is the height of the curve at x: the higher the curve, the faster the area is changing as x moves.

Proof. We go back to the definition of derivative, and prove that for all $x \in I$,

(5) $$\lim_{\Delta x \to 0} \frac{\Delta F}{\Delta x} \; = \; f(x) \; .$$

For this, we first need an expression for ΔF. Letting $\Delta x = t - x$,

$$\Delta F \; = \; F(x + \Delta x) - F(x) \; = \; \int_a^{x + \Delta x} f(t)dt - \int_a^x f(t) \, dt \; ;$$

using the Interval Addition Theorem 19.5, this becomes

$$\Delta F \; = \; \int_x^{x + \Delta x} f(t)dt \; .$$

At this point, the intuitive proof runs: ΔF is the shaded area, which is approximately the area of the rectangle having height $f(x)$ and width Δx:

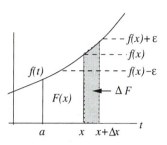

$$\Delta F \approx f(x)\Delta x .$$

Dividing this by Δx and letting $\Delta x \to 0$, we get (5).

To make this a real proof, we have to replace the vague symbol $\approx$ by upper and lower estimates for ΔF.

To estimate ΔF, we estimate the integrand $f(t)$. As $f(t)$ is continuous at x, the definition (11.1) of continuity shows that, given $\epsilon > 0$,

(6) $f(x) - \epsilon < f(t) < f(x) + \epsilon, \qquad \text{for } t \approx x .$

Assume first that $\Delta x > 0$; by the Comparison Theorem for integrals 19.4B,

$$\int_x^{x+\Delta x} (f(x) - \epsilon)\, dt \;\leq\; \int_x^{x+\Delta x} f(t)\, dt \;\leq\; \int_x^{x+\Delta x} (f(x) + \epsilon)\, dt, \quad \text{for } t \approx x^+ .$$

Since $f(x)$ and ϵ are constants in the integration, the above gives us

(7) $(f(x) - \epsilon)\Delta x \leq \Delta F \leq (f(x) + \epsilon)\Delta x , \qquad \text{for } \Delta x \approx 0^+ ;$

(8) $f(x) - \epsilon \leq \dfrac{\Delta F}{\Delta x} \leq f(x) + \epsilon , \qquad\qquad \text{for } \Delta x \approx 0^+ .$

Since $\epsilon > 0$ was arbitrary, this proves by the definition (11.2B) of right-hand limit that

$$\lim_{\Delta x \to 0^+} \frac{\Delta F}{\Delta x} = f(x) . \qquad\qquad \square$$

The argument is analogous if $\Delta x < 0$; we leave it for the Questions. It shows that the left-hand derivative is also $f(x)$; this proves (5). $\square\square$

If I is a closed interval and a or x is an endpoint, the theorem is still true, using one-sided continuity and one-sided derivatives.

Note that the theorem actually gives an explicit formula for the antiderivative, namely the function $F(x)$ defined by (4). In Example 10.1D for instance, we studied the error function

$$F(x) = \operatorname{erf} x = \int_0^x e^{-t^2/2}\, dt ;$$

it is an antiderivative of $e^{-x^2/2}$, and while not elementary, it is nonetheless a perfectly good function whose values can be calculated to as great an accuracy as desired and whose properties can be studied.

The antiderivative of $f(x)$ guaranteed to exist by Theorem 20.2A is not unique, for we can use another point b as the lower limit in (4), getting a new antiderivative

$$F_1(x) = \int_b^x f(t)\, dt .$$

However, $F_1(x)$ differs from $F(x)$ only by an additive constant, since by the Interval Addition Theorem 19.5,

$$F_1(x) = \int_b^x f(t)dt = \int_b^a f(t)dt + \int_a^x f(t)dt,$$

$$= F(x) + C, \qquad \text{where } C = \int_b^a f(x)dx .$$

In fact, the functions $F(x) + c$, where c is an arbitrary constant, give *all* the antiderivatives of $f(x)$; this is the next theorem.

The above argument does not prove this; it only shows that if we use a different lower limit for the integral, we get a function of the form $F(x) + c$. There remains the possibility that some entirely different approach could produce a genuinely different-looking antiderivative.

Theorem 20.2B Uniqueness theorem for antiderivatives.

Let $F(x)$ and $G(x)$ be differentiable on an interval I. Then on I,

$$G'(x) = F'(x) \;\Rightarrow\; G(x) = F(x) + c, \quad \text{for some constant } c .$$

Proof.

$$G'(x) = F'(x) \;\Rightarrow\; \big(G(x) - F(x)\big)' = 0;$$
$$\Rightarrow\; G(x) - F(x) = c, \quad \text{by Theorem 15.2 (5).} \qquad \square$$

The result is false if the domain is the union of two or more disjoint intervals, such as $\{x : x \neq 0\}$. See Question 4.

Corollary 20.2A Existence and uniqueness theorem for $y' = f(x)$.

Let $f(x)$ be continuous on an interval I, and $a \in I$. Then the differential equation with initial condition

(9) $$y' = f(x), \qquad y(a) = b,$$

has in I the unique solution $y = F(x)$, where

$$F(x) = b + \int_a^x f(t)\,dt .$$

Proof. $F(x)$ satisfies (9), by the Second Fundamental Theorem and 19.5 (24).

If $F_1(x)$ is any other function satisfying (9), then $F_1'(x) = f(x) = F'(x)$, so by Theorem 20.2B,

$$F_1(x) = F(x) + c, \qquad x \in I .$$

To evaluate c, since $y = F_1(x)$ satisfies (9), we have $F_1(a) = b = F(a)$; therefore $c = 0$ and so $F_1(x) = F(x)$; this proves uniqueness. $\qquad \square$

Since the proof of the Second Fundamental Theorem did not use the First Fundamental Theorem (1), it gives an alternative proof of that theorem; notice however that this proof requires that $f(x)$ be continuous, not merely integrable as required in the earlier proof.

Corollary 20.2B   The Second F.T. implies the First F.T.

Assume that on $[a, b]$, the function $F(x)$ has a continuous derivative $f(x)$. Then

(1)
$$\int_a^b f(t)\,dt \;=\; F(b) - F(a) \;.$$

Proof.   If we set $G(x) = \displaystyle\int_a^x f(t)\,dt$, then by the Second Fundamental Theorem 20.2A, we have

$$G'(x) = f(x) = F'(x) \qquad \text{on } [a, b] \;.$$

Therefore by the Uniqueness Theorem 20.2B, we have $G(x) = F(x) + c$, i.e.,

$$\int_a^x f(t)dt \;=\; F(x) + c \;, \qquad \text{for some constant } c.$$

Setting $x = a$ we get $c = -F(a)$; setting $x = b$ then gives (1). $\square$

Historically, the Second Fundamental Theorem was proved and the First Fundamental Theorem derived from it by the proof given in the corollary above. The overall argument is a bit elusive, depending as it does on introducing the function $G(x)$ below and proving its derivative is $f(x)$;

$$G(x) \;=\; \int_a^x f(t)\,dt \;.$$

Actually, all one wants is $G(b)$; the point is (and it is an important one in mathematical analysis) that paradoxically it is easier to determine the function $G(x)$ than the number $G(b)$. This is because the function is richer in properties—for instance, it can be differentiated.

Both fundamental theorems are stated in terms of a differentiation and an integration; they differ in the order in which the operations are performed:

⋆ First Fundamental Theorem: differentiate, then integrate;

⋆ Second Fundamental Theorem: integrate, then differentiate.

In both cases, the conclusion is that we return more or less to where we started from. In the same vein, we can say that the two proofs of the First Fundamental Theorem (in 20.1 and in Cor. 20.2B) both use the same essential ingredients—the definition of the integral and the Mean-value Theorem—but they use them in different ways.

Questions 20.2

1. In Theorem 20.2A, prove the left-hand limit (you will need to be careful with the inequalities): $\displaystyle\lim_{\Delta x \to 0^-} \frac{\Delta F}{\Delta x} = f(x)$.

2. Letting $F(x) = \operatorname{erf} x = \int_0^x e^{-t^2/2}\,dt$ (cf. p. 269), find
   (a) the critical points of $F(x)$;   (b) where $F(x)$ is convex and concave.

3. Let $u(x)$ be the unit step function $(= 1, \ x \geq 0; \ = 0, \ x < 0.)$
Show that Theorem 20.2A is not valid for $F(x) = \int_{-1}^{x} u(t) \, dt$: this function is not an antiderivative for $u(x)$ on $(-\infty, \infty)$.

4. Consider the function sgn x $(= 1, \ x > 0; \ = -1, \ x < 0)$.

(a) Find two antiderivatives $F_1(x)$ and $F_2(x)$ for sgn x on its domain D, such that on D, $F_2(x) \neq F_1(x) + c$ for any constant c.

(b) Show in the same vein that $y'(x) = $ sgn x, $y(1) = 1$, does not have a unique solution on D. Is Corollary 20.2A contradicted?

5. Streamline the proof that the Second Fundamental Theorem implies the First by using the Uniqueness Theorem for $y' = f(x)$, $y(a) = b$ (Cor. 20.2A).

6. Where is the Mean-value Theorem used in the proof of (1) via Cor. 20.2B?

20.3   Other relations between derivatives and integrals.

Besides the two fundamental theorems, there are two other important general theorems connecting differentiation and integration; they come from the formulas for differentiating products and composite functions. Though familiar (in a more elementary form) from calculus as "techniques of integration", they have an importance in analysis that goes far beyond this.

We will use the standard calculus notation: $F(x)\big]_a^b = F(b) - F(a)$.

Theorem 20.3A   Integration by parts.

If $u(x)$ and $v(x)$ have continuous derivatives on $[a, b]$, then

$$(10) \qquad \int_a^b u(x) v'(x) \, dx \ = \ u(x) v(x) \Big]_a^b - \int_a^b v(x) u'(x) \, dx \ .$$

Proof.   According to the formula for differentiating products,

$$(11) \qquad\qquad (u \cdot v)' \ = \ u' \cdot v + u \cdot v' \ .$$

Since u' and v' exist, u and v are continuous. Thus the two terms on the right in (11) are continuous, so the left side is also continuous. We can therefore integrate all three terms in (11) over $[a, b]$ and apply the First Fundamental Theorem to the term on the left. This gives

$$u(x) v(x) \Big]_a^b \ = \ \int_a^b u' \cdot v \, dx + \int_a^b u \cdot v' dx \ ,$$

which after transposing terms is the formula (10).   □

Equation (10) is more familiar in its form as a "technique of integration":

$$\int uv' dx = uv - \int vu' dx \ .$$

This however is not (10); it is nothing more than (11) in disguise, since it says

$$uv = \int (u'v + uv') dx,$$

i.e., that uv is an antiderivative for $u'v + uv'$; but this is (11).

By contrast, (10) involves integrability over $[a, b]$ and the First Fundamental Theorem: it is a theorem about definite integrals, not antiderivatives.

Theorem 20.3B   Change of variable rule.

Suppose $u(t)$ is a continuously differentiable function which maps the interval $[t_0, t_1]$ to the interval $[u_0, u_1]$, that is,

(a) $u : [t_0, t_1] \to [u_0, u_1]$ *and* $u(t_0) = u_0$, $u(t_1) = u_1$;

(b) $u'(t)$ *exists and is continuous on* $[t_0, t_1]$.

Assume $f(u)$ is a function continuous on $[u_0, u_1]$. Then

(12)
$$\int_{u_0}^{u_1} f(u)\, du = \int_{t_0}^{t_1} f(u(t)) \frac{du}{dt}\, dt \ .$$

Proof. Let $F(u)$ be an antiderivative for $f(u)$ on $[u_0, u_1]$. According to the chain rule for differentiation,

(13)
$$\frac{d}{dt} F(u(t)) = \frac{d}{du} F(u) \cdot \frac{du}{dt} = f(u(t)) \cdot \frac{du}{dt} \ .$$

The right side is continuous on $[t_0, t_1]$, therefore the left side is also. We may therefore integrate the left side of (12) and calculate:

$$\int_{u_0}^{u_1} f(u)\, du = F(u_1) - F(u_0) \ , \qquad \text{by the First Fundamental Theorem;}$$

$$= F(u(t_1)) - F(u(t_0)), \quad \text{by (a) above;}$$

$$= \int_{t_0}^{t_1} \frac{d}{dt} F(u(t))\, dt \ , \qquad \text{by the First Fundamental Theorem;}$$

$$= \int_{t_0}^{t_1} f(u(t)) \frac{du}{dt}\, dt \ , \qquad \text{by the chain rule (13).} \qquad \square$$

The last three sections are devoted to some applications of the theorems of these first three sections: the definition of $\ln x$ and e^x, Stirling's approximation for $n!$, and the use of l'Hospital's rule for evaluating indeterminate forms of the type $\frac{\infty}{\infty}$.

Questions 20.3

1. Suppose $f(x) = x\, g(x)$, where $g(x)$ has a continuous derivative for $x > 0$ and $g(a) = 0$, for some $a > 0$. Prove

$$\int_1^a f'(x) \ln x\, dx = -\int_1^a g(x)\, dx \ .$$

2. Evaluate $\displaystyle\int_0^1 \frac{dx}{(1 + x^2)^{3/2}}\, dx$, by making the substitution $x = \tan u$;

verify the hypotheses of Theorem 20.3B are satisfied.

20.4   The logarithm and exponential functions.

As an application of the theorems in the first three sections, we define the logarithm and exponential functions, and obtain their most important properties; proofs are sketched, or left as exercises.

Definition 20.4   The natural logarithm.   We define

$$\ln x = \int_1^x \frac{1}{t}\, dt, \qquad \text{for } x > 0.$$

Properties 20.4A

L1 $\ln x$ is defined, continuous, and differentiable;

L2 $\dfrac{d}{dx} \ln x = \dfrac{1}{x}$;

L3 $\ln x$ is strictly increasing and concave.

Proof. $\ln x$ is defined since $1/t$ is continuous for $t > 0$. It is differentiable with derivative $1/x$ by the Second Fundamental Theorem; therefore it is also continuous. It is strictly increasing since its derivative $1/x > 0$ for $x > 0$; it is concave since its second derivative is negative. □

L4 $\ln(ab) = \ln a + \ln b$;

L5 $\ln(1/a) = -\ln a$;

L6 $\ln(a^r) = r \ln a,$ for any rational number r.

Proofs.   To prove (L4), we use Theorem 20.3B (put $t = au$):

$$\ln(ab) = \int_1^{ab} \frac{dt}{t} = \int_1^a \frac{dt}{t} + \int_a^{ab} \frac{dt}{t} = \ln a + \int_1^b \frac{du}{u} = \ln a + \ln b.$$

From (L4) one deduces easily that $\ln 1 = 0$, and then (L5). As for (L6), if r is a positive integer n, it follows by applying (L4) repeatedly: if $r = 1/n$,

$$\ln a = \ln(a^{1/n})^n = n \ln a^{1/n}, \qquad \text{so that } \ln a^{1/n} = \tfrac{1}{n} \ln a.$$

If $r = m/n$ or $r = -m/n$, the result follows algebraically from the previously established cases and laws. □

L7 $\displaystyle \lim_{x \to \infty} \ln x = \infty, \qquad \lim_{x \to 0^+} \ln x = -\infty.$

Proof.   The first limit follows from (L6) and the fact that $\ln x$ is strictly increasing; the second limit from (L5) and the change of variable $u = 1/x$. □

L8 The range of $\ln x$ is $(-\infty, \infty)$.

L9 There is a unique number e such that $\ln e = 1$.

Proof.   The first follows from L7 and the Intermediate Value Theorem (Cor. 12.1), since $\ln x$ is continuous. The uniqueness of e follows from the fact that $\ln x$ is strictly increasing. □

The exponential function.

Since $\ln x$ is continuous and strictly increasing, its inverse function exists; we will temporarily give it the name $\exp x$, since the more usual designation e^x suggests "multiplying e by itself x times, which is not the basis for our definition. Since we are defining $\exp x$ as an inverse function, we will have to get its properties by making use of the various results about inverse functions scattered through some earlier chapters (Theorems 9.4, 12.4, and 14.2C).

According to the definition of inverse function (Section 9.4),

(14a) $$y = \exp x \quad \Leftrightarrow \quad \ln y = x \; ;$$

(14b) $$\exp(\ln x) = x, \qquad \ln(\exp x) = x \; .$$

The first group of properties derive from the general theorems cited.

E1 $\exp x$ is defined, continuous, and strictly increasing on $(-\infty, \infty)$; its range is $(0, \infty)$;

E2 $\exp x$ is differentiable for all x, and $D \exp x = \exp x$;

E3 $\exp x$ is convex for all x.

Proof.   The properties in E1 follow because $\exp x$ is the inverse of $\ln x$, which is continuous and strictly increasing on $(0, \infty)$ and has range $(-\infty, \infty)$; see Theorem 12.4 (continuity of inverse functions).

Since $\ln x$ is differentiable, so is $\exp x$, by Theorem 14.2C; the formula for its derivative follows from that of the derivative of $\ln x$; convexity is proved by the second derivative test (cf. Questions).

E4 $\exp (a + b) = \exp a \cdot \exp b$

E5 $\exp (-a) = 1/\exp a$

E6 $\exp ra = (\exp a)^r, \qquad$ if r is rational.

Proof.   These follow from the equations for $\ln x$. To show E4, for instance, we set $A = \exp a$, $B = \exp b$, and $C = \exp (a + b)$; using the right-hand side of (14b), we get

$$\ln C = a + b = \ln A + \ln B = \ln(A \cdot B), \quad \text{by L4};$$

applying exp to both sides and using (14b) again gives $C = A \cdot B$, which is E4. The others are proved similarly.

To relate $\exp x$ to powers, from (14a), L7, L8, and E6 we have

E7 $\exp 0 = 1, \; \exp 1 = e, \; \exp r = e^r$ if r is rational.

We therefore *define* the powers of e for all real numbers r by

E8 $e^r = \exp r$;

now at last we can write e^x instead of $\exp x$. The exponential function using other bases is defined in terms of e^x:

E9 $a^x = e^{x \ln a}, \qquad a > 0$.

A more naive way to define e^r for a general real number r would be to define it as the limit of a bounded increasing sequence; thus for instance,

$$e^{\sqrt{2}} = \lim \; < e^1, e^{1.4}, e^{1.41}, e^{1.414}, \ldots >,$$

each term of the sequence being already defined since it is a rational power of e. Though this definition is different from the one given above, by using continuity arguments it can be shown the two definitions give the same function.

Definition E9 is orderly and convenient, but it relies on $\exp x$ and $\ln x$. This means a statement like $\quad 3^2 = 9, \quad$ meaning: $f(2) = 9, \quad$ where $f(x) = 3^x$, becomes a deep theorem, if you take into account everything that has gone into the definition of the functions involved and the proofs of the relevant properties. It's like the express delivery services that simplify their operations by doing all the sorting centrally: your overnight package from Boston to Providence goes by way of St. Louis.

Questions 20.4

1. Prove: (a) the formula in E2;   (b) property E3;   (c) property E5.

2. Prove L6 by writing it $\ln x^r = r \ln x$, and showing both sides satisfy the same differential equation with initial conditions (cf. cor. 20.2A).

3. Prove using E9 as the definition that $3^2 = 9$.

20.5   Stirling's formula.

This is a famous formula which estimates $n!$. A refinement of it is used in hand calculators, which do not calculate $n!$ for large values of n by actually multiplying the first n integers together. The estimate given by Stirling's formula is an *asymptotic approximation*: one of the form $a_n \sim b_n$, meaning

$$\lim_{n \to \infty} \frac{a_n}{b_n} = 1 \; .$$

Recall that $a_n \sim b_n$ does not mean a_n and b_n are close for large n; as one sees from the above limit, it means they are *relatively close*; i.e., the relative error tends to zero:

$$\frac{a_n - b_n}{b_n} \to 0, \quad \text{as } n \to \infty.$$

Theorem 20.5   Stirling's formula.

(15) $$n! \; \sim \; \left(\frac{n}{e}\right)^n \sqrt{2\pi n} \; .$$

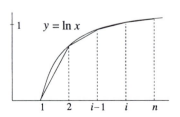

Partial proof.   We change $n!$ to a sum. Let

(16) $\quad S_n \; = \; \ln n! \; = \; \ln 1 + \ln 2 + \ldots + \ln n.$

We estimate the sum by relating it to the total area of the trapezoids shown. We have for one trapezoid

$$\text{area of trapezoid over } [i-1, i] \;=\; \tfrac{1}{2}(\ln(i-1) + \ln i).$$

Since $\ln x$ is concave, the trapezoids lie under $y = \ln x$, so

(17)
$$\tfrac{1}{2}\ln 1 + \ln 2 + \ldots + \tfrac{1}{2}\ln n \;\leq\; \int_{1}^{n} \ln x \, dx \;.$$

total area of the trapezoids   area under graph

Applying (16) to the left side, and integrating by parts on the right:

$$S_n - \tfrac{1}{2}\ln n \;\leq\; n \ln n - n + 1, \qquad \text{or}$$

(18)
$$S_n \;\leq\; (n + \tfrac{1}{2})\ln n - n + 1 \;=\; A_n \;,$$

where A_n is just an abbreviation for the quantity on the right.

The difference $A_n - S_n$ between the two sides of (18) is the same as the difference between the two sides of (17). The picture above shows that this difference is the sum of the little curved slivers lying above the trapezoids. This shows the sequence $A_n - S_n$ is increasing.

The sequence is bounded above, since all the slivers can be fitted as shown, without overlapping, into the region lying over $[1, 2]$. To see they do not overlap, we must show that

(*)   slope of chord over $[i - 1, \, i]$ > slope of $\ln x$ at i.

Apply the Mean-value Theorem (secant form 15.1(1a)) to the left side: the inequality (*) asserts that, for the Mean-value Theorem point $c \in (i - 1, i)$,

$$1/c \;>\; 1/i \;;$$

but this is indeed true, since $c < i$, so (*) is demonstrated.

Since we have shown $A_n - S_n$ is increasing and bounded above, it has by the Completeness Property a limit L :

$$A_n - S_n \;\to\; L; \quad \text{therefore}$$

(19)
$$e^{A_n - S_n} \;\to\; e^L, \quad \text{by sequential continuity (Cor. 11.5).}$$

Now $e^{S_n} = n!$, by (16); e^{A_n} can be gotten from (18); thus (19) turns into

$$\frac{n^{n+1/2} e^{-n} e}{n!} \;\to\; e^L, \quad \text{or} \quad n! \sim \left(\frac{n}{e}\right)^n \sqrt{n} K, \quad \text{where} \quad K = e^{1-L} \;. \quad \square$$

The exact value of K (it is $\sqrt{2\pi} \approx 2.5$) is found by another procedure, left for the Problems. From the picture one can estimate $2.2 < K < 2.8$, which isn't too bad—see the Questions.

Questions 20.5

1. Is it likely that an 8-digit calculator can display the exact value of 12!?

2. Taking as geometrically evident that the slivers take up less than half the area of the region under the graph of $\ln x$ and over $[1, 2]$, get the estimate given just above for K.

20.6   Growth rate of functions.

In analysis and its applications to the real world, it is often necessary to estimate the relative sizes of functions, when x is large or when $x \approx a$. To do this, a somewhat informal terminology is widely used.

To fix the ideas, suppose we wish to compare two functions f and g such that both $f(x) \to \infty$ and $g(x) \to \infty$ as $x \to \infty$. On the left below is the value of the limit; on the right is the terminology used.

$$\frac{f(x)}{g(x)} \to \infty : \quad \begin{cases} f & \text{tends to } \infty \text{ faster than } g, \\ f & \text{grows faster than } g; \end{cases}$$

$$\frac{f(x)}{g(x)} \to 0 \ : \quad f \text{ tends to } \infty \text{ more slowly than } g;$$

$$\frac{f(x)}{g(x)} \to 1 : \quad \begin{cases} f & \text{grows at the same rate as } g, \\ f & \text{is asymptotic to } g: \end{cases} \quad f(x) \sim g(x).$$

Note in particular the last of these, usually expressed by the symbol $\sim$ ("is asymptotic to"). We met it in the previous section, as well as in the asymptotic comparison test for series, 7.5B.

Repeating for functions what was said in the previous section about sequences, if f and g tend to ∞ and $f \sim g$, it does not mean that f and g are close for $x \gg 1$, only that they are relatively close. That is, the difference between them is negligible compared with the functions themselves; to put it another way, it is the percentage difference between them which is small.

To see this in symbols (note that $|g(x)| = g(x)$, since $g(x) \to \infty$):

$$f(x) \sim g(x) \quad \Leftrightarrow \quad \frac{f(x)}{g(x)} \to 1 \quad \Leftrightarrow \quad \left| \frac{f(x)}{g(x)} - 1 \right| < \epsilon \quad \text{for } x \gg 1;$$

$$\Leftrightarrow \quad \frac{|f(x) - g(x)|}{g(x)} < \epsilon \quad \text{for } x \gg 1 .$$

Examples 20.6A   As $x \to \infty$, the limit theorems show that (Question 1)

(a) $x^3 - 2x^2 - 1 \sim x^3$;

(b) $\sqrt{x^5 + 5x^3 + 2}$ grows more slowly than x^3;

(c) e^{ax} grows more rapidly than e^{bx}, if $a > b > 0$.

The above dealt with functions which tend to ∞ as $x \to \infty$; the same terminology would be used to express relative growth rate for those functions which tend to 0 as $x \to \infty$. As an example of the terminology:

$$\frac{f(x)}{g(x)} \to 0 : \quad \begin{cases} f & \text{tends to 0 more rapidly than } g, \\ g & \text{tends to 0 more slowly than } f. \end{cases}$$

(d) $\dfrac{1}{x^2 + 3x} \sim \dfrac{1}{x^2} \quad$ as $x \to \infty$;

(e) $\dfrac{1}{x^2}$ tends to 0 more rapidly than $\dfrac{1}{x}$, but more slowly than $\dfrac{1}{x^3}$.

The same terminology extends to limits as $x \to a$, $x \to a^+$, $x \to -\infty$, etc.

If f and g both tend to 0, deciding which goes to 0 faster can often be done by l'Hospital's rule 15.4B. In the same way, to compare the growth rates of functions which tend to ∞, one can use the version of l'Hospital's rule for indeterminate forms of the type ∞/∞, which we prove now.

A proof of this form of the rule, based on the Cauchy Mean-value Theorem, was given in the Problem section for Chapter 15. The idea here is to get from the right side of (20) to the left side by using integration; it gives a more straightforward proof, but needs somewhat stronger hypotheses.

Theorem 20.6   L'Hospital's rule for ∞/∞.

Suppose $f(x) \to \infty$ and $g(x) \to \infty$ (as $x \to \infty$, etc.).   Then

$$(20) \qquad \lim \frac{f(x)}{g(x)} = \lim \frac{f'(x)}{g'(x)} ,$$

assuming f' and g' exist and are continuous for $x \gg 1$, etc., and the limit on the right of (20) exists.

Proof.   Let L be the limit on the right of (20), and for definiteness, assume the limits are taken as $x \to \infty$. By definition of limit, given $\epsilon > 0$,

$$(21) \qquad \left| \frac{f'(x)}{g'(x)} - L \right| < \epsilon, \qquad \text{for } x \gg 1 .$$

Since the limit on the right of (20) exists, $g'(x) \neq 0$ for $x \gg 1$, and therefore $g'(x) > 0$ for $x \gg 1$ by Bolzano's Theorem 12.1, since $g(x) \to \infty$. Change the absolute values in (21) to the equivalent inequalities, and multiply through by $g'(x)$; the inequalities are preserved since $g' > 0$, and we get

$$-\epsilon g'(x) < f'(x) - Lg'(x) < \epsilon g'(x) \qquad \text{for } x \gg 1 .$$

Integrate everything from some big enough x-value a to a bigger x-value u, using the First Fundamental Theorem 20.1 to evaluate the integrals. The inequalities are preserved (by the Comparison Theorem for integrals), and we conclude there are constants b and c such that

$$(22) \qquad -\epsilon g(u) + b < f(u) - Lg(u) < \epsilon g(u) + c \qquad \text{for } u \gg 1.$$

Since $g(u) \to \infty$,

$$\frac{|b|}{g(u)} < \epsilon \qquad \text{and} \qquad \frac{|c|}{g(u)} < \epsilon \qquad \text{for } u \gg 1,$$

and therefore, dividing (22) through by $g(u)$ we can write the inequalities as

$$\left| \frac{f(u)}{g(u)} - L \right| < 2\epsilon, \qquad \text{for } u \gg 1 ,$$

which proves (20), according to the definition of limit. $\square$

Note that the rule can be applied even if the limit on the right of (20) is ∞: just turn both fractions upside down.

Examples 20.6B Verify that

(23) $\displaystyle\lim_{x\to\infty}\frac{\ln x}{x^k}=0,$ for all $k>0$; $\quad\left(\begin{array}{l}\ln x \text{ grows more slowly than any}\\ \text{power of } x, \text{ no matter how small;}\end{array}\right)$

(24) $\displaystyle\lim_{x\to\infty}\frac{e^{ax}}{x^m}=\infty,$ for $a,m>0$; $\quad\left(\begin{array}{l}e^{ax} \text{ grows more rapidly than any}\\ \text{power of } x, \text{ no matter how large.}\end{array}\right)$

Solution. Left as Questions.

Note that though (24) can be done by repeated use of l'Hospital's rule, it can also be done with a single use, by writing the fraction in a slightly different form. Please reread the cautionary notes at the end of Chapter 15 about common misuses of l'Hospital's rule. In general, to evaluate an indeterminate form, try algebraic transformations or substitutions first.

Questions 20.6

1. Verify the statements in Examples 20.6A.

2. Verify (23); verify (24).

3. Compare e^{x^2} and e^{2x} as $x\to\infty$, with proof. What happens when you use L'Hospital's rule?

Exercises

20.1

1. Suppose $F(x)$ is a function which is *smooth* (i.e., has a continuous derivative) on (a,b). By using the First Fundamental Theorem, prove that on (a,b),

(i) $F'\geq 0 \Rightarrow F$ is increasing;

(ii) $F'>0 \Rightarrow F$ is strictly increasing.

(This can also be proved using the Mean-value Theorem, as in 15.2. Note the stronger hypotheses being used here. Could they be weakened for this proof?)

20.2

1. Find $F'(x)$ by Theorem 20.2A and the chain rule, if $F(x)=$

(a) $\displaystyle\int_0^{x^2}\sqrt{1+t}\,dt$ (b) $\displaystyle\int_{x^3}^{1}e^{-t^2}\,dt$ (c) $\displaystyle\int_{x^2}^{x}e^{-t^2}\,dt$.

2. Evaluate the following limits, first by l'Hospital's rule, and then directly by considering the geometric meaning of the integrals.

(a) $\displaystyle\lim_{h\to 0}\frac{1}{h}\int_0^h e^{t^2+t}\,dt$ (b) $\displaystyle\lim_{\Delta x\to 0}\frac{1}{\Delta x}\int_1^{1+\Delta x}\sqrt{\sin t}\,dt.$

3. Rewrite the proof of the Second Fundamental Theorem, this time making use of the Maximum Theorem 13.3. Show first that

$$m \Delta x \leq \Delta F \leq M \Delta x \, ,$$

and then apply the Squeeze Theorem 11.3B. Prove carefully that the limits exist and have the value you want.

4. Prove that $y(x) = \displaystyle\int_0^x \frac{g(x)}{g(t)} r(t) \, dt$ is a solution to $y' + p(x)y = r(x)$, if the function $g(x)$ is a solution to $y' + p(x)y = 0$.

(Assume $p(x)$ and $r(x)$ are continuous and $g(t)$ is differentiable and non-zero on $\mathbb{R}$. Note that $g(x)$ can be moved outside the integral sign.)

20.3

1. (a) Let $I_k = \displaystyle\int_0^{\pi/2} \sin^k \theta \, d\theta$. Prove $I_k = (k-1)(I_{k-2} - I_k)$.

(Use integration by parts and trigonometric identities.)

(b) Deduce that $I_k = \dfrac{k-1}{k} I_{k-2}$, and hence $I_k = \dfrac{(k-1)!!}{k!!} c$, where

$$c = \begin{cases} 1, & \text{if } k \text{ is odd,} \\ \pi/2, & \text{if } k \text{ is even;} \end{cases} \qquad k!! = k(k-2)(k-4) \cdot \ldots \cdot m, \quad m = 1 \text{ or } 2.$$

2. Evaluate $\displaystyle\int_{-1}^1 (1 - x^2)^k \, dx$, where k is a positive integer, by making a change of variable which relates this integral to the one in the preceding exercise. Verify that the hypotheses for the Change-of-variable Theorem are satisfied.

3. A continuous periodic function $f(x)$ with period 2π can be represented as the sum of a series (called a *Fourier series*) of functions of the form:

$$f(x) = a_0 + a_1 \cos x + \ldots + a_n \cos nx + \ldots, \quad \text{if } f(x) \text{ is even;}$$
$$f(x) = b_1 \sin x + \ldots + b_n \sin nx + \ldots, \quad \text{if } f(x) \text{ is odd.}$$

The formulas for the *Fourier coefficients* are, for $n \geq 1$,

$$a_n = \frac{2}{\pi} \int_0^\pi f(x) \cos nx \, dx, \qquad b_n = \frac{2}{\pi} \int_0^\pi f(x) \sin nx \, dx \ .$$

Suppose $f(x)$ is continuously differentiable and even; then its derivative $f'(x)$ is odd. Show that the Fourier series for $f'(x)$ can be obtained by formally differentiating term-by-term the Fourier series for $f(x)$.

4. Suppose that $f(x)$ has a continuous n-th derivative on an open interval containing 0 and a, and that $f(x)$ has an n-fold zero at 0 (cf. Problem 17-1).

Evaluate $\displaystyle\int_0^a (x-a)^n f^{(n)}(x) \, dx$ in terms of $\displaystyle\int_0^a f(x) \, dx$.

5. (a) Integration by parts between limits is less familiar than the usual indefinite integration by parts done in calculus.

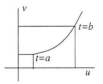

The curve shown is the parametric locus of $(u(t), v(t))$, as t ranges over the interval $a \leq t \leq b$. The picture illustrates Theorem 20.3A geometrically. How?

(b) In the picture, label the u-interval $[a_1, x]$ and the v-interval $[a_2, y]$. If a continuous strictly increasing elementary function $v = f(u)$ has an antiderivative that is an elementary function, so does its inverse function $u = g(v)$. Explain how the picture shows this.

(c) Using your explanation in (b), find elementary integrals for

(i) $\int \ln x \, dx$ (ii) $\int \sin^{-1} x \, dx$ (iii) $\int \tan^{-1} x \, dx$.

20.4

1. Give a proof of L4 by writing it in the form $\ln ax = \ln x + \ln a$ and using the Uniqueness Theorem (Cor. 20.2A).

2. Prove E6.

3. Prove that the alternative definition of e^x for x irrational, given in the remark at the end of this section, coincides with the definition given in E8.

20.5

1. By using Stirling's formula, obtain asymptotic estimates for:

(a) the number of different sets consisting of n positive integers $\leq 2n$;

(b) $(2n - 1)!! = 1 \cdot 3 \cdot \ldots \cdot (2n - 1)$.

2. By using the procedure in the proof of Stirling's formula (including the use of trapezoids), find with proof an estimate $f(n)$ within 1 for the sum C_n:

$$C_n = \sum_1^n \sqrt[3]{k} \underset{1}{\approx} f(n).$$

20.6

1. Let m and n be positive integers, c a positive real number.

The following functions all tend to ∞ as $x \to \infty$ and they are arranged in order of increasing growth rate:

$$\ln x \qquad x^{1/m} \qquad x^n \qquad e^{cx} \qquad e^{x^n} .$$

(a) Where in the list would you place x^x and $x/\ln x$?

(b) Find a function which grows faster than any on the list.

(c) Find a function which grows more slowly than any on the list.

2. Suppose f, g, and h all tend to ∞ as $x \to \infty$. Prove:

(a) if $f \sim g$, and $g \sim h$, then $f \sim h$;

(b) if f grows faster than g, then $f \sim f + g$.

3. Indeterminate forms of the type $0 \cdot \infty$ can be converted to the form $0/0$ or ∞/∞ by writing $f \cdot g$ as either $f/(1/g)$ or $g/(1/f)$. Knowing which form to use requires judgment, based in the beginning on trial-and-error.

Try both forms for each of the following, completing the calculation whenever possible.

$$\text{(a)} \quad \lim_{x \to 0^+} x \ln x \;; \qquad \text{(b)} \quad \lim_{x \to \infty} x\left(\tfrac{\pi}{2} - \tan^{-1} x\right) .$$

4. A function which gives an asymptotic estimate for the number of primes less than x is the "logarithmic integral" $\operatorname{Li} x$ defined below. Prove the asymptotic relation given for it. (The function $x/\ln x$ is elementary and simpler, but the non-elementary $\operatorname{Li} x$ gives a more accurate estimate.)

$$\operatorname{Li} x = \int_2^x \frac{dt}{\ln t} \;; \qquad\qquad \operatorname{Li} x \sim \frac{x}{\ln x}.$$

Problems

20-1 One way of rigorously defining the trigonometric functions is to start with the definition of the arctangent function. (This is the route used for example in the classic text *Pure Mathematics* by G. H. Hardy.)

So, assume amnesia has wiped out the trigonometric functions (but the rest of your knowledge of analysis is intact). Define

$$T(x) \;=\; \int_0^x \frac{dt}{1 + t^2} \;.$$

(a) Prove $T(x)$ is defined for all x and odd.

(b) Prove $T(x)$ is continuous and differentiable, and find $T'(x)$.

(c) Prove $T(x)$ is strictly increasing for all x; find where it is convex, where concave, and its points of inflection.

(d) Show $T(x)$ is bounded for all x, and $|T(x)| < 2.5$, using comparison of integrals. Can you get a better bound?

(e) Prove $u = T(x)$ has a differentiable inverse function $x = t(u)$ over some interval containing 0, and find a formula for $t'(u)$.

(f) Starting from $T(x)$, how could you define the number π, and the functions $\sin x$ and $\cos x$ (on the interval $[0, \pi/2]$, say)?

20-2 The proof of the Second Fundamental Theorem requires that $f(x)$ be continuous. Suppose that f is only piecewise continuous; we ask if the Second Fundamental Theorem will be true for such a function. Consider an example first.

(a) Let $f(x) = \begin{cases} x, & 0 \leq x \leq 1; \\ 1 - x, & 1 < x \leq 2; \end{cases}$ $F(x) = \int_0^x f(t)\, dt$.

Is $F(x)$ continuous at $x = 1$? Differentiable? Show the Second Fundamental Theorem does not hold for $F(x)$.

(b) Suppose in general $f(x)$ is piecewise continuous and has a jump discontinuity at the point x_0. On the basis of the preceding example, what statement about the derivative at x_0 of

$$F(x) = \int_a^x f(t)\, dt$$

should replace the Second Fundamental Theorem?

You could try getting more experimental evidence by calculating $F(x)$ for the function $\operatorname{sgn} x$ (cf. Exercises 9).

20-3 Mean-value Theorem for integrals.

This is the theorem in part (a); a generalization of it is given in part (b). Prove them both.

(a) Let $f(x)$ be continuous on $[a, b]$. Then for some $c \in [a, b]$,

$$\int_a^b f(x)\, dx = f(c)(b - a) .$$

(b) Let $f(x)$ and $g(x)$ be continuous and $g(x) \geq 0$ on $[a, b]$. Then for some point $c \in [a, b]$,

$$\int_a^b f(x)g(x)\, dx = f(c) \int_a^b g(x)\, dx .$$

(Get inequalities for $f(x)g(x)$ and integrate them. Why is this theorem a generalization of the one in part (a)?)

20-4 Taylor's Theorem.   Theorem 17.2 can be proved very elegantly using integration by parts. Suppose $f(t)$ has a continuous $(n + 1)$-st derivative on an open interval I containing a and x; say $a < x$.

By the First Fundamental Theorem, $f(x) - f(a) = \int_a^x f'(t)\, dt$.

(a) Apply integration by parts n times to the integral, differentiating f' and using $-(x - t)$ as the integral of dt (watch the signs carefully!); show that one gets Taylor's Theorem with the remainder term in the form

$$R_n = \int_a^x f^{(n+1)}(t)\, \frac{(x - t)^n}{n!}\, dt .$$

(b) Then use the Generalized Mean-value Theorem for integrals (Problem 20-3b) to show that the remainder can be expressed in the Lagrange form given in Theorem 17.2.

20-5 Generalizing the argument given for Stirling's formula, suppose we have a function $f(x)$ which is defined and non-negative for $x \geq 1$, say, and we wish to estimate $\sum_1^n f(k)$.

In the argument, a crucial step is the boundedness of the sum of the slivers lying between the curve and the trapezoidal approximation to it. Prove that this sum is bounded for all n whenever $f(x)$ is strictly increasing, differentiable, and concave, for $x \geq 1$.

20-6 Completion of Stirling's formula.   Here is an outline of the classical argument which shows $K = \sqrt{2\pi}$ in Stirling's formula 20.5. It refers to the notation of Exercise 20.3/1 and the formula for I_k given there.

(a) Prove $I_{2n+2} \leq I_{2n+1} \leq I_{2n}$, directly from the definition of I_n.

(b) Prove that $\lim\limits_{n \to \infty} \dfrac{I_{2n+1}}{I_{2n}} = 1$, by using the Squeeze Theorem 5.2 .

(c) Using the formula for I_n given in Exercise 20.3/1, and Stirling's formula to replace the factorials by asymptotic approximations, deduce from part (b) that $K = \sqrt{2\pi}$.   (Use the Squeeze Theorem.)

Answers to Questions

20.1

1. For the first line, the sum expression is a Riemann sum for $\mathcal{P}^{(n)}$, the standard n-partition. That means the c_i and Δx_i depend on n; when you go from n to $n + 1$, you don't just add on another term; every term of the sum changes. So it is not the type of infinite sequence for which you can use the $\sum_1^\infty$ notation: it's not an infinite series, in other words.

As to the second line: for each partition $\mathcal{P}^{(n)}$, the special Riemann sum constructed for the proof has the constant value $F(b) - F(a)$, independent of n. Thus it is a constant sequence, whose limit is $F(b) - F(a)$ as $n \to \infty$.

2. Assume $b < a$. Then by (1), $\int_b^a f(x)\,dx = F(a) - F(b)$. Multiply both sides by -1, and you get $\int_a^b f(x)\,dx = F(b) - F(a)$, according to 19.5 (24).

20.2

1. As a preliminary, note that by the Comparison Theorem,

$$(*) \qquad a < b \quad \text{and} \quad f(t) \leq g(t) \quad \Rightarrow \quad \int_b^a f(t)\,dt \geq \int_b^a g(t)\,dt ,$$

since reversing the limits changes the sign, and therefore reverses the inequality.

For the left-hand limit, start from (6), assume $\Delta x < 0$ and integrate (note that $x + \Delta x < x$, so you must use $(*)$); you get

$$\big(f(x) - \epsilon\big)\Delta x \ \geq \ \Delta F \ \geq \ \big(f(x) + \epsilon\big)\Delta x, \qquad \Delta x \approx 0^- ;$$

when you divide by Δx, the inequalities reverse again, giving

$$f(x) - \epsilon \ \leq \ \frac{\Delta F}{\Delta x} \ \leq \ f(x) + \epsilon , \qquad \Delta x \approx 0^- ,$$

which proves the left-hand limit is $f(x)$, by definition of left-hand limit 11.2B.

2. $F'(x) = e^{-x^2/2}$; $F''(x) = -xe^{-x^2/2}$; no critical points; convex if $F'' > 0$, therefore when $x < 0$; concave if $F'' < 0$, therefore when $x > 0$.

3. Explicitly, $F(x) = \begin{cases} 0, & x \leq 0, \\ x, & x \geq 0. \end{cases}$

Thus $F(x)$ is not differentiable at 0, therefore not an antiderivative for $u(x)$.

4. (a) For example, $F_1 = \begin{cases} -x, & x < 0; \\ x, & x > 0; \end{cases}$ and $F_2 = \begin{cases} -x + 1, & x < 0; \\ x, & x > 0. \end{cases}$

(b) F_1 and F_2 are an example: both satisfy the differential equation and the initial condition $y(1) = 1$ on $D = \{x : x \neq 0\}$. No: D is not an interval.

5. $F(x)$ and $F(a) + \int_a^x f(t)\, dt$ both satisfy $y' = f(x)$, $y(a) = F(a)$. Therefore $F(x) = F(a) + \int_a^x f(t)\, dt$, by Cor. 20.2A; now put $x = b$.

6. Theorem 15.2 (5), needed for Theorem 20.2B (Uniqueness Theorem for antiderivatives), depends on the Mean-value Theorem.

20.3

1. Since $g'(x)$ is continuous for $x > 0$, so is $f'(x)$; thus we can use integration by parts. Noting that $f(a) = 0$ and $\ln 1 = 0$, we get

$$\int_1^a f'(x) \ln x \, dx = f(x) \ln x \Big]_1^a - \int_1^a \frac{f(x)}{x} \, dx = 0 - \int_1^a g(x) \, dx \ .$$

2. The function $x = \tan u$ is continuous and strictly increasing on the interval $[0, \pi/4]$, and maps this interval to $[0, 1]$. Since the integrand is continuous on this interval, the change-of-variable formula applies, and we get

$$\int_0^1 \frac{dx}{(1 + x^2)^{3/2}} = \int_0^{\pi/4} \frac{\sec^2 u \, du}{\sec^3 u} = \int_0^{\pi/4} \cos u \, du = \frac{\sqrt{2}}{2} \ .$$

20.4

1. (a) Since $\ln x$ is differentiable, so is $\exp x$, by Theorem 14.2C. To find its derivative:

$$y = \exp x \implies x = \ln y \implies dx/dy = 1/y, \quad \text{by L2};$$
$$\implies dy/dx = y = \exp x, \quad \text{by Theorem 14.2C.}$$

(b) The second derivative test (Theorem 16.2B) shows $\exp x$ is convex, since $D^2 \exp x = \exp x$, by E2, and $\exp x > 0$, by E1 (the range is $(0, \infty)$).

(c) $A = \exp(-a) \implies \ln A = -a$ by (14a) $\implies \ln(1/A) = a$, by L5
$$\implies 1/A = \exp a, \text{ by (14a)} \implies A = 1/\exp a \ .$$

2. Let $f(x) = \ln x^r$ and $g(x) = r \ln x$, where r is a rational number; then
$$f'(x) = (1/x^r)rx^{r-1} = r/x, \quad \text{by the chain rule.}$$
Both f and g satisfy $y' = r/x$ and $y(1) = 0$, so $f(x) = g(x)$ by Cor. 20.2A.

3.   Let $a = 3^2 = e^{2\ln 3}$,   (definition of 3^x);
then $\ln a = 2\ln 3 = \ln(3\cdot 3)$,   by (14a) and L4;
   so $a = 3\cdot 3 = 9$,   by L3 or (14b).

20.5

1. Using Stirling's formula, and the estimates $e < 3$, $2^{10} = 1028 > 10^3$:

$$12! \sim \left(\frac{12}{e}\right)^{12}\sqrt{2\pi\cdot 12} > 4^{12}\cdot 8 = 2^{27} > 10^6\cdot 2^7 > 10^8,$$

so the answer is: probably not. (To be certain, we would need an error estimate for Stirling's formula.)

2. $A_n - S_n = $ sum of the slivers; estimating visually (using $\ln 2 \approx .7$),

$$0 < A_n - S_n < \tfrac{1}{2}(\text{area under }\ln x,\text{ over }[1,2]) = \ln 2 - \tfrac{1}{2} < .2;$$

Therefore, $A_n - S_n \to L \ \Rightarrow\ 0 < L < .2$, by the Limit Location Theorem 5.3A;

$$\Rightarrow\ .8 < 1 - L < 1 \ \Rightarrow\ e^{.8} < K < e \ \Rightarrow\ 2.2 < K < 2.8.$$

20.6

1. (a) $\displaystyle\lim_{x\to\infty}\frac{x^3 - 2x^2 - 1}{x^3} = \lim_{x\to\infty}\left(1 - \frac{2}{x} - \frac{1}{x^3}\right) = 1$

(b) $\displaystyle\lim_{x\to\infty}\frac{\sqrt{x^5 + 5x^3 + 2}}{x^3} = \lim_{x\to\infty}\sqrt{\frac{1}{x} + \frac{5}{x^3} + \frac{2}{x^6}} = 0$

(c) $\displaystyle\lim_{x\to\infty}\frac{e^{ax}}{e^{bx}} = \lim_{x\to\infty}e^{(a-b)x} = \infty$,   if $a > b$.

2. (a) (23): $\displaystyle\lim_{x\to\infty}\frac{\ln x}{x^k} = \lim_{x\to\infty}\frac{1}{kx^k} = 0$ if $k > 0$.

(b) (24): $\displaystyle\lim_{x\to\infty}\frac{e^{ax}}{x^m} = \lim_{x\to\infty}\left(\frac{e^{ax/m}}{x}\right)^m = \infty$,

since the quantity in parentheses has limit ∞ by l'Hospital's rule.

3. $\displaystyle\lim_{x\to\infty}\frac{e^{x^2}}{e^{2x}} = \lim_{x\to\infty}e^{x^2-2x} = \lim_{u\to\infty}e^u = \infty$, since $\lim_{x\to\infty}x^2 - 2x = \infty$.

Using l'Hospital's rule formally leads perpetually to an infinite sequence of indeterminate forms of the type ∞/∞: $\displaystyle\lim\frac{e^{x^2}}{e^{2x}} = \lim\frac{2xe^{x^2}}{2e^{2x}} = $???

21

Improper Integrals

21.1   Basic definitions.

Though they occur frequently in the applications, such expressions as

$$\int_1^\infty \frac{dx}{x^2} \qquad \text{or} \qquad \int_0^1 \frac{dx}{\sqrt{1-x^2}}$$

are not Riemann integrals—the first because the interval $[1,\infty)$ is not finite, the second because the integrand is not bounded on $[0,1]$.

Definition 21.1   An **improper integral of the first kind** is an integral over a semi-infinite interval; we define it to be a limit:

$$(1) \qquad \int_a^\infty f(x)\,dx \;=\; \lim_{R\to\infty} \int_a^R f(x)\,dx; \qquad \int_{-\infty}^a f(x)\,dx \;=\; \lim_{R\to\infty} \int_{-R}^a f(x)\,dx \ .$$

An **improper integral of the second kind** is one over a finite interval, such that the integrand becomes infinite at one of the endpoints; we define:

$$(2) \qquad \int_a^{b^-} f(x)\,dx \;=\; \lim_{u\to b^-} \int_a^u f(x)\,dx; \qquad \int_{a^+}^b f(x)\,dx \;=\; \lim_{u\to a^+} \int_u^b f(x)\,dx \ .$$

In each case, we say the improper integral **converges** if the limit exists; if it does not, the integral **diverges**.

For integrals of the second kind, we assume $f(x)$ is defined respectively on $[a,b)$ and $(a,b]$. Normally the $+$ and $-$ are dropped, and must be inferred from the integrand or the general context; we will often use them in this chapter, however, as an aid to understanding.

Examples 21.1A   The following are respectively of the first and second kind:

$$(3) \qquad \int_1^\infty \frac{dx}{x^p} \quad \text{converges if } p > 1, \ \text{diverges if } p \le 1 \ ;$$

$$(4) \qquad \int_{0^+}^1 \frac{dx}{x^p} \quad \text{converges if } p < 1, \ \text{diverges if } p \ge 1 \ .$$

Proof.   Use the definition; the integrals are readily evaluated using the First Fundamental Theorem 20.1, and the limits then follow (cf. Question 1).   □

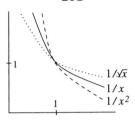

Remembering which of (3) and (4) converges for $p > 1$ and which for $p < 1$ is easy if you have a mental picture of how the graph of $1/x^p$ looks for different values of p : $1, 2$, and $1/2$ are good and typical choices.

Comparing $p = 2$ and $p = 1/2$, the graph of the first is closer to the x-axis over $(1, \infty)$, but further away from it over $(0, 1)$.

If an integral is improper at both ends, we break it in two, writing

$$\int_{a+}^{\infty} f(x)\, dx \;=\; \int_{a+}^{b} f(x)\, dx + \int_{b}^{\infty} f(x)\, dx \;,$$

where b is any convenient point $> a$, and we say the integral converges if each of the improper integrals on the right converges. (This does not depend on the choice of b.) In terms of limits, the usual notation for this is

$$\int_{a+}^{\infty} f(x)\, dx \;=\; \lim_{R\to\infty,\, u\to a+} \int_{u}^{R} f(x)\, dx \;,$$

by which we mean

$$\lim_{R\to\infty} \int_{b}^{R} f(x)\, dx \;+\; \lim_{u\to a+} \int_{u}^{b} f(x)\, dx \;.$$

That is, the two limits are to be taken independently. The following example will make this clearer.

Example 21.1B Does $\displaystyle\int_{-\infty}^{\infty} \frac{t}{1+t^2}\, dt$ converge?

Solution. Breaking up the integral as described above,

$$\int_{-\infty}^{\infty} \frac{t}{1+t^2}\, dt \;=\; \lim_{S\to\infty} \int_{0}^{S} \frac{t}{1+t^2}\, dt + \int_{-R}^{0} \frac{t}{1+t^2}\, dt,$$

$$=\; \lim_{S\to\infty} \tfrac{1}{2}\ln(1+S^2) - \lim_{R\to\infty} \tfrac{1}{2}\ln(1+R^2) \;;$$

and neither limit exists, so the integral does not converge. □

On the other hand, if we set $R = S$, so that the two endpoints no longer move apart independently but rather in tandem, we get (using the above)

$$(5) \qquad \int_{-\infty}^{\infty} \frac{t}{1+t^2}\, dt \;=\; \lim_{R\to\infty} \int_{-R}^{R} \frac{t}{1+t^2}\, dt \;=\; \lim_{R\to\infty} \tfrac{1}{2}\ln(1+t^2) \Big]_{-R}^{R} \;=\; 0 \;.$$

Though the integral diverges, the value 0 calculated by the above procedure is called its **Cauchy principal value** (CPV) and is occasionally useful. Some divergent integrals of mixed or second kind also have CPV's: see Exercises 21.1/3,4.

Questions 21.1

1. (a) Verify (3)   (b) Verify (4).

(c) Using (a) and (b), show: for $0 < p < 1$, $\displaystyle\int_{0+}^{1} \frac{dx}{x^{1/p}} = 1 + \int_{1}^{\infty} \frac{dx}{x^p}$,

and explain it geometrically.

2. Find all k for which $\displaystyle\int_0^\infty e^{kt}\, dt$ converges, and evaluate it.

3. Do the following converge or diverge? Evaluate, if possible.

(a) $\displaystyle\int_{2+}^4 \frac{dx}{\sqrt{x-2}}$ (b) $\displaystyle\int_0^{1^-} \frac{dt}{\sqrt{1-t^2}}$ (c) $\displaystyle\int_0^\infty \frac{t^2\, dt}{1+t^3}$

21.2   Comparison theorems.

Improper integrals of the first kind are the continuous analog of infinite series. This is easiest to see if we use computer-code notation, writing $\sum a_n$ as $\sum a(n)$. The analogy is then between

$$\sum_0^\infty a(n) \qquad \text{and} \qquad \int_0^\infty f(x)\, dx\ ;$$

in one case we sum the values of a discrete function $a(n)$ for $0 \le n < \infty$; in the other we "sum" a continuous function $f(x)$ for $0 \le x < \infty$.

Because of this analogy, some of the theorems and techniques for infinite series carry over to the study of improper integrals; for example, we have the following, whose proof is elementary and left for the Questions.

Theorem 21.2A   Tail-convergence.   *If $f(x)$ is integrable on $I = [x_0, \infty)$, (i.e., integrable on every compact subinterval of I), and $a, b \in I$, then*

(6) $\displaystyle\int_a^\infty f(x)\, dx$ converges $\Leftrightarrow$ $\displaystyle\int_b^\infty f(x)\, dx$ converges .

(There are similar statements for the other kinds of improper integrals.)

The proof of the comparison test for series (7.2) used the Completeness Property for sequences; the comparison test for improper integrals will use the following proposition, part (b) of which is the Completeness Property adapted to functions.

Proposition 21.2

(a) *$f(x)$ increasing for $x \gg 1$, $\displaystyle\lim_{x\to\infty} f(x) = L \Rightarrow f(x) \le L$ for $x \gg 1$;*

(b) *$f(x)$ increasing, $f(x) \le B$ for $x \gg 1$*

$$\Rightarrow \lim_{x\to\infty} f(x)\ \text{exists, and}\ \lim_{x\to\infty} f(x) \le B\ .$$

Proof.   See the Questions. There are similar statements for $\displaystyle\lim_{x\to a} f(x)$ etc., and for decreasing functions.

Just as the most fundamental test for convergence of a series is the comparison test, so too the convergence or divergence of an improper integral is often established by comparing it with a simpler integral. Two typical comparison theorems are these—others would be similar, and we will use them without comment.

Theorem 21.2B   Comparison theorems for improper integrals.

Assume $f(x)$ and $g(x)$ integrable and $0 \leq f(x) \leq g(x)$, for $x \geq a$. Then if the integral on the right below converges, so does the integral on the left, and

$$\int_a^\infty f(x)\,dx \ \leq \ \int_a^\infty g(x)\,dx.$$

Assume $f(x)$ and $g(x)$ integrable and $0 \leq f(x) \leq g(x)$, on $[a, b)$. Then if the integral on the right below converges, so does the integral on the left, and

$$\int_a^{b^-} f(x)\,dx \ \leq \ \int_a^{b^-} g(x)\,dx.$$

Proof. We prove the first; the second and all others would go similarly. We have

(7) $$\int_a^R f(x)\,dx \ \leq \ \int_a^R g(x)\,dx \ \leq \ \int_a^\infty g(x)\,dx;$$

the first inequality by the Comparison Theorem for integrals (19.4B), the second inequality by part (a) of the preceding proposition, applied to the increasing function of R defined by $\int_a^R g(x)dx$ (note that $g(x) \geq 0$). The theorem now follows from part (b) of the proposition, applied to the bounded increasing function

$$F(R) \ = \ \int_a^R f(x)\,dx \ \leq \ \int_a^\infty g(x)\,dx.$$

It tells us that $\int_a^\infty f(x)\,dx$ exists and $\int_a^\infty f(x)\,dx \ \leq \ \int_a^\infty g(x)\,dx.$ $\qquad\square$

Example 21.2A Does $\displaystyle\int_0^\infty \frac{e^{-x^2}}{\sqrt{x}}\,dx$ converge?

Solution. The integral is improper at both ends, so both must be studied separately, as explained in the preceding section. Using comparison (21.2B),

$$\frac{e^{-x^2}}{\sqrt{x}} < e^{-x} \quad \text{for } x > 1 \ \Rightarrow \ \int_1^\infty \frac{e^{-x^2}}{\sqrt{x}}\,dx \leq \int_1^\infty e^{-x}\,dx \quad \text{convergent;}$$

$$\frac{e^{-x^2}}{\sqrt{x}} < \frac{1}{\sqrt{x}} \quad \text{for } 0 < x < 1 \ \Rightarrow \ \int_{0+}^1 \frac{e^{-x^2}}{\sqrt{x}}\,dx \leq \int_{0+}^1 \frac{dx}{\sqrt{x}} \quad \text{convergent.}$$

Thus the given integral converges at both endpoints, and so converges. $\qquad\square$

As you can see from the previous example,

$$\int_0^\infty e^{kx}\,dx, \quad \int_0^1 \frac{dx}{x^p}, \quad \text{and} \quad \int_1^\infty \frac{dx}{x^p}$$

are useful comparison integrals. By making a change of variable, an improper integral can often be transformed so it can be compared with one of them. The next example illustrates.

Example 21.2B   Does $\displaystyle\int_0^1 \frac{\ln^3 x}{\sqrt{x}}\, dx$ converge?

Solution.   Put $\ln x = u$, so $x = e^u$, and then put $u = -t$; we get, in order:

$$\int_0^1 \frac{\ln^3 x}{\sqrt{x}}\, dx = \int_{-\infty}^0 u^3 e^{u/2}\, du = -\int_0^\infty t^3 e^{-t/2}\, dt \ .$$

Now we can use a standard comparison integral. We have

$$\int_a^\infty t^3 e^{-t/2} < \int_a^\infty e^{-t/4} dt \qquad \text{for } a \gg 1,$$

since (cf. 20.6 (24))

$$\lim_{t \to \infty} \frac{t^3}{e^{t/4}} = 0 \quad \Rightarrow \quad \frac{t^3}{e^{t/4}} < 1 \quad \text{for } t \gg 1 \ .$$

Since the integral on the right converges, so does the integral on the left by the Comparison Theorem, and therefore also the original integral, by the Tail-convergence Theorem. $\square$

A useful form of the comparison test compares the integrands asymptotically, rather than by means of inequalities.

Theorem 21.2C   Asymptotic comparison test.

Assume $f(x)$ and $g(x)$ are continuous and positive for $x \geq a$, and $f(x) \sim g(x)$ as $x \to \infty$ (cf. p. 280). Then

$$\int_a^\infty f(x)\, dx \ \text{converges} \quad \Leftrightarrow \quad \int_a^\infty g(x)\, dx \ \text{converges}.$$

This is the analog of the asymptotic comparison test for series, 7.5B. In most examples, both functions $\to 0$ as $x \to \infty$ (otherwise the integral would not be of any interest).

Note that the theorem is also true if "converges" is replaced by "diverges": this is just the contrapositive of the theorem.

There are analogous results for improper integrals of the second kind: for instance, if $f(x)$ and $g(x)$ tend to ∞ as $x \to a^+$, and $f \sim g$ at a^+ (i.e., $f/g \to 1$ as $x \to a^+$), then

$$\int_{a^+}^b f(x)\, dx \ \text{converges} \quad \Leftrightarrow \quad \int_{a^+}^b g(x)\, dx \ \text{converges}.$$

The proof is left as an Exercise.

Example 21.2C   Does $\displaystyle\int_0^\infty \frac{dx}{\sqrt{x(1+x)}}$ converge?

Solution.   Both endpoints are improper. Calling the integrand $f(x)$, we have:

at 0, $f(x) \sim 1/\sqrt{x}$,   and $\int_0^1 dx/\sqrt{x}$ converges, by 21.1 (4);

at ∞, $f(x) \sim 1/x$,   and $\int_1^\infty dx/x$ diverges, by 21.1 (3).

Therefore by the asymptotic comparison test, the integral converges at 0, but diverges at ∞; hence it is divergent.

> If we had tested the endpoints in the opposite order, we could have stopped after testing the endpoint ∞, since the integral diverges there.

Questions 21.2

1. Determine whether the following integrals converge by using the comparison test 21.2B with the suggested function (for a suitable k).

(a) $\displaystyle\int_0^\infty t^2 e^{-t}\, dt$ (compare with e^{kt})   (b) $\displaystyle\int_{0+}^1 \frac{\ln x}{\sqrt{x}}\, dx$ (compare with x^k)

2. Determine whether the following integrals converge or diverge by using the asymptotic comparison test 21.2C.

(a) $\displaystyle\int_0^\infty \frac{1+x^2}{1+x^4}\, dx$ (b) $\displaystyle\int_0^1 \frac{x\, dx}{\sqrt{1-x^3}}$ (put $1 - x = u$)

3. Prove the Tail-convergence Theorem (6).

4. (a) Prove Proposition 21.2(a) indirectly.

(b) Prove Proposition 21.2(b) by deducing it from the analogous theorem for sequences, as follows. Show the sequence $f(n)$, $n = 1, 2, 3, \ldots$ has a limit L, and $L \le B$; then show $L = \lim\limits_{x\to\infty} f(x)$, using the ϵ-definition of limit.

21.3   The Gamma function.

An important role in analysis is played by a function $\Gamma(x)$ whose value for positive integers n is given by $\Gamma(n) = (n-1)!$. We are considering it now because the study of this function gives a good example of how improper integrals and properties of definite integrals are used in analysis.

The definition of $\Gamma(x)$ is motivated by a famous improper integral value:

(8) $\displaystyle\int_0^\infty t^n e^{-t}\, dt \;=\; n!, \quad n = 0, 1, 2, \ldots$ (note $0! = 1$).

Proof.   We use mathematical induction on n. (See Appendix A.4.)

For $n = 0$, the integral is readily evaluated, and we get $1 = 1$.

For higher values of n, we integrate by parts between limits (20.3A):

$$\int_0^R t^n e^{-t}\, dt \;=\; -R^n e^{-R} + n \int_0^R t^{n-1} e^{-t}\, dt, \quad n = 1, 2, \ldots .$$

Taking limits of both sides as $R \to \infty$, the first term on the right has 0 as its limit (for example, by repeated applications of l'Hospital's rule, or by 20.6 (24)), while the second term has the limit $n(n-1)!$, by our induction assumption. Thus, if (8) is true for $n - 1$, it is true for n, and we are done. $\square$

The idea now is to read (8) backwards, viewing it not as an evaluation of an integral, but as a formula for $n!$. This is not as silly as it sounds, since there are methods (called "asymptotic methods") for estimating improper integrals like the one in (8); they can therefore estimate $n!$.

Here, however, our purpose in reading formula (8) backwards is to use it to define $n!$ for non-integer values of n, by replacing n by the continuous variable $x - 1$; we use this rather than x so that the resulting function will have $(0, \infty)$ as its natural domain, rather than $(-1, \infty)$, which is less convenient.

Definition 21.3   The Gamma function.   We define

$$(9) \qquad\qquad \Gamma(x) \; = \; \int_0^\infty t^{x-1} e^{-t} \, dt \; , \qquad x > 0.$$

We list some of the properties of this function accessible to us now.

G1 $\Gamma(n+1) = n!$ for all integers $n \geq 0$;   (this is (8)).

G2 $\Gamma(x)$ is defined for $x > 0$.

To show this, since the integral is improper at both ends, according to the remarks in 21.1 we have to break it into two parts, say at $t = 1$:

$$(10) \qquad\qquad \Gamma(x) \; = \; \int_{0+}^1 t^{x-1} e^{-t} \, dt + \int_1^\infty t^{x-1} e^{-t} \, dt \; .$$

The convergence of both of these integrals for $x > 0$ is established by the comparison test. See Questions 1 and 2.

G3 $\Gamma(x+1) \; = \; x\Gamma(x) \; , \qquad$ for $x > 0$.

To prove this, as in (10) we break up the integral into two pieces, and use integration by parts to calculate $\Gamma(x+1)$:

$$\int_1^R t^x e^{-t} \, dt \; = \; -t^x e^{-t} \Big]_1^R + x \int_1^R t^{x-1} e^{-t} \, dt \; ,$$

$$\int_u^1 t^x e^{-t} \, dt \; = \; -t^x e^{-t} \Big]_u^1 + x \int_u^1 t^{x-1} e^{-t} \, dt \; .$$

We now evaluate the middle terms in each of the above equations, and pass to the limit; we have

$$\lim_{R \to \infty} (-R^x e^{-R} + 1/e) \; = 1/e; \quad \lim_{u \to 0+} (-1/e + u^x e^{-u}) \; = \; -1/e, \quad \text{if } x > 0.$$

Thus after passing to the limit, the sum of the two equations is (G3). $\square$

G4 $\Gamma(1/2) \; = \; \sqrt{\pi}$.

To prove this, we start with the definition:

$$\Gamma(1/2) \; = \; \int_0^\infty \frac{e^{-t}}{\sqrt{t}} \, dt \; = \; \lim_{R \to \infty, u \to 0+} \int_u^R \frac{e^{-t}}{\sqrt{t}} \, dt.$$

Setting $t = s^2$, this becomes

$$\Gamma(1/2) = \lim_{R\to\infty, u\to 0^+} \int_{\sqrt{u}}^{\sqrt{R}} \frac{e^{-s^2}}{s}\, 2s\, ds = 2\int_0^\infty e^{-s^2}\, ds$$

$$= \sqrt{\pi}\,, \quad \text{as is well-known.} \qquad \square$$

G5 $\displaystyle\lim_{x\to 0^+} \Gamma(x) = \infty$.

G6 $\Gamma(x)$ is continuous, for all $x > 0$.

The proofs of these last two facts are left for the Exercises.

To go further, we would have to be able to differentiate $\Gamma(x)$ with respect to x. It is tempting to try to do this by just ignoring the integral sign and differentiating the integrand with respect to x, a procedure usually called "differentiating under the integral sign". In fact, this can be done, but the justification for it is quite subtle, and it will have to wait until Chapter 27. Then we can show that $\Gamma(x)$ is convex and infinitely differentiable.

Because $\Gamma(n+1) = n!$, we can think of $\Gamma(x+1)$ as the solution to a curve-fitting problem: find a function which interpolates between all the points $(n, n!)$, for $n \geq 0$, i.e., one whose graph passes through all these points. Obviously there can be many such graphs; from this viewpoint, definition (9) seems rather arbitrary. However, it can be proved that $\Gamma(x)$ is the only such interpolating function which is convex and infinitely differentiable.

If $x \leq 0$, the integral (9) does not converge at its lower end. However, the definition of $\Gamma(x)$ is extended to non-integral $x \leq 0$ in the following way. Using the functional equation G3 repeatedly, we get (for $x > 0$),

$$\Gamma(x+1) = x\Gamma(x),$$
$$\Gamma(x+2) = (x+1)\Gamma(x+1) = (x+1)x\Gamma(x),$$

and in general, for $n \geq 0$,

(11) $\Gamma(x+n) = (x+n-1)\cdots\cdots(x+1)x\Gamma(x).$

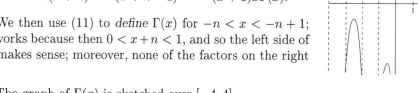

We then use (11) to *define* $\Gamma(x)$ for $-n < x < -n+1$; this works because then $0 < x+n < 1$, and so the left side of (11) makes sense; moreover, none of the factors on the right is 0.

The graph of $\Gamma(x)$ is sketched over $[-4, 4]$.

Questions 21.3

1. Prove the second integral in (10) converges for all x, using $e^{-t/2}$ as comparison integrand.

2. For what values of x is the first integral in (10) improper? Prove it converges for all $x > 0$. (Note that $e^{-t} \leq 1$ for $t \geq 0$.)

21.4   Absolute and conditional convergence.

The analogy with infinite series suggests that also for improper integrals there might be a useful distinction between the two kinds of convergence. We define them, then show by a proof very much like the one we gave for series (Theorem 7.3) that absolute convergence implies convergence.

Definition 21.4 $\displaystyle\int_a^\infty f(x)\,dx$ converges **absolutely** if $\displaystyle\int_a^\infty |f(x)|\,dx$ converges; it converges **conditionally** if it converges, but not absolutely.

Theorem 21.4   *If $f(x)$ is integrable on $[a,\infty)$ and the integral $\displaystyle\int_a^\infty f(x)\,dx$ is absolutely convergent, then it is convergent.*

Proof.   As in the proof of Theorem 7.3, we express $f(x)$ as the difference of two non-negative integrable functions: $f(x) = f^+(x) - f^-(x)$, where

$$f^+(x) = \frac{|f(x)| + f(x)}{2}, \qquad f^-(x) = \frac{|f(x)| - f(x)}{2},$$
$$= \max\{f(x), 0\}; \qquad\qquad = \max\{-f(x), 0\}.$$

Both functions are integrable, since $f(x)$ and $|f(x)|$ are (Theorem 18.4B), and

$$0 \le f^+(x) \le |f(x)|, \qquad 0 \le f^-(x) \le |f(x)|.$$

Therefore, by the Comparison Theorem 21.2B, the integrals of f^+ and f^- converge, and so therefore does the integral of $f = f^+ - f^-$, by the Linearity Theorem 11.3A for limits of functions, applied to the integrals from a to R.   □

Example 21.4   Show the integral below converges conditionally:

(12) $$\int_0^\infty \frac{\sin x}{x}\,dx\ .$$

Solution.   The graph of the integrand is shown at the right. Its value at 0 is defined as 1; this makes it a continuous function for all x.

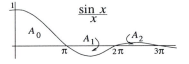

Let us set

$$A_n = \int_{n\pi}^{(n+1)\pi} \frac{\sin x}{x}\,dx.$$

The integral (12) is not absolutely convergent; this follows from the following two estimations which Exercise 21.4/2 asks you to prove:

(13) $$\int_0^R \frac{|\sin x|}{x}\,dx > \sum_0^N |A_n|\,, \qquad \text{if } R > (N+1)\pi;$$

(14) $$|A_n| > \frac{1}{2n+1}\ .$$

To show the integral (12) does converge conditionally, one idea would be to use Cauchy's test for series with alternating sign (7.6) to show that $\sum A_n$ converges; then show that (12) must converge because it stays close to the series:

$$(15) \qquad \int_0^R \frac{\sin x}{x}\, dx \;\approx\; \sum_0^\infty A_n, \quad \text{for } R \gg 1 \;.$$

Problem 21-2 asks you to carry out this program, and gives some help.

It is easier, however, to use integration by parts, which converts a tail of the integral (12) into one which is absolutely convergent:

$$(16) \qquad \int_1^R \frac{\sin x}{x}\, dx \;=\; \frac{-\cos x}{x}\Big]_1^R \;-\; \int_1^R \frac{\cos x}{x^2}\, dx \;.$$

As $R \to \infty$ in (16), the first term on the right $\to \cos 1$; the second term is an integral which converges absolutely, and therefore converges (Theorem 21.4). So the left side has a limit as $R \to \infty$, i.e., it converges, and therefore (12) converges by the Tail-convergence Theorem. $\qquad\qquad\square$

Questions 21.4

1. For each of the following, tell if it converges or diverges; if it converges, is the convergence absolute or conditional? (Prove divergence or absolute convergence; you need not prove conditional convergence.)

$$\text{(a)} \quad \int_0^\infty \sin x \, dx \qquad \text{(b)} \quad \int_0^\infty e^{-x} \sin x \, dx \qquad \text{(c)} \quad \int_1^\infty \frac{\sin x}{x^2} \, dx$$

2. For which values of k will $\displaystyle\int_1^\infty x^k \sin x \, dx$ be divergent? absolutely convergent? conditionally convergent? (Prove only absolute convergence.)

Exercises

21.1

1. Prove that $\displaystyle\int_2^\infty \frac{dx}{x \ln^k x}$ converges if $k > 1$, diverges if $k \le 1$.

2. Does $\displaystyle\int_{0+}^1 \ln x \, dx$ converge? Prove your answer.

3. (a) Show that if $n \in \mathbb{N}$, $\displaystyle\int_{-1}^1 \frac{1}{t^n} \, dt$ does not converge.

   (b) Define its Cauchy principal value, and show it is $0 \iff n$ is odd.

4. How would you define the Cauchy principal value of $\displaystyle\int_{0+}^\infty f(x) \, dx$? Apply your definition to $\displaystyle\int_0^\infty \frac{\ln x}{x} \, dx$.

21.2

1. Test each of the following improper integrals for convergence or divergence. Indicate the reasoning. If the integral is improper at both ends, then each end must be tested separately. As is customary in the literature, we omit the $+$ and $-$ on the endpoints of improper integrals of the second kind.

If the integral contains k, the answer will usually depend on k. For some of them, a change of variable might be helpful.

(a) $\displaystyle\int_0^\infty \frac{x\,dx}{1+x^3}$
(b) $\displaystyle\int_0^\infty \frac{dx}{\sqrt{1+x^3}}$
(c) $\displaystyle\int_1^\infty \frac{x^k\,dx}{\sqrt{x^2-1}}$

(d) $\displaystyle\int_2^\infty \frac{x\,dx}{\sqrt{x^4-1}}$
(e) $\displaystyle\int_0^\infty x^k e^{-x^2}\,dx$
(f) $\displaystyle\int_0^1 \frac{dx}{\sqrt{x-x^3}}$

(g) $\displaystyle\int_0^1 x^k \ln x\,dx$
(h) $\displaystyle\int_1^\infty \sin\frac{1}{x}\,dx$

2. Suppose that $f(x)$ is continuous on $[1,\infty)$, and for some $k > 1$, we have that $\lim\limits_{x\to\infty} x^k f(x)$ exists and is positive. Prove that

$$\int_1^\infty f(x)\,dx \quad \text{converges.}$$

Use the comparison test 21.2B, to get practice using it in theoretical arguments; this also gives insight into how to prove the asymptotic comparison test 21.2C (the exercise is too easy using the asymptotic test).

3. Prove the asymptotic comparison test 21.2C.

4. For improper integrals of the first kind, is there an analog of the n-th term divergence test for series 7.2A? It would say:

$$\int_a^\infty f(x)\,dx \text{ converges} \implies \lim_{x\to\infty} f(x) = 0.$$

Give a proof or counterexample. (Assume $f(x)$ positive and continuous.)

5. Assume f and g are continuous and positive. Prove that if $\int_0^\infty f(x)\,dx$ converges and $g(x)$ is bounded, then $\int_0^\infty f(x)g(x)\,dx$ converges.

6. Assume f and g are continuous and positive. Show that if $\int_0^\infty f(x)\,dx$ and $\int_0^\infty g(x)\,dx$ converge, this does not imply $\int_0^\infty f(x)g(x)\,dx$ converges.

21.3

1. Prove G5: $\Gamma(x) \to \infty$ as $x \to 0^+$, in two ways:

(a) using the other five listed properties of $\Gamma(x)$;

(b) directly from the definition of $\Gamma(x)$, by estimating the integral.

21.4

1. Determine whether $\displaystyle\int_0^\infty \frac{\cos x}{\sqrt{1+x^3}}\,dx$ converges absolutely, conditionally, or diverges, with indication of reasoning.

2. Fill in the argument that $\int_0^R \frac{\sin x}{x}\,dx$ is not absolutely convergent by proving (13) and (14), and then drawing the conclusion.

3. (a) Prove that $\int_0^\infty \sin(x^2)\,dx$ converges by making the change of variable $u = x^2$, and applying integration by parts to the resulting integral.

(b) Prove the integral does not converge absolutely. (Use the u form.)

Problems

21-1 Imagine we are back in prehistoric times when the average person on the trail didn't know about $\sin x$ and $\sin^{-1} x$. Define a function by

$$x(u) = \int_0^u \frac{dt}{\sqrt{1 - t^2}}\ .$$

(a) Show the natural domain of $x(u)$ is $[-1, 1]$, and give a reasonable upper estimate for $x(1)$. (What is its true value, from a 21-st century perspective?)

(b) Show $x(u)$ is continuous, including the endpoints.

(c) Show $x(u)$ is monotonic.

(d) Where is $x(u)$ convex, and where concave?

(e) Show it has a twice-differentiable inverse function $u(x)$, which satisfies the differential equation $u'' + u = 0$, on its domain.

(f) Show that if you try to make the unit circle definition of $\sin x$ rigorous, the preceding is what you are led to: the basic function is actually $\sin^{-1} x$, not $\sin x$. (You will need the usual calculus integral for arclength.)

21-2 Fill in the details of the straightforward approach sketched in the text to proving that (12) converges, essentially by comparing it to the series $\sum_0^\infty A_n$.

(a) Show the series $\sum A_n$ converges by showing it satisfies the hypotheses of Cauchy's test (7.6):

(i) the A_n alternate in sign: $A_{2n} > 0$, $A_{2n+1} < 0$;

(ii) $|A_n|$ is a decreasing sequence, and $|A_n| \to 0$.

(b) Then prove (15) by estimating the difference between the two sides: break the difference into two parts, both of which can be shown to be small.

21-3 Suppose $\lim_{x \to \infty} f(x) = 0$, $\int_a^\infty f'(x)\,dx$ is absolutely convergent, and $f'(x)$ is continuous for $x \geq a$. Prove that $\int_a^\infty f(x) \sin x\,dx$ converges.

21-4 Prove that $\Gamma(x)$ is continuous at $x = 1^+$.

(Method: consider $|\Gamma(1 + h) - \Gamma(1)|$. To estimate it, break up the interval $[0, \infty)$ into two parts. Remember that you can estimate differences of the form $|f(b) - f(a)|$ by using the Mean-value Theorem, if $f(x)$ is differentiable on the relevant interval.)

21-5 Prove that $\Gamma(x)$ is continuous for $x > 0$ (cf. Problem 21-4).

Answers to Questions

21.1

1. (a) and (b): $\displaystyle \int \frac{dx}{x^p} = \frac{x^{1-p}}{1-p}$, if $p \neq 1$;

For (3): $\displaystyle \lim_{R \to \infty} \frac{x^{1-p}}{1-p}\bigg]_1^R = \lim_{R \to \infty} \frac{R^{1-p} - 1}{1 - p} = \begin{cases} 1/(p-1), & \text{if } p > 1; \\ \infty, & \text{if } p < 1. \end{cases}$

For (4): $\displaystyle \lim_{u \to 0^+} \frac{x^{1-p}}{1-p}\bigg]_u^1 = \lim_{u \to 0^+} \frac{-u^{1-p} + 1}{1 - p} = \begin{cases} 1/(1-p), & \text{if } p < 1; \\ \infty, & \text{if } p > 1. \end{cases}$

In both cases, for $p = 1$ we get divergence, because

$$\int \frac{dx}{x} = \ln x \quad \text{and} \quad \lim_{R \to \infty} \ln R = \infty, \quad \lim_{u \to 0^+} - \ln u = \infty.$$

(c) Since $1/x^{1/p}$ and $1/x^p$ are inverse functions, a reflection about the diagonal exchanges the two graphs; the two areas represented by the integrals are interchanged, except for a square of side one that must be added to one of them.

2. $\displaystyle \int_0^R e^{kt}\, dt = \frac{e^{kt}}{k}\bigg]_0^R = \frac{e^{kR} - 1}{k}, \quad \text{and}$

$$\lim_{R \to \infty} \frac{e^{kR} - 1}{k} = \begin{cases} -1/k, & \text{if } k < 0; \\ \infty, & \text{if } k > 0. \end{cases}$$

For $k = 0$, the integral diverges, by inspection.

3. (a) $\displaystyle \int_u^4 \frac{dx}{\sqrt{x - 2}} = 2\sqrt{x - 2}\bigg]_u^4 \to 2\sqrt{2} \;\text{ as } u \to 2^+.$

(b) $\displaystyle \int_0^u \frac{dt}{\sqrt{1 - t^2}} = \sin^{-1} t\bigg]_0^u \to \frac{\pi}{2} \;\text{ as } u \to 1^-.$

(c) $\displaystyle \int_0^R \frac{t^2\, dt}{1 + t^3} = \tfrac{1}{3} \ln(1 + t^3)\bigg]_0^R \to \infty \;\text{ as } R \to \infty; \text{ divergent.}$

21.2

1. (a) Using the comparison test 21.2B for convergence:
$$t^2 < e^{t/2}, \qquad\qquad \text{for } t \gg 1, \ (\text{cf. } 20.6,(24)); \quad \text{thus}$$
$$\int_a^\infty t^2 e^{-t}\, dt \ < \ \int_a^\infty e^{-t/2}, \qquad \text{for } a \gg 1.$$
Since the right-hand integral converges, so does the left; and therefore, by the Tail-convergence Theorem 21.2A, even when $a = 0$.

(Any k value in the range $-1 < k < 0$ would do just as well.)

(b) Using the comparison test 21.2B for convergence:
$$|\ln x| < x^{-1/4}, \qquad\qquad \text{for } x \underset{\neq}{\approx} 0^+ \ (\text{cf. Note below}); \text{ thus}$$
$$\int_{0+}^b |\ln x|/\sqrt{x}\, dx \ < \ \int_{0+}^b dx/x^{3/4}, \qquad \text{for } b \underset{\neq}{\approx} 0^+.$$
Since the right-hand integral converges, so does the left; and therefore, by the Tail-convergence Theorem 21.2A, it also converges when $b = 1$.

Note: any k value such that $-1/2 < k < 0$ would work. The first inequality is equivalent to $x^{1/4}|\ln x| < 1$ for $x \approx 0$, $x > 0$, which by the Sequence Location Theorem for functions follows from $\lim_{x\to 0+} x^{1/4}|\ln x| = 0$, which can in turn be justified by l'Hospital's rule applied to $\lim \ln x / x^{-1/4}$ (or use 20.6 (23)).

The absolute value is needed, since as it is stated the Comparison Theorem requires non-negative integrands.

2. (a) Calling the integrand $f(x)$, we have $f(x) \sim 1/x^2$. Since $\int_1^\infty dx/x^2$ converges, so does $\int_0^\infty f(x)\, dx$ by the asymptotic comparison test 21.2C and the Tail-convergence Theorem 21.2A.

(b) Since the given integral is improper at $x = 1^-$, we make the change of variable $x = 1 - u$ so the integral becomes improper at $u = 0^+$ instead, which is an easier point to work with. The new integral is
$$-\int_{0+}^1 f(u)\, du, \quad \text{where} \quad f(u) = \frac{1-u}{\sqrt{u(3 - 3u + u^2)}} \ \sim \ \frac{1}{\sqrt{3u}}, \ \text{as } u \to 0^+;$$
since $\int_0^1 du/\sqrt{u}$ converges (Ex. 21.1A), so does the original integral, by the asymptotic comparison test 21.2C.

3. By the Interval Addition Theorem 19.5,
$$\int_a^R f(x)\, dx = \int_a^b f(x)\, dx + \int_b^R f(x)\, dx;$$
take limits of both sides as $r \to \infty$: by the Sum Theorem for limits of functions (of the variable R), the left-hand integral has a limit if and only if the second integral on the right does; so one integral converges if and only if the other does.

4. (a) Let $L = \lim_{x \to \infty} f(x)$. We use indirect reasoning.

If for some x_0, we have $f(x_0) > L$, then $f(x) \geq f(x_0) > L$ for $x \geq x_0$, hence $\lim_{x \to \infty} f(x) \geq f(x_0) > L$ by the Limit Location Theorem for functions; this is a contradiction. $\quad\square$

(b) For $n \gg 1$, the sequence $f(n)$ is increasing and $f(n) \leq B$. Thus by the Completeness Property (p.30), it has a limit L, and $L \leq B$ by the Limit Location Theorem for sequences.

To finish, we show $L = \lim_{x \to \infty} f(x)$. Namely, given $\epsilon > 0$, we have
$$L - \epsilon \leq f(n) \leq L, \qquad \text{for } n > \text{ some } N .$$
Now if $x > N$, let n' be some integer $> x$; then since f is increasing,
$$L - \epsilon \leq f(N) \leq f(x) \leq f(n') \leq L ,$$
which shows that $f(x) \underset{\epsilon}{\approx} L$ for $x > N$, i.e., that $\lim_{x \to \infty} f(x) = L$. $\quad\square$

21.3

1. The proof is identical to that of Question 21.2/1a, substituting $x - 1$ for 2 throughout.

2. The integral $\int_{0+}^{1} t^{x-1} e^{-t}\, dt$ is improper only if $x < 1$. In this case, we use the comparison
$$t^{x-1} e^{-t} \leq t^{x-1} \qquad \text{for } t \geq 0 ;$$

the comparison test 21.2B then shows the integral converges if $\int_{0+}^{1} t^{x-1}\, dt$ converges, which by (4) is so if $x - 1 > -1$, i.e., if $x > 0$.

21.4

1. (a) Divergent: since $\int_{0}^{R} \sin x\, dx = \begin{cases} 2, & R = \pi, 3\pi, 5\pi, \ldots; \\ 0, & R = 2\pi, 4\pi, 6\pi, \ldots, \end{cases}$ it follows

that $\lim_{R \to \infty} \int_{0}^{R} \sin x\, dx$ does not exist.

(b) Absolutely convergent, since $\int_{0}^{\infty} e^{-x} |\sin x|\, dx \leq \int_{0}^{\infty} e^{-x}\, dx.$

(c) Absolutely convergent, since $\int_{1}^{\infty} \frac{|\sin x|}{x^2}\, dx \leq \int_{1}^{\infty} \frac{dx}{x^2}.$

2. (a) diverges for $k \geq 0$, since the successive arched areas are not decreasing in size;

(b) absolutely convergent for $k < -1$, by the same reasoning as in (1c) above (which was the case $k = -2$);

(c) conditionally convergent for $-1 \leq k < 0$ (the reasoning would be as in Example 21.4).

22

Sequences and Series of Functions

22.1 Pointwise and uniform convergence.

In the early part of the book we defined and studied the convergence of infinite sequences $\{x_n\}$ and infinite series $\sum a_n$ of real numbers. In this chapter, we will generalize this work by considering the convergence of sequences and series of *functions*.

Actually, we have already gotten a good start on this in our work with power series $\sum a_n x^n$, since these can be thought of as an infinite sum of the simple functions $a_n x^n$. In Chapter 8 we considered the convergence and algebraic properies of such series, and in Chapter 17 the problem of representing a given function $f(x)$ by a power series:

$$f(x) = \sum a_n x^n .$$

We need to extend this work in two ways. First, by connecting it with differentiation and integration; for instance, can we differentiate or integrate a power series term-by-term? That is, if $f(x)$ is as above, will the equations

$$f'(x) = \sum n a_n x^{n-1}$$

and

$$\int_a^b f(x)\, dx = \sum \int_a^b a_n x^n dx$$

also be valid? Many applications of analysis to the real world use these.

Second, we need to consider more general series of functions. For example, periodic functions are usually represented not by power series but by Fourier series, which have the form

$$f(x) = a_0 + \sum a_n \cos nx + b_n \sin nx .$$

Similarly, in numerical analysis, one often represents a function by a series of Tchebyshev or Legendre polynomials. The solutions to the most common partial differential equations of physics are often expressed as series of Bessel functions, or Legendre polynomials composed with trigonometric functions.

All this suggests that we should be studying infinite series $\sum u_n(x)$ whose terms are just functions of some unspecified type. This will cover all the series of the past millenium, as well as those of the next.

The analogy with our earlier work with real numbers suggests that we should begin by first studying the convergence of a *sequence* of functions $f_n(x)$ to its limit:

$$f_n(x) \to f(x) \, .$$

This will be our starting point. But going from sequences of numbers to sequences of functions brings surprises; we illustrate.

Example 22.1A   Let $f_n(x) = x^n$. What is $\lim_{n\to\infty} f_n(x)$ on $I = [0,1]$?

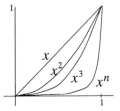

Solution.   By Section 3.4, $\lim_{n\to\infty} x_0^n = \begin{cases} 0, & \text{if } 0 \le x_0 < 1 \,; \\ 1, & \text{if } x_0 = 1 \,. \end{cases}$

So it is natural to say that on $[0,1]$,

$$\lim_{n\to\infty} f_n(x) = f(x), \quad \text{where } f(x) = \begin{cases} 0, & \text{if } 0 \le x < 1 \,; \\ 1, & \text{if } x = 1 \,. \end{cases}$$

The sequence of continuous functions converges to a discontinuous function! This suggests we had better not rely too heavily on naive intuition. We begin therefore by writing down the definition of convergence implicitly used in the above example; since there is more than one notion of convergence, for clarity we will often refer to this one by the adjective *pointwise*; the literature also uses "ordinary" convergence, or most frequently, no adjective at all.

Definition 22.1A   Let $f_n(x)$, $n = 0, 1, 2, \ldots$ be a sequence of functions, all defined on the interval I. By definition,

$f_n(x)$ **converges (pointwise) on** I $\Leftrightarrow$ $f_n(x_0)$ converges for all $x_0 \in I$.

If the sequence $f_n(x)$ converges, its *limit* is the function $f(x)$ defined by

$$f(x_0) = \lim_{n\to\infty} f_n(x_0) \, , \qquad x_0 \in I \, .$$

The notation is similar to that for sequences of real numbers; for example:

$$f(x) = \lim_{n\to\infty} f_n(x), \quad x \in I; \qquad f_n(x) \to f(x) \quad \text{on } I.$$

Example 22.1B   Determine $\lim_{n\to\infty} \dfrac{n}{1 + nx}$, on the interval $(0, \infty)$.

Solution.   For every $x_0 > 0$, we have

$$\frac{n}{1 + nx_0} = \frac{1}{1/n + x_0} \to \frac{1}{x_0}, \quad \text{since } 1/n \to 0.$$

Therefore by the definition of convergence, $\lim_{n\to\infty} \dfrac{n}{1 + nx} = \dfrac{1}{x}$, on $(0, \infty)$.   □

There is another notion of convergence for a sequence of functions—*uniform convergence*—which is more demanding than pointwise convergence; it is the main new idea in this chapter.

Definition 22.1B Let $f_n(x)$ be a (pointwise) convergent sequence of functions on I, and let
$$f(x) = \lim_{n \to \infty} f_n(x) \quad \text{on } I .$$

We say $f_n(x)$ **converges uniformly** on I to $f(x)$, (noted : $f_n(x) \rightrightarrows f(x)$ on I) if,

(1) $\qquad\qquad$ given $\epsilon > 0$, $\quad f_n(x) \underset{\epsilon}{\approx} f(x)$ on I, $\quad$ for $n \gg 1$,

where "for $n \gg 1$" means: for $n >$ some N_ϵ depending only on ϵ, *and not on x*.

Remarks.

1. Despite appearances, to prove $f_n(x) \rightrightarrows f(x)$, it is not necessary to give a separate proof that $f_n(x) \to f(x)$, since the approximation (1) automatically implies that $f(x)$ is the pointwise limit of the $f_n(x)$.

Namely, given $\epsilon > 0$, there is an N depending only on ϵ such that
$$f_n(x) \underset{\epsilon}{\approx} f(x) \text{ on } I \quad \text{for } n > N;$$
$$\Rightarrow \quad f_n(x_0) \underset{\epsilon}{\approx} f(x_0), \text{ for } x_0 \in I, \; n > N;$$
$$\Rightarrow \quad \lim_{n \to \infty} f_n(x_0) = f(x_0), \quad x_0 \in I .$$

2. If we defined pointwise convergence from scratch by an ϵ-N statement, as in (1), with the N made explicit by replacing "for $n \gg 1$" with the following line using N, it would look exactly the same; the difference would be that N would depend not only on ϵ, but on x as well.

What distinguishes therefore the two types of convergence is their answer to this question: how far out in the sequence must you go for the approximation $f_n(x) \underset{\epsilon}{\approx} f(x)$ to be valid?

For pointwise convergence, the answer depends on x and ϵ; for uniform convergence an answer can be given which does not depend on x. To emphasize this, one often says the series "converges uniformly with respect to x", or that "the convergence is uniform over x".

Uniform convergence is a stronger hypothesis than pointwise convergence, and if it holds, we will be able to draw stronger conclusions.

Example 22.1C Show that $\displaystyle \lim_{n \to \infty} \frac{n}{1 + nx} \rightrightarrows \frac{1}{x}$ on $[1, \infty)$, but not on $(0, \infty)$.

Solution. To show uniform convergence, we show (1) holds on $[1, \infty)$.
$$\left| \frac{n}{1 + nx} - \frac{1}{x} \right| = \frac{1}{(1 + nx)x} < \frac{1}{n}, \quad \text{since } x \geq 1;$$
$$< \epsilon \quad \text{if } n > 1/\epsilon .$$

To show the convergence is not uniform on $(0, \infty)$, choose say $\epsilon = 1$. Then referring to the above calculation of the difference, it is not true that
$$\frac{1}{(1 + nx)x} < \epsilon = 1 \quad \text{on } (0, \infty) \text{ for } n \gg 1,$$
since no matter how large n is, the inequality fails if say $x = 1/n$. $\qquad\square$

Example 22.1D   Show the convergence $x^n \to 0$ on $[0, 1)$ is not uniform.

Solution.   Choose say $\epsilon = \frac{1}{2}$. Then referring to (1), it is *not* true that

(2) $$x^n \underset{\epsilon}{\approx} 0 \quad \text{on } [0, 1), \text{ for } n \gg 1;$$

namely, since $x^n \to 1$ as $x \to 1^-$, it follows that no matter what n is,

$$x^n > \epsilon = \tfrac{1}{2}, \quad \text{for } x \approx 1^-,$$

which contradicts (2).   □

We now translate the two definitions of convergence from the language of sequences of functions to infinite series of functions.

Given such an infinite series $\sum u_k(x)$, we let as in Chapter 7

(3) $$s_n(x) = u_0(x) + u_1(x) + \ldots + u_n(x)$$

denote the n-th partial sum of the series.

Definition 22.1C   Let $u_k(x)$ be defined on I, for $k = 0, 1, 2, \ldots$, and let $s_n(x)$ be as in (3). Then by definition,

$$\sum u_k(x) \text{ converges (resp. *pointwise, uniformly*)}$$

$$\Leftrightarrow \quad s_n(x) \text{ converges (resp. pointwise, uniformly).}$$

If the series converges, its *sum* is the function $f(x)$ defined by

$$f(x_0) = \lim_{n \to \infty} s_n(x_0) = \sum u_k(x_0), \quad x_0 \in I.$$

Again, insofar as it is ordinary convergence we are dealing with, there is no change from what we did when working with power series: to say $\sum u_k(x) = f(x)$ on I simply means according to the above definition that this relation is true for every $x_0 \in I$. But the convergence of $\sum u_k(x)$ may not be uniform on I.

Example 22.1E   Show that the power series $1 + x + \dfrac{x^2}{2!} + \ldots + \dfrac{x^n}{n!} + \ldots$

(a) converges uniformly to e^x on any interval $[-R, R]$;

(b) converges pointwise to e^x but not uniformly on $(-\infty, \infty)$.

Solution.   (a) By Taylor's Theorem with remainder (17.2), we know that for any given $x \in [-R, R]$, there is a c such that

$$e^x = 1 + x + \frac{x^2}{2!} + \ldots + \frac{x^n}{n!} + \frac{e^c x^{n+1}}{(n+1)!}, \quad -R \le x \le R, \ 0 < |c| < |x|.$$

Therefore, since $0 < e^c < e^R$ when $|c| < R$, we have

$$\left| e^x - \left(1 + x + \frac{x^2}{2!} + \ldots + \frac{x^n}{n!} \right) \right| \le \frac{e^R R^{n+1}}{(n+1)!} < \epsilon, \quad \text{for } n \ge \text{ some } N_\epsilon,$$

since for any R,

$$\lim_{n \to \infty} \frac{e^R R^{n+1}}{(n+1)!} = 0 \quad \text{(cf. Example 17.4).} \qquad \square$$

(b)   According to the remark following Definition 22.1B, since we know the series converges uniformly on $[-R, R]$, it also converges pointwise there (which we already knew from Chapter 17), and hence it converges pointwise on $(-\infty, \infty)$. However it does not converge uniformly on $(-\infty, \infty)$. For, reasoning indirectly, if it did, then given $\epsilon > 0$, we would have for large n a relation

(4)
$$e^x \underset{\epsilon}{\approx} 1 + x + \frac{x^2}{2!} + \ldots + \frac{x^n}{n!} \quad \text{for all } x \in \mathbb{R}.$$

But this is impossible, no matter how large n is, since the right side of (4) is like $x^n/n!$ when $x \gg 1$, whereas the left side is like e^x, which is much greater. To say this more formally, if (4) were true for some n, then for $x > 1$, say, we can divide through by x^n (which is also > 1) and preserve the approximation:

$$\frac{e^x}{x^n} \underset{\epsilon}{\approx} \frac{1}{x^n} + \frac{1}{x^{n-1}} + \frac{1}{2!x^{n-2}} + \ldots + \frac{1}{n!};$$

but this is impossible, since the left side tends to ∞ as $x \to \infty$, whereas the right side has the limit $1/n!$.   □□

It may seem paradoxical, that a property of a function or sequence of functions can hold for all intervals $[-R, R]$, yet not hold on $(-\infty, \infty)$. But consider the simple property of boundedness: the function x is bounded on every $[-R, R]$, but not on $(-\infty, \infty)$. This is typical behavior for global properties, and boundedness and uniformity of convergence are global; by contrast, ordinary convergence is a pointwise property, so that if it holds over all $[-R, R]$, it automatically holds for $(-\infty, \infty)$ as well.

Questions 22.1

1.   Let $f_n(x) = \dfrac{1}{x + n}$, $x \geq 0$, $n \geq 1$. Find $f(x) = \lim\limits_{n \to \infty} f_n(x)$, and show the convergence is uniform on $[0, \infty)$.

2.   Let $f_n(x) = \sin nx$. Show that $f_n(x)$ does not converge pointwise on $\mathbb{R}$.

3.   Let $f_n(x) = e^{-nx}$;   find $f(x) = \lim\limits_{n \to \infty} f_n(x)$ on $[0, \infty)$,   and show the convergence is uniform on $[R, \infty)$ for any $R > 0$.

4.   Let $f_n(x) = x/n$. Find $f(x) = \lim\limits_{n \to \infty} f_n(x)$, and prove the convergence is uniform on $[-R, R]$ for every $R > 0$, but not on $(-\infty, \infty)$.

5.   Prove: if a tail $\sum_N^\infty u_k(x)$ converges uniformly, then $\sum_0^\infty u_k(x)$ does also.

22.2   Criteria for uniform convergence.

We begin with an almost trivial test for uniform convergence of a sequence.

Theorem 22.2A   Elementary criterion for uniform convergence.

On an interval I, suppose that for all n the functions $f_n(x)$ and $f(x)$ are defined, and there is a sequence of constants B_n such that

(5) $|f(x) - f_n(x)| \leq B_n,$ and $B_n \to 0.$

Then $f_n(x) \rightrightarrows f(x)$ on $I.$

Proof.   Since $\lim B_n = 0 \Rightarrow B_n < \epsilon$ for $n \gg 1$, it follows that (1) holds.   □

Example 22.2A Show $\dfrac{x+n}{1+nx} \rightrightarrows \dfrac{1}{x}$ on $[1, a].$

Solution. $\left| \dfrac{x+n}{1+nx} - \dfrac{1}{x} \right| = \dfrac{x^2 - 1}{x(1+nx)} < \dfrac{a^2}{n},$ since $1 \leq x \leq a.$

We can take $B_n = a^2/n$ in (5), showing the convergence is uniform.   □

Example 22.2B   Show $e^{x/n} \rightrightarrows 1$ on $[0, 1]$.

Solution.   We use the Mean-value Theorem 15.1(1) to estimate $|f_n(x) - f(x)|$:

$$\left| e^{x/n} - e^0 \right| = e^c |x/n|,   \text{for some } 0 < c < x/n ;$$

$$< e/n ,   \text{since } 0 \leq x \leq 1 .$$

We can take $B_n = e/n$ in (5), which proves the uniform convergence.   □

The most frequently used criterion for uniform convergence is the next one for the uniform convergence of a series. It is easy to remember and easy to apply.

Theorem 22.2B   Weierstrass M-test (uniform convergence of series).

If for $k \geq 0$, we have $|u_k(x)| \leq M_k$ on I, and the series $\displaystyle\sum_0^k M_k$ converges,

then $\displaystyle\sum_0^\infty u_k(x)$ converges uniformly on $I.$

Proof.   For each $x_0 \in I$, the series $\sum u_k(x_0)$ is absolutely convergent, by the comparison test 7.2D; thus it is convergent by Theorem 7.3. The series $\sum_0^\infty u_k(x)$ therefore has a sum $f(x)$ in I, to which it converges pointwise:

$$f(x) = \sum_0^\infty u_k(x) ,   x \in I .$$

Let $s_n(x)$ be the sequence of partial sums. To show $s_n(x) \rightrightarrows f(x)$, we consider the difference, and use the infinite triangle inequality (Exercise 7.3/1),

(6)
$$\left| \sum_0^\infty a_k \right| \leq \sum_0^\infty |a_k| \; .$$

Following this plan, we have

$$|f(x) - s_n(x)| \;=\; \left| \sum_{n+1}^\infty u_k(x) \right| \; ;$$

$$\leq \; \sum_{n+1}^\infty |u_k(x)| \; , \qquad \text{by (6);}$$

$$\leq \; \sum_{n+1}^\infty M_k \; , \qquad \text{by the Comparison Theorem 7.2D.}$$

Finally, we have by definition of the sum of a convergent series,

$$\lim_{n \to \infty} \sum_{n+1}^\infty M_k \;=\; \lim_{n \to \infty} \left(\sum_1^\infty M_k - \sum_1^n M_k \right) \;=\; 0 \; .$$

So the elementary criterion 22.2A shows that $s_n(x) \rightrightarrows f(x)$.   □

Example 22.2C   Show $\displaystyle\sum_1^\infty \frac{\cos nx}{n^2}$ converges uniformly on $(-\infty, \infty)$.

Proof.   We have $\left| \dfrac{\cos nx}{n^2} \right| \leq \dfrac{1}{n^2}$ for all $n \geq 1$ and all x, and $\sum \dfrac{1}{n^2}$ converges. The Weierstrass M-test conditions are satisfied, so the convergence is uniform. □

Theorem 22.2C   Uniform convergence of power series

If $\sum a_n x^n$ has radius of convergence R, then the series converges uniformly on every interval $[-L, L]$, where $0 \leq L < R$.

Proof.   We use the Weierstrass M-test. We have
$$|a_n x^n| \leq |a_n| L^n, \quad \text{for } |x| \leq L \; ;$$
and we know that $\sum |a_n| L^n$ converges, because by the Radius-of-convergence Theorem 8.1, a power series is absolutely convergent for $|x| < R$. So the M-test applies, and $\sum a_n x^n$ converges uniformly on $[-L, L]$.   □

For any $L < R$, the power series converges uniformly on the compact symmetric interval $[-L, L]$, but this does not mean it converges uniformly on $(-R, R)$. The geometric series provides a counterexample, since its convergence on $(-1, 1)$ is not uniform (see Exercise 22.2/3b).

Note that in the proof above one cannot take $L = R$, since the series might not converge for $x = R$.

Questions 22.2

1. Redo Questions 22.1/1, 3, 4 using the elementary criterion 22.2A; that is, use it to prove the uniform convergence of the sequences:

(a) $\dfrac{1}{x+n}$ on $[0, \infty)$;   (b) e^{-nx} on $[R, \infty)$, $R > 0$;   (c) $\dfrac{x}{n}$ on $[-R, R]$.

2. Prove the following series converge uniformly on the given interval:

(a) $\sum x^n \sin nx$, $[-R, R]$, $0 < R < 1$;   (b) $\sum \dfrac{\sin nx}{n!}$, $(-\infty, \infty)$.

22.3   Continuity and uniform convergence.

The rest of this chapter is devoted to three theorems in analysis which are used all the time and whose hypotheses require uniform convergence.

We take up the Continuity Theorem first. Exs. 22.1A,D show $\{x^n\}$ converges to a discontinuous limit, but not uniformly; it cannot be uniform, because

   the uniform limit or sum of continuous functions is continuous.

Theorem 22.3   Continuity of uniform limits and sums.

(a) Let $f_n(x)$ be continuous on I for $n \geq 0$; suppose $f_n(x) \rightrightarrows f(x)$ on I. Then $f(x)$ is continuous on I.

(b) Let $u_k(x)$ be continuous on I for $k \geq 0$, and suppose $\sum u_k(x)$ converges uniformly on I. Then the sum $f(x) = \sum u_k(x)$ is continuous on I.

Proof. Statement (b) follows immediately from (a) by applying (a) to the sequence (3) of partial sums $s_n(x)$ of the series; these are continuous since they are *finite* sums of continuous functions.

To prove (a), we show $f(x)$ is continuous at an arbitrary point x_0 of I (assume x_0 is not an endpoint of I; if it is, the proof that follows must be modified by using $x \approx x_0^+$ or $x \approx x_0^-$).

Given $\epsilon > 0$, choose a value N large enough so that

(7) $f(x) \underset{\epsilon}{\approx} f_N(x)$,   for all $x \in I$.

This can be done since $f_n \rightrightarrows f$ on I. Since $f_N(x)$ is continuous at x_0,

$$f_N(x) \underset{\epsilon}{\approx} f_N(x_0), \qquad \text{for } x \approx x_0 .$$

Combining this with (7) we get a chain of approximations

$$f(x) \underset{\epsilon}{\approx} f_N(x) \underset{\epsilon}{\approx} f_N(x_0) \underset{\epsilon}{\approx} f(x_0), \quad \text{for } x \in I, \ x \approx x_0.$$

Therefore

$$f(x) \underset{3\epsilon}{\approx} f(x_0) \qquad \text{for } x \approx x_0 ,$$

which shows that $f(x)$ is continuous at x_0.   □

Note how the proof gets from $f(x)$ to $f(x_0)$ by using $f_N(x)$ to make a short detour: it's like crossing a river by stepping on two rocks a little upstream. It will be be a model for similar arguments involving uniformity. Study it carefully.

Remarks.   The proof shows that if we only assume the $f_N(x)$ or $u_k(x)$ are continuous at x_0, it still follows that the limit $f(x)$ is at least continuous at x_0.

The statement (b) should be looked on as the correct generalization to infinite sums of the fact that the sum of a finite number of continuous functions is continuous.

Corollary 22.3   *A power series $\sum a_n x^n$ represents a continuous function inside its radius of convergence R:*
$$f(x) \;=\; \sum a_n x^n \;\; \text{is continuous on } (-R, R)\ .$$

Proof.   Given any $x_0 \in (-R, R)$, choose L so that $|x_0| < L < R$.

The power series converges uniformly on $[-L, L]$ by Theorem 22.2C, so its sum $f(x)$ is continuous on $[-L, L]$, therefore at $x_0 \in [-L, L]$. $\square$

> The argument is simple, but a bit elusive; study it. Note that we cannot apply the preceding Continuity Theorem directly to $(-R, R)$, since the power series does not converge uniformly on this interval. Instead, we show that the sum $f(x)$ is continuous on $[-L, L]$ for every $L < R$; therefore it is continuous on $(-R, R)$.
>
> Unlike continuity, uniform convergence on $[-L, L]$ for every $L < R$, does not imply it on $(-R, R)$. The difference is that continuity is local, while uniform convergence is a global property.

Students in elementary courses often take Cor. 22.3 for granted as "obvious", because "each term $a_k x^k$ is continuous". But Fourier series can very well have discontinuous sums, even though each term $a_n \sin nx$ is continuous. The laws of finite sums do not extend automatically to infinite sums.

Still, the students are in good company: the early analysts also thought that the infinite sum of continuous functions would be continuous. The discovery in the early 1800's that this was not so sent shock waves through the community; it forced mathematicians to extend the notion of function to include discontinuous functions, and in general to give a stronger foundation to analysis by making careful definitions and giving proofs of the theorems.

Questions 22.3

1. Where does $\displaystyle\sum_{1}^{\infty} \frac{x^{2n}}{2^n n}$ converge uniformly? Where is its sum continuous?

2. Prove that $\displaystyle\sum_{1}^{\infty} \frac{\cos nx}{n\sqrt{n+1}}$ represents a continuous function for all x.

3. Prove that $\displaystyle\sum_{0}^{\infty} \frac{1}{1 + n^2 x}$ represents a continuous function for $x > 0$.

22.4   Integration term-by-term.

We study the integration of series; as usual, we begin with sequences.

Theorem 22.4A   Integration of a uniform limit.

If on a finite interval $[a, b]$, the functions $f_n(x)$ are continuous for every n, and $f_n(x) \rightrightarrows f(x)$, then

$$(8) \qquad \int_a^b f_n(x)\, dx \;\to\; \int_a^b f(x)\, dx \; .$$

Proof.   The integrals on the left exist since the $f_n(x)$ are continuous; the integral on the right exists since $f(x)$ is the uniform limit of continuous functions, therefore itself continuous, by Theorem 22.3.

To prove (8), we are given $\epsilon > 0$, and estimate the difference.

$$\left| \int_a^b f(x)\, dx - \int_a^b f_n(x)\, dx \right| = \left| \int_a^b \big(f(x) - f_n(x) \big)\, dx \right|$$

$$\leq \int_a^b |f(x) - f_n(x)|\, dx, \quad \text{by 19.4C;}$$

$$\leq \epsilon\,(b - a), \qquad \text{if } n \gg 1 \; ,$$

since the hypothesis $f_n(x) \rightrightarrows f(x)$ on $[a, b]$ implies by (1) that

$$|f(x) - f_n(x)| \; < \; \epsilon \quad \text{on } [a, b], \quad \text{for } n \gg 1 \; .$$

This shows the two integrals in (8) are within $\epsilon(b - a)$, so we are done.   □

Theorem 22.4B   Term-by-term integration of a series.   *Assume*

  (i) *for each k, the function $u_k(x)$ is continuous on $[a, b]$;*

  (ii) $\sum_0^\infty u_k(x) = f(x)$,   *uniformly on $[a, b]$.*

Then the series can be integrated term-by-term over $[a, b]$, i.e.,

$$(9) \qquad \sum_0^\infty \int_a^b u_k(x)\, dx \; = \; \int_a^b f(x)\, dx \; .$$

Proof.   Use first the definition of uniform convergence of a series and then Theorem 22.4A; as $n \to \infty$, we get, in turn,

$$\sum_0^n u_k(x) \; \rightrightarrows \; f(x) \quad \text{in } [a, b];$$

$$\int_a^b \sum_0^n u_k(x)\, dx \; \to \; \int_a^b f(x)\, dx, \quad \text{by 22.4A;}$$

$$\sum_0^n \int_a^b u_k(x)\, dx \; \to \; \int_a^b f(x)\, dx,$$

by the Linearity Theorem for integrals 19.5A. By definition, this last line is what is meant by (9).   □

Corollary 22.4   Term-by-term integration of power series.

A power series can be integrated term-by-term inside its open interval of convergence $(-R, R)$; that is, if

$$\sum_{0}^{\infty} a_n x^n = f(x) , \qquad -R < x < R ,$$

then
$$\sum_{0}^{\infty} \frac{a_n}{n+1} x^{n+1} = \int_{0}^{x} f(t)\, dt , \qquad -R < x < R .$$

Proof.   The power series converges uniformly on $[0, x]$ or $[x, 0]$, according to Theorem 22.2C; by the preceding theorem therefore it can be integrated term-by-term on $[0, x]$ or $[x, 0]$.   □

Note that the corollary only claims that the integrated series still converges on $(-R, R)$; we have not proved that R is its exact radius of convergence. This is true, however; it will follow from the theorems about term-by-term differentiation.

Examples 22.4   Apply term-by-term integration if possible to

(a) $\dfrac{1}{1 + x} = 1 - x + x^2 - \ldots + (-1)^n x^n + \ldots ,\quad |x| < 1;$

(b) $x = \dfrac{\pi}{2} - \dfrac{4}{\pi} \left(\cos x + \dfrac{\cos 3x}{3^2} + \dfrac{\cos 5x}{5^2} + \ldots \right) ,\quad 0 \le x \le \pi .$

Solution.   (a) By Corollary 22.4, term-by-term integration is possible; we get
$$\ln(1 + x) = x - \tfrac{1}{2}x^2 + \tfrac{1}{3}x^3 - \ldots, \quad |x| < 1.$$

To do this previously, we had to estimate the remainder term, either by Taylor's formula, as in 17.2, or by integrating the remainder term for the geometric series, as in 4.2. This explicit estimating of remainders is avoided now, because the notion of uniform convergence gives us more powerful theorems.

(b)   Though not a power series, it converges uniformly by the Weierstrass M-test (see Example 22.2C), so it can be integrated term-by-term. We get

$$\frac{x^2}{2} = \frac{\pi x}{2} - \frac{4}{\pi} \left(\sin x + \frac{\sin 3x}{3^3} + \frac{\sin 5x}{5^3} + \ldots \right) ,\quad 0 \le x \le \pi .$$

General theorems about Fourier series show that (b) is valid, but only on the interval $[0, \pi]$. But since the series converges uniformly for all x by the M-test, Theorem 22.3 shows its sum $f(x)$ is defined and continuous for all x; moreover, its sum is even and periodic with period 2π, since the series itself is. Together, these facts show the sum $f(x)$ of the series (b) must be the sawtooth function whose graph is illustrated.

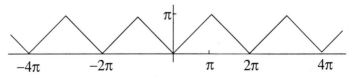

Questions 22.4

1.   Prove Theorem 22.4A, assuming only that $f_n(x)$ is integrable; you will need an extra hypothesis, however. Why wasn't it needed for Theorem 22.4A?

2   Prove $\tan^{-1} x \; = \; x - \frac{1}{3}x^3 + \frac{1}{5}x^5 - \ldots$ over a suitable interval; which?

22.5   Differentiation term-by-term.

The theorems on differentiating sequences and series are a little trickier than those on integration, and the hypotheses are a little more awkward to use. The essential point to bear in mind is that it is the *differentiated* sequence or series that must converge uniformly, not the original one.

To see why, consider the basic question: on an interval I,
$$\text{if } \; f_n(x) \to f(x), \;\; \text{does} \;\; f_n'(x) \to f'(x) \; ?$$
The answer is certainly *no*, for even if we assume that $f_n(x)$ is uniformly close to $f(x)$ over the whole interval I, this says nothing about how the derivative behaves. If the function $f_n(x)$ lies within an ϵ-band of $f(x)$, but its graph wiggles up and down rapidly, then $f_n'(x)$ will not be close to $f'(x)$ (see Question 1.) We must control how the sequence of derivatives $f_n'(x)$ converges.

Theorem 22.5A   Differentiation of a limit of functions.

Suppose that for all $n \geq 0$ the functions $f_n(x)$ have continuous derivatives on a given interval I, and on this interval,
$$f_n(x) \; \to \; f(x), \quad \text{and} \quad f_n'(x) \; \rightrightarrows \; g(x) \; .$$
Then $f(x)$ is differentiable, and $f'(x) = g(x)$ on I.

Proof.   On the interval I we have by hypothesis $f_n'(x) \rightrightarrows g(x)$. We fix some arbitrarily chosen point $a \in I$ and integrate these functions from a to x; then by Theorem 22.4A on term-by-term integration of sequences, we see that
$$\int_a^x f_n'(t)\, dt \; \to \; \int_a^x g(t)\, dt \; .$$
Applying the First Fundamental Theorem 20.1 to the left-hand side gives
$$f_n(x) - f_n(a) \; \to \; \int_a^x g(t)\, dt \; .$$
On the other hand, by the Linearity Theorem for limits of sequences 5.1(1),
$$f_n(x) - f_n(a) \; \to \; f(x) - f(a) \; .$$
Since the limit of a sequence is unique, comparing the two limits above shows
$$f(x) - f(a) \; = \; \int_a^x g(t)dt \; .$$
Since $g(t)$ is continuous by Theorem 22.3, the Second Fundamental Theorem 20.2A can be applied to this last equality; it shows that $f(x)$ is differentiable on the interval I and $f'(x) = g(x)$. □

This is a significant argument; you should write it out yourself. Start with the differentiated sequence so you can make use of the uniform convergence hypothesis. Note how the argument uses both the First and Second Fundamental Theorems of calculus, as well as the two preceding theorems of this chapter.

Try formulating by yourself and proving the corresponding statement for series before you read the next theorem.

Theorem 22.5B   Term-by-term differentiation of series.

Let $\sum u_k(x)$ be a series of functions. Assume that on an interval I,

(a)   each $u_k(x)$ is defined and has a continuous derivative;

(b) $\sum u_k(x)$ converges;

(c) $\sum u_k'(x)$ converges uniformly.

Then on I, the sum $f(x) = \sum u_k(x)$ is differentiable and $f'(x) = \sum u_k'(x)$.

Proof.   Let $s_n(x) = \sum_0^n u_k(x)$ and $g(x) = \sum_0^\infty u_k'(x)$; then our hypotheses (b) and (c) translate into

$$s_n(x) \to f(x) \quad \text{and} \quad s_n'(x) \rightrightarrows g(x) .$$

The hypotheses of the previous theorem are satisfied, and we conclude that

$$s_n'(x) \to f'(x), \qquad \text{i.e.,} \qquad f'(x) = \sum_0^\infty u_k'(x) . \qquad \square$$

In the two preceding theorems, if x_0 is an endpoint of the interval, then the derivatives are to be interpreted as the left- or right-hand derivatives at the point.

Example 22.5   Recalling Example 22.4(b), discuss the applicability of term-by-term differentiation to the sawtooth function pictured on page 315.

$$f(x) = \frac{\pi}{2} - \frac{4}{\pi} \left(\cos x + \frac{\cos 3x}{3^2} + \frac{\cos 5x}{5^2} + \ldots \right) .$$

Solution.   Differentiating term-by-term, we get formally the Fourier series

$$(10) \qquad f'(x) \stackrel{?}{=} \frac{4}{\pi} \left(\sin x + \frac{\sin 3x}{3} + \frac{\sin 5x}{5} + \ldots \right) .$$

To establish the equality, we must know the series on the right converges uniformly. We cannot use the Weierstrass M-test, since the only reasonable choice for a comparison series is

$$\sum M_k = \frac{4}{\pi} \sum \frac{1}{2n+1} ,$$

and this series fails to converge. While this failure doesn't prove anything in itself, it should make us suspicious.

Indeed, the differentiated series does *not* converge uniformly. If it did, we could apply Theorem 22.5B, equality would hold in (10), and the sum $f'(x)$ would be a continuous function, by Theorem 22.3. But $f'(x)$ has jump discontinuities at $0, \pm\pi, \pm 2\pi, \ldots$, since the graph of $f(x)$ has a sawtooth shape.

Even though we are not able to use Theorem 22.5B, it is nonetheless true that (10) is valid at all places where $f'(x)$ exists. This follows from the theory of Fourier series, but is too delicate to be handled here; our general theorem about differentiating series isn't strong enough to handle those Fourier series which don't converge very well. It is however good enough to handle power series, as we will see in the next section.

Questions 22.5

1. Let $f_n(x) = \frac{1}{n}\sin nx$, $\quad f(x) = \lim f_n(x)$. Show that $f_n(x) \rightrightarrows f(x)$, but $\lim f_n'(x) \neq f'(x)$.

2. Let $f(x) = \sum_1^\infty (\cos nx)/n^3$. Prove $f(x)$ and $f'(x)$ exist, and find the series for $f'(x)$.

22.6   Power series and analytic functions.

The theorem on differentiating a power series term-by-term is a special case of Theorem 22.5B, but is so important we give it a label of its own.

Theorem 22.6   Differentiation of power series.

A power series $\sum a_n x^n$ can be be differentiated term-by-term within its radius R of convergence. That is, if

$$f(x) \;=\; \sum_0^\infty a_n x^n \;, \qquad |x| < R,$$

then $f(x)$ is differentiable and

(11) $$f'(x) \;=\; \sum_1^\infty n a_n x^{n-1} \;, \qquad |x| < R \;.$$

Proof.   Let $0 < L < R$; we prove first the series (11) converges for $x = L$. This will show its radius of convergence is at least R, by Theorem 8.1.

Choose any M such that $0 < L < M < R$. We claim that for some large N, we have term-by-term

(12) $$\sum_N^\infty n|a_n|L^{n-1} \;<\; \sum_N^\infty |a_n|M^n \;;$$

namely, (12) is equivalent to

$$n\left(\frac{L}{M}\right)^n \;<\; L \;, \quad \text{for } n \gg 1,$$

and since $L/M < 1$, this follows by the Sequence Location Theorem 5.3B from

(13) $$\lim_{n\to\infty} n\left(\frac{L}{M}\right)^n \;=\; 0 \;;$$

(13) is true because in general if $0 \le a < 1$, then $xa^x \to 0$ as $x \to \infty$ (write $xa^x = xe^{-kx}$, $(k > 0)$, and use 20.6 (24) upside down). This proves (12).

The right side of (12) converges, since $\sum a_n x^n$ converges absolutely inside its radius of convergence (Theorem 8.1) and $M < R$. Therefore the left side converges, by the Comparison Theorem 7.2D. Thus (11) converges for $x = L$.

Since the power series $\sum n a_n x^{n-1}$ has radius of convergence at least R, by Theorem 22.2C it converges uniformly for $|x| \leq L$. Therefore the hypotheses of Theorem 22.5B on term-by-term differentiation of series are satisfied, and we thus know (11) holds for $|x| \leq L$. But since L was any number satisfying $0 < L < R$, the equality (11) holds for $|x| < R$. $\square$

Example 22.6A Evaluate $f(x) = \dfrac{x^2}{1 \cdot 2} + \dfrac{x^3}{2 \cdot 3} + \ldots + \dfrac{x^n}{(n-1) \cdot n} + \ldots$.

Solution. The radius of convergence is 1 by the ratio test. Two term-by-term differentiations show that

$$f''(x) \;=\; 1 + x + x^2 + x^3 + \ldots \;=\; \frac{1}{1-x}, \quad |x| < 1 .$$

Now integrate the right-hand side twice; to evaluate the two constants of integration, use the additional data $f(0) = 0$, $f'(0) = 0$, which come from the original series. The final result is

$$f(x) \;=\; x + (1-x)\ln(1-x), \quad \text{for } |x| < 1 .$$ $\square$

Corollary 22.6A   Taylor series theorem.

Suppose $\sum_0^\infty a_n x^n$ has radius of convergence $R > 0$. Then the function

$$(14) \qquad\qquad f(x) \;=\; \sum_0^\infty a_n x^n$$

is infinitely differentiable in $(-R, R)$, and $\sum a_n x^n$ is its Taylor series around the point $x = 0$, that is, $a_n = \dfrac{f^{(n)}(0)}{n!}$.

Proof. We apply Theorem 22.6 repeatedly, getting

$$f'(x) \;=\; \sum_1^\infty n\, a_n x^{n-1} , \qquad |x| < R ;$$

$$f''(x) \;=\; \sum_2^\infty n(n-1) a_n x^{n-2}, \qquad |x| < R ;$$

and so on. Since the derivative of a power series is again a power series, the function has derivatives of all orders.

To calculate the a_n, we observe that after differentiating (14) a total of n times to get $f^{(n)}(x)$, the differentiated series has the form

$$(15) \qquad\qquad f^{(n)}(x) \;=\; n!\, a_n + c_1 x + c_2 x^2 + \ldots ,$$

where the constant term comes from differentiating $a_n x^n$ a total of n times. Substituting $x = 0$ in (15) gives

$$(16) \qquad\qquad a_n \;=\; \frac{f^{(n)}(0)}{n!} .$$ $\square$

Corollary 22.6B   Zero theorem for power series.   *Let $R > 0$.*

If $\sum a_n x^n = 0$ for $|x| < R$, then $a_n = 0$ for all n.

Corollary 22.6C   Uniqueness of power series.   *Let $R > 0$.*

$$f(x) = \sum a_n x^n = \sum b_n x^n \text{ for } |x| < R \Rightarrow a_n = b_n \text{ for all } n \ .$$

Proofs.   The first is just (16) applied to the function $f(x) = 0$; for the second, we have, again by (16), $a_n = \dfrac{f^{(n)}(0)}{n!} = b_n$ for every n. $\square$

The following example illustrates how these theorems are used in finding series solutions to a first-order differential equation. The same principles apply to higher-order equations and nonlinear equations (see the Questions and Exercises).

Example 22.6B   Find a series solution $y(x)$ to $y' + xy = 0$, satisfying $y(0) = 1$.

Solution.   Assume that for some $R > 0$, $y(x)$ has a series representation:

$$y = \sum_{0}^{\infty} a_n x^n , \qquad \text{for } |x| < R.$$

Then by using term-by-term differentiation (Theorem 22.6),

$$y' + xy = \sum_{0}^{\infty} n\, a_n x^{n-1} + \sum_{0}^{\infty} a_n x^{n+1}, \qquad \text{for } |x| < R.$$

Rewriting the series in terms of x^n (we define $a_{-1} = 0$),

$$y' + xy = \sum_{0}^{\infty} (n+1)a_{n+1} x^n + \sum_{0}^{\infty} a_{n-1} x^n \ .$$

Since $y' + xy = 0$, we get

$$0 = \sum_{0}^{\infty} \big((n+1)a_{n+1} + a_{n-1} \big) x^n , \qquad \text{for } |x| < R.$$

The Zero Theorem for power series (Cor. 22.6B) tells us then that
$$(n+1)a_{n+1} + a_{n-1} = 0 \qquad \text{for } n \geq 0.$$
So we get a recursion formula for the a_n (replacing n by $n+1$):

(17) $$a_{n+2} = \frac{-a_n}{n+2}, \qquad \text{for } n \geq -1 \ .$$

The starting values are

$$a_{-1} = 0 \ \text{(by definition)}, \qquad a_0 = 1 \ \text{(since } y(0) = 1\text{)}.$$

From (17) it then follows that

$$a_{-1} = a_1 = a_3 = a_5 = \ldots = 0;$$
$$a_0 = 1, \ a_2 = -\frac{1}{2}, \ a_4 = -\frac{1}{4} \cdot -\frac{1}{2}, \ a_6 = -\frac{1}{6} \cdot -\frac{1}{4} \cdot -\frac{1}{2} ,$$

and in general

$$a_{2n} = \frac{(-1)^n}{2^n n!} \; .$$

The resulting series solution converges for all x and is readily summed:

(18) $$y = \sum_0^\infty \frac{(-1)^n}{2^n n!} x^{2n} = \sum_0^\infty \Big(\frac{-x^2}{2}\Big)^n \cdot \frac{1}{n!} = e^{-x^2/2} \; .$$ □

All the above really shows is that *if* there is a series solution with a positive radius of convergence, it is given by (18). So we ought to substitute (18) into the differential equation and show it actually is a solution. This is not really necessary, however, since the steps leading up to (18) are all reversible.

The main point of the example is to observe the use of Theorem 22.6 and the key role played by the Zero Theorem for power series (Cor.22.6B). In differential equations courses, term-by-term differentiation of power series is usually taken for granted without proof, and the Zero Theorem often used implicitly without comment.

Definition 22.6   Analytic functions.

A function $f(x)$ which is represented by a power series with a positive radius of convergence

$$f(x) = \sum a_n x^n \; , \qquad |x| < R, \quad R > 0,$$

is said to be *real analytic at the origin*, or simply **analytic**. The Corollaries show

(19) $f(x)$ *is analytic* $\Leftrightarrow$ *the Taylor series for* $f(x)$ *converges to* $f(x)$.

(20)   *An analytic function has a unique power series representation.*

An analytic function is infinitely differentiable, by Corollary 22.6A, but the converse is not true: the following elementary function is the best-known example of a function which is infinitely differentiable but not analytic at the origin.

(21) $$f(x) = \begin{cases} e^{-1/x^2}, & \text{if } x \neq 0; \\ 0, & \text{if } x = 0. \end{cases}$$

This function is "infinitely flat" at the origin, since it can be proved (see Problem 22-4) that all its derivatives are zero at the origin. Thus its Taylor series is $\sum 0 \cdot x^n$; since the Taylor series does not converge to $f(x)$, according to (19) the function cannot be analytic.

Questions 22.6

1. Prove that in Theorem 22.6, the differentiated series actually has R as its radius of convergence. (Use the information in the proof, plus Cor. 22.4.)

2. Find explicitly the sum $f(x) = x + x^3/3 + x^5/5 + \ldots$, using the ideas of this section. For what x is your answer valid?

3. Find a series solution $y(x)$ to $y'' - y = 0$, satisfying the initial conditions $y(0) = 1$, $y'(0) = 0$; justify the steps using the theorems of this section.

4. Find an infinitely differentiable function $f(x)$ whose Taylor series at the origin is $\sum_0^\infty x^n/n!$, yet $f(x) \neq e^x$ for $x \approx 0$. Explain how you know that $f(x)$ cannot be analytic.

Exercises

22.1

1. For each of the following sequences of functions $f_n(x)$,

(i) find $f(x) = \lim_{n\to\infty} f_n(x)$;

(ii) determine directly from the definition whether $f_n(x) \rightrightarrows f(x)$ over the given interval.

(a) $\dfrac{x}{1+nx}$, $[0,\infty)$ (b) $\dfrac{\sin nx}{n}$, $(-\infty,\infty)$

(c) $\dfrac{nx^2}{1+nx}$, $[0,\infty)$ (d) $\dfrac{nx}{1+n^2x^2}$, $[0,1]$

2. Prove that the Taylor series for $\sin x$ converges uniformly in any interval $[-R, R]$, but does not converge uniformly on $(-\infty, \infty)$.

(Don't quote any theorems presented in later sections of this chapter; do it directly from the definition of uniform convergence. Model the argument after that in Example 22.1E, using the remainder term in the Taylor series.)

3. For every integer k, let $u_k = \begin{cases} 1, & \text{if } x \in [k, k+1); \\ 0, & \text{otherwise.} \end{cases}$

Find $f(x) = \sum_0^\infty u_k(x)$, and show directly from the definition that the series converges uniformly on $[-R, R]$ for all $R > 0$, but does not on $(-\infty, \infty)$.

22.2

1. Prove that $f_n \rightrightarrows f$ on I $\Leftrightarrow$ $\sup_I |f(x) - f_n(x)| \to 0$.

2. Using the criteria given in this section, prove the following sequences and series converge uniformly on the indicated interval.

(a) $\dfrac{n}{x+n}$, $[0, R]$ for any $R > 0$ (b) $\cos \dfrac{x}{n}$, $|x| < R$;   use the Mean-value Theorem

(c) $\sum_1^\infty \dfrac{x^n}{n^2}$, $[-1,1]$ (d) $\sum_1^\infty \dfrac{\sin nx}{x^2 + n^2}$, $(-\infty, \infty)$

3. (a) Show the geometric series $\sum x^n$ converges uniformly in every interval $[-L, L]$, for $0 < L < 1$.

(b) Show that it does not converge uniformly in $(-1, 1)$, by considering the explicit remainder term for the series (Section 4.2 (4)).

4. Prove that if $\sum a_n$ is absolutely convergent, then $\sum a_n \sin nx$ is uniformly convergent on $(-\infty, \infty)$.

5. Prove the Cauchy criterion for uniform convergence, given below. Unlike the elementary criterion (Theorem 22.2A), it does not require advance knowledge of the limit $f(x)$.

Cauchy criterion for uniform convergence.

Suppose that on an interval I, the functions $f_n(x)$ are defined, and

$$\text{given } \epsilon > 0, \quad f_m(x) \underset{\epsilon}{\approx} f_n(x) \quad \text{for } m, n \gg 1,$$

i.e., for $m, n >$ some N_ϵ depending only on ϵ. Then the sequence $f_n(x)$ converges uniformly on I.

(The Cauchy criterion for convergence of numerical sequences (6.4) will be a big help in the argument, but its proof will not be.)

22.3

1. Prove that the function $\displaystyle\sum_{2}^{\infty} \frac{\sin nx}{n(n-1)}$ is continuous for all x.

2. Prove that $\sum_0^\infty e^{-nx} \sin nx$ converges to a continuous function on $(0, \infty)$. (Warning: the convergence is not uniform on $(0, \infty)$.)

3. Prove that the function $\displaystyle\sum_{1}^{\infty} \frac{x}{n(x+n)}$ is continuous on $[0, \infty)$.

4. (a) Prove that if the functions $f_n(x)$ are uniformly continuous on I, and $f_n(x) \rightrightarrows f(x)$ on I, then $f(x)$ is uniformly continuous on I.

(b) How do the functions $f_n(x) = \begin{cases} n, & 0 < x < 1/n; \\ 1/x, & 1/n \le x \le 1; \end{cases}$ for $n \in \mathbb{N}$,

illustrate the theorem in part (a)? (Do they satisfy the hypotheses? Does their limit $f(x)$ satisfy the conclusion?)

22.4

1. Prove that if on an interval $[a, b]$ the functions $g_n(x)$ are continuous and $g_n(x) \rightrightarrows g(x)$, then on $[a, b]$,

$$\int_a^x g_n(t)\, dt \;\rightrightarrows\; \int_a^x g(t)\, dt\;.$$

2. As usual, let $\;\text{erf}\, x = \displaystyle\int_0^x e^{-t^2/2} dt$. Find an infinite power series which represents this function, justifying your steps. Over what interval does it converge, and how do the theorems of this section predict this?

3. Let $f(x) = \displaystyle\sum_{1}^{\infty} \frac{\cos nx}{(n-1)!}\;$. Prove carefully that $\displaystyle\int_0^{\pi/2} f(x)\, dx$ exists and evaluate it. (Indicate which theorems you are using.)

4. (a) Prove $f(x) = \sum\limits_{1}^{\infty} \dfrac{x^{n-1}}{n^2(1+x^n)}$ is continuous for $x \geq 0$.

(b) Evaluate $\displaystyle\int_{0}^{1} f(x)\,dx$, justifying all steps of your work, and expressing your answer in terms of values of the function $\zeta(s) = \sum\limits_{1}^{\infty} \dfrac{1}{n^s}$.

5. Let $Z_p(u)$ be the function defined by the series $Z_p(u) = \sum\limits_{1}^{\infty} \dfrac{u^n}{n^p}$.

Evaluate $\displaystyle\int_{0}^{1} Z_2(e^{-x})\,dx$ in terms of values of $Z_p(u)$, justifying all steps.

22.5

1. Prove that if $f(x) = \sum a_n \sin nx$, and $\sum na_n$ is absolutely convergent, then $f(x)$ is differentiable and $f'(x) = \sum na_n \cos nx$.

2 Let $f_n(x) = x^n/n$; discuss this sequence from the standpoint of Theorem 22.5A, over the interval $[0,1]$. Does it satisfy the hypotheses? the conclusion? (As usual, the derivative at 1 means the left-hand derivative.)

22.6

1. Let $n!! = n(n-2)(n-4)\cdot\ldots\cdot k$, where $k = 1$ or 2, depending on whether n is odd or even. (We define $0!! = 1$.)

Evaluate the sum $f(x) = \sum_{0}^{\infty} x^n/n!!$, using the methods of this section, and justifying your steps.

2. Find the sum of $\dfrac{x^3}{1\cdot 3} + \dfrac{x^4}{2\cdot 4} + \dfrac{x^5}{3\cdot 5} + \dfrac{x^6}{4\cdot 6} + \ldots$ by using term-by-term differentiation.

3. Let $\sum a_n x^n$ be the power series solution to the differential equation $y' - y = e^x$, satisfying the initial condition $y(0) = 0$.

(a) Use the Uniqueness Theorem (Cor. 22.6C) to find the recursion relation connecting a_{n+1} and a_n.

(b) Calculate the first few values of a_n, guess what a_n will be in general, and prove it by induction (Appendix A.4).

(c) Find the sum of the power series.

4. Show that for any power series $\sum a_n x^n$ with a positive radius of convergence, there is a function $f(x)$ whose Taylor series is $\sum a_n x^n$, yet for which $f(x) \neq \sum a_n x^n$ for $x \approx 0$.

5. Find the sum of the following, justifying the steps.

(a) $1 + 2x + 3x^2 + 4x^3 + \ldots$

(b) $1 + 4x + 9x^2 + 16x^3 + \ldots + n^2 x^{n-1} + \ldots$

Problems

22-1 Find the sum of $\sum_1^\infty x^n/n^2$, expressed as the integral of an elementary function. Justify the steps, including the existence of the integral. On what interval is the representation valid, and why?

22-2 Prove that the initial-value problem (Bessel's equation of order 0)
$$x^2 y'' + xy' + x^2 y = 0, \qquad y(0) = 1, \ y'(0) = 0 \ ,$$
has a unique real-analytic solution (the zero-th order Bessel function)
$$J_0(x) \ = \ \sum_0^\infty \frac{(-1)^n x^{2n}}{4^n (n!)^2} \ .$$
For what x is this solution valid? Justify the steps, as in Example 22.6B.

22-3 There is an elegant definite integral expression for the Bessel function of the previous problem, which you are now in a position to derive. (Part (a) repeats part of Exercise 20.3/1, without the hints.)

(a) Let $I_k = \int_0^\pi \sin^k \theta d\theta$, $k = 0, 1, 2, \ldots$. Prove that
$$I_k = (k-1)(I_{k-2} - I_k), \quad k = 2, 3, \ldots \ ,$$
and deduce from this that for $m = 0, 1, 2, \ldots$, (cf. Exercise 22.6/1 for $n!!$)
$$\int_0^\pi \sin^{2m} \theta \, d\theta \ = \ \frac{(2m-1)!!}{(2m)!!} \, \pi \ .$$

(b) Using part (a) and the theorems of this chapter, prove by using term-by-term integration that the Bessel function can also be represented as the following definite integral (this was the form in which Bessel originally discovered the function):
$$J_0(x) \ = \ \frac{1}{\pi} \int_0^\pi \cos(x \sin \theta) \, d\theta \ .$$

22-4 Prove that if $f(x) = e^{-1/x^2}$, (and defining $f(0) = 0$), then $f(x)$ has derivatives of all orders at $x = 0$, and $f^{(n)}(0) = 0$ for all n.

(Try the first few derivatives, to see the pattern. In addition to the usual differentiation rules, you will have to use the definition of derivative, as well as l'Hospital's rule.)

22-5

(a) Define the notion "the sequence $f_n(x)$ is locally uniformly convergent on I".

(b) Show that $\dfrac{nx}{1 + nx^2}$ is convergent on $(-\infty, \infty)$; show it is not locally uniformly convergent however, by two methods: directly from the definition, and by using theorems of this chapter.

(c) Give a sequence which is locally uniformly convergent on an interval I, but not uniformly convergent on I.

Answers to Questions

22.1

1. For any fixed $x \geq 0$, $\lim\limits_{n \to \infty} \dfrac{1}{x+n} = 0$. Therefore $f(x) = 0$ for all x.

Given $\epsilon > 0$, $\left| \dfrac{1}{x+n} - 0 \right| \leq \dfrac{1}{n}$, which is $< \epsilon$ if $n > 1/\epsilon$; since $1/\epsilon$ does not depend on x, this shows the convergence is uniform.

2.   For example, if $x_0 = \pi/2$, then the sequence $f_n(x_0) = \sin(n\pi/2)$ is $0, 1, 0, -1, 0, 1, 0, -1, \ldots$, which has no limit.

In general, $\lim\limits_{n \to \infty} \sin n x_0$ exists only for $x_0 = 0, \pm\pi, \pm 2\pi, \ldots$.

3. We have $f(x) = \lim\limits_{n \to \infty} (e^{-x})^n = \begin{cases} 1, & \text{if } x = 0; \\ 0, & \text{if } x > 0, \end{cases}$ by Theorem 3.4,

since $0 < e^{-x} < 1$ if $x > 0$.

To show the convergence is uniform for $x \geq R > 0$, we have:

$$\text{given } \epsilon > 0, \quad |e^{-nx}| \leq e^{-nR} < \epsilon, \quad \text{for } n > -(\ln \epsilon)/R ,$$

and $(\ln \epsilon)/R$ does not depend on x.

4. For any fixed x, we have $\lim\limits_{n \to \infty} x/n = 0$. Therefore $f(x) = 0$ for all x.

Given $\epsilon > 0$, we have $|x/n| \leq |R/n| < \epsilon$, if $|x| \leq R$ and $n > R/\epsilon$. Since R/ϵ does not depend on x, this shows the convergence is uniform on $[-R, R]$, i.e., for $|x| \leq R$.

To see the convergence is not uniform on $(-\infty, \infty)$, take for example $\epsilon = 1$; it is not true that $|x/n| < 1$ for all x and all $n > $ some N, where N does not depend on x. For no matter what n we use, it is false when $|x| > n$.

5. Let the tail be $f(x) = \sum_M^\infty u_k(x)$ and let $S(x) = \sum_0^\infty u_k(x)$.

Then given $\epsilon > 0$, we have for $n > M$

$$\left| S(x) - \sum_0^n u_k(x) \right| = \left| f(x) - \sum_M^n u_k(x) \right|,$$

$$< \epsilon, \quad \text{for } n > \text{ some } N_\epsilon > M,$$

which shows $\sum_0^\infty u_k(x)$ converges uniformly to $S(x)$.

22.2

1. (a) $\left| \dfrac{1}{x+n} - 0 \right| \leq \dfrac{1}{n}$ on $[0, \infty)$, and $\lim\limits_{n \to \infty} 1/n = 0$.

(b) $e^{-nx} \leq e^{-nR}$ on $[R, \infty)$, and $\lim\limits_{n \to \infty} e^{-nR} = 0$, since $R > 0$..

(c) $|x/n - 0| \leq R/n$ on $[-R, R]$, and $\lim\limits_{n \to \infty} R/n = 0$.

2. (a) M-test: $|x^n \sin nx| \leq R^n$ if $|x| \leq R$, and $\sum R^n$ converges if $R < 1$.

(b) M-test: $|(\sin nx)/n!| \leq 1/n!$ for all x, and $\sum 1/n!$ converges.

22.3

1. Using the ratio test to find the radius of convergence,

$$\frac{x^{2(n+1)}}{2^{n+1}(n+1)} \cdot \frac{2^n n}{x^{2n}} = \frac{n\, x^2}{2(n+1)} \rightarrow \frac{x^2}{2} \, , \text{ which is } < 1 \iff |x| < \sqrt{2} \, .$$

So the series represents a continuous function in $(-\sqrt{2}, \sqrt{2})$, by Cor. 22.3, and converges uniformly in every interval $[-L, L]$, where $L < \sqrt{2}$, by Theorem 22.2C.

2. Using the M-test, we have

$$\left| \frac{\cos nx}{n\sqrt{n+1}} \right| \leq \frac{1}{n^{3/2}}, \text{ for all } x, \text{ and } \sum \frac{1}{n^{3/2}} \text{ converges;}$$

therefore the series converges uniformly on $(-\infty, \infty)$, and so it represents a continuous function on $(-\infty, \infty)$, by Theorem 22.3.

3 Using the M test, we have

$$\left| \frac{1}{1+n^2 x} \right| \leq \frac{1}{n^2 a}, \text{ if } x \in [a, \infty), \ a > 0; \text{ and } \sum \frac{1}{n^2 a} \text{ converges.}$$

Therefore the series converges uniformly in every interval $[a, \infty), a > 0$; by Theorem 22.3, its sum is continuous on every such $[a, \infty)$.

To show it is continuous on $(0, \infty)$, let x_0 be any point in $(0, \infty)$; the sum is continuous on the interval $(x_0/2, \infty)$ by the above, thus it is continuous at x_0.

This argument mimics the proof of Corollary 22.3. The point of it is that that the series does *not* converge uniformly on $(0, \infty)$, and therefore Theorem 22.3 is not directly applicable.

22.4

1. We must also assume $f(x)$ is integrable, since we do not know that the uniform limit of integrable functions is integrable. The corresponding hypothesis is not necessary in Theorem 22.4A as stated, since we know that the uniform limit of continuous functions is continuous.

If we do assume that $f(x)$ is integrable, the proof goes through exactly as before; no changes are necessary.

2. We have $\dfrac{1}{1+x^2} = 1 - x^2 + x^4 - x^6 + \dots, \ |x| < 1$, by replacing x by $-x^2$ in the geometric series (see Example 22.4(a)). Integration term-by-term between 0 and x gives the series for $\tan^{-1} x$, which is also valid in the same interval $(-1, 1)$.

22.5

1. We have $f_n(x) \rightrightarrows 0$, by Theorem 22.2A, since $|\sin nx/n| \leq 1/n$, and $1/n \to 0$.

On the other hand, $f_n'(x) = \cos nx$, and $\lim_{n\to\infty} \cos nx$ does not exist (for example, it does not exist when $x = \pi$, since $\cos n\pi = (-1)^n$).

Thus $\lim_{n\to\infty} f_n'(x) \neq f'(x)$.

2. $f(x)$ exists for all x, since the series is absolutely convergent by comparison:

$$\sum |\cos nx|/n^3 \le \sum 1/n^3.$$

The terms of the series have continuous derivatives, and the derived series $\sum \dfrac{-\sin nx}{n^2}$ converges uniformly on $(-\infty, \infty)$ by the M-test:

$$\left|\frac{\sin nx}{n^2}\right| \le \frac{1}{n^2}, \quad \text{and} \quad \sum \frac{1}{n^2} \text{ converges.}$$

Thus by Theorem 22.5B, $f'(x) = -\sum \dfrac{\sin nx}{n^2}$.

22.6

1. Let S be the radius of convergence for the differentiated series. The proof shows the series for $f'(x)$ converges in $(-R, R)$, therefore $R \le S$. Corollary 22.4 shows $f(x)$ converges in $(-S, S)$, therefore $S \le R$. So $R = S$.

2. By Theorem 22.6, $f'(x) = 1 + x^2 + x^4 + \ldots = \dfrac{1}{1 - x^2}$, $|x| < 1$.

Integrating, $f(x) = \dfrac{1}{2} \ln \left(\dfrac{1 + x}{1 - x}\right) + c$; since $f(0) = 0$, we get $c = 0$.

3. Following the steps of Example 22.6B, using the same two theorems as there, the recursion formula is

$$a_{n+2} = \frac{a_n}{(n + 2)(n + 1)}; \quad a_0 = 1, \ a_1 = 0;$$

using this, one gets as the series $y = 1 + x^2/2! + x^4/4! + \ldots + x^{2n}/(2n)! + \ldots$.

4. An example would be $f(x) = e^x + e^{-1/x^2}$; (see 22.6 (21)). It is not analytic, since its Taylor series $\sum x^n/n!$ converges to e^x, not to $f(x)$.

23

Infinite Sets
and the Lebesgue Integral

23.1   Introduction. Infinite sets.

Because of the importance of the Lebesgue integral in analysis, there is some point even in an elementary course in seeing how it differs from the Riemann integral and what it can do for you. That is the purpose of this chapter. The integral will not be defined, but you will learn some of its most important properties; many who use the integral all the time scarcely know much more about it than that.

This material is not used in any later chapters; you can skip it if you wish. The set terminology we will need for it is all in Appendix A.0.

Infinite sets.

To describe the properties of the Lebesgue integral, we need some background material about infinite sets; the first section is devoted to this.

A set S is said to be **finite**, with **cardinality n**, if its elements can be arranged without repetition in a finite sequence $< x_1, x_2, \ldots, x_n >$. Such a sequence is called an *enumerating sequence* for S; it is a function (or map)

$$f : \{1, 2, \ldots, n\} \; \to \; S \qquad f(i) = x_i \; ,$$

in which distinct numbers are assigned to distinct elements of S, and all the elements of S are used. In the language of Appendix A.0, the map f is a *bijection*, and it establishes a *one-one correspondence* between S and the set $\{1, \ldots, n\}$.

The definition of cardinality is completed by assigning to the empty set the cardinality 0; it is the only set with this cardinality.

If a set is not finite, it is called **infinite**. In the theory of sets, a cardinality is also assigned to infinite sets, in order to give a measure of how big they are, but it is of course not a positive integer. Here we will be primarily interested in the infinite sets having the lowest cardinality; they are the "smallest" kind of infinite set and are called *countable* according to the following definition.

Definition 23.1   Countable sets.

A set is called **countable** if it is infinite and its elements can be arranged without repetition in a sequence $< x_1, x_2, \ldots, x_n, \ldots >$; such a sequence is called an **enumerating sequence** for S.

Equivalently, an infinite set S is *countable* if there is a bijection

$$f : \mathbb{N} \to S ; \qquad f(n) = x_n .$$

The enumerating sequence is really just another way of looking at the bijective map, but psychologically it feels different; most of the proofs of countability work with the sequence rather than with f. Here is a simple example.

Example 23.1A   Show that the integers $\mathbb{Z}$ are a countable set.

Solution.   An enumerating sequence is $0, 1, -1, 2, -2, \ldots, n, -n, \ldots$.   □

One's first try might be just $\ldots, -n, \ldots, -2, -1, 0, 1, 2, \ldots, n, \ldots$, but this is not a sequence—it has no first element. Note that it is easier to understand the countability by seeing the sequence; giving the bijection explicitly by a formula for $f(n)$ would be tedious and unilluminating.

Before going on to give more examples of countable sets, we give some principles which will make it easier to prove various sets are countable. Try proving them yourself, before reading the arguments.

Theorem 23.1A   Countability principles.

(a)   *A set S is finite or countable if there is a sequence $< x_1, x_2, \ldots, x_n, \ldots >$ of elements of S in which each element of S occurs at least once.*

(b)   *A subset S of a countable set T is either finite or countable.*

(c)   *If $S_1, S_2, \ldots, S_n, \ldots$ is a countable collection of finite sets, then their union $\bigcup\limits_{n \in \mathbb{N}} S_n$ is finite or countable.*

Proof.

(a)   Examine each term x_n in order and delete it if it occurs earlier in the sequence; this turns the given sequence into an enumerating sequence, if the set is infinite.

(b)   Take an enumerating sequence for T and delete from it all elements not in S; the result is an enumerating sequence for S.

(c)   Make a sequence of elements of $\bigcup S_n$ by first listing the finitely many elements of S_1, then the elements of S_2, and so on. It is clear this sequence contains all elements of $\bigcup S_n$, so by (a) above, the union is finite or countable. (Note that if two of the S_n overlap, the sequence will have repetitions.)   □

Example 23.1B   Show that the positive rationals $\mathbb{Q}^+$ are countable.

Solution.

$$
\begin{array}{ccccc}
1 & 2 & 3 & 4 & \cdots \\
\frac{1}{2} & \frac{2}{2} & \frac{3}{2} & \frac{4}{2} & \cdots \\
\frac{1}{3} & \frac{2}{3} & \frac{3}{3} & \frac{4}{3} & \cdots \\
\frac{1}{4} & \frac{2}{4} & \frac{3}{4} & \frac{4}{4} & \cdots \\
\cdots & \cdots & \cdots & \cdots & \cdots
\end{array}
$$

The infinite square array shown contains all positive rationals, with repetitions. Let S_n be the finite square array using the first n rows and columns. Then $\mathbb{Q}^+ = \bigcup_{n \in \mathbb{N}} S_n$, so $\mathbb{Q}^+$ is countable by Theorem 23.1A(c) above.

The positive rationals in their natural order on the number line do not form a sequence; they are dense on the line: there is never a "next" rational. So the idea that they can be enumerated one-by-one goes against common sense, and was not well-received when Cantor first announced it.

Theorem 23.1B   If $S_1, S_2, \ldots, S_n, \ldots$ is a countable collection of countable sets, then the union $S = \bigcup_{n \in \mathbb{N}} S_n$ is also countable.

Proof.   Imitate the method of the previous example, making an infinite matrix whose first row is an enumerating sequence for S_1, second row a sequence for S_2 and so on. S is the countable union of the finite square $n \times n$-matrices in the upper left corner, and so it is countable by Theorem 23.1A(c).   ☐

Theorem 23.1C   The set $[0, 1)$ of real numbers is not countable.

Proof.   By contradiction. Suppose there were a sequence $x_1, x_2, \ldots, x_n, \ldots$ enumerating all the reals in $[0, 1)$. Write out the decimal expansion of each x_i; to make it unique, do not use any decimal ending with all 9's—use the equivalent one ending with all 0's. (We indicate by a_i, b_i, etc., the decimal digits; they lie between 0 and 9.)

$$
\begin{array}{rcllll}
x_1 & = & . \, a_1 & a_2 & a_3 & a_4 & \ldots \\
x_2 & = & . \, b_1 & b_2 & b_3 & b_4 & \ldots \\
x_3 & = & . \, c_1 & c_2 & c_3 & c_4 & \ldots \\
x_4 & = & . \, d_1 & d_2 & d_3 & d_4 & \ldots \\
& & \vdots & \vdots & \vdots & \vdots & \vdots
\end{array}
$$

Now form a decimal number $x' = . \, t_1 t_2 t_3 t_4 \ldots$ in $[0, 1)$ according to the rule:

$$
t_i = \begin{cases} 1, & \text{if the } i\text{-th decimal digit of } x_i \text{ is } 0; \\ 0, & \text{otherwise.} \end{cases}
$$

Then $x' \neq x_i$ for all i, since x' and x_i have different i-th decimal places. Thus our enumerating sequence omits x', so it is not an enumerating sequence. This contradiction completes the indirect proof that $[0, 1)$ is uncountable.   ☐

Since $[0, 1) \subset \mathbb{R}$, this also shows that $\mathbb{R}$ is uncountable, by Theorem 23.1A(b).

The indirect argument above is known as *Cantor's diagonal argument*, since the new number x' is constructed so that each digit differs from the corresponding digit in the "diagonal" sequence $a_1 b_2 c_3 \ldots$ Generalizations of the argument are at the core of many proofs in modern logic—one of them is Gödel's proof that there are mathematical statements that are true but unprovable.

Remarks about infinite cardinals.

The cardinal number of a finite set is the number of elements in it. Cantor invented a corresponding notion for infinite sets, assigning to each such set S a "cardinal number" $N(S)$, with the rules

$$N(S) = N(T) \quad \Leftrightarrow \quad \text{there is a bijection } S \to T;$$

$$N(S) < N(T) \quad \Leftrightarrow \quad S \text{ is bijective with a subset of } T, \text{ but not vice-versa.}$$

With these rules, the set of cardinal numbers is totally ordered, just like the positive integers: given any two, one is larger. There is no largest cardinal number, for if S is any infinite set and $\mathcal{P}(S)$ the set of all subsets of S, one can prove by Cantor's diagonal argument (see the Exercises) that

$$N(S) < N(\mathcal{P}(S)) .$$

The usual notation for the two smallest known infinite cardinal numbers is ($\aleph$="aleph")

$$\aleph_0 = N(\mathbb{Z}), \qquad \aleph_1 = N(\mathbb{R}) .$$

It is unknown if there is any cardinal number N between these two:

$$\aleph_0 < N < \aleph_1 .$$

The conjecture that no such N exists is called the *continuum hypothesis*. (It is equivalent to the assertion that any uncountable infinite subset of $\mathbb{R}$ can be put in 1-1 correspondence with $\mathbb{R}$.)

> Both the continuum hypothesis and its negation are known to be unprovable with the set-theoretic tools currently available. The hypothesis is probably false; that is, there probably does exist an uncountable S which is "smaller" than the set $\mathbb{R}$ of reals; it's just that we cannot construct it.
>
> To get some feeling why we cannot, we ask what it means to "know" a real number. A reasonable answer would be that a real number is regarded as known if we can give a finite algorithm for calculating its n-th decimal place. But since it is not hard to see that all the possible finite algorithms form a countable set (Exercise 23.1/9), it follows that the "knowable" real numbers form a countable set. In other words, most real numbers are "unknowable", which makes it hard to use them to construct an uncountable set S of reals smaller in cardinality than $\mathbb{R}$ itself.

Questions 23.1

1. Let T be the set of numbers in $[0, 1)$ representable as terminating decimals. Prove two ways that T is countable:

    (a) directly, by describing an enumerating sequence;

    (b) as briefly as possible, by using the results of this section.

2. Prove the set of power series with integer coefficients is uncountable.

3. In the proof of Theorem 23.1A(a), does the procedure produce an enumerating sequence if the set is finite?

23.2   Sets of measure zero.

In considering how one might extend the notion of integral to functions which are not Riemann-integrable (19.2), it helps to think of the definite integral of a positive function as giving the area of the plane region under its graph. Defining the integral is really the same problem as defining the area, or as one says, the *measure*, of this region. To do this, Lebesgue began by defining the measure of a one-dimensional region, i.e., a set of points on the line.

For the rest of this chapter, all sets S will be subsets of $\mathbb{R}$.

Definition 23.2A   Measure of intervals.

By the *measure* of any type of interval, we mean its length $|I|$. (If I is an infinite interval, we say its measure is ∞.)

If S is a finite or countable union of intervals I_k, any two of which overlap at most at an endpoint, we define the **measure** $|S|$ of S to be the sum

$$|S| = \sum |I_k| \ .$$

(If S is a countable union of I_k, this is an infinite series whose terms are non-negative; if it converges, $|S|$ is its sum; if it diverges, we say $|S| = \infty$.)

Assigning a measure to more general sets which are not the countable union of almost non-overlapping intervals is problematical. It turns out most sets have no measure, in any reasonable sense. Lebesgue was able however to describe a class of sets called *measurable* and define their measure. Here we will consider only a special case of these, but an important special case.

Definition 23.2B   A set $S \subseteq \mathbb{R}$ has **measure zero** if, given $\epsilon > 0$, there is a finite or countable collection of intervals $I_1, I_2, \ldots$ (which may overlap), such that

(1) $$S \subseteq \bigcup I_k \quad \text{and} \quad \sum |I_k| \le \epsilon \ .$$

In words, the set S can be covered by a finite or countable collection of intervals having arbitrarily small total length.

From the definition, we see immediately that a point has measure zero, as well as any finite or countable set $S = \{x_1, x_2, \ldots\}$, since we can take the I_k to be the intervals $[x_k, x_k]$ of zero length. More generally,

Theorem 23.2A   Let $S = \bigcup_k S_k$, a finite or countable union of sets, where each S_k has measure 0. Then S has measure 0.

Proof.   Leaving the case of finite union for Question 2, assume the union is countable. Since each S_k has measure zero, it follows that given $\epsilon > 0$, we can find for each S_k a collection of intervals I_{km} such that

$$S_k \subseteq I_{k1} \cup I_{k2} \cup I_{k3} \cup \ldots \quad \text{and} \quad \sum_{m=1}^{\infty} |I_{km}| \le \frac{\epsilon}{2^k} \ .$$

The set of intervals $\{I_{km} : k, m \in \mathbb{N}\}$ is a countable union of countable sets; it is therefore countable, by Theorem 23.1B. Moreover

$$\sum_{k,m\leq N}\left|I_{km}\right| = \sum_{m\leq N}\left|I_{1m}\right| + \sum_{m\leq N}\left|I_{2m}\right| + \ldots + \sum_{m\leq N}\left|I_{Nm}\right|$$

$$\leq \frac{\epsilon}{2} + \frac{\epsilon}{4} + \ldots + \frac{\epsilon}{2^N} < \epsilon \,.$$

Thus we have

$$S = \bigcup_{k\in\mathbb{N}} S_k \subseteq \bigcup_{k,m\in\mathbb{N}} \left|I_{km}\right| \quad\text{and}\quad \sum_{k,m\in\mathbb{N}}\left|I_{km}\right| = \lim_{N\to\infty}\sum_{k,m\leq N}\left|I_{km}\right| \leq \epsilon\,,$$

which proves that S has measure zero. $\qquad\square$

So far, the only sets of measure zero we have seen have been the finite or countable ones. We now construct a famous uncountable set of measure zero, the Cantor set C.

Definition 23.2B   The Cantor set.

Starting with $C_0 = [0,1]$, get a sequence of sets C_n as follows:

from C_0 remove the middle third $(\frac{1}{3}, \frac{2}{3})$ to get C_1;

from C_1 remove the two middle thirds $(\frac{1}{9}, \frac{2}{9})$ and $(\frac{7}{9}, \frac{8}{9})$ to get C_2;

from C_2 remove the middle third of each of the four intervals to get C_3, etc.

The *Cantor set* C is the limit of these sets C_n, defined to be

(2)
$$C = \bigcap_{n\in\mathbb{N}} C_n \,.$$

Theorem 23.2B   *The Cantor set has measure zero, and is uncountably infinite.*

Proof. Since C_n is a sum of non-overlapping intervals which arises from C_{n-1} by removing the middle third of each interval,

(3)
$$\left|C_n\right| = \left(\tfrac{2}{3}\right)\left|C_{n-1}\right| = \left(\tfrac{2}{3}\right)^2\left|C_{n-2}\right| = \ldots = \left(\tfrac{2}{3}\right)^n\left|C_0\right| = \left(\tfrac{2}{3}\right)^n \,.$$

It follows that C has measure zero, since (2) shows that $C \subset C_n$ for all n, where C_n is a finite union of intervals, and (3) shows that $\left|C_n\right| < \epsilon$, for $n \gg 1$.

To see that C is infinite and not countable, we represent the numbers in $[0,1]$ to the base 3, i.e., writing them as ternary decimals, just using the digits $0, 1$, and 2. The representation is

$$x = .a_1 a_2 a_3 \ldots a_n \ldots = \frac{a_1}{3} + \frac{a_2}{3^2} + \frac{a_3}{3^3} + \ldots + \frac{a_n}{3^n} + \ldots \,.$$

In this representation, we avoid using 1 as far as possible, by replacing any finite ternary decimal ending with 1 by the equivalent ternary decimal ending with an infinity of 2's; for example,

$$1 = .22222\ldots, \quad 1/3 = .1 = .022222\ldots, \quad 4/9 = .11 = .102222\ldots \,.$$

Given $x \in [0,1]$, its first ternary decimal digit a_1 is then given by the rule:

$$a_1 = \begin{cases} 0, & \text{if } 0 \le x \le \frac{1}{3}; \\ 1, & \text{if } \frac{1}{3} < x < \frac{2}{3}; \\ 2, & \text{if } \frac{2}{3} \le x \le 1 \, . \end{cases}$$

It follows that

$$C_1 = \{x \in [0,1] : a_1 \ne 1\} \, .$$

In the same way, we find a_2 by taking the interval of length $1/3$ containing x and dividing it into three subintervals of length $1/9$; then $a_2 = 0, 1$, or 2 according to which of these subintervals contains x, as above. This shows that

$$C_2 = \{x \in [0,1] : a_1 \ne 1, a_2 \ne 1\} \, .$$

Continuing in this way, we see that since $C = \bigcap C_n$,

$$C = \{x \in [0,1] : x \text{ has a base 3 representation using only 0 and 2}\} \, .$$

C is obviously infinite, and, suitably modified, Cantor's diagonal argument (proof of Theorem 23.1C) then shows that C is not countable.   □

One can avoid having to repeat the diagonal argument by instead using the base 2 representation of the numbers in $[0,1]$ to show there is a bijection from C to the uncountable interval $[0,1]$; thus C is also uncountable. (See the Exercises.)

Cantor's set is an example of what is called a "fractal set": if you magnify any part of it, the magnified part looks like the original set.

Questions 23.2

1. Show the set of discontinuities of $\sec(1/x)$ on $[0,1]$ has measure zero.

2. Prove a finite union of sets of measure zero has measure zero.

3. Would the Cantor construction work if one used "quaternary decimal" representation (using just the digits 0,1,2,3), and removing each time say the second quarter of each interval? if one used "binary decimal" representation, removing each time say the second half of each interval ?

23.3   Measure zero and Riemann-integrability.

Which functions are Riemann-integrable? So far, we know from Section 18.3 that monotone functions and continuous functions are Riemann-integrable, and Section 19.6 extends this to functions which are bounded and piecewise continuous or piecewise monotone — those which have only a *finite* number of discontinuities or changes of direction on any finite interval.

On the other hand, if a bounded function has a *countable* number of discontinuities on $[a,b]$, the question of its integrability remains. The complete answer was first provided by the Lebesgue theory of measure, and it was viewed as one of the early triumphs of the theory.

Theorem 23.3A   Characterization of Riemann-integrability.

Let f be defined on $I = [a, b]$, and S be the set of points in I where f is not continuous. Then on I,

   f is Riemann-integrable $\Leftrightarrow$ f is bounded and S has measure zero.

We are not in a position to prove this theorem; it is best done as a corollary of Lebesgue integration. But that shouldn't stop us from using it. The theorem shows for example that any bounded function with at most a countable number of discontinuities on $[a, b]$ is Riemann-integrable. The theorem gives another proof that monotone functions are Riemann-integrable, since a monotone function can have only a countable number of discontinuities (see Problem 23-1).

To give other illustrations of the theorem, it is convenient to introduce the notion of "characteristic function" of a set.

Definition 23.3   Let $S \subseteq \mathbb{R}$ be a subset of the reals.

We define the **characteristic function** $f_S(x)$ of S: $f_S(x) = \begin{cases} 1, & \text{if } x \in S \text{ ;} \\ 0, & \text{if } x \notin S \text{ .} \end{cases}$

Example 23.3A   Let C be the Cantor set. Then $f_C(x)$ is Riemann-integrable.

Proof.   In outline, the argument runs:

(a) $f_C(x)$ is continuous at all points not in C (Problem 23-2); therefore its set S of discontinuities is contained in C.

(b) Since C has measure zero (Theorem 23.2B), so does S (Exercise 23.2/2).

(c) Thus $f_C(x)$ is Riemann-integrable, by Theorem 23.3A.   □

As an example of how the characterization of Riemann-integrability given in Theorem 23.3A can be used to get stronger results, we improve our earlier Theorem 22.4A, which dealt with the integral of a uniform limit of functions.

Theorem 23.3B   On the compact interval $[a, b]$, if the functions $f_n(x)$ are Riemann-integrable and $f_n(x) \rightrightarrows f(x)$, then

(4) $f(x)$ is Riemann-integrable ;

(5) $\lim_{n \to \infty} \int_a^b f_n(x)\, dx \; = \; \int_a^b f(x)\, dx$.

Remarks.   The difference between this and Theorem 22.3A is that now we are not assuming the f_n are continuous, only Riemann-integrable.

In fact, the only reason for assuming them continuous before was to guarantee that their uniform limit f would be continuous and therefore Riemann-integrable. What the argument given there actually shows is that, with the current hypotheses, (4) $\Rightarrow$ (5); therefore it suffices now to prove (4).

Proof of (4).   We prove first that f is bounded. Each f_n is bounded, since it is Riemann-integrable; since f is close to the f_n, it will be bounded too. To see this formally, choose a suitable large n and write, using the error-form for the limit,

$$f = f_n + (f - f_n);$$

Then the triangle inequality shows f is bounded:

(6) $$|f| \leq |f_n| + |f - f_n| \leq K + 1;$$

here K is a bound for $|f_n|$, where n has been chosen so large (using the uniformity of the convergence) that

$$|f - f_n| < 1 \quad \text{on } [a, b].$$

To prove (4) now, use the remark at the end of the proof of Theorem 22.3A (a uniform limit f of continuous functions $f_n(x)$ is continuous), which observes that the proof shows that if all the $f_n(x)$ are continuous at x_0, then $f(x)$ is continuous at x_0. Put another way, $f(x)$ can only be discontinuous at a point c where at least one of the $f_n(x)$ is discontinuous. But each f_n is discontinuous at most on a set S_n of measure zero, since it is Riemann-integrable (Theorem 23.3A). Therefore $f(x)$ is discontinuous at most on the set

$$S = \bigcup_1^\infty S_n \, ,$$

which still has measure zero according to Theorem 23.2A, since it is a countable union of sets of measure zero. Since the set of discontinuities of $f(x)$ is thus contained in the set S, it too has measure zero (Exercise 23.2/2), and therefore $f(x)$ is Riemann-integrable. This proves (4), and therefore (5) as well, according to the remarks following the statement of the theorem. $\qquad\square$

Questions 23.3

1. Let $\mathbb{Q}$ denote the rational numbers; it is a countable set (Example 23.1B), and therefore has measure 0. In Question 18.2/3, it is shown that the characteristic function $f_\mathbb{Q}(x)$ is not Riemann-integrable; why doesn't this contradict Theorem 23.3A?

2. Let $< a_1, a_2, a_3, \ldots >$ be the rational numbers in $[0, 1]$, arranged in a sequence. Let $f_n(x)$ be the characteristic function of the first n of them, i.e., of the set $\{a_1, \ldots, a_n\}$.

(a) Prove that $\lim_{n \to \infty} f_n(x) = f_\mathbb{Q}(x)$.

(b) Why doesn't this prove by Theorem 23.3B (4) that $f_\mathbb{Q}(x)$ is Riemann-integrable?

23.4   Lebesgue integration.

The Lebesgue integral is, like the Riemann integral, defined only for certain functions (the *Lebesgue-integrable* functions), over an interval I. Unlike the Riemann integral, the functions need not be bounded, nor does the interval I have to be finite, but as compensation, we will allow the integral to have ∞ as a "value". There is no simple charecteization of the functions that are Lebesgue-integrable; however all the Riemann-integrable functions certainly are—we make this our first theorem about the Lebesgue integral.

Theorem 23.4A   *If $f(x)$ is Riemann-integrable on $[a, b]$, it is also Lebesgue-integrable, and*

$$(7) \qquad \int_a^b f(x)\, dx \;=\; \int_a^b f(x)\, dx \; .$$
$$\qquad\qquad \text{(Riemann)} \qquad \text{(Lebesgue)}$$

Definition 23.4   A statement $P(x)$ is said to hold **almost everywhere** on an interval I if the set $\{x \in I : P(x) \text{ is false}\}$ has measure zero.

Examples 23.4A

(a) $\tan x$ is continuous almost everywhere on $\mathbb{R}$;

(b) the characteristic function of the Cantor set (Ex. 23.3A) has the value zero almost everywhere;

(c) Assume $f(x)$ is bounded on an interval I; then on I ,

$f(x)$ is Riemann-integrable $\Leftrightarrow$ $f(x)$ is continuous almost everywhere.

(This is just a restatement of Theorem 23.3A.)

Theorem 23.4B   *Suppose $f(x) = g(x)$ almost everywhere in $[a, b]$.*
   Then if $g(x)$ is Lebesgue-integrable on $[a, b]$, so is $f(x)$, and

$$(8) \qquad \int_a^b f(x)\, dx \;=\; \int_a^b g(x)\, dx \qquad\qquad \text{(Lebesgue integrals)} \; .$$

In other words, sets of measure zero are negligible as far as the Lebesgue integral is concerned—you can change the values of $f(x)$ on a set of measure zero without changing the value of its Lebesgue integral.

Example 23.4B   Show $\displaystyle\int_a^b f_Q(x)\, dx$ (Lebesgue) is defined, and evaluate it..

Solution.   Since the set of rationals $\mathbb{Q}$ is countable and therefore has measure zero, its characteristic function $f_Q(x) = 0$ almost everywhere. The function 0 is Lebesgue-integrable since it is Riemann-integrable. Therefore by (8), the characteristic function $f_Q(x)$ is also Lebesgue-integrable and by (8) and then (7), as Lebesgue integrals,

$$\int_a^b f_Q(x)\, dx \;=\; \int_a^b 0\, dx \;=\; 0.$$

Note that $f_{\mathbb{Q}}(x)$ is not Riemann-integrable; cf. Question 23.3/1.

The Lebesgue integral has all the standard properties given in Section 19.4 for the Riemann integral. The most important one for making estimations is the absolute value property; we state it for reference.

Theorem 23.4C   Absolute value property.

If $f(x)$ is Lebesgue-integrable on an interval I, so is $|f(x)|$, and as Lebesgue integrals,

$$(9) \qquad \left| \int_I f(x)\, dx \right| \leq \int_I |f(x)|\, dx .$$

The main power of the Lebesgue integral comes from two convergence theorems Lebesgue proved, which we describe now. The essential question is::

$$(10) \qquad \text{if} \quad f_n(x) \to f(x) , \quad \text{does} \quad \int_I f_n(x)\, dx \to \int_I f(x)\, dx \; ?$$

This is a basic question, since in the applications of analysis, often a solution $f(x)$ to a problem is known only as the limit of a sequence of functions $f_n(x)$ which approximate it, and its integral can only be found as the limit of the approximating integrals.

The answer we have given so far to (10) is in Theorem 22.4A and its extension 23.3B. There we said the answer is yes, if the $f_n(x)$ are Riemann-integrable, and the convergence $f_n(x) \to f(x)$ is uniform. This last is the rub, because in practice either the convergence is not uniform, or you don't know whether it is or not.

Here the Lebesgue theorems come to the rescue—by using the Lebesgue integral, they make it possible to assert (10) under very general hypotheses, even though the $f_n(x)$ may not converge uniformly to $f(x)$.

In stating them, we will abbreviate Lebesgue-integrable to "L-integrable".

Theorem 23.4D   Monotone convergence theorem.

Suppose on an interval I, the functions f_n are all L-integrable,

$$0 \leq f_1 \leq f_2 \leq \dots \leq f_n \leq \dots , \quad \text{and} \quad f_n \to f .$$

Then f is also L-integrable (allowing the value ∞ for the integral) and

$$(11) \qquad \int_I f_n(x)\, dx \to \int_I f(x)\, dx .$$

Corollary 23.4D   *Suppose on an interval I,*

(a) *the functions $u_n(x)$ are defined, L-integrable, and $u_n(x) \geq 0$;*

(b) *the series $\sum u_n(x)$ converges (pointwise).*

Then its sum $\sum u_n(x)$ is L-integrable, and as Lebesgue integrals,

$$(12) \qquad \int_I \sum u_n(x)\, dx = \sum \int_I u_n(x)\, dx .$$

The important word here is *pointwise*; the series $\sum u_n(x)$ is not assumed to converge uniformly. Again, the value ∞ is allowed for the integral on the left, in which case the numerical series on the right will diverge to ∞. The corollary is proved by taking as $f_n(x)$ the partial sums of $\sum u_n(x)$.

Outside of the situation in the corollary, one does not usually have a sequence $f_n(x)$ which converges monotonically to its limit. It is the following theorem therefore which is used most often, in which the functions $f_n(x)$ do not have to be non-negative, nor do they have to form a monotone sequence.

Theorem 23.4E   Dominated convergence theorem.

Suppose on an interval I on which the functions $f_n(x)$ are defined,

(a) $|f_n(x)| \le g(x)$ for all n, where $\int_I g(x)\,dx$ exists and is finite;

(b) $f_n(x)$ are all L-integrable, and $f_n(x) \to f(x)$ almost everywhere.

Then $f(x)$ is L-integrable, and

(13)
$$\int_I f_n(x)\,dx \;\to\; \int_I f(x)\,dx \;.$$

Corollary 23.4E   Bounded convergence theorem.

Suppose on a finite interval I on which the functions $f_n(x)$ are defined,

(a) $|f_n(x)| \le K$ for some constant K and all n;

(b) $f_n(x)$ are all L-integrable, and $f_n(x) \to f(x)$ almost everywhere.

Then $f(x)$ is L-integrable, and (13) holds.

Exercises and problems illustrate these two theorems and their corollaries.

As far as the connection between differentiation and integration goes, for Lebesgue integration, the First Fundamental Theorem (20.1) remains true:

(14)   If $F'(x)$ is L-integrable on $[a,b]$,
$$\int_a^b F'(x)\,dx = F(b) - F(a) \;.$$

On the other hand, we have seen that the Second Fundamental Theorem fails when the integrand is not continuous. Thus, the best one can do is:

Theorem 23.4F   *If $f(x)$ is L-integrable on I, and*
$$F(x) = \int_a^x f(t)\,dt, \quad a, x \in I,$$
then almost everywhere on I, its derivative $F'(x)$ exists and
$$F'(x) = f(x).$$

Questions 23.4

1. As in Question 23.1/1, let T be the countable set of terminating decimals in $[0,1]$, arranged in an enumerating sequence $< x_1, x_2, x_3, \ldots >$. Let $u_n(x)$ be the characteristic function of the set $\{x_n\}$: $u_n(x_n) = 1, u_n(x) = 0$ elsewhere.

(a) Show that Cor. 23.4D applies to $\sum u_n(x)$, and tell what it says.

(b) Show that it would not be true for Riemann integrals.

Exercises

23.1

1. (a) Let $S_2 = \{(x, y) : x, y \in \mathbb{Z}\}$ be the set of points in $\mathbb{R}^2$ having integer coordinates. Prove S_2 is countable.

(b) Prove the same thing for the analogous set S_n of n-tuples in $\mathbb{R}^n$.

2. Prove the set consisting of all the numbers $\sqrt[n]{a}$, where a and n are positive integers, is countable.

3. (a) Fix a positive integer n. Prove that the binary sequences (that is, sequences of 0's and 1's) of length n form a finite set.

(b) Prove that the binary sequences of finite (but arbitrary) length form a countable set.

(c) Given an alphabet of k symbols, prove that the set of possible words formed from the alphabet is a countable set. (Parts (a) and (b) were the case $k = 2$.)

4. (a) Prove that the polynomials which are simple sums of powers, i.e., of the form $x^a + x^b + \ldots + x^c$, where the $a, b, c \ldots$ are distinct non-negative integers, form a countable set.

(b) Prove that the set of all finite subsets of $\mathbb{N}_0$ (the non-negative integers) is a countable set.

5. Prove the set of all monomials $x_{i_1}^{n_1} x_{i_2}^{n_2} \ldots x_{i_k}^{n_k}$ (k can vary) in infinitely many variables $x_1, x_2, \ldots$ is a countable set.

6. (a) Prove the polynomials $f(x)$ whose coefficients are rational and whose degree $\leq k$ (where k is fixed) form a countable set.

(b) Prove the polynomials $f(x)$ with rational coefficients are countable.

7. Prove the infinite binary sequences (cf. Exercise 3) form an uncountable set. Do this two ways:

(a) by Cantor's diagonal argument;

(b) by relating the sequences to the points in $[0, 1]$, carefully!

8. Do the real-analytic functions whose Taylor series at the origin use only 0 and 1 as coefficients form a countable or uncountable set? Prove your answer.

9. Define a finite algorithm to be one describable by a computer program of finite length. Prove the finite algorithms form a countable set. (This is the basis for saying the set of real numbers that are "known" is countable.)

10. (a) Fix an $n \in \mathbb{N}$. Prove that the set of $n \times n$ matrices having integer entries is countable. (Begin by considering the matrices with entries from an interval $[-N, N]$, $N \in \mathbb{N}$.)

(b) Prove the set of all square matrices with integer entries is countable.

23.2

1. Suppose S is the union of a countable set of intervals I_n which overlap at most at their endpoints and which all lie in $[a, b]$. Prove that $|S| \leq b - a$.

2. Prove that if S is a set of measure zero, then any subset $T \subseteq S$ also has measure zero.

3. Complete the proof that the Cantor set is uncountable by using the ternary decimal representation of the numbers in C to show there is a bijection from C to the set $[0, 1]$ (cf. Exercise 23.1/7(b)).

4. Prove the numbers in $[0, 1]$ whose decimal representation does not contain any 7 form a set of measure zero.

5. Let us say temporarily that a set of reals has *strict* measure zero if in the definition of measure zero, the intervals I_k must be *open* intervals.

   (a) Prove that a countable set has strict measure zero.

   (b) Prove that every set of measure zero has strict measure zero. (Since the converse is trivial, this shows the two concepts are equivalent.)

23.3

1. Let $\operatorname{sgn} x = \begin{cases} 1, & x > 0; \\ -1, & x < 0; \\ 0, & x = 0; \end{cases}$ $\qquad f(x) = \begin{cases} \sin 1/x, & x \neq 0; \\ 0, & x = 0. \end{cases}$

Is $\operatorname{sgn}(f(x))$ Riemann-integrable on $[-1, 1]$? Prove your answer.

2. Prove: on an interval $[a, b]$, if $f(x)$ is Riemann-integrable and $f(x) > 0$, then $\displaystyle\int_a^b f(x)\, dx > 0$.

23.4

1. Let $f_n(x) = \begin{cases} 1, & \text{if } x \text{ can be written } i/2^n \text{ for some integer } i; \\ 0, & \text{otherwise.} \end{cases}$

Answer the following, with proofs.

   (a) Let $f(x) = \lim f_n(x)$. What is $f(x)$ explicitly?

   (b) Is the convergence $f_n \to f$ uniform on $I = [0, 1]$?

   (c) Is f_n Riemann-integrable? Is f?

   (d) Show that f is L-integrable and evaluate $\int_0^1 f(x)\, dx$. Do this three different ways, by using Theorems 23.4B, D, and E.

2. Let $f_n(x) = \begin{cases} 1/n, & \text{on } [0, n]; \\ 0, & \text{otherwise.} \end{cases}$ $\qquad$ Take $I = [0, \infty)$.

Answer the following, with proofs.

   (a) Let $f = \lim f_n$. What is f, explicitly? Is the convergence uniform on the interval I?

   (b) Show that $\lim \int_I f_n(x)\, dx = \int_I f(x)\, dx$ is false.

   (c) Show how the hypotheses of both the Monotone and Bounded Convergence Theorems are not satisfied.

3. Let $f_n(x) = \begin{cases} n, & 0 < x \le 1/n; \\ 0, & \text{elsewhere.} \end{cases}$

Take $I = [0, 1]$. Answer the same three questions as in Exercise 2.

4. Let $f_n(x) = \begin{cases} 1, & 0 \le x \le n; \\ 0, & \text{elsewhere.} \end{cases}$

What does the Monotone Convergence Theorem say about this sequence?

5. Arrange the countable set of rational numbers in $[0, 1]$ in an enumerating sequence $< a_1, a_2, a_3, \ldots >$. Let S_n be $[0, 1)$ with the subset $\{a_1, \ldots, a_n\}$ removed, and $f_n(x)$ the characteristic function of S_n.

Show the hypotheses of the Bounded Convergence Theorem are satisfied for the f_n; what does the conclusion of the theorem say?

6. Non-Theorem.   Let $f(x)$ be Riemann-integrable, $F(x) = \displaystyle\int_a^x f(t)\, dt$. Then $F(x)$ is differentiable, and $F'(x) = f(x)$ almost everywhere ("a.e.").

Non-Proof.

$f(x)$ Riemann-integrable $\Rightarrow$ $f(x)$ continuous a.e.   (Ex.23.4A(c);

$\qquad\qquad\qquad\Rightarrow f(x) = g(x)$ a.e.,   where g is continuous;

$\qquad\qquad\qquad\Rightarrow \displaystyle\int_a^x f(t)\, dt = \int_a^x g(t)\, dt$ as Lebesgue integrals;

$\qquad\qquad\qquad\Rightarrow \displaystyle\int_a^x f(t)\, dt = \int_a^x g(t)\, dt$ as Riemann integrals;

$\qquad\qquad\qquad\Rightarrow \qquad F(x) = \displaystyle\int_a^x g(t)\, dt$.

By the Second Fundamental Theorem 20.2, $F'(x) = g(x) = f(x)$ a.e.. $\square$

Find the error. (If stuck, trace a counterexample through the proof.)

Problems

23-1

(a) Let S be a set of non-overlapping intervals of positive length. Prove that S is a countable set.

(b) Deduce that a function which is monotone on an interval I has at most a countable number of discontinuities on I.

(Draw a picture. Begin by establishing what kind of discontinuities they must be.)

23-2 Let C be the Cantor set, and $f(x)$ its characteristic function. Prove that $f(x)$ is continuous almost everywhere on $[0, 1]$, by showing that its discontinuities lie entirely inside C. Deduce that f is Riemann-integrable on $[0, 1]$.

23-3 With $f(x)$ as in the previous problem, prove directly from the definition of integrability that $f(x)$ is Riemann-integrable on $[0, 1]$.

(Begin by estimating, for a partition $\mathcal{P}$ of $[0, 1]$ of mesh ϵ, and for an interval $I \subset [0, 1]$, the total length of those subintervals of the partition which are contained in I and do not contain an endpoint of I. Then estimate the total length of the subintervals not containing any point of C_k, where C_k are the sets whose limit defines the Cantor set.)

23-4 Prove that $\mathcal{P}(\mathbb{N})$, the set of all subsets of the positive integers $\mathbb{N}$, is uncountable. Do this two different ways.

(a) Show there is a 1-1 correspondence between the subsets and the infinite binary sequences.

(b) Suppose by contradiction that the subsets of $\mathbb{N}$ can all be enumerated: $A_1, A_2, \ldots$. Construct a subset that cannot be in this list. (You can use the representation in part (a), combined with Cantor's diagonal argument, or you can do it directly.)

23-5 Generalize the argument in Problem 23-4(b) above so that it proves that for any infinite set S there is no 1-1 correspondence between S and its power set $\mathcal{P}(S)$ (the set of all subsets of S).

Does this hold for finite sets? How about the empty set $\emptyset$?

23-6 Prove that the graph of a continuous function on $(-\infty, \infty)$ is completely determined once one knows a certain countable set of points on it. (By contrast, the graph of an arbitrary function is not determined until one knows every point on it, and these form an uncountable set.)

Answers to Questions

23.1

1. (a) Let $T_n = \{$terminating decimals of length exactly $n\}$. Then T_n is finite (in fact, it has $10^{n-1} \cdot 9$ elements, since there are 10 ways to fill each of the first $n-1$ places, but only 9 ways to fill the last).

Make an enumerating sequence by first listing the elements of T_1, then T_2, and so on. Every terminating decimal occurs exactly once in this listing.

(b) The numbers in T are all rational (their denominators being a power of 10), so $T \subset \mathbb{Q}$; since we know $\mathbb{Q}$ is countable, it follows by Theorem 23.1A(b) that also T is countable.

2. (a) One way would be to repeat the diagonal argument of Theorem 23.1C (uncountability of $[0,1)$): assume the power series countable, list them in some order, then use the diagonal argument to construct a new power series with integer coefficients which is not on the list, by making its n-th coefficient different from the n-th coefficient of the n-th power series on the list.

(b) Another way: let S be the subset consisting of all power series whose coefficients just use the 10 numbers $0, 1, \ldots, 9$, and which do not end with coefficients all 9. Then S is in one-one correspondence with $[0,1)$ via the map

$$a_0 + a_1 x + a_2 x^2 + \ldots \rightarrow .a_0 a_1 a_2 \ldots .$$

Since $[0,1)$ is uncountable, so is S, and since S is a subset of the power series with integer coefficients, this latter set is also uncountable (Theorem 23.1A(b)).

3. No, for since the original sequence is presumed infinite (like all sequences in this book), the procedure keeps removing terms forever, and at no stage does the sequence become finite.

23.2

1. Since $\sec(1/x) = \dfrac{1}{\cos(1/x)}$, the function is discontinuous in the interval $[0,1]$ exactly where $\cos(1/x) = 0$, i.e., at $x = 2/\pi, 2/3\pi, 2/5\pi, \ldots$.

Since this is an enumerating sequence, the set of discontinuities is countable, and therefore has measure zero.

2. Let $S_1, \ldots, S_n$ be the sets of measure zero, and $S = \bigcup_1^n S_k$.

Given $\epsilon > 0$, cover the set S_k by a countable collection of intervals $I_{ki}, i \in \mathbb{N}$, such that $\sum_{i=1}^{\infty} |I_{ki}| = \epsilon/n$. Then

$$S = \bigcup_1^n S_k \subseteq \bigcup_{k,i} I_{ki}, \qquad \sum_{k,i} |I_{ki}| = \epsilon .$$

The set of intervals $\{I_{ki}\}$ is still countable, since it is a finite union of countable sets (Theorem 23.1B), so this proves S has measure zero.

3. (a) It does work for quaternary decimals — the intervals remaining at the n-th stage have total length $(3/4)^n$, whose limit is 0 as $k \to \infty$; what's left is an uncountable set, since it consists of all the quaternary decimals not containing 1.

(b) It does not work for binary decimals, since all that remains at the end are the binary decimals not using 1, i.e., just the single point 0.

23.3

1. This is a common error. Though $\mathbb{Q}$ is countable, it is not the set S of discontinuities of $f_{\mathbb{Q}}(x)$: this function is discontinuous at every point.

For if say a is irrational, we may (as on page 1) approximate it by a sequence of terminating decimals x_n; then the x_n are rational, and we have

$$x_n \to a, \quad \text{but} \quad \lim f_{\mathbb{Q}}(x_i) \neq f_{\mathbb{Q}}(a), \quad \text{since } 1 \neq 0 .$$

By the Sequential Continuity Theorem (11.5), this shows that $f_{\mathbb{Q}}$ is not continuous at a. A similar argument shows it is not continuous at the rational numbers.

2. (a) If a is irrational, then $f_n(a) = 0 = f_{\mathbb{Q}}(a)$ for all n ; if a is rational then $a = a_k$ for some k; thus $f_n(a) = 1 = f_{\mathbb{Q}}(a)$ for all $n \geq k$.

(b) The convergence is not uniform, so Theorem 23.3B does not apply: choose say $\epsilon = 1/2$; then

$$|f_n(x) - f_{\mathbb{Q}}(x)| \not< \epsilon \quad \text{for } n \gg 1,$$

since no matter what n is,

$$|f_n(a_{n+1}) - f_{\mathbb{Q}}(a_{n+1})| = 1 .$$

23.4

1. (a) The series $\sum u_n(x)$ converges pointwise; its sum is the characteristic function $f_T(x)$ of the set T. (The proof is the same as the one for Question 23.3/2a above, applied to the partial sums of the series.)

Therefore its sum $f_{\mathbb{Q}}(x)$ is L-integrable. Since $\int_0^1 u_n(x)dx = 0$, Cor. 23.4D tells us that

$$\int_0^1 f_T(x) \, dx = \sum \int_0^1 u_n(x)dx = 0.$$

(b) The theorem does not hold if R-integrable is substituted everywhere for L-integrable, since while the hypotheses would still be valid, the conclusion that $f_T(x)$ is R-integrable is false.

24

Continuous Functions on the Plane

24.1 Introduction. Norms and distances in $\mathbb{R}^2$.

To go further in analysis, we have to work with functions of several variables. Two variables is a good place to start; you avoid a lot of mess with subscripts and superscripts and can draw pictures; yet all the essential new ideas appear, so that extending the work later to n variables is mostly just a matter of getting used to the notation.

Even those who only care about $f(x)$ cannot avoid functions of two variables, since they appear as the integrands of definite integrals depending on a parameter. Such integrals are used to solve differential equations, to define integral transforms such as the Laplace transform

$$\int_0^\infty e^{-st} g(t)dt \ ,$$

and to represent functions, like the Γ-function (21.4) or Bessel function (Problem 22-3). We will study these integrals in Chapters 26 and 27.

By the **two-dimensional real Euclidean space** $\mathbb{R}^2$ we mean the product set $\mathbb{R} \times \mathbb{R}$, i.e., the set of all ordered pairs of real numbers:

$$\mathbb{R}^2 \ = \ \{(x,y) : x,y \in \mathbb{R}\} \ .$$

We call the pair (x,y) a **point** of $\mathbb{R}^2$, and denote it by $\mathbf{x}$ (handwritten: $\underline{x}$); the origin $(0,0)$ is $\mathbf{0}$.

If we identify $\mathbf{x}$ with the origin vector having its head at $\mathbf{x}$, we can use vector operations: if $\mathbf{x} = (x,y)$ and $\mathbf{x}' = (x',y')$, we define

(1) $$\mathbf{x} \pm \mathbf{x}' = (x \pm x', y \pm y'); \qquad c\mathbf{x} = (cx, cy), \quad c \in \mathbb{R} \ .$$

Norms and Distances.

On the line, we measure the size of x by $|x|$, its distance from 0.

In the plane, there are two common measures (or **norms**, as one calls them) for the size of an origin vector $\mathbf{x} = (x,y)$, i.e., for the distance of the point $\mathbf{x}$ from $\mathbf{0}$. They are

(2)
$$|\mathbf{x}| \ = \ \sqrt{x^2 + y^2} \qquad \qquad \textit{(Euclidean norm)};$$
$$\|\mathbf{x}\| \ = \ \max(|x|, |y|) \qquad \textit{(uniform norm)}.$$

The first is more familiar — it is the magnitude (length) of the vector $\mathbf{x}$ — but the second is more convenient for the work of this chapter. The two norms give different ways of measuring the **distance** between two points $\mathbf{x}$ and $\mathbf{x}'$, which we define to be the norm of $\mathbf{x} - \mathbf{x}'$:

(3)
$$|\mathbf{x} - \mathbf{x}'| = \sqrt{(x - x')^2 + (y - y')^2} \qquad \textit{(Euclidean distance)};$$
$$\|\mathbf{x} - \mathbf{x}'\| = \max(|x - x'|, |y - y'|) \qquad \textit{(uniform distance)}.$$

One can do all the work of this chapter using either norm, i.e., either notion of distance. We will use the uniform distance, since this makes it easy to estimate distances between points, once you know their coordinates. Namely, by (2) and by (3), (we could also substitute $<$ throughout), we have

(4) $$\|\mathbf{x}\| \leq K \iff |x| \leq K \quad \text{and} \quad |y| \leq K \; ;$$
(4a) $$\|\mathbf{x} - \mathbf{x}'\| \leq K \iff |x - x'| \leq K \text{ and } |y - y'| \leq K \; .$$

To get some feeling for the difference between the two norms, we ask where the points lie whose distance is less than r from some fixed point $\mathbf{a}$. They lie in the sets which are the analog in $\mathbb{R}^2$ of a symmetric interval around $a \in \mathbb{R}$. Depending on which norm is used, they are

(5)
$$D(\mathbf{a}, r) = \{\mathbf{x} : |\mathbf{x} - \mathbf{a}| < r\} \qquad \text{(the \textit{open r-disk around} } \mathbf{a});$$
$$B(\mathbf{a}, r) = \{\mathbf{x} : \|\mathbf{x} - \mathbf{a}\| < r\} \qquad \text{(the \textit{open r-box around} } \mathbf{a});$$

the first picture is well-known, and the second follows immediately from (4a). The word "open" is used by analogy with the open interval on the line, since these sets do not include their boundary points. If those points are included we get instead

(5a)
$$\overline{D}(\mathbf{a}, r) = \{\mathbf{x} : |\mathbf{x} - \mathbf{a}| \leq r\} \qquad \text{(the \textit{closed r-disk around} } \mathbf{a});$$
$$\overline{B}(\mathbf{a}, r) = \{\mathbf{x} : \|\mathbf{x} - \mathbf{a}\| \leq r\} \qquad \text{(the \textit{closed r-box around} } \mathbf{a}).$$

We give some examples of the use of the uniform norm and distance in $\mathbb{R}^2$; note how convenient the equivalent forms (4) and (4a) are to use.

Approximations. Boundedness.

Once we have a way to measure the distance between points in the plane, we can say two points are close and use the language of approximations. We write $\mathbf{x} \underset{\epsilon}{\approx} \mathbf{a}$ if $\mathbf{x}$ and $\mathbf{a}$ are within ϵ of each other:

(6) $$\mathbf{x} \underset{\epsilon}{\approx} \mathbf{a} \iff \|\mathbf{x} - \mathbf{a}\| < \epsilon \iff \mathbf{x} \in B(\mathbf{a}, \epsilon) \; .$$

If we write this using the coordinates of the points, the use of the uniform norm makes it very simple; for according to (4a) (in the $<$ version),

(6a) $$(x, y) \underset{\epsilon}{\approx} (a, b) \iff x \underset{\epsilon}{\approx} a \quad \text{and} \quad y \underset{\epsilon}{\approx} b \; .$$

Definition 24.1A A set $S \subset \mathbb{R}^2$ is **bounded** if there is a K such that

(7) $$\|\mathbf{x}\| \leq K \qquad \text{for all } \mathbf{x} \in S \; ; \qquad \text{equivalently,}$$
(7a) $$|x| \leq K \text{ and } |y| \leq K \qquad \text{for all } (x, y) \in S \; .$$

The equivalence of (7) and (7a) follows immediately from (4).

Properties of the two norms.

Both norms are positive, symmetric, and satisfy the triangle inequality; this is what we ask of something measuring size and distance. The statements are the same for either norm; we give them here for the uniform norm.

(8a)   *positivity* $\|\mathbf{0}\| = 0;$ $\|\mathbf{x}\| > 0,$ if $\mathbf{x} \neq \mathbf{0};$

(8b)   *symmetry* $\|\mathbf{x}\| = \| - \mathbf{x}\|;$

(8c)   *triangle inequality* $\|\mathbf{x}_1 + \mathbf{x}_2\| \leq \|\mathbf{x}_1\| + \|\mathbf{x}_2\|.$

The proof of (8c) for the uniform norm is left as a Question. For the Euclidean norm, (8c) actually is the classical Euclidean triangle inequality, as the picture illustrates: the sum of the lengths of two sides of an honest triangle is greater than the length of the third side. (This explains why the name is used for the inequality in $\mathbb{R}^1$, where no triangles are in sight.)

Questions 24.1

1. Find the distance between $(1, 1)$ and $(4, 5)$, using (i) $|\ |$ (ii) $\|\ \|$.

2. Show $\|\mathbf{a}_1 - \mathbf{a}_2\| \leq |\mathbf{a}_1 - \mathbf{a}_2|$ always; how do $\mathbf{a}_1$ and $\mathbf{a}_2$ look if equality holds?

3. What is the analog of (4) for the Euclidean norm? (Note: $\Rightarrow$ and $\Leftarrow$ must be separated.)

4. Prove the triangle inequality (8c) for the uniform norm. (Use (2) and the triangle inequality for $\mathbb{R}$.)

24.2   Convergence of sequences.

Here is the generalization to $\mathbb{R}^2$ of our earlier work with sequences. We can be brief, because almost no new ideas are involved; everything will look just as before, except that $\|\ \|$ replaces $|\ |$ and $\mathbf{a}$ replaces a everywhere.

Definition 24.2A   Limit of a convergent sequence.

A **sequence** of points in $\mathbb{R}^2$ is an infinite list

(9) $\{\mathbf{x}_n\} \ = \ \mathbf{x}_1, \mathbf{x}_2, \ldots, \mathbf{x}_n, \ldots \ ;$ $\mathbf{x}_n \in \mathbb{R}^2$.

It **converges** if there is a point $\mathbf{a} \in \mathbb{R}^2$, the *limit* of the sequence, such that

(10)   given $\epsilon > 0,$ $\mathbf{x}_n \underset{\epsilon}{\approx} \mathbf{a}$ for $n \gg 1$.

As before, the notation is: $\lim_{n \to \infty} \mathbf{x}_n = \mathbf{a},$ or $\mathbf{x}_n \to \mathbf{a}.$

The only new idea is that convergence of a sequence in $\mathbb{R}^2$ can be checked one coordinate at a time, according to the following theorem.

Theorem 24.2A   Coordinate-wise convergence.

(11) $(x_n, y_n) \to (a, b) \quad \Leftrightarrow \quad x_n \to a$ and $y_n \to b$.

Proof.   If we translate both sides into approximation statements, using the definition of limit, we get two statements which are equivalent by (6a): given any $\epsilon > 0$,

$$(x_n, y_n) \underset{\epsilon}{\approx} (a, b) \text{ for } n \gg 1 \quad \Leftrightarrow \quad x_n \underset{\epsilon}{\approx} a \text{ and } y_n \underset{\epsilon}{\approx} b \text{ for } n \gg 1. \qquad \square$$

Definition 24.2B   A **subsequence** $\{x_{n_i}\}$ of the sequence $\{x_n\}$ is a selection of infinitely many members of $\{x_n\}$, listed in the original order, that is,

$$n_1 < n_2 < n_3 < \dots .$$

Theorem 24.2B   Subsequence theorem for $\mathbb{R}^2$.

(12)   If $\mathbf{x}_n \to \mathbf{a}$, then $\mathbf{x}_{n_i} \to \mathbf{a}$ for every subsequence $\{\mathbf{x}_{n_i}\}$ of $\{\mathbf{x}_n\}$.

Proof. For $\mathbb{R}^1$, this is the Subsequence Theorem 5.3. One can either repeat verbatim its proof, just changing x_n to $\mathbf{x}_n$ throughout, or one can use the Coordinate-wise Convergence Theorem:

$$\mathbf{x}_n \to \mathbf{a} \Rightarrow x_n \to a \text{ and } y_n \to b, \qquad \text{by (11)};$$
$$\Rightarrow x_{n_i} \to a \text{ and } y_{n_i} \to b , \quad \text{by Theorem 5.3};$$
$$\Rightarrow \mathbf{x}_{n_i} \to \mathbf{a}, \quad \text{by (11) again.} \qquad \square$$

Theorem 24.2C   Bolzano-Weierstrass Theorem for $\mathbb{R}^2$.

Let $S \subset \mathbb{R}^2$ be bounded. Then any sequence $\mathbf{x}_n$ of points in S has a convergent subsequence.

Proof.   Let the sequence be (x_n, y_n). Since S is bounded, (7a) above shows that x_n is bounded. Thus by the Bolzano-Weierstrass Theorem (6.3) for $\mathbb{R}$, it has a convergent subsequence x_{n_k}; let $\lim_{n \to \infty} x_{n_k} = a$.

Using (7a) again, the corresponding sequence y_{n_k} is bounded, therefore it has a convergent subsequence $y_{n_{k_i}}$; let $\lim_{i \to \infty} y_{n_{k_i}} = b$.

Since $x_{n_k} \to a$, we know $x_{n_{k_i}} \to a$, by the Subsequence Theorem (5.3).

Therefore we have $\lim_{i \to \infty} (x_{n_{k_i}}, y_{n_{k_i}}) = (a, b)$, by the Coordinate-wise Convergence Theorem, and this is a convergent subsequence of (x_n, y_n). $\qquad \square$

The preceding proof uses coordinate-wise convergence to deduce the theorem from the Bolzano-Weierstrass Theorem for $\mathbb{R}^1$. One could also give a direct proof of the theorem for $\mathbb{R}^2$ by adapting to the plane the bisection argument which proved it in $\mathbb{R}^1$. Exercise 24.2/5 asks you to do this.

We could continue developing the theory of sequences in the plane, but most questions can be reduced to what we have done in $\mathbb{R}^1$, and so we leave them for the Exercises.

Questions 24.2

1. Find $\lim \mathbf{x}_n$, if $\mathbf{x}_n = \left(\dfrac{1}{n}, \dfrac{n+1}{n}\right)$.

2. If $\mathbf{x}_0 = \mathbf{a}$ and $\mathbf{x}_{n+1} = \mathbf{x}_n/2$, prove $\lim \mathbf{x}_n = \mathbf{0}$ both directly and by using coordinate-wise convergence.

3.   Show the sequence $\mathbf{x}_n = (\sin n\pi/5, \cos n\pi/7)$ is bounded, and find a convergent subsequence.

4.   In the proof of the Bolzano-Weierstrass Theorem for $\mathbb{R}^2$, it looks like one could simplify it by also using the B-W Theorem for $\mathbb{R}^1$ to select from y_n a convergent subsequence y_{n_k}; then, letting $b = \lim y_{n_k}$, we would have the limit $(x_{n_k}, y_{n_k}) \to (a, b)$ by coordinate-wise convergence. What do you think?

24.3   Functions on $\mathbb{R}^2$.

We turn now to functions of two variables. Much of what we say will echo the corresponding statements for functions of one variable; we will try to emphasize what is different when you work with more variables.

Speaking informally at first, a (real-valued) function f on $\mathbb{R}^2$ is a rule assigning a single real number $f(\mathbf{a})$ to each point $\mathbf{a}$; in coordinates,

$$(a, b) \;\rightharpoonup\; f(a, b) .$$

Such functions are usually given by expressions in two variables, like

$$\sin(xy + 3), \quad \sqrt{1 - x - y}, \quad \ln(x^2 + y^2) .$$

The general notation for such functions is $f(x, y)$, $f(\mathbf{x})$, or simply f; one can also introduce a dependent variable, writing for example

$$z = f(x, y) .$$

For a function $f(x, y)$ given by an expression in x and y, we say it is *defined* at $\mathbf{a} = (a, b)$ if the expression $f(a, b)$ makes sense; its *domain* $\mathcal{D}_f$ is then the set of all $\mathbf{x} \in \mathbb{R}^2$ for which it is defined.

Examples 24.3   Describe in symbols and geometrically the domains of

$$\text{(a)} \; \sqrt{1 - x^2 - y^2} \qquad \text{(b)} \; \frac{1}{x^2 + y^2} \qquad \text{(c)} \; \sin^{-1}(y/x)$$

Solution. (a) $\{(x, y) : x^2 + y^2 \leq 1\}$; the closed unit disk $\overline{D}(\mathbf{0}, 1)$;

(b) $\{(x, y) : (x, y) \neq (0, 0)\}$; the plane with the origin $\mathbf{0}$ removed;

(c) $\{(x, y) : -1 \leq y/x \leq 1\}$; the two infinite horizontal regions between the lines $y = x$ and $y = -x$, including the lines themselves.

The *graph* G_f of f is the set $\{(\mathbf{x}, w) : w = f(\mathbf{x})\}$, in $\mathbb{R}^2 \times \mathbb{R}$. It is not easily pictured and not always intelligible when computer-drawn on the screen. But it gives the best way to give a formal definition of function; we do that now.

Definition 24.3   A **function** f on $\mathbb{R}^2$ is a set of ordered pairs in $\mathbb{R}^2 \times \mathbb{R}$,
$$G_f = \{(\mathbf{x}, w) : \mathbf{x} \in \mathbb{R}^2, \ w \in \mathbb{R}\},$$
which contains for each $\mathbf{a} \in \mathbb{R}^2$ at most one pair of the form $(\mathbf{a}, c)$.

If $(\mathbf{a}, c) \in G_f$, we say f is **defined** at $\mathbf{a}$, and write $c = f(\mathbf{a})$. Then the **domain** of f is the set
$$\mathcal{D}_f = \{\mathbf{a} \in \mathbb{R}^2 : f \text{ is defined at } \mathbf{a}\} \ .$$

Note that the domain of the function is part of its definition. If the domain is unspecified, it is the *natural domain* — wherever the expression defining the function makes sense. But for some purposes, one may want to restrict the domain to some subset $S \subset \mathcal{D}_f$; this is common in science and engineering, where variables often are restricted to be positive or non-negative, or one may be describing a property of some physical object (the temperature of a metallic plate, say), whose physical boundaries provide natural limits for the variables.

In such cases, we speak of the *restriction of f to the set S*, and if necessary for clarity, use the notation $f|_S$.

This restriction $f|_S$ is a "new" function; it agrees with f on S, and is undefined outside of S. Its properties may differ from those of f. For example, $x^2 + y^2$ is unbounded, but $x^2 + y^2|_R$, where $R = B(\mathbf{0}, r)$, is bounded.

Questions 24.3

1. Give the domain of

(a) $\ln(x^2 + y^2)$ (b) $\dfrac{1}{\|\mathbf{x} - \mathbf{a}\| \ \|\mathbf{x} - \mathbf{a}'\|}$ (c) $\tan(x^2 + y^2)\big|_R$, $R = D(\mathbf{0}, 2)$.

24.4   Continuous functions.

The basic idea of continuity for functions on $\mathbb{R}^2$ is the same as for functions of one variable — if $\mathbf{x}$ changes only a little, then $f(\mathbf{x})$ changes only a little. Here is the formal definition.

Remark.   The statement "P is true for $\mathbf{x} \approx \mathbf{a}$" means that there is some unspecified $\delta > 0$ such that P is true for $\mathbf{x} \underset{\delta}{\approx} \mathbf{a}$.

Definition 24.4A   We say $f(\mathbf{x})$ is **continuous at a point** $\mathbf{a} \in \mathcal{D}_f$ if

(13)   given $\epsilon > 0$, $f(\mathbf{x}) \underset{\epsilon}{\approx} f(\mathbf{a})$ for $\mathbf{x} \approx \mathbf{a}$, $\mathbf{x} \in \mathcal{D}_f$.

We say $f(\mathbf{x})$ is **continuous** if it is continuous at each point of $\mathcal{D}_f$.

Notice in the definition that if the domain is a closed box, say, and $\mathbf{a}$ is on its boundary, then we only look at the behavior of $f(\mathbf{x})$ at those points near to $\mathbf{a}$ which also lie in $\mathcal{D}_f$. This is analogous to the definition of continuity at an endpoint, for a function $f(x)$ which is defined on a compact interval $[a, b]$.

Actually, we need a slightly more general concept; consider this example.

Example 24.4A   In $\mathbb{R}^2$, let $f(\mathbf{x}) = \begin{cases} 1, & \|\mathbf{x}\| \le 1; \\ 0, & \text{otherwise.} \end{cases}$

Is $f(\mathbf{x})$ continuous on the closed box $\overline{B}(\mathbf{0}, 1) = \{x : \|\mathbf{x}\| \le 1\}$?

It is a matter of definition. The function $f(\mathbf{x})$ is discontinuous at the points on the square $\|\mathbf{x}\| = 1$, and these are in the closed unit box. So the answer ought to be "no".

But this would be inconvenient for applications — if the box represented a metal plate, for instance, and $f(\mathbf{x})$ its temperature function, then the plate would be at the constant temperature 1 everywhere, so that its temperature ought to be considered to be a continuous function. In other words, we should _first_ restrict the domain of $f(\mathbf{x})$ to the closed box, and <u>then</u> answer the question for the restricted function: "yes".

As a result of this discussion, we extend Definition 24.4A as follows.

Definition 24.4B   Let $S \subset \mathcal{D}_f$. We say $f(\mathbf{x})$ is **continuous on** S if the restricted function $f|_S$ is continuous on its domain S, i.e., at every $\mathbf{a} \in S$:

(14)   given $\epsilon > 0, \quad f(\mathbf{x}) \underset{\epsilon}{\approx} f(\mathbf{a}) \text{ for } \mathbf{x} \approx \mathbf{a}, \mathbf{x} \in S$.

The algebraic theorems about continuous functions are the two-dimensional analogues of the ones in Chapter 11, and they are proved the same way. In general, if $f(\mathbf{x})$ and $g(\mathbf{x})$ are continuous on some set S where both are defined, so are $f + g$, fg, and also f/g provided $g(\mathbf{x}) \ne 0$ on S. If $g(u)$ is continuous on $\mathbb{R}^1$, then $g(f(\mathbf{x}))$ is continuous wherever $f(\mathbf{x})$ is continuous.

In particular, since the coordinate functions x and y on $\mathbb{R}^2$, defined by

$$x(a, b) = a, \qquad y(a, b) = y, \quad \text{for all } (a, b) \in \mathbb{R}^2$$

are continuous (see the Questions), it follows that any polynomial $p(x, y)$ is continuous on $\mathbb{R}^2$, any rational function $p(x, y)/q(x, y)$ is continuous wherever its denominator is nonzero, and composite functions like

$$\sqrt{x^2 + y^2}, \quad e^{ax+by}, \quad J_0(x/y), \quad \ln(x - y), \quad |\mathbf{x} - \mathbf{a}|,$$

are continuous at all points where they are defined.

Questions 24.4

1.   Prove that the coordinate functions x and y are continuous on $\mathbb{R}^2$.

2.   Let $f(\mathbf{x})$ be as in Example 24.4B. On which of the following sets S is $f(\mathbf{x})$ continuous, according to Definition 24.4B?

(a) $x^2 + y^2 \le 1$ (b) $x + y = 1$, $x \ge 0$, $y \ge 0$ (c) $\mathbf{0}$

(d) the square $0 \le x \le 1$, $0 \le y \le 1$ (e) $x = 1$ (f) $x = 1$, $y > 1$

24.5   Limits and continuity.

The definitions and results look a lot like they do for functions of one variable — all one does is replace x by $\mathbf{x}$.

Definition 24.5   Limit of a function.   We say $\lim\limits_{\mathbf{x} \to \mathbf{a}} f(\mathbf{x}) = L$ if

$$(15) \qquad\qquad \text{given } \epsilon > 0, \quad f(\mathbf{x}) \underset{\epsilon}{\approx} L \quad \text{for } \mathbf{x} \underset{\neq}{\approx} \mathbf{a}, \; \mathbf{x} \in \mathcal{D}_f .$$

Notice that just as for functions of one variable, the function $f(\mathbf{x})$ need not be defined at $\mathbf{a}$ to have a limit there. To express continuity in terms of limits, we have, as in $\mathbb{R}^1$,

Theorem 24.5A   Limit form for continuity.

$$(16) \qquad\qquad f(\mathbf{x}) \text{ is continuous at } \mathbf{a} \quad \Leftrightarrow \quad \lim\limits_{\mathbf{x} \to \mathbf{a}} f(\mathbf{x}) = f(\mathbf{a}) .$$

Proof.   This follows immediately by comparing the definition (13) of limit with the definition (15) of continuity at $\mathbf{a}$. $\square$

There can be many different types of behavior at a point where $f(\mathbf{x})$ is not continuous, so there is no simple classification of points of discontinuity. We do have two recognizable types however:

the *removable* discontinuity, where the limit L exists, but is not equal to $f(\mathbf{a})$; the discontinuity is removed by redefining $f(\mathbf{a})$ to be L;

the *infinite* discontinuity, where the limit (suitably defined) is ∞ or $-\infty$.

The difficulty with jump discontinuities is that we do not have just two directions of appoach (as in $\mathbb{R}^1$), but infinitely many possible directions.

Theorem 24.5B   Sequential continuity.   *If $f(\mathbf{x})$ is continuous at $\mathbf{a}$, and defined for all x_n, then*

$$\mathbf{x}_n \to \mathbf{a} \;\Rightarrow\; f(\mathbf{x}_n) \to f(\mathbf{a}) .$$

Proof.   The proof is the same as that for the one-variable Theorem 11.5 . $\square$

As is true for one-variable functions, the contrapositive version of the theorem is a useful way of showing $f(\mathbf{x})$ has a discontinuity at $\mathbf{a}$: produce a sequence $\mathbf{x}_n$ which converges to $\mathbf{a}$, but for which $\lim f(\mathbf{x}) \neq f(\mathbf{a})$.

Example 24.5   Let $f(\mathbf{x}) = x/y$. Show that $f(\mathbf{x})$ has a discontinuity at $\mathbf{0}$ which is not removable.

Solution.   The natural domain of f is the plane with the x-axis removed: $\{(x, y) : y \neq 0\}$. We must show that no definition for $f(\mathbf{0})$ will make f continuous at $\mathbf{0}$.

Given any real number c, consider the sequence $\mathbf{x}_n = (c/n, 1/n)$. Then we have $\mathbf{x}_n \to \mathbf{0}$, and $f(\mathbf{x}_n) = c$ for all n. Thus if $f(x)$ is to be continuous at $\mathbf{0}$, the Sequential Continuity Theorem requires that we define $f(\mathbf{0}) = c$. This is a contradiction, since c can be any real number.   □

Note that the difficulty is not that $f(x, y)$ is undefined at points on the x-axis, since these points would be excluded anyway in testing for continuity — they are not in $\mathcal{D}_f$. The trouble is rather that x/y approaches different values according to what direction you approach $(0, 0)$ from.

Theorem 24.5C   Positivity theorem for continuous functions

(17) $f(\mathbf{x})$ continuous at $\mathbf{a}$, $f(\mathbf{a}) > 0$ $\Rightarrow$ $f(\mathbf{x}) > 0$ for $\mathbf{x} \approx \mathbf{a}$, $\mathbf{x} \in \mathcal{D}_f$.

Proof.   The proof is easy, and left as an Exercise.   □

Questions 24.5

1.   Let $f(x, y) = a_0 + a_1 x + a_2 y + \dots$ be a polynomial in x and y, with $a_0 \neq 0$. Show that $f(x, y) \neq 0$ for $x \approx 0$, $y \approx 0$.

2.   Let $\mathbf{a} = (a, a)$. Prove from the definition that $\displaystyle\lim_{\mathbf{x} \to \mathbf{a}} \frac{\sin(x - y)}{x - y} = 1$.

(You can use $\displaystyle\lim_{u \to 0} \frac{\sin u}{u} = 1$.)

3.   Assume $f(t)$ continuous. By using sequential continuity, show that

$$\lim_{n \to \infty} \int_{1/n}^{1 - 1/n} f(t)\, dt = \int_0^1 f(t)\, dt .$$

24.6   Compact sets in $\mathbb{R}^2$.

In the next two sections, we want to extend to continuous functions on $\mathbb{R}^2$ our earlier results about continuous functions on compact intervals.

As the analog of an interval, we will mostly use rectangles R in the plane having their sides parallel to the coordinate axes:

$$R = \{(x, y) : a \le x \le b, \ c \le y \le d\} .$$

Such a rectangle can be looked on as the product of the two intervals $[a, b]$ and $[c, d]$ on the x-axis and y-axis respectively:

$$R = \{(x, y) : x \in [a, b], \ y \in [c, d]\} ;$$
$$= [a, b] \times [c, d] .$$

We can classify such rectangles according to the kinds of intervals which form their sides. The two most important kinds are:

$$[a, b] \times [c, d] \qquad \text{finite closed rectangle;}$$
$$(a, b) \times (c, d) \qquad \text{finite open rectangle.}$$

If one of the intervals is semi-infinite, the important kinds for us are:

$$[a, b] \times [c, \infty) \qquad \text{semi-infinite closed rectangle,}$$
$$(a, b) \times (c, \infty) \qquad \text{semi-infinite open rectangle.}$$

The analog in $\mathbb{R}^2$ of a compact interval in $\mathbb{R}^1$ is a *compact set*: one having the property called sequential compactness which makes the proofs work.

Definition 24.6 Sequential compactness property.

A set $S \subset \mathbb{R}^2$ is **compact** if it has the sequential compactness property:

(18)   *any sequence $\mathbf{x}_n \in S$ has a subsequence $\mathbf{x}_{n_i}$ converging to an $\mathbf{a} \in S$.*

Theorem 24.6 Compactness Theorem for finite closed rectangles.

If I and J are compact intervals, then $S = I \times J$ is compact.

Proof. We show S has the sequential compactness property (18).

Let $\mathbf{x}_n$ be a sequence in S; since S is bounded, the Bolzano-Weierstrass Theorem for $\mathbb{R}^2$ shows that $\mathbf{x}_n$ has a subsequence $\mathbf{x}_{n_i}$ converging to a point $\mathbf{a} = (a, b)$. We prove $\mathbf{a} \in S$, as follows.

$$(x_{n_i}, y_{n_i}) \in I \times J \;\Rightarrow\; x_{n_i} \in I \text{ and } y_{n_i} \in J, \quad \text{for all } n_i \,;$$
$$(x_{n_i}, y_{n_i}) \to (a, b) \;\Rightarrow\; x_{n_i} \to a \text{ and } y_{n_i} \to b, \quad \text{by (11)}.$$

Since I is a closed interval, $x_{n_i} \in I \Rightarrow \lim x_{n_i} \in I$, by the Limit Location Theorem (5.3A). This shows $a \in I$; the proof that $b \in J$ is analogous. Thus $\mathbf{a} = (a, b) \in I \times J$.   □

There are many sets besides finite closed rectangles which are compact, i.e., have the sequential compactness property: any finite closed disk $\overline{D}(\mathbf{a}, r)$, for example. In the next chapter, Point-sets in $\mathbb{R}^2$, we shall describe all compact sets in the plane, as part of a brief introduction to what is called "point-set topology".

24.7   Continuous functions on compact sets in $\mathbb{R}^2$.

Just as for $\mathbb{R}^1$, there are three main theorems, and their proofs are almost identical to the corresponding proofs for $\mathbb{R}^1$ in Chapter 13. So this is a good opportunity to review those arguments.

Theorem 24.7A   Boundedness theorem.

If $f(\mathbf{x})$ is continuous on the compact set $S \subset \mathbb{R}^2$, then $f(\mathbf{x})$ is bounded on S, i.e., there is a constant K such that

(19) $$|f(\mathbf{x})| < K \quad \text{for all } \mathbf{x} \in S \,.$$

Proof. By indirect argument. If $f(\mathbf{x})$ is not bounded above on S, we can find for each $n \in \mathbb{N}$ a point $\mathbf{x}_n \in S$ such that

(20) $$f(\mathbf{x}_n) > n, \qquad n = 1, 2, 3, \ldots .$$

Since S has the sequential compactness property, the sequence $\mathbf{x}_n$ has a subsequence $\mathbf{x}_{n_i}$ converging to a point $\mathbf{a} \in S$, by (18). Since $f(\mathbf{x})$ is continuous on S, the Sequential Continuity Theorem 24.5B says

$$\lim_{i \to \infty} f(\mathbf{x}_{n_i}) = f(\mathbf{a}) .$$

On the other hand,

$$\lim_{i \to \infty} f(\mathbf{x}_{n_i}) = \infty , \qquad \text{by (20) above.}$$

This is a contradiction, for since $\mathbf{a} \in S$, the function f is defined at $\mathbf{a}$, i.e., $f(\mathbf{a})$ is a number in $\mathbb{R}$. $\qquad\square$

This shows $f(\mathbf{x})$ is bounded above on S.

By a similar argument, it is bounded below (or apply the above to $-f(\mathbf{x})$). Thus $f(\mathbf{x})$ is bounded on S. $\qquad\square\square$

Theorem 24.7B   Maximum theorem.

If $f(\mathbf{x})$ is continuous on the compact set S, then $f(\mathbf{x})$ attains its maximum and minimum on S, that is, for some $\mathbf{a}$ and $\mathbf{a}'$ in S,

$$f(\mathbf{a}) = M = \sup_S f(\mathbf{x}), \qquad f(\mathbf{a}') = m = \inf_S f(\mathbf{x}) .$$

Proof.   Since $f(\mathbf{x})$ is continuous on S, by the preceding theorem it is bounded on S. That is, the set $\{f(\mathbf{x}) : \mathbf{x} \in S\}$ is bounded; hence by the Completeness Property, it has a supremum and infimum; let

$$M = \sup_S f(\mathbf{x}) = \sup\{f(\mathbf{x}) : \mathbf{x} \in S\} \quad \text{and} \quad m = \inf_S f(\mathbf{x}) .$$

We now show there is a point $\mathbf{a} \in S$ such that $f(\mathbf{a}) = M$.

For each integer $n > 0$ we can find a point $\mathbf{x}_n \in S$ such that

$$f(\mathbf{x}_n) > M - \frac{1}{n} ,$$

for otherwise M would not be the <u>least</u> upper bound of $\{f(\mathbf{x}) : \mathbf{x} \in S\}$. Since S has the sequential compactness property, (18) says that from $\mathbf{x}_n$ we can select a subsequence $\mathbf{x}_{n_i}$ converging to a point $\mathbf{a} \in S$. We then get a diagram where the limits are taken as $i \to \infty$,

$$M - \frac{1}{n_i} \quad < \quad f(\mathbf{x}_{n_i}) \quad \leq \quad M ;$$
$$\downarrow \qquad\qquad \downarrow \qquad\qquad \downarrow$$
$$M \quad \leq \quad f(\mathbf{a}) \quad \leq \quad M .$$

Here the middle arrow is valid by the Sequential Continuity Theorem 24.5B, since the subsequence converges to $\mathbf{a}$ and $f(\mathbf{x})$ is continuous. The bottom line inequalities result from the Limit Location Theorem 5.3A. These last inequalities now show that $f(\mathbf{a}) = M$. $\qquad\square$

The proof that $f(\mathbf{a}') = m$ for some $\mathbf{a}'$ is similar, or one can apply the preceding to the function $-f(\mathbf{x})$. $\qquad\square\square$

Theorem 24.7C   Uniform continuity theorem.

    A continuous function on a compact set S is **uniformly continuous** *on S,* i.e.,

given any $\epsilon > 0$, there is a $\delta > 0$ (depending on ϵ) such that

$$f(\mathbf{x}) \underset{\epsilon}{\approx} f(\mathbf{x}') \qquad \text{whenever } \mathbf{x} \underset{\delta}{\approx} \mathbf{x}' .$$

Proof. Indirect argument. Suppose $f(\mathbf{x})$ is <u>not</u> uniformly continuous on S. Then there is an $\epsilon > 0$ such that for each value of n, we can find a pair of points $\mathbf{x}_n$, $\mathbf{x}'_n \in S$ such that

$$(21) \qquad\qquad |f(\mathbf{x}_n) - f(\mathbf{x}'_n)| > \epsilon, \qquad \text{yet } \mathbf{x}_n \underset{1/n}{\approx} \mathbf{x}'_n .$$

By (18), there is a subsequence $\mathbf{x}_{n_i}$ converging to a point $a \in S$. The right side of (21) shows $\mathbf{x}'_{n_i}$ also converges to $\mathbf{a}$; thus

$$\mathbf{x}_{n_i} \to \mathbf{a}, \qquad \mathbf{x}'_{n_i} \to \mathbf{a} .$$

Since $f(\mathbf{x})$ is continuous on S, sequential continuity (24.5B) shows that

$$|f(\mathbf{x}_{n_i}) - f(\mathbf{x}'_{n_i})| \to |f(\mathbf{a}) - f(\mathbf{a})| = 0,$$

which contradicts (21). $\qquad\qquad\qquad\qquad\qquad\qquad\qquad\qquad\qquad\qquad$ □

    Using the Euclidean norm rather than the uniform norm would not affect the results in this chapter: using either norm, the same sequences are convergent, the same functions continuous, and the same theorems about them are true. The two norms are *equivalent*; see three of the Exercises for more about this.

Questions 24.7

    1. Prove that a finite line segment in the plane is compact. (You can assume it is non-vertical; take it in the form $y = rx + s$, $x \in [a_1, a_2]$.)

    2. Suppose $f(x)$ is continuous and positive on $[a, b]$. Prove there is a pair of points x', x'', in $[a, b]$ for which the trapezoid having base $[x', x'']$ and parallel sides of heights $f(x')$ and $f(x'')$ has biggest area.

    3. Suppose $f(x, y)$ is a function of the form $g(x)h(y)$, where $g(x)$ and $h(y)$ are respectively continuous on the compact intervals I and J. Deduce from the one-variable Uniform Continuity Theorem 13.5 that $f(x, y)$ is uniformly continuous on $I \times J$. (Do not cite Theorem 24.7C; if stuck, cf. proof of the product rule for differentiation.)

    4. Could you prove the Maximum Theorem for the case of a continuous function on a compact rectangle $I \times J$, along the following lines, by using the one-variable Maximum Theorem 13.3?

    For each $y_0 \in J$, let $\overline{x}(y_0)$ be the maximum point of $f(x, y_0)$ for $x \in I$; and let $\overline{y}$ be the maximum point of $\overline{x}(y)$, for $y \in J$.

    Then $(\overline{x}, \overline{y})$ is the maximum point for $f(x, y)$ on $I \times J$.

    To find the position of the tallest soldier in a rectangular formation, first find the position of the tallest in each row, then find the row with the tallest of the tall.

Exercises

24.1

1. Show the point $(\mathbf{x} + \mathbf{y})/2$ is equally distant from $\mathbf{x}$ and $\mathbf{y}$, without using coordinates.

2. Prove (using the uniform norm) that $\mathbf{x} \underset{\epsilon}{\approx} \mathbf{y},\ \mathbf{y} \underset{\epsilon}{\approx} \mathbf{z}\ \Rightarrow\ \mathbf{x} \underset{2\epsilon}{\approx} \mathbf{z}$.

3. Prove that there are positive constants $c \leq d$ such that
$$c|\mathbf{x}| \leq \|\mathbf{x}\| \leq d|\mathbf{x}| \quad \text{for all } \mathbf{x} \in \mathbb{R}^2 .$$
(Two norms $|\ |$ and $\|\ \|$ related in this way are called **equivalent**.)

24.2

1. Prove directly from the definition of convergence that the sequence defined recursively by
$$\mathbf{x}_n = \tfrac{1}{2}(\mathbf{x}_{n-1} + \mathbf{a}) \qquad (\mathbf{x}_0 \text{ arbitrary})$$
converges to $\mathbf{a}$. (Hint: express $\mathbf{x}_n - \mathbf{a}$ in terms of $\mathbf{x}_{n-1} - \mathbf{a}$.)

2. Given a starting point $\mathbf{x}_1 = (x, y)$, construct the sequence from it defined by $\mathbf{x}_n = (x^n, y^n)$. Find the region $\mathcal{D} \subset \mathbb{R}^2$ having the property
$$\mathbf{x}_n \text{ converges} \quad \Leftrightarrow \quad \mathbf{x}_1 \in \mathcal{D} .$$
Draw $\mathcal{D}$, and divide it up into subregions so that for the $\mathbf{x}_1$ in any subregion, the corresponding sequences $\mathbf{x}_n$ all have the same limiting point.

3. A sequence $\mathbf{x}_n$ is **convergent in the Euclidean norm** if it satisfies Definition 24.2A, where $\mathbf{x} \underset{\epsilon}{\approx} \mathbf{a}$ means $|\mathbf{x} - \mathbf{a}| < \epsilon$ (Euclidean norm). Prove that

$\mathbf{x}_n$ is convergent (in the uniform norm)

$\Leftrightarrow \quad \mathbf{x}_n$ is convergent in the Euclidean norm.

(All you need is the equivalence of the two norms, i.e., the result in 24.1/3 above; you do not need the explicit definition of the two norms. This exercise shows that the notion of convergence does not depend on which norm is used.)

4. (a) Give a definition of *Cauchy sequence* in $\mathbb{R}^2$.

(b) Prove that a Cauchy sequence in $\mathbb{R}^2$ converges.

(Use the Coordinate-wise Convergence Theorem to reduce it to the same statement for $\mathbb{R}^1$.)

(c) Prove that a convergent sequence in $\mathbb{R}^2$ is a Cauchy sequence. Do it directly, without reducing it to $\mathbb{R}^1$.

5. Prove the Bolzano-Weierstrass Theorem for $\mathbb{R}^2$ by adapting the method of bisection used to prove it in $\mathbb{R}^1$.

24.3

    1. Describe the graph of $f(\mathbf{x}) = \|\mathbf{x}\|$. (Hint: divide its domain up into four regions.)

24.4

    1. Prove that $f(\mathbf{x}) = \|\mathbf{x}\|$ is continuous. (Hint: use the triangle inequality to relate $\|\mathbf{x} - \mathbf{a}\|$ and $\|\mathbf{x}\| - \|\mathbf{a}\|$, if $\|\mathbf{x}\| \geq \|\mathbf{a}\|$.)

    2. In the definition (24.4) of continuity at $\mathbf{a}$, the symbol $\approx$ refers to distance measured by the uniform norm. Prove that if the Euclidean norm is used instead, one gets the same notion of continuity; i.e., if f is continuous at $\mathbf{a}$ using one norm, it is also continuous at $\mathbf{a}$ using the other norm (cf. 24.2/3 for a similar exercise).

    3. We can generalize the idea of function by allowing the value of the function to be a point in $\mathbb{R}^n$, rather than just a single real number. For example, a vector-valued function of one variable would be a mapping

$$\mathbf{f} \; : \; \mathbb{R}^1 \to \mathbb{R}^2, \quad \text{where } \mathbf{f}(t) = (f_1(t), \; f_2(t)) \; .$$

The image of $\mathbb{R}^1$ under $\mathbf{f}$ is in general a curve in $\mathbb{R}^2$; t can be thought of as a time parameter. (The function $\mathbf{f}$ might be defined only on a subset $\mathcal{D}$ of $\mathbb{R}$.)

    We define $\mathbf{f}$ to be **continuous** at t_0 if

$$\text{given } \epsilon > 0, \quad \mathbf{f}(t) \underset{\epsilon}{\approx} \mathbf{f}(t_0) \;\; \text{for } t \approx t_0, \; t \in \mathcal{D} \; .$$

Prove: $\mathbf{f}(t)$ is continuous at t_0 $\;\Leftrightarrow\;$ each $f_i(t)$ is continuous at t_0.

    4. Suppose $f(x)$ has a continuous derivative for all x. Let

$$g(x,y) \; = \; \frac{f(x) - f(y)}{x - y} \; , \;\; x \neq y \; ; \qquad g(a,a) = f'(a), \; a \in \mathbb{R}.$$

Prove $g(x,y)$ is a continuous function on $\mathbb{R}^2$.

24.5

    1. Prove the function defined by

$$f(x,y) = \frac{xy}{x^2 + y^2} \; , \;\; \mathbf{x} \neq \mathbf{0}; \qquad f(\mathbf{0}) = 0$$

is not continuous at $\mathbf{0}$. (Use the Sequential Continuity Theorem.)

    2. Prove the Positivity Theorem (24.5C) for continuous functions.

    3. If $\mathbf{x}_n \to \mathbf{a}$, does it follow that $|\mathbf{x}_n| \to |\mathbf{a}|$? Proof or counterexample.

    4. Let $g(\mathbf{x})$ be a continuous function on $\mathbb{R}^2$, and let

$$\mathcal{D}^- = \{\mathbf{x} : g(\mathbf{x}) < 0\}, \qquad \mathcal{D}^+ = \{\mathbf{x} : g(\mathbf{x}) > 0\}, \qquad \mathcal{D}^0 = \{\mathbf{x} : g(\mathbf{x}) = 0\}.$$

Let $\mathbf{f}(t)$ be a continuous mapping from $[0, 1]$ to $\mathbb{R}^2$, as in Exercise 24.4/3 above; its image (or range) is in general a continuous curve in $\mathbb{R}^2$.

    Assume $\mathbf{f}(0) \in \mathcal{D}^-$ and $\mathbf{f}(1) \in \mathcal{D}^+$. Prove that the curve must cross the 0-contour set $\mathcal{D}^0$ at least once.

    5. Let $\mathbf{x}_n = (\cos n, \sin n)$. Prove the sequence has a subsequence which converges to a point on the unit circle. (Use sequential continuity.)

24.6

1. Prove the closed r-disk $\overline{D}(\mathbf{0}, r) = \{\mathbf{x} : |\mathbf{x}| \leq r\}$ is compact. (Hint: imitate the proof of Theorem 24.6; prove $\mathbf{a} \in S$ by using the function $|\mathbf{x}|^2$.)

2. As in Exercise 24.4/3, let $\mathbf{f} : [a, b] \to \mathbb{R}^2$, where $\mathbf{f}(t) = \big(f_1(t), f_2(t)\big)$, and f_1, f_2 are continuous on $[a, b]$.

The image of $\mathbf{f}$ is a curve in $\mathbb{R}^2$; prove it is a compact subset of $\mathbb{R}^2$. (Use the Bolzano-Weierstrass Theorem in $\mathbb{R}^1$.)

3. Prove the unit circle $\{(x, y) : x^2 + y^2 = 1\}$ is compact. (You can use methods suggested by either of the preceding exercises.)

24.7

1. Let S be a compact set in the plane. Prove there is a point of S which is closest to the origin (i.e., no point of S is closer, in the sense of the usual Euclidean distance).

2. Let S be a compact set in $\mathbb{R}^2$, and let $\mathbf{a}_1$ and $\mathbf{a}_2$ be two distinct points not in S.. Prove that there is an $\mathbf{x} \in S$ which subtends the largest angle between the points $\mathbf{a}_1$ and $\mathbf{a}_2$. (By "angle" we mean the angle $\leq \pi$ between the two vectors from $\mathbf{x}$ to $\mathbf{a}_1$ and from $\mathbf{x}$ to $\mathbf{a}_2$.)

3. Fix $a > 0$, and let $R = \{(x, y) : x \geq a, \ y \geq a\}$. Prove that the function $f(x, y) = \sqrt{x + y}$ is uniformly continuous on R.

(Note that R is not a compact set. Do it directly from the definition of uniform continuity. Use the identity $(A + B)(A - B) = A^2 - B^2$.)

4. Prove that $\|\mathbf{x}\|$ is uniformly continuous on $\mathbb{R}^2$ (cf. Exercise 24.4/1).

Problems

24-1 Let $x(t), y(t)$ be functions defined for $t \geq 0$, and suppose there is a number $r > 0$ such that

$$\|\big(x(t), y(t)\big)\| \leq r \quad \text{for all } t \geq 0 .$$

Prove there is at least one point $\mathbf{a} = (x_0, y_0)$ such that, given any $\epsilon > 0$, the point $(x(t), y(t))$ lies in the ϵ-disk around $\mathbf{a}$ for infinitely many values of t, separated from each other by at least unit intervals of t: $t_{n+1} \geq t_n + 1$.

Entomologically, this means a two-dimensional fly buzzing around inside the unit box forever must visit arbitrarily small neighborhoods of some point infinitely often — must, in other words, keep returning to the vicinity of some point in the box. As a special case, it might just get to the point right away and sit there forever.

Answers to Questions

24.1

1. (i) 5   (ii) 4

2. Let $\mathbf{a}_1 = (a_1, b_1)$, $\mathbf{a}_2 = (a_2, b_2)$.

Both $|a_1 - a_2|$ and $|b_1 - b_2|$ are $\leq \sqrt{(a_1 - a_2)^2 + (b_1 - b_2)^2}$; therefore
$$\|\mathbf{a}_1 - \mathbf{a}_2\| = \max(|a_1 - a_2|, |b_1 - b_2|) \leq |\mathbf{a}_1 - \mathbf{a}_2|.$$
Equality holds if $a_1 = a_2$ or $b_1 = b_2$; geometrically this means the two points are on a line parallel to the x-axis or y-axis.

3. $|\mathbf{x}| \leq K \Rightarrow |x| \leq K, |y| \leq K;$ $|x| \leq K, |y| \leq K \Rightarrow |\mathbf{x}| \leq K\sqrt{2}$.

4. $\|\mathbf{x}_1 + \mathbf{x}_2\| = |x_1 + x_2|$ or $|y_1 + y_2|$, by (2); say the latter. Then
$$\|\mathbf{x}_1 + \mathbf{x}_2\| = |y_1 + y_2| \leq |y_1| + |y_2| \leq \|\mathbf{x}_1\| + \|\mathbf{x}_2\| .$$

24.2

1. Since $1/n \to 0$ and $(n+1)/n \to 1$, we get $\lim \mathbf{x}_n = (0, 1)$ by coordinate-wise convergence (24.2A)

2. Directly: $\mathbf{x}_n = \mathbf{a}/2^n$; so, given $\epsilon > 0$, $\|\mathbf{x}_n\| = \|\mathbf{a}\|/2^n < \epsilon$ if $n \gg 1$. This shows that $\lim \mathbf{x}_n = \mathbf{0}$, by definition of convergence.

By (11): $\mathbf{x}_n = (a/2^n, b/2^n)$; $a/2^n \to 0$, $b/2^n \to 0$; thus $\mathbf{x}_n \to \mathbf{0}$.

3. $\|x_n\| \leq 1$ for all n, so it is bounded. The subsequence $\mathbf{x}_{70k}$ is the constant sequence $\{(0, 1)\}$, so it is convergent. (Other convergent subsequences could be produced.)

4. The subsequence x_{n_k} converges, but the subsequence y_{n_k} is mislabeled and must be called y_{m_k}, since it does not consist of the "partners" of the x_{n_k}. The rest of the proof is then invalidated.

24.3

1. (a) the plane with $\mathbf{0}$ removed   (b) the plane with the two points $\mathbf{a}$ and $\mathbf{a}'$ removed

(c) $\{(x, y) : x^2 + y^2 < r, \ x^2 + y^2 \neq \pi/2\}$: the open disk of radius 2 and center $\mathbf{0}$, with the circle of radius $\sqrt{\pi/2}$ and center at the origin removed.

24.4

1. Let $f(x, y) = y$ for example. Then,

given $\epsilon > 0$, $|f(\mathbf{x}) - f(\mathbf{a})| = |y - b|$, which is $< \epsilon$ if $\mathbf{x} \underset{\epsilon}{\approx} \mathbf{a}$.

2. (a) yes (b) no (c) yes (d) yes (e) no (f) yes

24.5

1. $f(x, y)$ is continuous, so the Positivity Theorem can be used:

if $a_0 > 0$, then $f(\mathbf{0}) > 0$, so $f(\mathbf{x}) > 0$ for $\mathbf{x} \approx \mathbf{0}$;

if $a_0 < 0$, then $-f(\mathbf{0}) > 0$, so $-f(\mathbf{x}) > 0$ and thus $f(\mathbf{x}) < 0$ for $\mathbf{x} \approx \mathbf{0}$.

2. Given $\epsilon > 0$, $\dfrac{\sin u}{u} \underset{\epsilon}{\approx} 1$ if $u \underset{\delta}{\approx} 0$.   Set $u = x - y$; the preceding implies

$$\dfrac{\sin(x-y)}{x-y} \underset{\epsilon}{\approx} 1 \quad \text{if } (x,y) \underset{\delta/2}{\approx} (a,a), \quad \text{since} \quad (x,y) \underset{\delta/2}{\approx} (a,a) \quad \Rightarrow \quad u \underset{\delta}{\approx} 0.$$

3. Apply the Sequential Continuity Theorem to $F(x,y) = \displaystyle\int_x^y f(t)dt$. First
we have to show $F(x,y)$ is continuous. Break it into two parts:

$$F(x,y) = \int_x^{1/2} f(t)dt + \int_{1/2}^y f(t)dt.$$

Both integrals are continuous functions (by the Second Fundamental Theorem 20.2A); therefore $F(x,y)$ is continuous; therefore by the Sequential Continuity
Theorem 24.5B, we get $\displaystyle\lim_{n\to\infty} F(1/n, 1-1/n) = F(0,1)$.

24.7

1. Given a sequence of points (x_n, y_n) on the line segment between (a_1, b_1)
and (a_2, b_2), the sequence x_n has a subsequence x_{n_i} which converges to some
point $a_0 \in [a_1, a_2]$, since the interval is compact. Then since the points lie on the
line, we have by the limit theorems (or sequential continuity),

$$y_{n_i} = rx_{n_i} + s \;\to\; b_0 = ra_0 + s; \qquad (x_{n_i}, y_{n_i}) \;\to\; (a_0, b_0) \, ,$$

which shows the original sequence has a subsequence converging to a point on the
line segment; therefore the line segment is compact, since it has the sequential
compactness property.

2. The area of the trapezoid is given by

$$F(x', x'') \;=\; \frac{1}{2}(x'' - x')\big(f(x') + f(x'')\big) \, .$$

Since $f(x)$ is continuous on $[a,b]$, so is $F(x', x'')$ on the rectangle $[a,b] \times [a,b]$.
Therefore by the Maximum Theorem 24.7B it has a maximum point (x_0', x_0'') in
the rectangle; these values x_0' and x_0'' correspond to the largest trapezoid.

3. In outline, we have, given ϵ, and "changing one thing at a time",

$$f(\mathbf{x}') - f(\mathbf{x}'') \;=\; g(x')h(y') - g(x'')h(y') + g(x'')h(y') - g(x'')h(y'')$$
$$|f(\mathbf{x}') - f(\mathbf{x}'')| \;\le\; |g(x') - g(x'')||h(y')| + |g(x'')||h(y') - h(y'')|$$
$$<\; \epsilon M_1 + M_2 \epsilon \, ,$$

if $|x' - x''| < \delta$ and $|y' - y''| < \delta$, for some δ not depending on ϵ; the mid-
dle inequality uses the triangle inequality; the last uses the fact that $g(x)$ and
$h(y)$ are continuous and therefore bounded and uniformly continuous on I and J
respectively.

4. The argument won't work; the first use of the Maximum Theorem is
all right, since $f(x, y_0)$ is a continuous function of x, but the second use is not,
because we do not know that $\bar{x}(y)$ is a continuous function of y.

25

Point Sets in the Plane

25.1   Closed sets in $\mathbb{R}^2$.

In the last chapter we talked about "closed" and "open" disks, boxes, and rectangles, While these particular sets of points will be all we need for the last two chapters (so this one could be omitted), as a foundation for further work in analysis it is important to know how these words — closed, open, — are applied to more general sets of points in the plane, as well as their connection to the "compact" sets for which the theorems about continuous functions hold. We describe this now, as an introduction to what is called *point-set topology*

The definitions, theorems, and proofs will be stated for $\mathbb{R}^2$, but they are the same for $\mathbb{R}^1$ as well, changing $\mathbf{x}$ to x and $\| \ \|$ to $| \ |$. They are also true for the analogous k-dimensional space
$$\mathbb{R}^k = \{(x_1, x_2, \ldots, x_k) : x_i \in \mathbb{R} \text{ for all } i\} \ ,$$
if we use as the distance between two points $\mathbf{x}$ and $\mathbf{x}'$ either the uniform or the Euclidean distance:

$$\|\mathbf{x} - \mathbf{x}'\| = \max_i |x_i - x_i'|, \quad i = 1, \ldots, k \ ; \qquad \text{(uniform distance)}$$

$$|\mathbf{x} - \mathbf{x}'| = \sqrt{(x_1 - x_1')^2 + \ldots + (x_k - x_k')^2} \ . \qquad \text{(Euclidean distance)}$$

Definition 25.1A   Cluster point.

In $\mathbb{R}^2$ (or $\mathbb{R}^n$), a point $\mathbf{a}$ is a **cluster point** of a set S if any of the following equivalent statements are true:

(1a)   given $\delta > 0$, $\mathbf{x} \underset{\delta}{\approx} \mathbf{a}$ for some point $\mathbf{x} \in S$, $\mathbf{x} \neq \mathbf{a}$;

(1b)   given $\delta > 0$, $\mathbf{x} \underset{\delta}{\approx} \mathbf{a}$ for infinitely many points $\mathbf{x} \in S$;

(1c)   there is a convergent sequence $\mathbf{x}_n \to \mathbf{a}$, where $\mathbf{x}_n \in S$, $\mathbf{x}_n \neq \mathbf{a}$.

The equivalence of these three statements is essentially what we called the Cluster-point Theorem 6.2, way back when we talked about cluster points on the line. The argument is the same here, so we leave it for the Questions.

Definition 25.1B   Closed set.

A set $\subset \mathbb{R}^2$ (or $\mathbb{R}^n$) is **closed** if it contains all its cluster points.

Examples 25.1A

(i) A finite set of points is closed: it has no cluster points, by (1b).

(ii) The open box $B(\mathbf{a}, r)$ is not closed: it does not contain the cluster point $\mathbf{a} + (r, 0)$, for instance. (See the picture, page 348.)

(iii) A finite closed rectangle is closed: this is proved in the second part of the proof of the Compactness Theorem 24.6 (another proof can be given using the next two theorems below).

(iv) $\mathbb{R}^2$ is a closed set: it contains all points, *a fortiori* all cluster points. The empty set $\emptyset$ is closed: it has no cluster points.

(v) The set $S = \{1/n : n \in \mathbb{N}\}$ in $\mathbb{R}^1$ is not closed: it has one cluster point 0, which is not in S.

To see more examples of closed sets, we need the standard tools for constructing and recognizing them which are provided by the next two theorems.

Theorem 25.1A   Union-intersection theorem for closed sets.

(2)   *The finite union and infinite intersection of closed sets is a closed set.*

In more detail: in $\mathbb{R}^2$ (or $\mathbb{R}^n$),

(i) *if $S_1, \ldots, S_n$ are closed sets, then $S = S_1 \cup \ldots \cup S_n$ is a closed set;*

(ii) *let $\{S_i\}$ be a collection of closed sets, indexed by some set I (finite or infinite); then their intersection $S = \bigcap_{i \in I} S_i$ is a closed set.*

Proof of (i).

Let $\mathbf{a}$ be a cluster point of S; we must show $\mathbf{a} \in S$.

By (1c), there is a sequence $\mathbf{x}_n \in S$ such that $\mathbf{x}_n \to \mathbf{a}$, $\mathbf{x}_n \neq \mathbf{a}$. Since there are only finitely many S_i, at least one, say S_1, contains an infinite subsequence $\mathbf{x}_{n_i}$, and $\mathbf{x}_{n_i} \to \mathbf{a}$ by the Subsequence Theorem 24.2B. Therefore $\mathbf{a}$ is a cluster point of S_1, by (1c), and since S_1 is closed, this means $\mathbf{a} \in S_1$, by definition of closed set. Therefore $\mathbf{a} \in S$. $\qquad\square$

Proof of (ii).

Since $S \subseteq S_i$ for every i, definition (1a) shows that

$\mathbf{a}$ is a cluster point of S $\Rightarrow$ $\mathbf{a}$ is a cluster point of every S_i;

$\Rightarrow$ $\mathbf{a} \in S_i$ for every i, since S_i is closed;

$\Rightarrow$ $\mathbf{a} \in S$. $\qquad\square\square$

Note that the infinite union of closed sets is not necessarily closed; for example a single point is closed set, but the infinite union of the points inside an open box is the open box itself, which is not a closed set, as we saw.

Theorem 25.1B If $f(\mathbf{x})$ is continuous on $\mathbb{R}^2$, then these three sets are closed:

(3) $$\overline{S}_f^+ = \{\mathbf{x} : f(\mathbf{x}) \geq 0\}; \qquad \overline{S}_f^- = \{\mathbf{x} : f(\mathbf{x}) \leq 0\};$$

(4) $$\overline{S}_f = \{\mathbf{x} : f(\mathbf{x}) = 0\}.$$

Proof. To prove $\overline{S}_f^+$ is closed, let $\mathbf{a}$ be a cluster point of $\overline{S}_f^+$. Then by (1c) there is a sequence $\mathbf{x}_n$ such that

$$\mathbf{x}_n \in \overline{S}_f^+, \quad \text{and} \quad \mathbf{x}_n \to \mathbf{a}.$$

Then by the definition of $\overline{S}_f^+$ and the Sequential Continuity Theorem 24.5B,

$$f(\mathbf{x}_n) \geq 0, \quad \text{and} \quad f(\mathbf{x}_n) \to f(\mathbf{a}),$$

Therefore by the Limit Location Theorem for sequences 5.3A,

$$f(\mathbf{a}) \geq 0, \quad \text{i.e.,} \quad \mathbf{a} \in \overline{S}_f^+.$$

This proves $\overline{S}_f^+$ is closed. □.

The proofs that $\overline{S}_f^-$ and $\overline{S}_f$ are closed can be done similarly.

Alternatively, one can remark that $\overline{S}_f^- = \overline{S}_{-f}^+$, so it is closed, and then $\overline{S}_f = \overline{S}_f^+ \cap \overline{S}_f^-$, so it is closed by the Union-intersection Theorem (2). □□

Questions 25.1

1. Use the preceding theorems to show all these are closed sets:

(a) lines, circles, ellipses, and in general any plane curve which is the graph of a polynomial equation $p(x, y) = 0$;

(b) the closed disk $\overline{D}(\mathbf{a}, r)$;

(c) the complement in $\mathbb{R}^2$ of the open disk $D(\mathbf{a}, r)$, that is, $\mathbb{R}^2$ with the open disk removed;

(d) any closed half-plane (the set consisting of a line $ax + by + c = 0$ and the part of the plane lying on one side of it);

(e) the finite closed rectangle $[a, b] \times [c, d]$.

2. Prove the equivalence of the three definitions of cluster point, by showing (note that (1b) $\Rightarrow$ (1a) is trivial)

(a) (1c) $\Rightarrow$ (1b) (yes, there actually is something to show);

(b) (1a) $\Rightarrow$ (1c).

25.2   The Compactness theorem in $\mathbb{R}^2$.

It is important to get as clear an idea as possible of how a compact set in $\mathbb{R}^2$ looks, because the three fundamental theorems about continuous functions given in Section 24.7 — that a continuous function is bounded, has a maximum point, and is uniformly continuous — are valid only when the variables are restricted to lie in a compact set.

Recall the definition: a compact set S is a set having the **sequential compactness property**:

(5)   *any sequence* $\mathbf{x}_n \in S$ *has a subsequence* $\mathbf{x}_{n_i}$ *converging to an* $\mathbf{a} \in S$.

The notion of closed set allows us to give the general version of the Compactness Theorem, which describes all the compact sets in $\mathbb{R}^2$ (and in $\mathbb{R}^k$ too).

Theorem 25.2   Compactness theorem.

A set $S \subset \mathbb{R}^2$ *is compact* $\Leftrightarrow$ *S is closed and bounded.*

Proof. $\Leftarrow$:   Let $\mathbf{x}_n$ be a sequence in S. Since S is bounded, the two-dimensional Bolzano-Weierstrass Theorem 24.2 shows $\mathbf{x}_n$ has a convergent subsequence:

$$\mathbf{x}_{n_i} \in S, \quad \mathbf{x}_{n_i} \to \mathbf{a} \ .$$

To prove (5), we will show $\mathbf{a} \in S$; reasoning indirectly, if $\mathbf{a} \notin S$, then $\mathbf{a} \neq \mathbf{x}_{n_i}$ for any i; thus by (1c), $\mathbf{a}$ is a cluster point of S; therefore, since S is closed, $\mathbf{a} \in S$; contradiction. Therefore $\mathbf{a} \in S$. $\qquad\qquad\square$

If the above reasoning made you unhappy, see the end of Appendix A.2, where a similar proof is given and discussed.

$\Rightarrow$:   To see first that S is closed, suppose $\mathbf{a}$ is a cluster point of S; then by (1c) there is a convergent sequence $\mathbf{x}_n \in S$ such that

$$\mathbf{x}_n \to \mathbf{a} \ .$$

Since we assume (5) is true for S, the sequence $\mathbf{x}_n$ has a subsequence $\mathbf{x}_{n_i}$ such that

$$\mathbf{x}_{n_i} \to \mathbf{a}', \quad \text{and} \quad \mathbf{a}' \in S \ .$$

By the Subsequence Theorem 24.2B, $\mathbf{x}_n$ and $\mathbf{x}_{n_i}$ have the same limit, so we have $\mathbf{a} = \mathbf{a}'$, which shows that $\mathbf{a} \in S$. This proves that S is closed.

To see that S is bounded, we note that the function $f(x, y) = x^2 + y^2$ is continuous, therefore by the Boundedness Theorem 24.7A, it is bounded on the compact set S. That is, there is a positive constant K^2 such that

$$x^2 + y^2 \leq K^2, \quad \text{for all } (x, y) \in S \ .$$

It follows that $|x| \leq K$ and $|y| \leq K$ for all $(x, y) \in S$. This shows S is bounded, and completes the proof of $\Rightarrow$. $\qquad\qquad\square\square$

The Compactness Theorem holds in $\mathbb{R}^k$; the proof is word-for-word the same, but it requires that the three theorems on which it rests: the Bolzano-Weierstrass Theorem, the Subsequence Theorem, and the Boundedness Theorem be proved for $\mathbb{R}^k$, that is, for sequences in $\mathbb{R}^k$ and continuous functions on $\mathbb{R}^k$. There is no difficulty in any of this — the proofs given for $\mathbb{R}^2$ go over almost unchanged to $\mathbb{R}^k$.

Note that the Compactness Theorem holds in $\mathbb{R}^1$ also, since we have proved in earlier chapters the three theorems in $\mathbb{R}^1$ on which it rests.

Questions 25.2

1. For which $a \in \mathbb{R}$ is the set $\{(x,y) : ax^4 + y^4 = 1\}$ a compact subset of $\mathbb{R}^2$?

2. Prove that the finite union and infinite intersection of compact sets in $\mathbb{R}^2$ is compact. (Reread the last bit of advice at the end of Section 5.5 first.)

3. Let $\{x_n\}$ be an increasing sequence in $\mathbb{R}^1$. Describe all such sequences for which the underlying set $\{x_1, x_2, \ldots\}$ is a compact subset of $\mathbb{R}^1$.

25.3   Open sets.

Roughly speaking, a set U of points in $\mathbb{R}^2$ is called "open" if it is two-dimensional everywhere (i.e., it is not a curve or a collection of isolated points), and it does not include any of its boundary points (i.e., its edges). The way to ensure that both of these will be true is to say:

every $\mathbf{a} \in U$ is the center of a δ-box lying entirely inside U.

(The size of the δ-box will depend on where its center $\mathbf{a}$ lies.)

For example, in $\mathbb{R}^2$ the open r-box $B(\mathbf{0}, r)$ is an open set. The picture shows some typical δ-boxes around points in a general open set U. The closer the point is to the boundary, the smaller δ has to be for the box to fit entirely inside U.

We will refer to the open box $B(\mathbf{a}, \delta)$ as the *δ-neighborhood of* $\mathbf{a}$.

Definition 25.3   Open set.

A set U in $\mathbb{R}^2$ is called an **open set** if it satisfies the following condition (three slightly different ways of saying it are given; cf. (6), p. 348 for (6c)):

(6a)   for every $\mathbf{a} \in U$, there is a $\delta > 0$ such that $B(\mathbf{a}, \delta) \subseteq U$;

(6b)   every point $\mathbf{a} \in U$ has a δ-neighborhood lying inside U;

(6c) $\mathbf{a} \in U \Rightarrow \mathbf{x} \in U$ for all $x \approx \mathbf{a}$.

The condition guarantees that U has to be two-dimensional everywhere, since every point is the center of a two-dimensional box lying entirely inside U. An open set cannot contain "edge" or "boundary" points, speaking loosely, since these cannot be surrounded by a box lying entirely inside the set.

Examples 25.3   Prove each of the following statements.

   (a) $\mathbb{R}^2$ and the empty set $\emptyset$ are open sets. (Both are also closed.)

   (b)   The open upper half-plane $U = \{(x, y) : y > 0\}$ is an open set.

   (c)   The closed box $\overline{B}(\mathbf{0}, a)$ is not an open set.

Solution.

   (a) $\mathbb{R}^2$ satisfies the definition, since for any $\mathbf{a}$, we have $B(\mathbf{a}, 1) \subset \mathbb{R}^2$; the empty set $\emptyset$ satisfies the definition vacuously, since there is no $\mathbf{a}$.

   (b)   Given any $\mathbf{a} = (a, b) \in U$, the open box $B(\mathbf{a}, b/2) \subset U$. This is clear from the picture; formally we can argue

$$
\begin{aligned}
\mathbf{x} \in B(\mathbf{a}, b/2) &\Rightarrow \|\mathbf{x} - \mathbf{a}\| < b/2 & &\text{(definition of } B(\mathbf{a}, b/2)); \\
&\Rightarrow |y - b| < b/2 & &(\|\mathbf{x} - \mathbf{a}\| = \max(|x - a|, |y - b|)); \\
&\Rightarrow y > b/2 & &(-b/2 < y - b < b/2); \\
&\Rightarrow \mathbf{x} \in U & &\text{(definition of } U). \quad \Box
\end{aligned}
$$

   (c)   Take $\mathbf{a} = (a, 0)$. Any δ-box $B(\mathbf{a}, \delta)$ centered at $\mathbf{a}$ contains the point $\mathbf{x} = (a + \delta/2, 0)$, which is not in $B(\mathbf{0}, a)$, since $\|\mathbf{x} - \mathbf{0}\| = a + \delta/2$.

Theorem 25.3A   Union-intersection theorem for open sets.

(7)   *The finite intersection and infinite union of open sets is open.*

   In more detail: in $\mathbb{R}^2$ (and in $\mathbb{R}^k$ as well):

   (i)   *if $U_1, \ldots, U_n$ are open sets, then $U_1 \cap \ldots \cap U_n$ is an open set;*

   (ii)   *if $\{U_i\}$ is a collection of open sets, indexed by some set I, finite or infinite, then their union $\bigcup_{i \in I} S_i$ is an open set.*

Proof of (i).

$$
\begin{aligned}
\mathbf{a} \in S &\Rightarrow \mathbf{a} \in U_i \quad \text{for all } i \; ; \\
&\Rightarrow B(\mathbf{a}, \delta_i) \subseteq U_i \text{ for some choice of } \delta_1, \ldots, \delta_n \text{ and all } i; \\
&\Rightarrow B(\mathbf{a}, \delta) \subseteq U_i \quad \text{for all } i, \text{ where } \delta = \min \delta_i \; ; \\
&\Rightarrow B(\mathbf{a}, \delta) \subseteq S \; . \hspace{4cm} \Box
\end{aligned}
$$

Proof of (ii).

$$
\begin{aligned}
\mathbf{a} \in S &\Rightarrow \mathbf{a} \in U_{i_0}, \quad \text{say}; \\
&\Rightarrow B(\mathbf{a}, \delta) \subseteq U_{i_0} \quad \text{for some } \delta > 0, \text{ since } U_{i_0} \text{ is open}; \\
&\Rightarrow B(\mathbf{a}, \delta) \subseteq S \; . \hspace{3.5cm} \Box\Box
\end{aligned}
$$

Theorem 25.3B *Let $f(\mathbf{x})$ be continuous on $\mathbb{R}^2$. These sets are open:*

(8) $$S_f^+ = \{\mathbf{x} : f(\mathbf{x}) > 0\} \quad \text{and} \quad S_f^- = \{\mathbf{x} : f(\mathbf{x}) < 0\}.$$

Proof. This follows immediately from the Positivity Theorem for continuous functions, since by that Theorem (24.5C), there is a $\delta > 0$ such that

$$f(\mathbf{a}) > 0 \;\Rightarrow\; f(\mathbf{x}) > 0 \quad \text{for all } x \underset{\delta}{\approx} \mathbf{a}, \quad \text{i.e.,}$$

$$\mathbf{a} \in S_f^+ \;\Rightarrow\; B(\mathbf{a}, \delta) \subseteq S_f^+ .$$

For S_f^-, one uses $S_f^- = S_{-f}^+$ to reduce it to the case above. $\square$

Theorem 25.3C Complementation theorem. *Let $S \subseteq \mathbb{R}^2$, and set*

$$S' = \{\mathbf{x} \in \mathbb{R}^2 : \mathbf{x} \notin S\} \quad \text{(the complement of S)}.$$

Then: $\qquad\qquad$ *S is a closed set $\;\Leftrightarrow\;$ S' is an open set.*

You can show a set S is closed, by showing its complement is open. Often this is easier than seeing whether S contains all its cluster points.

WARNING: People like dichotomies: winners and losers, haves and have-nots, you're part of the solution or part of the problem. Analysis students often seem to think every set is either open or closed — at any rate, they try to prove a set is open by showing it is not closed. You can't do this: most sets are neither open nor closed. The only useful relation between open and closed sets is the one in the Complementation Theorem.

Proof. By contraposition: S is not closed $\;\Leftrightarrow\;$ S' is not open.

S is not closed

$\quad \Leftrightarrow \quad$ there is an $\mathbf{a} \in S'$ such that $\mathbf{a}$ is a cluster point of S;

$\quad \Leftrightarrow \quad$ $\mathbf{a} \in S'$ and every δ-box $B(\mathbf{a}, \delta)$ has points of S;

$\quad \Leftrightarrow \quad$ $\mathbf{a} \in S'$ and no δ-box $B(\mathbf{a}, \delta)$ lies entirely in S';

$\quad \Leftrightarrow \quad$ S' is not open. $\square$

By using the Complementation Theorem, and elementary set theory, the two other theorems of this section can be deduced from the two theorems of Section 25.1 on closed sets (see Exercises 25.3). Other ideas connected with open and closed sets are in the Problems; we won't need them for the rest of this book, but they are important in further study of analysis of functions of several variables.

Questions 25.3

1. Tell for which functions the domain is an open set in $\mathbb{R}^2$ or $\mathbb{R}^1$; give reason.

    (a) $1/(x^2 + y^2)$ (b) $\sqrt{x + y}$ (c) $\sec x$

2. Prove two ways that if the function $f(x, y)$ is continuous on $\mathbb{R}^2$ then the set $S = \{(x, y) : f(x, y) \neq 0\}$ is an open set.

Exercises

Note: For Exercises in $\mathbb{R}^1$, you can use the definitions and results of this chapter, which are valid for $\mathbb{R}^1$ (and $\mathbb{R}^n$ as well).

25.1

1. For each of these sets, tell if it is closed or not in $\mathbb{R}^1$ or $\mathbb{R}^2$, using just the definition, not the theorems of this section. Give reasons.

(a) $\mathbb{N} = \{1, 2, 3, \ldots\} \subset \mathbb{R}^1$ (b) $\mathbb{Q} = \{$rationals $a/b\} \subset \mathbb{R}^1$

(c) $\{\mathbf{x} \in \mathbb{R}^2 : 1/\|\mathbf{x}\|$ is defined$\}$ (d) $\{\mathbf{x} \in \mathbb{R}^2 : \|\mathbf{x}\| \geq 1\}$

2. Prove that a closed subset S of $\mathbb{R}^1$ (thought of as the x-axis in $\mathbb{R}^2$), is also a closed set when viewed as a subset of $\mathbb{R}^2$.

3. (a) (If you have read Chapter 23): prove the Cantor set (23.3) is a closed subset of $\mathbb{R}^1$.

(b) Which are the cluster points of the Cantor set? Prove it.

4. For each locus below, prove it is or is not a closed set in $\mathbb{R}^2$.

(a) $xy = 1$, $x > 0$, $y > 0$ (b) $y = x^2$, $x > 0$, $y > 0$

25.2

1. For which a and b is the set $S = \{(x, y) : x, y \geq 0, y \leq ax + b\}$ compact?

2. Do Exercise 24.7/1, assuming only that S is closed, not necessarily compact. (Make a compact set somehow.)

3. Given two compact curves in $\mathbb{R}^2$, defined by

$$\mathbf{f}(t) : I \to \mathbb{R}^2, \qquad \mathbf{g}(u) : J \to \mathbb{R}^2;$$

where I and J are respectively compact intervals on the t and u lines, and the mappings are continuous (cf. Exercise 24.6/2).

Prove there are two points — one on each curve — which are closest together, and two points which are furthest apart.

4. Prove that if S and T are both compact subsets of $\mathbb{R}^1$, then $S \times T$ is a compact subset of $\mathbb{R}^2 = \mathbb{R}^1 \times \mathbb{R}^1$. (This is true if S and T are compact intervals, by Theorem 24.6, say, but how about in general?)

5. Suppose S is a compact subset of $\mathbb{R}^2$, and $f(\mathbf{x})$ a continuous function on S. Think of f as a mapping from $\mathbb{R}^2$ to $\mathbb{R}^1$. Prove its image

$$f(S) \;=\; \{f(\mathbf{x}) : \mathbf{x} \in S\}$$

is a compact subset of $\mathbb{R}^1$. (Use the definition of compactness.)

This can be thought of as a partial generalization of the Continuous Mapping Theorem 13.4, which says that for a continuous function $f(x) : \mathbb{R}^1 \to \mathbb{R}^1$, the image $f(I)$ of a compact interval I is itself a compact interval.

6. Repeat the previous exercise for a continuous mapping $\mathbf{f} : \mathbb{R}^2 \to \mathbb{R}^2$, defined by a pair of continuous functions $f(x,y), g(x,y)$:

$$(f,g)(x,y) = (f(x,y), g(x,y)).$$

That is, prove that if S is a compact subset of $\mathbb{R}^2$, then the image $\mathbf{f}(S)$ is a compact subset of $\mathbb{R}^2$. (Go back to the definition of compactness.)

25.3

1. For each of the following sets in $\mathbb{R}^2$, tell whether it is open, closed, compact, or none of these, indicating a brief reason. Use any of the theorems in this chapter.

(Notation: $S \setminus R = \{x : x \in S,\ x \notin R\}$; cf. Appendix A.0 .)

(a) $\{(x,y) : 1 < x^2 + y^2 < 2\}$ (b) $\{(x,y) : x^2 + y^2 = 1\}$

(c) $\mathbb{R}^2$ (d) $\{(x,y) : x^2 - y^2 = 1\}$

(e) $\{(x,y) : 1 \le x^2 + y^2 < 2\}$ (f) $\mathbb{R}^2 \setminus \mathbf{0}$

(g) $\{(x,y) : x \ge 0, y \ge 0, x^2 + y^2 < 1\}$ (h) $\mathbb{R}^2 \setminus$ (the positive x-axis)

(i) $\mathbb{R}^2 \setminus$ (the non-negative x-axis)   (j) $\{(x,y) : x, y$ are integers$\}$

2. Let S be a closed set and U an open set, in $\mathbb{R}^2$ say. Prove that

  (a) $S \setminus U$ is a closed set;   (b) $U \setminus S$ is an open set.

3. What kind of a set is the domain of $\sec(1/x)$ in $\mathbb{R}^1$? Indicate reason.

4. Show that the finite intersection property of open sets (25.3A(i)) does not extend to an infinite family of open sets.

5. By using Theorems 25.3A and 25.3B, give proofs that an open disk is an open set and a finite open rectangle is an open set.

6. Prove: for any finite or infinite collection of sets S_i (here $'$ denotes complementation with respect to some larger set T in which all the S_i lie),

    (a) $\left(\bigcup_i S_i\right)' = \bigcap_i S_i'$ (b) $\left(\bigcap_i S_i\right)' = \bigcup_i S_i'$

7. Using the previous exercise and the Complementation Theorem,

    (a) show that Theorem 25.3A follows from Theorem 25.1A;

    (b) show that Theorem 25.3B follows from Theorem 25.1B.

Problems

25-1 It is natural to try to define what it means for a point to be "on the edge of" a set S in $\mathbb{R}^2$.

Definition. We call $\mathbf{x}$ a **boundary point** of S if every ϵ-disk $D(\mathbf{x}, \epsilon)$ contains points of S and S' (the complement of S). The set of all boundary points of S is called the **boundary** of S (notation: $\partial(S)$).

For each set S in Exercise 25.3/1, describe its boundary.

25-2 An important operation in point-set topology is one which turns an arbitrary set into a closed set by adding its boundary points (cf. Problem 25-1).

Definition. The **closure** of a set S is the set $\overline{S} = S \cup \partial(S)$.

(a) Describe the closure of each set in Exercise 25.3/1.

(b) Prove that $\overline{S}$ is a closed set.

(c) Prove that $\overline{S}$ is the smallest closed set containing S, by showing that if T is a closed set containing S, then $T \supseteq \overline{S}$.

25-3 Here is another way of describing the closure of S (see the preceding problem): define

$$cl(S) = \{\mathbf{a} \in \mathbb{R}^2 : \mathbf{a} = \lim \mathbf{x}_n \text{ for some sequence } \mathbf{x}_n \text{ in } S\}.$$

(a) Prove that $cl(S)$ is a closed set by using Definition 25.1.

(b) Prove that $cl(S)$ is the smallest closed set containing S, by showing that if T is a closed set containing S, then $T \supseteq cl(S)$.

This shows that $\overline{S} \supseteq cl(S)$, and 25-2(c) above shows the reverse inclusion; hence the two sets are equal, and the two notions of closure coincide. In one view, we add boundary points to S, in the other view, we add cluster points (or limit points).

25-4 Definition.   The **interior** of S is the set $\mathrm{int}\,(S) = \overline{S} - \partial(S)$.

(a) Describe the interior of each of the sets of Exercise 25.3/1.

(b) Prove that $\mathrm{int}\,(S)$ is an open set.

(c) Prove that $\mathrm{int}\,(S)$ is the largest open set contained in S, by showing that if U is an open set such that $U \subseteq S$, then $U \subseteq \mathrm{int}\,(S)$.

(d) Give an example in $\mathbb{R}^2$ of a set S such that $\mathrm{int}\,(S)$ is empty and $\overline{S}$ is all of $\mathbb{R}^2$.

25-5 If $h : S \to T$ is a map between two sets, and $X \subseteq T$, we define

$$h^{-1}(X) = \{s \in S : h(s) \in X\} .$$

Prove that if $\mathbf{f} = (f, g) : \mathbb{R}^2 \to \mathbb{R}^2$ is continuous (see Exercise 25.2/6), then

(a) if V is closed, $\mathbf{f}^{-1}(V)$ is closed;   (b) if U is open, $\mathbf{f}^{-1}(U)$ is open.

Answers to Questions

25.1

1.

(a) This follows from (4), since polynomials are continuous functions.

(b) By (3): set $f(x, y) = (x - a)^2 + (y - b)^2 - r^2$; then $\overline{D}(\mathbf{a}, r) = \overline{S}_f^{\,-}$.

(c) By (3): it is $\overline{S}_f^{\,+}$, where $f(x, y)$ is as in (b).

(d) It is either $\overline{S}_f^{\,+}$ or $\overline{S}_f^{\,-}$, where $f(x, y) = ax + by + c$.

(e) By (2), since it is the intersection of four closed half-planes.

2. (a) The sequence $\mathbf{x}_n$ contains an infinity of distinct points, otherwise $\mathbf{x}_n = \mathbf{a}'$ for $n \gg 1$, and the sequence would converge to $\mathbf{a}'$, not $\mathbf{a}$. By definition of convergence, any δ-box centered at $\mathbf{a}$ contains all $\mathbf{x}_n$, for $n \gg 1$, therefore an infinity of distinct points $\mathbf{x}_n$.

(b) For each $n \in \mathbb{N}$ let $\delta = 1/n$, and choose $\mathbf{x}_n \in S$, $\mathbf{x}_n \neq \mathbf{a}$ so that $\mathbf{x}_n \underset{\delta}{\approx} \mathbf{a}$. Then $\mathbf{x}_n \to \mathbf{a}$, by definition of convergence.

25.2

1. Answer: for $a > 0$. The set S is closed, by Theorem 25.1B (4); therefore by the Compactness Theorem 25.2, it is compact if and only if it is bounded. We have

$a > 0 \Rightarrow S$ is bounded, since $(x, y) \in S \Rightarrow |x| \le a^{-1/4}$, $|y| \le 1$.

$a = 0 \Rightarrow S$ is the pair of horizontal lines $y = \pm 1$, which is unbounded.

$a < 0 \Rightarrow S$ is unbounded, since it contains $(n, \sqrt[4]{1 - an^4})$ for all n.

2. (a) Let $S = S_1 \bigcup \ldots \bigcup S_n$ and $T = \bigcap T_i$.

If each of the S_i and T_i is compact, it is closed and bounded, by the Compactness Theorem. Therefore S and T are closed, by Theorem 25.1A. They are also bounded, for since S_i and T_1 are bounded, there are numbers r_i and s such that $S_i \subset B(\mathbf{0}, r_i)$ for all i, and $T_1 \subset B(\mathbf{0}, s)$; then

$$S \subset \bigcup_1^n B(\mathbf{0}, r_i) \subset B(\mathbf{0}, r), \quad \text{where} \quad r = \max r_i; \qquad T \subset T_1 \subset B(\mathbf{0}, s).$$

Since S and T are closed and bounded, they are compact.

3. Since the set is compact, the sequence must be bounded, by the Compactness Theorem. Therefore the sequence has a limit (Completeness Property), which must be in the set (since it is closed, by the Compactness Theorem).

This proves the limit is one of the x_n, say x_N. Then since the sequence is increasing, $x_n = x_N$ for all $n \ge N$; i.e., the sequence has only a finite number of distinct points, and has a tail which is a constant sequence.

25.3

1. (a) $\mathbb{R} \setminus \mathbf{0}$, open (its complement is $\{\mathbf{0}\}$, which is closed);

(b) $\{(x, y) : x + y \ge 0\}$; not open, since say $\mathbf{0}$ cannot be surrounded by a box lying entirely inside;

(c) $\bigcup_{n \in \mathbb{Z}} \left(\frac{\pi}{2} + n\pi, \ \frac{\pi}{2} + (n + 1)\pi\right)$; open (infinite union of open intervals).

2. (a) The complement of S is the set where $f(x, y) = 0$, and this is closed by Theorem 25.1B (4). Hence S is open, by the Complementation Theorem 25.3C.

(b) $S = S_f^+ \bigcup S_f^-$, both of which are open by Theorem 25.3B; therefore S is open by the Union-intersection Theorem 25.3A.

26

Integrals with a Parameter

26.1 Integrals depending on a parameter.

Suppose $f(x,t)$ is defined on some rectangle $[a,b] \times [c,d]$. In this chapter we study the continuity and differentiability of the function

$$(1) \qquad \phi(x) = \int_c^d f(x,t)\, dt \ .$$

Functions defined in this way occur frequently in analysis, and are a major tool in real-world problem-solving. For example, if $r(x)$ is continuous, then a solution $y(x)$ to the differential equation with initial conditions

$$y'' + y = r(x), \qquad y(0) = 0,\ y'(0) = 0,$$

is given explicitly by

$$y(x) = \int_0^x \sin(x - t) r(t)\, dt \ .$$

More generally, a particular solution $y_p(x)$ to a linear differential equation

$$y^{(n)} + a_1(x) y^{(n-1)} + \ldots + a_n(x) y = r(x)$$

can be expressed in the form

$$y_p(x) = \int_a^x G(x,t) r(t)\, dt \ ,$$

where $G(x,t)$ is a function depending only on the coefficients $a_i(x)$.

Other examples of the integral (1) are provided by the integral transforms (like the Laplace transform), though in these the interval $[c,d]$ is usually infinite. We will study them in the next chapter, but must first start with the simpler case where the interval is finite.

To verify that (1) solves a differential equation, we need to be able to differentiate it. The main goal of this chapter is to see under what circumstances it is permissible to calculate $\phi'(x)$ by "differentiating under the integral sign" with respect to the parameter x. In other words, to determine when

$$(2) \qquad \phi'(x) = \int_c^d \frac{\partial f(x,t)}{\partial x}\, dt \ .$$

We begin first with the simpler question as to whether $\phi(x)$ is continuous. That is, if we think of (1) as a definite integral whose value depends on the parameter x, does its value depend continuously on x? If we vary x a little, does the value of the integral vary only by a small amount?

The key point of the continuity proof will be the fact that $f(x, t)$ is uniformly continuous in a compact rectangle; you should first review the meaning of this together with the Uniform Continuity Theorem 24.7C.

Theorem 26.1   Continuity theorem.

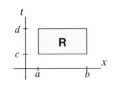

For any x-interval I, if $f(x, t)$ is continuous in $R = I \times [c, d]$, then $\phi(x)$ is continuous on I, where

$$(1) \qquad \phi(x) \; = \; \int_c^d f(x, t) \, dt \; .$$

Proof.   Assume that $I = [a, b]$; the general case follows easily and is left as Question 1b.

We show that $\phi(x)$ is continuous at any point x_0 in $[a, b]$, by showing the difference $\phi(x) - \phi(x_0)$ is small, if $x \approx x_0$.

$$(3) \quad |\phi(x) - \phi(x_0)| \; = \; \left| \int_c^d \left(f(x, t) - f(x_0, t) \right) dt \right|, \quad \text{by linearity (19.4A);}$$

$$(4) \qquad\qquad\qquad \leq \; \int_c^d \left| f(x, t) - f(x_0, t) \right| dt, \qquad \text{by Theorem 19.4C .}$$

We now obtain an upper estimate for the integrand in (4).

Since $f(x, t)$ is continuous on the compact rectangle R, by Theorem 24.7C it is uniformly continuous on R. According to the definition, this means that, given $\epsilon > 0$, there is a $\delta > 0$ such that on R we have

$$\left| f(x, t) - f(x_0, t) \right| \; < \; \epsilon, \qquad \text{for } (x, t) \underset{\delta}{\approx} (x_0, t).$$

We can change the last clause to "for $x \underset{\delta}{\approx} x_0$", since

$$(x, t) \underset{\delta}{\approx} (x_0, t) \quad \Leftrightarrow \quad x \underset{\delta}{\approx} x_0 \; .$$

Using the above estimate for the integrand in (4), we obtain from (3) and (4),

$$|\phi(x) - \phi(x_0)| \; \leq \; \epsilon(d - c), \qquad \text{for } x \underset{\delta}{\approx} x_0 \; ,$$

which proves $\phi(x)$ continuous at x_0, by the K-ϵ principle. $\square$.

(As usual, if x_0 is one of the endpoints, slight modifications must be made — for instance, if $x_0 = a$, one should write $x \approx x_0^+$ throughout.)

Questions 26.1

1. (a) Where does the proof break down if $I = [0, \infty)$ or $I = (a, b)$?

   (b) Use the case $I = [a, b]$ to show the theorem is true for any interval I.

2. (a) Prove that $\phi(x) = \displaystyle\int_0^1 \frac{x - t}{x + t} \, dt$ is continuous on $[1, 2]$ by following the argument given in the proof; estimate $\left| f(x, t) - f(x_0, t) \right|$ by algebra.

   (b) Predict the largest interval on which $\phi(x)$ is continuous, and verify this by calculating $\phi(x)$ explicitly.

26.2   Differentiating under the integral sign.

The main theorem will say that, on an x-interval I, the derivative formula

$$(2) \qquad\qquad \phi'(x) \;=\; \int_c^d \frac{\partial f(x,t)}{\partial x}\, dt\,, \qquad x \in I,$$

is valid if $f(x,t)$ and the integrand $f_x(x,t)$ are continuous in $I \times [c,d]$.

Example 26.2A   Verify (2), if $\phi(x) = \displaystyle\int_0^1 e^{xt}\, dt, \quad I = (0,\infty)$.

Solution.   Calculating $\phi(x)$ explicitly and then differentiating it, we get

$$\phi(x) \;=\; \frac{e^{xt}}{x}\Big|_0^1 = \frac{e^x}{x} - \frac{1}{x}; \qquad \phi'(x) \;=\; e^x\left(\frac{1}{x} - \frac{1}{x^2}\right) + \frac{1}{x^2}.$$

We get the same answer using (2) instead: since $\dfrac{\partial e^{xt}}{\partial x} = te^{xt}$, we get

$$\phi'(x) \;=\; \int_0^1 te^{xt}\, dt; \quad \text{now use integration by parts:}$$

$$=\; \frac{te^{xt}}{x}\Big|_0^1 - \int_0^1 \frac{e^{xt}}{x}\, dt = \frac{e^x}{x} - \frac{e^{xt}}{x^2}\Big|_0^1 = \frac{e^x}{x} - \frac{e^x}{x^2} + \frac{1}{x^2}\,. \qquad\qquad \square$$

Example 26.2B   Find the derivative of $\phi(x) = \displaystyle\int_0^\pi \frac{\sin xt}{t}\, dt$.

Solution.   This is a well-known function in physics; $\phi(x)$ is not an elementary function, but its derivative is, as we shall see. We take I to be (∞,∞).

The integrand $f(x,t)$ is by convention defined for all t by

$$(*) \qquad\qquad f(x,t) = \frac{\sin xt}{t}, \quad t \neq 0; \qquad f(x,0) = x\,.$$

It then becomes a continuous function; the best way to see this is to write

$$f(x,t) = \frac{\sin xt}{xt}\, x = \frac{\sin u}{u}\, x\,, \quad \text{where } u = xt\,,$$

and we make the usual convention that $(\sin u)/u$ is the continuous function whose value at $u = 0$ is 1. The above line shows that $f(x,t)$ is the composition of two continuous functions, therefore continuous.

Since the partial derivative $f_x(x,t)$ is calculated by holding t constant and taking the ordinary derivative with respect to x, we get from $(*)$:

$$f_x(x,t) = \cos xt,\ t \neq 0; \qquad f_x(x,0) = 1\ ; \quad \text{i.e.,}$$
$$f_x(x,t) = \cos xt \quad \text{for all } (x,t).$$

Since $f(x,t)$ and $f_x(x,t)$ are continuous everywhere, we can use (2):

$$\phi'(x) \;=\; \int_0^\pi \cos xt\, dt \;=\; \begin{cases} (\sin \pi x)/x, & \text{if } x \neq 0; \\ \pi, & \text{if } x = 0. \end{cases}$$

By the definition convention in $(*)$, therefore, (changing t to x and x to π) :

$$\phi'(x) \;=\; \frac{\sin \pi x}{x}, \qquad \text{for all } x. \qquad\qquad \square$$

Theorem 26.2A   Differentiating under the integral sign.

Let I be any x-interval. If $f(x,t)$ and $f_x(x,t)$ are continuous in the rectangle $I \times [c,d]$, then for all $x \in I$, the function $\phi(x)$ below is differentiable, and its derivative is given by (2):

$$(1) \qquad\qquad \phi(x) \;=\; \int_c^d f(x,t)\,dt$$

$$(2) \qquad\qquad \phi'(x) \;=\; \int_c^d f_x(x,t)\,dt \;.$$

Proof.   Let $x_0 \in I$. We show ϕ is differentiable at x_0, and that $\phi'(x_0)$ is given by the integral on the right of (5), with integrand $f_x(x_0, t)$.

Choose a compact interval $[a,b] \subseteq I$ which contains x_0; then it suffices to prove the theorem when $R = [a,b] \times [c,d]$, since differentiability is a local notion (cf. the proof of Question 26.1/1b); note also that if x_0 is say the right-hand endpoint of I, then the derivatives should be left-hand derivatives at x_0 and slight modifications to the proof are needed.)

We begin by calculating the difference quotient which approximates $\phi'(x_0)$; by the definition (1) of $\phi(x)$ and the linearity property of integrals (19.4A),

$$(6) \qquad\qquad \frac{\phi(x) - \phi(x_0)}{x - x_0} \;=\; \int_c^d \frac{f(x,t) - f(x_0,t)}{x - x_0}\,dt \;.$$

The left sides of (6) and (2) are close to each other; we show the right sides are also close. To do this, we transform the integrand by using the secant form of the Mean-value Theorem (15.1); it is applicable since for each $t \in [c,d]$, $f(x,t)$ is a differentiable function of x. It tells us that for each $t \in [c,d]$, we can select a point $\bar{x}$ (this point depends on t, so we should really write $\bar{x} = \bar{x}(t)$) such that

$$(7) \qquad \frac{f(x,t) - f(x_0,t)}{x - x_0} \;=\; f_x(\bar{x},t), \qquad \begin{cases} x_0 < \bar{x} < x, \quad \text{or} \\ x < \bar{x} < x_0. \end{cases}$$

We do not know how $\bar{x}$ depends on t; nonetheless, the right side of (7) is a continuous function of t because our hypotheses tell us the left side is. So we can integrate both sides of (7) with respect to t; in view of (6), we get

$$(8) \qquad\qquad \frac{\phi(x) - \phi(x_0)}{x - x_0} \;=\; \int_c^d f_x(\bar{x},t)\,dt \;.$$

The integrals in (8) and (2) are close because their integrands are. To see this, note that $f_x(x,t)$ is continuous on the compact rectangle $R = [a,b] \times [c,d]$; therefore it is uniformly continuous on R, by the Uniform Continuity Theorem 24.7C. This shows that, given $\epsilon > 0$, there is a $\delta > 0$ such that in R,

$$f_x(x,t) \underset{\epsilon}{\approx} f_x(x_0,t) \quad \text{for } (x,t) \underset{\delta}{\approx} (x_0,t), \quad \text{i.e., for } x \underset{\delta}{\approx} x_0;$$

since $\bar{x}(t)$ lies between x_0 and x, $\; x \underset{\delta}{\approx} x_0 \;\Rightarrow\; \bar{x} \underset{\delta}{\approx} x_0$ also;   therefore

$$(9) \qquad\qquad f_x(\bar{x}(t),t) \underset{\epsilon}{\approx} f_x(x_0,t) \quad \text{for } x \underset{\delta}{\approx} x_0.$$

If we integrate both sides of (9) over $[c, d]$, the resulting integrals are within $\epsilon(d - c)$ of each other. Therefore from (8) we get

$$\frac{\phi(x) - \phi(x_0)}{x - x_0} \underset{\epsilon(d-c)}{\approx} \int_c^d f_x(x_0, t)\, dt , \qquad \text{for } x \underset{\delta}{\approx} x_0 .$$

By the definition of limit and the K-ϵ principle, this shows the limit of the quotient on the left exists as $x \to x_0$, and equals the quantity on the right, i.e.

$$\phi'(x_0) \;=\; \int_c^d f_x(x_0, t)\, dt , \qquad a \le x_0 \le b . \qquad \qquad \square$$

In using Theorem 26.2A, often the integral on the right will have variable upper limits, so we need to extend it to this case. This will require the chain rule for functions of three variables, which we state below and use without proof.

Chain Rule. *Suppose $F(u, v, w)$ has continuous partial derivatives on an open rectangular box in uvw-space containing a smooth parametrized curve (here x is the parameter, and u, v, w are assumed differentiable on $[a, b]$):*

$$\{ \, (u(x), v(x), w(x)) \; : \; a \le x \le b \, \} ;$$

then the composite function $G(x) = F(u(x), v(x), w(x))$ is differentiable and

$$(10) \qquad\qquad \frac{dG}{dx} \;=\; \frac{\partial F}{\partial u}\frac{du}{dx} + \frac{\partial F}{\partial v}\frac{dv}{dx} + \frac{\partial F}{\partial w}\frac{dw}{dx} .$$

Theorem 26.2B *Let I be an x-interval, and suppose that for $x \in I$,*
  (a) $u(x)$ *and* $v(x)$ *are differentiable, and* $c \le u(x) \le d$, $c \le v(x) \le d$;
  (b) $f(x, t)$ *and* $f_x(x, t)$ *are continuous on an open rectangle* $\supset I \times [c, d]$.

Then we have, for $x \in I$, **Leibniz' Formula**:

$$(11) \qquad \frac{d}{dx} \int_{u(x)}^{v(x)} f(x, t)\, dt \;=\; -f(x, u)\frac{du}{dx} + f(x, v)\frac{dv}{dx} + \int_{u(x)}^{v(x)} f_x(x, t)\, dt .$$

Proof. We apply the chain rule to the function

$$F(u, v, w) \;=\; \int_u^v f(w, t)\, dt, \qquad \text{where } u = u(x), \; v = v(x), \; w = x.$$

By the Second Fundamental Theorem 20.2A, we have

$$F_v \;=\; f(w, v), \quad \text{and}$$

$$F_u \;=\; \frac{\partial}{\partial u} \int_v^u -f(w, t)\, dt \;=\; -f(w, u), \quad \text{while}$$

$$F_w \;=\; \int_u^v f_w(w, t)\, dt, \qquad \text{by Theorem 26.2A.}$$

The hypotheses guarantee that the conditions for the chain rule are fulfilled; using the chain rule (10), and substituting $w = x$ at the end, we get (11). $\qquad \square$

Questions 26.2

1. Let $\phi(x) = \displaystyle\int_0^1 \ln(x-t)\,dt$.

    (a) Determine the natural domain I of $\phi(x)$.

    (b) Find $\phi'(x)$ by the methods of this section; your answer should not contain a definite integral. (Verify that the hypotheses of any theorems used are satisfied.)

2. Let $\phi(x) = \displaystyle\int_0^{x^2} (x^2 + t^2)\,dt$. Evaluate $\phi'(x)$ two ways, and check that you get the same answer both ways; on what interval I are the results valid? Verify hypotheses as needed.

26.3   Changing the order of integration.

The analog of "differentiating under the integral sign" is "integrating under the integral sign". If we try integrating in this way the function

(1) $$\phi(x) \;=\; \int_c^d f(x,t)\,dt,$$

we are led to equation (12) below: the left side is the integral of $\phi(x)$ over $[a,b]$, while the right side is the result of integrating over $[a,b]$ "under the integral sign".

A more familiar way however of looking at equation (12) is to view it as saying that the order of integration can be changed in a double iterated integral. The general case is called *Fubini's Theorem*; we will consider the special case where the function is continuous and the region is a finite rectangle.

Theorem 26.3   Fubini's theorem (special case).

Let $f(x,t)$ be continuous on the rectangle $R = [a,b] \times [c,d]$. Then as iterated integrals,

(12) $$\int_a^b \int_c^d f(x,t)\,dt\,dx \;=\; \int_c^d \int_a^b f(x,t)\,dx\,dt \ .$$

Note that both sides make sense. The inner integral on the left in (12) defines a continuous function of x by Theorem 26.1, so it can be integrated over $[a,b]$. The same is true on the right in (12), reversing the roles of x and t.

Proof.   We prove (12) for an integrand having the form

(13) $$f(x,t) \;=\; g_x(x,t) \ .$$

In fact, this is no restriction, for, given f, if we define g on R by

$$g(x,t) \;=\; \int_a^x f(u,t)\,du \ ,$$

then since f is continuous, the Second Fundamental Theorem 20.2A shows that g is differentiable with respect to x and (13) is true.

To prove Fubini's Theorem, with f taken to be in the form (13), we have, by the First Fundamental Theorem 20.1,

$$\int_c^d \int_a^b g_x(x,t)\,dx\,dt \;=\; \int_c^d \big(g(b,t) - g(a,t)\big)\,dt;$$

$$= \psi(b) - \psi(a), \qquad\qquad \text{where } \psi(x) = \int_c^d g(x,t)\,dt;$$

$$= \int_a^b \psi'(x)\,dx, \qquad\qquad \text{by the FFT 20.1;}$$

note $\psi'(x)$ exists by Theorem 26.2; using the formula there for it, the above

$$= \int_a^b \int_c^d g_x(x,t)\,dt\,dx. \qquad\qquad\qquad \square$$

Remarks.

If you want to weaken the continuity hypothesis, you have to assume that $f(x,y)$ is integrable (as a function of y) for each fixed x in $[a,b]$, and also integrable (as a function of x) for each fixed y in $[c,d]$; also, you must assume $f(x,y)$ is integrable as a double integral over the rectangle R.

We haven't discussed double integrals, but the definition of integrability is the natural one: in outline, the upper and lower sums for a sequence of partitions of R into small rectangles should get arbitrarily close to each other, as the mesh of the partitions tends to 0.

More general bounded regions S can be handled by a trick. Draw a rectangle R around S and change the definition of f by keeping it the same on S, but making it 0 outside of S. Then the integral of f over R will be the same as its integral over S, and one needs therefore only a theorem about rectangular regions. Unfortunately Theorem 26.3 as stated won't do, since the altered function f is now discontinuous on R.

Questions 26.3

1. Recall the definition of the error function: $\operatorname{erf}(x) = \displaystyle\int_0^x e^{-t^2/2}\,dt$.

   (a) Show first that $\displaystyle\int_0^1 e^{-a^2 t^2/2}\,dt = \frac{1}{a}\operatorname{erf}(a)$.

   (b) Let $\phi(x) = \displaystyle\int_0^1 t^2 e^{-x t^2/2}\,dt$.   Evaluate $\displaystyle\int_1^4 \phi(x)\,dx$ in terms of values of the error function. Verify hypotheses.

Exercises

26.1

1. Evaluate the following, with proof; the integrals are not elementary. In each case, predict the largest x-interval on which the integral will be continuous.

(a) $\lim\limits_{x \to 0} \int_0^\pi \dfrac{\sin t}{1 + xt}\, dt$;   (b) $\lim\limits_{x \to 0} \int_0^1 \dfrac{e^{x+t}}{1 + x^2 t^2}\, dt$.

26.2

1. Let $\phi(x) = \displaystyle\int_{x^2}^{x^3} \cos xt\, dt$. Calculate $\phi'(x)$ two ways:

(a) finding $\phi(x)$ explicitly;   (b) using Leibniz' formula (verify hypotheses).

2. Let $\phi(x) = \displaystyle\int_a^x g(x - t)\, dt$, where $g'(u)$ is assumed to exist and be continuous for all u. Give with proof a formula for $\phi'(x)$ which does not contain any definite integrals.

3. In each of the following, calculate $\phi'(x)$. Verify the hypotheses, indicating for what x your result is valid.

(a) $\phi(x) = \displaystyle\int_{x^2}^x \dfrac{\sin xt}{t}\, dt$ (b) $\phi(x) = \displaystyle\int_{1/\sqrt{x}}^{\sqrt{x}} \dfrac{e^{-xt^2}}{t}\, dt$

4. Consider the differential equation $y'' + y = r(x)$, where $r(x)$ is continuous, together with the initial conditions

$$y(0) = 0, \qquad y'(0) = 0 .$$

Prove that the function

$$y(x) = \int_0^x \sin(x - t) r(t)\, dt$$

satisfies the differential equation and the given initial conditions.

This expresses the solution to the differential equation as a definite integral. The function $\sin(x - t)$ is called the *Green's function* for the operator $D^2 + 1$ which defines the equation: $(D^2 + 1)y = r(x)$.

Notice that this way of writing the solution allows for any function $r(t)$ which is continuous; suitably interpreted, it works for many discontinuous or unbounded functions $r(t)$ as well.

5. Generalizing the preceding exercise, prove that the differential equation and initial conditions (a, b constants, $r(x)$ continuous):

$$y'' + ay' + by = r(x), \qquad y(0) = 0, \ y'(0) = 0$$

has as its solution the function

$$y(x) = \int_0^x g(x - t) r(t)\, dt ,$$

where $g(x)$ is the unique solution to the same differential equation with $r(x)$ taken to be 0, and with initial conditions $y(0) = 0$, $y'(0) = 1$.

26.3

1. Let $f(x) = \int_0^x \sin(t^2)\, dt$,   and $\phi(x) = \int_0^\pi t^2 \cos(xt^2)\, dt$.

These integrals cannot be expressed in terms of elementary functions.

Evaluate $\int_0^a \phi(x)\, dx$, for $a > 0$, using values of $f(x)$;   verify hypotheses.

2. Deduce Theorem 26.2A from Fubini's Theorem 26.3, in the following way. Let $\phi(x)$ and $f(x,t)$ be as in in (1), and consider

$$\int_c^d \int_a^x f_t(t,y)\, dt\, dy \ .$$

First evaluate it as it stands, expressing your answer in terms of $\phi(x)$.   Then reverse the order of integration, and differentiate the resulting iterated integral to get an expression for $\phi'(x)$.

State the hypotheses you are using, and indicate where in the proof they are used.

Problems

26-1 Prove the Bessel equation of order 0, with initial conditions:

$$x^2 y'' + xy' + x^2 y = 0; \qquad y(0) = 1,\ y'(0) = 0,$$

is solved by the Bessel function of zero-th order:

$$J_0(x) \ = \ \frac{1}{\pi} \int_0^\pi \cos(x \sin \theta)\, d\theta \ .$$

We now have three descriptions of $J_0(x)$:

> the unique function satisfying
> $x^2 y'' + xy' + x^2 y = 0,$
> $y(0) = 1,\ y'(0) = 0$

$$J_0(x) = \frac{1}{\pi} \int_0^\pi \cos(x \sin \theta)\, d\theta$$

$$J_0(x) = \sum_0^\infty \frac{(-1)^n x^{2n}}{4^n (n!)^2}$$

Problem 22-2 deals with the right side of the triangle: using term-by-term differentiation to verify that the unique power series satisfying the differential equation with initial conditions is the one given above.

Problem 22-3 deals with the bottom of the triangle, using term-by-term integration to show that the definite integral can be converted into the power series.

Problem 26-1 above completes the picture with the left side of the triangle, verifying that the definite integral also satisfies the differential equation with initial condition.

26-2 Generalize Exercises 26.2/4,5 in the following way. Consider the general linear second-order differential equation with continuous coefficients; take initial conditions at a given point a to be zero:

$$y'' + p(x)y' + q(x)y = r(x), \qquad y(a) = 0, \ y'(a) = 0 .$$

Let y_1 and y_2 be two independent solutions to the associated homogeneous equation, i.e., the one in which $r(x)$ has been replaced by 0. Assume y_1'' and y_2'' continuous. Define the Green's function for the operator $D^2 + pD + q$ by

$$G(x,t) \ = \ \frac{\begin{vmatrix} y_1(t) & y_2(t) \\ y_1(x) & y_2(x) \end{vmatrix}}{\begin{vmatrix} y_1(t) & y_2(t) \\ y_1'(t) & y_2'(t) \end{vmatrix}} .$$

Prove $y(x) = \displaystyle\int_a^x G(x,t)r(t)\,dt$ solves the differential equation with the given initial conditions. (For the denominator, cf. Theorem D.3A.)

Answers to Questions

26.1

1. (a) Since in either case I is not a compact interval, $I \times [c,d]$ will not be a compact rectangle, so that in general, $f(x,t)$ will not be uniformly continuous on it.

(b) Let $x_0 \in I$; assume it is not an endpoint of I. For some sufficiently small $\delta > 0$, the interval $I' = [x_0 - \delta, x_0 + \delta] \subset I$; since I' is compact, the case already proved shows $\phi(x)$ is continuous on I', therefore at $x_0 \in I'$.

If x_0 is say the left endpoint of I, use $I' = [x_0, x_0 + \delta]$ instead; then $\phi(x)$ will be continuous at x_0, (i.e., right-hand continuous), by Theorem 26.1.

2. (a) Using algebra and simple estimations, we have

$$\left| \frac{x-t}{x+t} - \frac{x_0-t}{x_0+t} \right| = \left| \frac{2(x-x_0)t}{(x+t)(x_0+t)} \right| \le 2|x-x_0| \quad \text{on } R;$$

namely, for $(x,t) \in R$, we have $0 \le t \le 1$ and $1 \le x \le 2$, so that

$$|2(x-x_0)t| \le 2|x-x_0|, \qquad |(x+t)(x_0+t)| \ge 1 .$$

Now, given $\epsilon > 0$, for $x \underset{\epsilon}{\approx} x_0$ we have (using the linearity and absolute value properties of the definite integral)

$$|\phi(x) - \phi(x_0)| \le \int_0^1 \left| \frac{x-t}{x+t} - \frac{x_0-t}{x_0+t} \right| dt$$

$$\le \int_0^1 2|x-x_0|\,dt \ \le \ 2\epsilon ;$$

this proves that $\phi(x)$ is continuous at x_0, by the K-ϵ principle ($K = 2, \ \delta = \epsilon$).

(b) Since the integral is discontinuous when $x = -t$, the largest rectangle $R = I \times [0, 1]$ on which the hypotheses are satisfied (and therefore $\phi(x)$ will be continuous) is $I = (0, \infty)$.

Calculating $\phi(x)$ explicitly, we get — note that $\dfrac{x - t}{x + t} = -1 + \dfrac{2x}{x + t}$ —

$$\phi(x) = -1 + 2x \ln\left(1 + \frac{1}{x}\right) ,$$

which is continuous for $x > 0$.

(This can be extended to $x \geq 0$ since $\lim\limits_{x \to 0^+} \phi(x)$ exists.)

26.2

1. (a) $\ln(x - t)$ is continuous where $x > t$; since $0 \leq t \leq 1$, the admissible x-values are $x > 1$. Therefore $I = (1, \infty)$.

(b) $\ln(x - t)$ is differentiable with respect to x in $R = (1, \infty) \times [0, 1]$. Therefore Theorem 26.2A applies, and we get

$$\phi'(x) = \int_0^1 \frac{dt}{x - t} = -\ln(x - t)\Big|_{t=0}^{t=1} = -\ln(x - 1) + \ln x = \ln \frac{x}{x - 1} .$$

2. Since the integrand is differentiable for all x and t, the results will be valid on the whole x-axis; the other hypotheses are trivial to verify.

Using the Leibniz formla,

$$\phi'(x) = \left(x^2 + (x^3)^2\right) \cdot 3x^2 + \int_0^{x^3} 2x \, dt = 5x^4 + 3x^8.$$

Directly, we get by evaluating the integral,

$$\phi(x) = x^2 t + t^3/3 \Big|_0^{x^3} = x^5 + x^9/3 ;$$

from which it is clear that $\phi'(x)$ agrees with the results of Leibniz' formula.

26.3

1. (a) Make the change of variable $u = at$; then $dt = (1/a)du$ and the u-integral goes from 0 to a; and the result follows.

(b) Since the function $f(x, t)$ is continuous for all x and t, the hypotheses of Fubini's theorem are satisfied; we get

$$\int_1^4 \phi(x) \, dx = \int_0^1 \int_1^4 t^2 e^{-xt^2/2} \, dx \, dt$$

$$= \int_0^1 -2e^{-xt^2/2} \Big|_1^4 \, dt$$

$$= -2 \int_0^4 \left(e^{-4t^2/2} - e^{-t^2/2}\right) dt$$

$$= -\mathrm{erf}\,(2) + 2 \, \mathrm{erf}\,(1) .$$

27

Differentiating Improper Integrals

27.1 Introduction.

We would like to extend the results of the preceding chapter to improper integrals. That is, we now suppose our function $\phi(x)$ is defined by an improper integral depending on a parameter:

$$(1) \qquad \phi(x) \; = \; \int_c^\infty f(x,t)\, dt \; ,$$

and we ask if $\phi(x)$ is continuous, and whether it can be differentiated and integrated "under the integral sign". Typical examples of such a function are

$$(2) \qquad H(x) \; = \; \int_0^\infty e^{-xt} h(t)\, dt \qquad\qquad \text{(Laplace transform of } h\text{)};$$

$$(3) \qquad \Gamma(x) \; = \; \int_0^\infty t^{x-1} e^{-t} dt \qquad\qquad \text{(Gamma function)}.$$

For concreteness, we will work with improper integrals of the first kind (1), taken over a semi-infinite interval. Results for other types of improper integrals are similar and can be obtained *mutatis mutandis*. (Note that the Γ-function is also improper of the second kind at the lower end if $0 < x < 1$.)

Though we shall of course make use of the analogous results for integrals over finite intervals in the previous chapter, the work in this chapter can be best thought of as generalizing the work in Chapter 22 on differentiating and integrating series of functions. To make this clearer, use computer notation, writing subscripts as variables:

$$u_n(x) \; = \; u(x,n) \; ;$$

that is, the family of functions $u_n(x)$ forming the terms of the series is now thought of as a single function $u(x,n)$ of two variables: a continuous real-valued variable x, and a discrete variable n whose values are the non-negative integers. We are guided by the analogy between

$$s(x) \; = \; \sum_{n=0}^{\infty} u(x,n) \qquad \text{and} \qquad \phi(x) \; = \; \int_c^\infty f(x,t)\, dt \; .$$

On the left is the discrete summation over n; on the right is the continuous summation over t. Also, the partial sums and the total sum of the series,

$$s_N(x) \;=\; \sum_0^N u_n(x), \qquad s(x) = \lim_{N\to\infty} s_N(x),$$

have, according to the definition of improper integral, their analog for integrals:

(4) $$\phi_R(x) \;=\; \int_c^R f(x,t)\,dt\,, \qquad \phi(x) = \lim_{R\to\infty} \phi_R(x).$$

27.2   Pointwise vs. uniform convergence of integrals.

We will assume throughout that $f(x,t)$ is continuous in a semi-infinite strip $S = I \times [c,\infty)$, where I is any kind of interval, finite or infinite.

Definition 27.2A   We say the integral (1) **converges** (pointwise) on I to the function $\phi(x)$, if for each $x_0 \in I$,

$$\phi(x_0) \;=\; \int_c^\infty f(x_0,t)\,dt \;=\; \lim_{R\to\infty} \int_c^R f(x_0,t)\,dt\,.$$

That is, given $\epsilon > 0$, for each $x_0 \in I$ there is an R_0 *(depending on ϵ and x_0)* such that

(5) $$\left| \phi(x_0) \;-\; \int_c^R f(x_0,t)\,dt \right| < \epsilon \qquad \text{for } R > R_0\,,$$

or equivalently,

(5a) $$\left| \int_R^\infty f(x_0,t)\,dt \right| < \epsilon \qquad \text{for } R > R_0\,.$$

Definition 27.2B   We say the integral (1) **converges uniformly** on I to the function $\phi(x)$ if, given $\epsilon > 0$, there is an R_0 *(depending only on ϵ)* such that

(6) $$\left| \phi(x_0) \;-\; \int_c^R f(x_0,t)\,dt \right| < \epsilon \qquad \text{for } R > R_0 \text{ and all } x_0 \in I$$

or equivalently,

(6a) $$\left| \int_R^\infty f(x_0,t)\,dt \right| < \epsilon \qquad \text{for } R > R_0 \text{ and all } x_0 \in I.$$

The difference between the definitions is that for uniform convergence, the number R_0 is to depend only on the choice of ϵ, and not on what particular x_0 you are considering; that is, one value R_0 must do for all the points in I.

Compare this with the definitions we gave in Chapter 22 for the convergence of series of functions. It will look simpler if use the partial sum notation given in (4) to express the statements in (5) and (6): given $\epsilon > 0$,

$$|s(x) - s_N(x)| < \epsilon \;\; \text{if } N > N_0; \quad \begin{cases} N_0 & \text{depends on } \epsilon \text{ and } x \text{ (ptwise conv.)} \\ N_0 & \text{depends only on } \epsilon \text{ (uniform conv.)} \end{cases}$$

$$|\phi(x) - \phi_R(x)| < \epsilon \;\; \text{if } R > R_0; \quad \begin{cases} R_0 & \text{depends on } \epsilon \text{ and } x \text{ (ptwise conv.)} \\ R_0 & \text{depends only on } \epsilon \text{ (uniform conv.)} \end{cases}$$

To test the improper integral (1) for uniform convergence, we use the analog of the M-test for series (Theorem 22.2C). We recall it now, so you can have the two tests side-by-side for comparison. (After reading the version for series, try conjecturing the analogous version for integrals yourself before reading further.)

If $|u_n(x)| \leq M_n$ for $x \in I$ and $n \geq n_0$,   and $\displaystyle\sum_{n_0}^{\infty} M_n < \infty$, then

$$\sum_{n_0}^{\infty} u_n(x) \text{ is uniformly convergent on } I. \qquad \text{(M-test for series)}$$

Theorem 27.2   Analog of the M-test for improper integrals.

Assume $f(x,t)$ is continuous in the infinite strip $S = I \times [c, \infty)$.

If $|f(x,t)| \leq g(t)$ for $(x,t) \in S$, and $\displaystyle\int_c^{\infty} g(t)\, dt$ converges, then

$$\int_c^{\infty} f(x,t)\, dt \text{ is uniformly convergent in } I \ .$$

Proof.   Using (6a) as the definition of uniform convergence, we have (see footnote below for justification of the first inequality)(*) :   given $\epsilon > 0$,

(7)
$$\left| \int_R^{\infty} f(x,t)\, dt \right| \ \leq \ \int_R^{\infty} |f(x,t)|\, dt$$

$$\leq \ \int_R^{\infty} g(t)\, dt \qquad \text{by Theorem 21.2B;}$$

$$< \ \ \epsilon, \qquad\qquad \text{for } R \geq R_0,$$

since the integral is the tail of a positive convergent improper integral,   □

Example 27.2A   Show $\displaystyle\int_0^{\infty} \frac{\sin xt}{1+t^2}\, dt$ converges uniformly in $I = (-\infty, \infty)$.

Solution.   Using the analog of the M-test, $\left| \dfrac{\sin xt}{1+t^2} \right| \leq \dfrac{1}{1+t^2}$ for all (x,t),

and $\displaystyle\int_0^{\infty} \frac{dt}{1+t^2}$ converges (to $\tan^{-1}(\infty) = \pi/2$).   □

(*) To see this, we have by the Absolute Value Theorem for integrals (19.4C),

$$\left| \int_R^S f(x,t)\, dt \right| \ \leq \ \int_R^S |f(x,t)|\, dt, \quad \text{for all } S > R, \ R \text{ fixed.}$$

Let $g(S) = $ [right side] $-$ [left side]. Then $g(S) \geq 0$ for $S > R$, so that by the Limit Location Theorem for functions (11.3C), $\displaystyle\lim_{S \to \infty} g(S) \geq 0$; this proves (7).

Laplace transform of functions of exponential type.

We give another illustration of the use of the M-test analog. Suppose $h(t)$ is defined and continuous on $[0, \infty)$. We say the function is of *exponential type* if there is a constant k such that

(8) $$\left| h(t) \right| < e^{kt}, \qquad \text{for } t \gg 1 .$$

Such a function does not grow faster than a simple exponential, as $t \to \infty$. By contrast, a function like $h(t) = e^{t^2}$ grows too rapidly to belong to this class.

Functions of exponential type are the simplest general class of functions which have a Laplace transform. Recall the definition:

(2) $$H(x) = \int_0^\infty e^{-xt} h(t)\, dt \qquad\qquad \text{(Laplace transform of h).}$$

Example 27.2B Assume $h(t)$ is of exponential type, with k as in (8). On what x-interval is its Laplace transform known to exist (i.e., where does the integral converge)? Where does it converge uniformly?

Solution. According to (8), we have for some sufficiently large t_0,

$$\int_{t_0}^\infty e^{-xt} \left| h(t) \right| dt \ \le\ \int_{t_0}^\infty e^{-xt} e^{kt} dt \ =\ \int_{t_0}^\infty e^{(k-x)t} dt ,$$

and this last integral converges for $k - x < 0$, i.e., for $x \in (k, \infty)$. Thus the comparison test for improper integrals shows the integral on the left converges for $x > k$; by the Tail-convergence Theorem, it still converges when the lower limit is 0, and since for an improper integral, absolute convergence implies convergence (Theorem 21.4), it follows that the Laplace transform

(2) $$H(x) = \int_0^\infty e^{-xt} h(t)\, dt$$

exists for $x > k$, i.e., on the interval (k, ∞).

Uniform convergence is a bit more subtle. Using the analog of the M-test, the best estimation independent of x and giving a convergent integral is

$$\left| e^{-xt} h(t) \right| \ \le\ e^{-k't} e^{kt}, \qquad \text{for } (x, t) \in [k', \infty) \times [t_0, \infty), \quad k' > k .$$

(We cannot take $k' = k$, since $\int_0^\infty e^{(k-k')t} dt$ does not converge when $k' = k$. We cannot include x on the right side of the inequality as we did before, since the M-test does not permit this.)

We conclude therefore that the convergence is uniform on every x-interval $[k', \infty)$, where $k' > k$. It will generally not be uniform on (k, ∞), however.

Questions 27.2

1. Where does $\displaystyle\int_0^\infty \frac{e^{-t}}{1 + xt}\, dt$ converge pointwise? uniformly?

2. Which of the following is of exponential type? For each that is, tell where its Laplace transform converges uniformly.   (a) $\sin t$ (b) t^t (c) te^{-t}

27.3 Continuity theorem for improper integrals.

We consider now the continuous analog of the Continuity Theorem 22.3, which tells when a convergent series of functions has a continuous sum. We will use the Continuity Theorem 26.1 for finite integrals, but the method of proof will be the exact analog of the method used to prove Theorem 22.3.

Theorem 27.3 Continuity theorem. *If $f(x,t)$ is continuous in the strip $I \times [c, \infty)$, and the integral converges uniformly on I, then the function*

$$\phi(x) = \int_c^\infty f(x,t)\,dt$$

is continuous on I.

Proof. Given $\epsilon > 0$, we wish to show that for an arbitrary point $x_0 \in I$,

$$\phi(x) \underset{\epsilon}{\approx} \phi(x_0) \qquad \text{for } x \approx x_0 .$$

Since $\int_c^\infty f(x,t)\,dt$ converges uniformly on I, we may choose an R so

(9) $\qquad \phi(x) \underset{\epsilon}{\approx} \phi_R(x)$ for $x \in I$, where $\phi_R(x) = \int_c^R f(x,t)\,dt.$

The integral on the right in (9) is over a finite interval $[c, R]$, and $\phi_R(x)$ is therefore continuous on I, by the Continuity Theorem 26.1 of the previous chapter. Therefore

$$\phi_R(x) \underset{\epsilon}{\approx} \phi_R(x_0), \qquad \text{for } x \approx x_0 .$$

Combining this with (9), we get for this value of R the approximations

$$\phi(x) \underset{\epsilon}{\approx} \phi_R(x) \underset{\epsilon}{\approx} \phi_R(x_0) \underset{\epsilon}{\approx} \phi(x_0), \quad \text{if } x \approx x_0 .$$

This shows $\phi(x)$ is continuous at x_0, since it shows that

$$\phi(x) \underset{3\epsilon}{\approx} \phi(x_0) \qquad \text{for } x \approx x_0. \qquad \square$$

If x_0 is say the left endpoint of the interval I, one should in the proof use $x \approx x_0^+$; similarly if it is the right endpoint.

Example 27.3 Assume $h(t)$ is continuous and of exponential type, with k as in (8). Show its Laplace transform (2) is continuous for $x > k$.

Solution. By Example 27.2B, the integral converges uniformly on $[k', \infty)$ when $k' > k$, so by Theorem 27.3, the Laplace transform $H(x)$ is continuous on $[k', \infty)$ for all $k' > k$, and therefore is continuous on (k, ∞) as well. $\square$

Questions 27.3

1. Referring to Question 27.2/1, for what x is $\int_0^\infty \dfrac{e^{-t}}{1+xt}\,dt$ continuous?

2. For what x is $\int_0^\infty e^{-xt}\ln(1+t)\,dt$ continuous?

27.4   Integrating and differentiating improper integrals.

If we try to integrate over $[a, b]$ the function

$$(1) \qquad \phi(x) \; = \; \int_c^\infty f(x, t)\, dt$$

by "integrating under the integral sign", we are led to the equality (10) below. Just as in the corresponding Theorem 26.3 for finite integrals, it can also be interpreted as a Fubini Theorem about interchanging the order of integration in an iterated integral; the difference is that now the rectangular region is an infinite strip.

Theorem 27.4A   Fubini theorem.

If $f(x, t)$ is continuous in the strip $S = [a, b] \times [c, \infty)$, and the integral (1) above is uniformly convergent on $[a, b]$, then

$$(10) \qquad \int_a^b \int_c^\infty f(x, t)\, dt\, dx \; = \; \int_c^\infty \int_a^b f(x, t)\, dx\, dt \; .$$

Proof.   We use the analogous finite Fubini Theorem 26.3, according to which

$$(11) \qquad \int_a^b \int_c^R f(x, t)\, dt\, dx \; = \; \int_c^R \int_a^b f(x, t)\, dx\, dt \; .$$

From the definition of improper integral, to prove (10), we must show

$$\int_a^b \int_c^\infty f(x, t)\, dt\, dx \; = \; \lim_{R \to \infty} \int_c^R \int_a^b f(x, t)\, dx\, dt \; .$$

We do this by showing the difference between the above two iterated integrals has size $< \epsilon$, if $R \gg 1$. Calculating this difference, and using (11) above,

$$\left| \int_a^b \int_c^\infty f(x, t)\, dt\, dx - \int_c^R \int_a^b f(x, t)\, dx\, dt \right| = \left| \int_a^b \int_c^\infty f(x, t)\, dt\, dx - \int_a^b \int_c^R f(x, t)\, dt\, dx \right| \; ;$$

$$= \left| \int_a^b \int_R^\infty f(x, t)\, dt\, dx \right| \; ,$$

by the Linearity Theorem 19.4A; by the Absolute Value Theorem 19.4C, this is

$$\leq \int_a^b \left| \int_R^\infty f(x, t)\, dt \right| dx \; ;$$

$$\leq \quad \epsilon(b - a), \qquad \text{if } R \gg 1 \; ,$$

since according to the definition (6a) of uniform convergence, we have

$$\left| \int_R^\infty f(x, t)\, dt \right| \leq \epsilon \; , \qquad \text{for } R \geq \text{ some } R_0 \; .$$

This completes the proof, by the K-ϵ principle. $\square$

Theorem 27.4B   Differentiation of improper integrals. *Assume that*

(a) $f(x,t)$ *and* $f_x(x,t)$ *are continuous on* $S = I \times [c, \infty)$;

(b) $\phi(x) = \displaystyle\int_c^\infty f(x,t)\, dt$ *converges for each* $x \in I$;

(c) $\psi(x) = \displaystyle\int_c^\infty f_x(x,t)\, dt$ *converges uniformly on* I;   *then*

(12) $$\phi'(x) = \int_c^\infty f_x(x,t)\, dt \qquad \text{for } x \in I \ .$$

Once again, there are two interpretations for (12). It is Theorem 26.2A about differentiating under the integral sign, extended to improper integrals. It is also the continuous analog of Theorem 22.5 about differentiating a series term-by-term. The proof below imitates the proof of Theorem 22.5 .

Proof.   Let x_0 be a point of I. We integrate $\psi(x)$, switching to the dummy variable of integration u to avoid confusion:

$$\int_{x_0}^x \psi(u)\, du = \int_{x_0}^x \int_c^\infty f_u(u,t)\, dt\, du \ ;$$

$$= \int_c^\infty \int_{x_0}^x f_u(u,t)\, du\, dt \ ,$$

by the the uniform convergence and the Fubini Theorem 27.4A;

$$= \int_c^\infty \big(f(x,t) - f(x_0,t)\big)\, dt \ ,$$

by the First Fundamental Theorem;

$$= \phi(x) - \phi(x_0), \qquad\qquad \text{by definition of } \phi(x).$$

Now differentiate both sides with respect to x, using the Second Fundamental Theorem to differentiate the left-hand side; this gives (12), in the form

$$\psi(x) = \phi'(x) \ . \qquad\qquad\qquad \square$$

Questions 27.4

1. Let $\phi(x) = \displaystyle\int_0^\infty e^{-xt} t\, dt$, the Laplace transform of t.

(a) Evaluate $\displaystyle\int_1^\infty \phi(x)\, dx$ by Fubini's theorem, verifying the hypotheses. (Integrate up to R, then let $R \to \infty$.)

(b) Verify your answer by using the explicit value of $\phi(x)$.

2. Let $\phi(x) = \displaystyle\int_1^\infty \frac{\sin xt}{t(1+t^2)}\, dt$. Find $\phi'(0)$, verifying the hypotheses.

27.5 Differentiating the Laplace transform.

There are many tricky applications of these theorems, especially to the evaluation of certain non-elementary integrals that appear in applied mathematics. A couple are in the Exercises and Problems. but these things are mostly for specialists. Instead, let's justify one of the standard formulas of the Laplace transform.

Example 27.5 Derivative of the Laplace transform.

Assume $h(t)$ is continuous, and of exponential type: there is a k such that
$$|h(t)| \leq e^{kt}, \qquad \text{for } t \gg 1 .$$
Show that the derivative of its Laplace transform

(2) $$H(x) = \int_0^\infty e^{-xt} h(t)\, dt = \mathcal{L}\big(h(t)\big), \quad x > k,$$

is given by

(13) $$H'(x) = -\int_0^\infty e^{-xt} t\, h(t)\, dt = \mathcal{L}\big(-t\, h(t)\big), \quad x > k .$$

Solution.   The definition (2) of the Laplace transform shows that, in the notation of Theorem 27.4B,
$$f(x,t) = e^{-xt} h(t); \qquad\qquad f_x(x,t) = e^{-xt}(-t\, h(t)) .$$
Thus, to see over what x-interval the hypotheses of Theorem 27.4B hold, we have to study the uniform convergence of the integral on the right side of (13) above.

We claim that $t\, h(t)$ is also of exponential type: namely, given $\epsilon > 0$, we get successively,

$$\lim_{t \to \infty} \frac{t}{e^{\epsilon t}} = 0, \qquad \text{by l'Hospital's rule;}$$

$$\frac{t}{e^{\epsilon t}} \leq 1, \qquad \text{by the Function Location Theorem 11.3D;}$$

$$|t\, h(t)| \leq e^{(k+\epsilon)t}, \qquad \text{by our hypothesis on } h(t).$$

By the result in Example 27.2B, we can conclude that the convergence is uniform on any $[k', \infty)$, where $k' > k + \epsilon$; therefore when $k' > k$, since ϵ can be arbitrarily small.

Theorem 27.4B then tells us that (13) above is valid for all $x \in [k', \infty)$, where k' is any number such that $k' > k$. Since the derivative is a local property, it follows that (13) is valid for $x \in (k, \infty)$. □

Other improper integrals

The definitions and theorems of this chapter also hold for improper integrals of the second kind, that is, those for which the integrand is unbounded. For example, assume f is continuous in the rectangle $S = I \times [c, d)$, but unbounded as $t \to d^-$, and let

$$\phi(x) = \int_c^{d^-} f(x, t)\, dt .$$

If the integral converges (pointwise) for all $x \in I$, then $\phi(x)$ is defined on I. The definition of *uniform convergence on I* would be:

given $\epsilon > 0$, there is a $\delta > 0$ (depending on ϵ, but not on x), such that

$$\left| \phi(x) - \int_c^u f(x,t)\, dt \right| < \epsilon \qquad \text{for all } u \underset{\delta}{\approx} d^- .$$

All the theorems of this chapter are then true, replacing ∞ by d^- in the statements, and R by u in the proofs.

Exercises

27.2

1. 1. Prove that $\displaystyle\int_0^\infty e^{-xt^2} \sin t\, dt$ converges pointwise on $(0, \infty)$ and uniformly on $[a, \infty)$, for any $a > 0$.

2. Prove that $\displaystyle\int_0^\infty \frac{t \sin xt}{1 + t^3}$ converges uniformly on $(-\infty, \infty)$.

27.3

1. Prove that $\displaystyle\phi(x) = \int_0^\infty \frac{e^{-xt}}{1+t}\, dt$ is continuous for $x > 0$.

2. Let $\displaystyle\phi(x) = \int_0^\infty \frac{\sin xt}{1 + t^k}\, dt$. Prove $\phi(x)$ is continuous for all x, if $k > 1$. What is the difficulty if $k \leq 1$?

27.4

1. Let $b > a > 0$. Prove that $\displaystyle\int_0^\infty \frac{e^{-at} - e^{-bt}}{t}\, dt = \ln\left(\frac{b}{a}\right)$.

Method: set $\displaystyle\phi(x) = \int_0^\infty e^{-xt}\, dt$, and apply Fubini's Theorem to $\displaystyle\int_a^b \phi(x)\, dx$. (Note that $\phi(x)$ can be calculated explicitly.) Verify the hypotheses. How should the integrand be defined at $t = 0$ so as to make it continuous for all t?

2. Let $\displaystyle\phi(x) = \int_0^\infty \frac{\sin xt}{1 + t^k}\, dt$. Prove that the function $\phi(x)$ is differentiable for all x, if $k > 2$. What is the difficulty if $k \leq 2$?

3. Prove the differential equation

$$y' - y = -1/x, \qquad x > 0,$$

has the solution

$$y(x) = \int_0^\infty \frac{e^{-xt}}{1+t}\, dt , \qquad x > 0.$$

Begin by proving that $y(x)$ is continuous and differentiable for $x > 0$. Verify hypotheses.

27.5

1. Evaluate the Laplace transform $\int_0^\infty e^{-xt}t^n\,dt$, $n \in \mathbb{N}$ by using the results of this section, and mathematical induction. (For $n = 0$ you'll have to do it directly.) For what x are the results valid?

2. From the Laplace transform formula $\mathcal{L}(\sin t) = \dfrac{1}{1+x^2}$, $x > 0$, deduce the formula for $\mathcal{L}(t\sin t)$.

3. Show that, with reasonable assumptions, if $F(x) = \mathcal{L}(f(t))$, then

$$\int_x^\infty F(u)du = \mathcal{L}\left(\frac{f(t)}{t}\right).$$

State explicitly what you are assuming.

4. (a) Prove $\displaystyle\int_0^\infty \frac{dt}{x+t^2} = \frac{\pi}{2\sqrt{x}}$, $\quad x > 0$.

    (b) Deduce that $\displaystyle\int_0^\infty \frac{dt}{(x+t^2)^{n+1}} = \frac{(2n-1)!!}{(2n)!!\,x^n\sqrt{x}}\cdot\frac{\pi}{2}$, for $n \geq 1$.

(See Exercise 20.3/1 for $n!!$.)

Problems

27-1 Prove that $\phi(x) = \displaystyle\int_0^\infty \frac{t^{x-1}}{1+t}\,dt$ is continuous for $0 < x < 1$.

27-2 Evaluate the integral $\phi(x) = \displaystyle\int_0^\infty \frac{e^{-t}(1-\cos xt)}{t}\,dt$, by calculating the derivative. Verify all hypotheses.

(Note that from the resulting formula $\phi'(x)$ can be calculated explicitly — look it up in a table of integrals or any calculus text.) How should one define the original integrand at 0 to make it continuous for all x and t?

27-3 Prove the Γ-function (3) is continuous for $x > 0$.

(Show it is continuous on any interval $[a, b]$, $a > 0$. Break up the integral into two parts.)

27-4 The value of $\Gamma'(1)$ is a classical result.

    (a) Give an improper integral whose value should be $\Gamma'(1)$.

    (b) Prove your improper integral $= \Gamma'(1^+)$.

    (c) Prove your improper integral $= \Gamma'(1)$.

It turns out that $\Gamma'(1) = -\gamma$, the Euler constant described at the very beginning of this book in Prop. 1.5B. This can be proved by transforming the limit there into the negative of the improper integral of part (a), but a number of intermediate steps are required.

27-5 An integral which generalizes the ones in the previous three problems is the *Mellin transform*:

$$F(x) = \int_0^\infty t^{x-1} f(t) \, dt \; .$$

(a) Assume that $f(t)$ is continuous for $t \geq 0$, and that for some fixed constant $a > 1$, we have $0 \leq f(t) < 1/t^a$ for $t \gg 1$. Prove that $F(x)$ exists on $(0, a)$ and is continuous on say $[1, a)$ (so you won't have to work too hard).

(b) Calculate the Mellin transform of e^{-kt}, for $x > 1$, in terms of $\Gamma(x)$ and k. Justify the steps.

Answers to Questions

27.2

1. If $x < 0$, then the integral $\displaystyle\int_0^\infty \frac{e^{-t}}{1 + xt} \, dt$ is improper of the second kind, since the denominator is 0 when $t = -1/x$; it does not converge.

The integral is however uniformly convergent (and therefore pointwise convergent) on the x-interval $[0, \infty)$. Namely, using the analog of the M-test, we have

$$\left| \frac{e^{-t}}{1 + xt} \right| \leq e^{-t} , \qquad \text{if } x \geq 0, t \geq 0 \; ;$$

and $\displaystyle\int_0^\infty e^{-t} \, dt$ converges.

2. (a) $\sin t$ is of exponential type: $|\sin t| \leq e^{0t}$ for all t. Its Laplace transform is therefore convergent on $(0, \infty)$, and uniformly convergent on every interval $[a, \infty), a > 0$.

(b) $t^t = e^{t \ln t}$; since $\lim \ln t = \infty$, it ultimately becomes larger than any given k, and therefore t^t is not of exponential type.

(c) Given any $\epsilon > 0$, we have $t < e^{\epsilon t}$ for $t \gg 1$.
Therefore $te^{-t} < e^{(\epsilon - 1)t}$ for $t \gg 1$, which shows that te^{-t} is of exponential type with exponential constant $\epsilon - 1$, for any $\epsilon > 0$. Therefore its Laplace transform is uniformly convergent on any interval $[k, \infty)$, where $k > -1$. (See Example 27.5, where the same reasoning is used, and spelled out in more detail.)

27.3

1. It is continuous where it is uniformly convergent, therefore on the interval $[0, \infty)$.

2. Since $\left| \ln(1+t) \right| \leq e^{kt}$ for any $k > 0$, and $t \gg 1$, (by l'Hospital's rule and the Function Location Theorem; cf. Example 27.5), the integral is by Example 27.2B uniformly convergent on $[k', \infty)$, for any $k' > k > 0$, i.e., for any $k' > 0$. Therefore it is continuous on $[k', \infty)$, and since continuity is a local property, it is continuous on $(0, \infty)$.

27.4

1. (a) We check the hypotheses for Fubini's Theorem 27.4A. First of all, $e^{-xt}t$ is continuous for all x and t. Moreover, by l'Hospital's rule, the function t is of exponential type:

$$t < e^{\epsilon t} \qquad \text{for } t \gg 1, \text{ and any } \epsilon > 0.$$

Therefore, by Example 27.2B, its Laplace transform integral converges uniformly on $[k, \infty)$, where $k > \epsilon > 0$, or simply $k > 0$, since ϵ can be arbitrarily small.

We can therefore apply Fubini's Theorem; since the interval I is supposed to be finite, we integrate only from 1 to R. We get

$$\int_1^R \phi(x)\,dx = \int_0^\infty \int_1^R e^{-xt}t\,dx\,dt$$

$$= \int_0^\infty -e^{-xt}\Big|_1^R\,dt$$

$$= \int_0^\infty (e^{-t} - e^{-Rt})\,dt = 1 - \frac{1}{R}.$$

Therefore, taking the limit as $R \to \infty$, we get $\displaystyle\int_1^\infty \phi(x)\,dx = 1$.

(b) Integration by parts (or Laplace transform tables) produces the value $\phi(x) = \dfrac{1}{x^2}$, $x > 0$, and by an easy calculation, $\displaystyle\int_1^\infty \frac{dx}{x^2} = 1$, verifying our calculation in part (a).

2. We have $\phi(x) = \displaystyle\int_1^\infty \frac{\sin xt}{t(1+t^2)}\,dt$; and formally, $\phi'(x) = \displaystyle\int_1^\infty \frac{\cos xt}{1+t^2}\,dt$.
We verify the hypotheses of Theorem 27.4B, to justify the expression for $\phi'(x)$.

1. The integrands are continuous for $t > 0$ and all x, so the integrals exist.

2. For all x, the integral for $\phi(x)$ converges (pointwise) absolutely, by comparison with $\displaystyle\int_1^\infty \frac{dt}{t(1+t^2)}$ (which converges since the integrand is asymptotic to $1/t^3$). Since it converges absolutely, it converges (Theorem 21.4).

3. The integral for $\phi'(x)$ converges uniformly for all x by the M-test analog (Theorem 27.2), comparing it with the convergent integral $\displaystyle\int_1^\infty \frac{dt}{1+t^2}\,dt$.

4. Therefores the integral for $\phi'(x)$ is valid for all x; put $x = 0$: we get

$$\phi'(0) = \int_1^\infty \frac{dt}{1+t^2}\,dt$$

$$= \lim_{R\to\infty} \tan^{-1}(R) - \tan^{-1}(1)$$

$$= \pi/2 - \pi/4 = \pi/4.$$

Appendix A

Sets, Numbers, and Logic

The first section establishes some of the set notation used in the book and briefly reviews the number systems we need; the rest deals with logical implication and proofs—if-then statements. indirect proof, proof by contraposition, and what a counterexample is are all discussed, followed by a section on mathematical induction. Read this appendix alongside Chapter 1 if these are unfamiliar ideas; or use it as a reference when puzzled by something in the main text, which in the first few chapters cites it frequently.

A.0 Sets and Numbers.

First a quick mention about how sets are described; then we will comment on the numbers we will use.

Set notation.

Sets in general will be denoted here by capital letters: $S, T, \ldots$; in this book they will almost always be sets of numbers, or toward the end, of points in the plane, i.e., ordered pairs of numbers (x, y).

The notation $a \in S$ means "the number a is an element of the set S", or said more simply, "a is in S".

Sets can be defined either by listing their elements between braces:
$$S = \{1, \ 2, \ 3, \ 4, \ 5\}$$
or by describing the criteria for membership in the set:
$$S = \{x : x \text{ is an integer, } 1 \leq x \leq 5\},$$
read, "S is the set of all x such that x is an integer and..."

Set equality and inclusion.   The notation is

$S = T :$ the two sets have the same elements;

$S \subseteq T, \ T \supseteq S :$ S is a subset of T (every $x \in S$ is also in T);

$S \subset T, \ T \supset S :$ S is a proper subset of T ($S \subseteq T$ but $S \neq T$).

You can prove $S = T$ by proving $S \subseteq T$ and $T \subseteq S$; here is a simple example.

Example A.0A.   Let $S = \{x : x^3 = x\}$. Prove that $S = \{-1, 0, 1\}$.

Solution. $\{-1, 0, 1\} \subseteq S :$ $0^3 = 0, \ 1^3 = 1, \ (-1)^3 = -1.$

$S \subseteq \{-1, 0, 1\} :$ $x^3 = x \ \Rightarrow \ x^3 - x = x(x+1)(x-1) = 0$

$\Rightarrow \ x = 0, -1, \text{ or } 1. \quad \square$

Number systems. The evolution of our number system can be summarized roughly as the series of set inclusions
$$\emptyset \subset \mathbb{N} \subset \mathbb{N}_0 \subset \mathbb{Z} \subset \mathbb{Q} \subset \mathbb{R} \subset \mathbb{C}.$$
Let's talk briefly about each of these in turn.

In the beginning there was
$$\emptyset = \text{the empty set} : \text{the set with no elements}.$$
The empty set is a subset of every other set, but there is only one empty set — the set with no integers is the same as the one with no apples. Out of the empty set, man learned (only in the 20th century, actually) to construct
$$\mathbb{N} = \{1, 2, 3, \ldots, n, \ldots\} : \text{the natural numbers}.$$
These have been around a long time, since they are needed for counting and describing the size of finite sets; it's just that they weren't defined until recently.

The set $\mathbb{N}$ was expanded a thousand years ago to
$$\mathbb{N}_0 = \{0, 1, 2, 3, \ldots\} : \text{the non-negative integers};$$
with the addition of zero, now the empty set also had a size. Zero made decimal notation and therefore effective calculation possible. (Try calculating with Roman numerals!)

But $\mathbb{N}_0$ is defective: you can add and multiply without leaving the system, but not always subtract: to solve $x + 5 = 3$, the system must be expanded to
$$\mathbb{Z} = \{\ldots, -2, -1, 0, 1, 2, \ldots\} : \text{the integers}.$$
This causes some problems: negative numbers are traumatic — they don't count anything, which is what numbers are for, and the law $(-1)(-1) = 1$, passed so that $(a+b)c = ab+ac$ would always be true, has led one generation after another over the years to decide that mathematics is gibberish.

Even $\mathbb{Z}$ is defective: it doesn't contain $1/n$, a number that n-person families with pie for dessert find indispensable. In fact, the ancient Egyptians managed to do all their needed arithmetic using just the integers and the numbers $1/n$. To include them, one has to include their multiples, i.e., expand the system to
$$\mathbb{Q} = \{\tfrac{m}{n} : m, n \in \mathbb{Z}, \ n \neq 0\} : \text{the rational numbers}.$$
With the rational numbers we have finally what is called a *field*: a system in which one can add, subtract, multiply, and divide without ever leaving the system (and where these operations satisfy some well-known laws like $ab = ba$ and the distributive law alluded to above).

Unfortunately, many m/n can represent the same rational number. We can make the quotient m/n unique by specifying that $n > 0$ and that m and n have no common factors, i.e., that the representation is in *lowest terms*. You can't always write every fraction in lowest terms, however: it makes calculation too hard. For instance, any two rational numbers can be added only because they are "commensurable", that is, can be written so they have a common denominator; but when you do this, they won't be in lowest terms any more: $\frac{1}{2} + \frac{2}{3} = \frac{3}{6} + \frac{4}{6} = \frac{7}{6}$.

Reals and irrationals   As we discuss at the end of Chapter 1, the system $\mathbb{Q}$ is still too small — it's not complete. Pythagoras discovered that the diagonal of the unit square was incommensurable with its side: in other words, that $\sqrt{2}$ was not rational. (Section A.2 gives his proof.) Suddenly there were a lot of new numbers, *irrational* numbers. There was no way of representing them except as lengths, that is, as points on a line, a representation not well-suited to calculation. But then, no one really needed them. (In a sense, it is only mathematicians who do.) At any rate, to include them, the number system had to be expanded to

$$\mathbb{R} \;=\; \text{the real numbers,}$$

thought of first as the points on a line, then many centuries later, after decimal notation had been invented, also as infinite decimals.

Like the smaller set of rational numbers, the real numbers also form a field: arithmetic operations on real numbers always lead to real numbers. They were constructed rigorously for the first time in the 19th century, in two different ways. The proofs that the so-constructed numbers have the right properties (including the Completeness Property of Chapter 1) take time and effort.

The *irrational* numbers are, in set notation,

$$\mathbb{R} \setminus \mathbb{Q}: \text{ everything in } \mathbb{R} \text{ that's not in } \mathbb{Q}.$$

They most emphatically do *not* form a field, since the arithmetic operations on irrationals do not necessarily lead to irrationals: for instance $\sqrt{2} \cdot \sqrt{2} = 2$. An irrational is either *algebraic* — like $\sqrt{2}$, a zero of a polynomial with coefficients in $\mathbb{Q}$ — or if not, *transcendental*. Proving a number is transcendental is hard.

> Though we shall not make use of them in this book, one should mention the still bigger field $\mathbb{C}$ of *complex numbers* in which all polynomial equations $p(x) = 0$ have solutions; full recognition of these as a valid number system took hundreds of years, and many are still a little uncomfortable with them, since they don't measure lengths on a line. They bear the stigma of being called "imaginary", but of course in a sense they are.

Rational numbers and infinite decimals   The beginning of Chapter 1 gives a brief sketch of the real numbers, thought of as infinite decimals: how you add and multiply them, and why the Completeness Property holds for them. Where do the terminating decimals fit in, and what's their relation to rational numbers?

(i) *A terminating decimal represents a rational number.*

Namely, it can be represented in the form $m/10^n$, if it has n decimal places:

$$3.141 = \tfrac{3141}{1000}, \qquad\qquad -1.42 = -\tfrac{142}{100}.$$

The rationals represented by terminating decimals are very special, however, since the above shows they are the ones writable (not in lowest terms, of course) with only powers of 10 in the denominator. Most rationals are not of this form.

(ii) *An infinite decimal is a rational number* $\Leftrightarrow$ *it is a repeating decimal.*

That is, after some point, it contains a group of digits which repeats for the rest of the decimal expansion, like $2.1333\ldots, -6.366014014014\ldots$.

Question 1 below illustrates why statement (ii) is true: briefly, a repeating decimal represents a geometric series, which can be summed to a rational number. Going the other way, in long division of n into m there are only a finite number of possibilities for the intermediate steps, so at some point, the process will start repeating.

Fact (i) above is the basis for the engineer's claim (very irritating to mathematicians) that "all numbers are rational": for in fact all experimental work and all calculations, whether done by hand or computer, of necessity use only terminating decimals.

Operations on sets.   After this somewhat breezy account, let's return to more general facts about sets. Four operations produce new sets from old ones:

$$S \cup T = \{x : x \in S \text{ or } x \in T\} \qquad \text{(union)};$$
$$S \cap T = \{x : x \in S \text{ and } x \in T\} \qquad \text{(intersection)};$$
$$S \times T = \{(x, y) : x \in S \text{ and } y \in T\} \qquad \text{(product)};$$
$$S \setminus T = \{x : x \in S \text{ but } x \notin T\} \qquad \text{(difference)}.$$

The word "or" in mathematics always has the inclusive sense: that is, "A or B" is rendered in ordinary speech by "A or B or both"; if the other (exclusive) sense is wanted, it would be written in mathematics as "A or B, but not both".

The notation (x, y) in the definition of $S \times T$ stands for the ordered pair of numbers x and y; the same notation is used for the open interval, but the context will always make it clear which is meant.

The definition of the union and intersection (and product as well, though we shall not need it) can be extended from two sets to a finite or infinite collection of sets $\{S_i\}$, where i runs over some set I of indices: the notation for this is

$$\bigcup_{i \in I} S_i = \{x : x \in S_i \text{ for at least one } i\} \qquad \text{(union)}$$

$$\bigcap_{i \in I} S_i = \{x : x \in S_i \text{ for every } i\} \qquad \text{(intersection)}$$

If the collection is finite, the set of indices I is usually $\{1, 2, \ldots, n\}$, and if infinite, it is usually the natural numbers $\mathbb{N}$, although there are occasional exceptions. As examples, using the usual notation for intervals:

$$[a, b] = \{x \in \mathbb{R} : a \le x \le b\}, \qquad (a, b) = \{x \in \mathbb{R} : a < x < b\},$$

we have

$$\bigcap_{n \in \mathbb{N}} (0, 1/n) = \emptyset; \qquad \bigcup_{a \in \mathbb{R}} (0, a^2) = \mathbb{R}^+ .$$

Functions on sets.

Functions of one and two real variables are discussed in detail in Chapters 9 and 24 respectively. The words introduced below to describe functions are used very briefly in Chapter 9 and in Chapter 23, so there's no urgency in reading this quick account now. It's put here just to have as a reference.

A function f from a set S to a set T is given by a rule associating with each element $s \in S$ a corresponding element of T, denoted $f(s)$; in notation:

$$f : S \to T , \qquad s \to f(s) .$$

It is called

injective, if it sends distinct elements of S into distinct elements of T:

$$s_1 \neq s_2 \quad \text{implies} \quad f(s_1) \neq f(s_2) ;$$

surjective, if for each $t \in T$ there is an $s \in S$ such that $f(s) = t$;

bijective, if it is both injective and surjective.

Examples A.0B. Let $f_1(x) = x^2$ and $f_2(x) = 2x$; classify them as injective, surjective, or bijective on the following sets (here $\mathbb{R}^+ = \{x \in \mathbb{R} : x > 0\}$) :

(a) $f_1 : \mathbb{R}^+ \to \mathbb{R}$ (b) $f_1 : \mathbb{R} \to \mathbb{R}^+ \cup \{0\}$ (c) $f_1 : \mathbb{R}^+ \to \mathbb{R}^+$

(d) $f_2 : \mathbb{Z} \to \mathbb{Z}$ (e) $f_2 : \mathbb{Q} \to \mathbb{Q}$

Solution. (a) injective, since if $a > b > 0$, then $a^2 > b^2$; (b) surjective, since every non-negative real number has a square root; (c) bijective, since every positive real has a unique positive square root; (d) injective; (e) bijective.

If $f : S \to T$ is bijective, it has a unique *inverse* $g : T \to S$ defined by

$$g(t) = s \quad \text{if} \quad f(s) = t ;$$

given any t, such an s exists since f is surjective; the s is unique since f is injective. The association $s \to f(s)$ gives what is called a *one-one correspondence* between the elements of S and T: each element of S is paired with one and only one element of T, and vice-versa.

In the examples above, the inverse of $f_1 : \mathbb{R}^+ \to \mathbb{R}^+$ is the function g_1 given by $g_1(x) = \sqrt{x}$; the inverse of $f_2 : \mathbb{Q} \to \mathbb{Q}$ is the function g_2, where $g_2(x) = \frac{1}{2}x$.

The rest of this appendix is devoted to reviewing some points of logic.

Questions A.0 (Answers at end of Appendix A.)

1. (a) Show that $1.1232323\ldots$ is a rational number. (Use Section 4.2 (2).)

   (b) Write $\frac{3}{7}$ as an infinite decimal, using division; see why it repeats.

2. Express these sets in terms of the standard sets and intervals, and the four finite operations.

(a) $\{x \in \mathbb{R} : x \geq 0\}$ (b) $\{x \in \mathbb{Q} : a \leq x \leq b\}$

(c) $\{\text{pairs } (x, y) : a < x < b,\ c < y < d\}$ (d) $\{x \in \mathbb{Q} : x > 0\}$

(e) $\bigcup_{a \in \mathbb{N}_0} [a, a + 1]$ (f) $\bigcup_{a \in \mathbb{Z}} (a, a + 1)$

3. Let $S = \{x \in \mathbb{R} : x(1 - x) > 0\}$. Prove $S = (0, 1)$, by proving $\subseteq$ and $\supseteq$.

A.1   If-then statements

In mathematics, statements to be proved can often be put in the form

(1)   if A, then B; $A \Rightarrow B$ (read: "A implies B").

The two forms say the same thing; the second form uses a symbol called the "forward implication arrow". The A and B represent simpler statements:

> A is the **hypothesis**: "what's given", "what's known";
>
> B is the **conclusion**: "what follows", "what's to be proved".

Here are some examples. Note that often a preliminary statement must be made, explaining what the symbols in the $A \Rightarrow B$ statement stand for.

Examples A.1A

(i) A, B, C are the vertices of a triangle; a, b, c are the non-zero lengths of the opposite sides, respectively.

$$ACB \text{ is a right angle } \Rightarrow a^2 + b^2 = c^2 \ . \qquad\qquad \textit{(true)}$$

(ii) Let a be a real number.

$$2a^6 + a^4 + 3a^2 = 0 \Rightarrow a = 0 \ . \qquad\qquad \textit{(true)}$$

(iii) $f(x)$ is differentiable $\Rightarrow$ $f(x)$ is continuous. $\qquad\qquad$ *(true)*

(iv) Let a, b, n be positive integers.

$$n \text{ divides } ab \Rightarrow n \text{ divides } a \text{ or } b \ . \qquad\qquad \textit{(false)}$$

We shall use where possible the arrow notation, since it allows the hypothesis and conclusion to stand out clearly. But with the if-then form one can avoid a preliminary sentence:

> "If a is a real number such that $2a^6 + a^4 + 3a^2 = 0$, then $a = 0$."

However, the problem with all of this is that in ordinary mathematical writing, the hypothesis and conclusion may not be spelled out so clearly; it is you that has to extract them from the prose sentence. For instance, (ii) and (iii) would probably appear in the form:

(ii) 0 is the only real root of $2x^6 + x^4 + 3x^2 = 0$;

(iii) a differentiable function is continuous.

Thus, if a statement is given in the form $A \Rightarrow B$, some of the work has already been done for you.

In bad mathematical writing, the ambiguities of English may make it impossible to decide what the implication is; for example,

> "$f(x)$ has a relative maximum or minimum at a point a where $f'(a) = 0$."

Does this say (in rough outline) "relative max/min at a $\Rightarrow$ $f'(a) = 0$" or "$f'(a) = 0 \Rightarrow$ relative max/min at a"? Your guess is as good as mine.

Converse.   If we interchange hypothesis and conclusion in $A \Rightarrow B$, we get

(2) $B \Rightarrow A$ (or $A \Leftarrow B$) ,

which is called the *converse* to the statement (1).

Examples A.1B   The converses to A.1A are (omitting the preliminaries):

(i) $a^2 + b^2 = c^2 \Rightarrow ACB$ is a right angle.   (*true*)

(ii) $a = 0 \Rightarrow 2a^6 + a^4 + 3a^2 = 0$.   (*true*)

(iii) $f(x)$ continuous $\Rightarrow$ $f(x)$ differentiable.   (*false*)

(iv) n divides a or b $\Rightarrow$ n divides ab .   (*true*)

As one can see from these examples, the truth or falsity of the converse is unrelated to the truth or falsity of the original statement.

Equivalent statements.

We can combine the two implication arrows into one double-ended arrow:

(3) , $A \Leftrightarrow B$

which is a true statement if both $A \Rightarrow B$ and $A \Leftarrow B$ are true. If this is so, we say A and B are *equivalent* statements. To give our examples one last time:

Examples A.1C

(i) $a^2 + b^2 + c^2 \Leftrightarrow ACB$ is a right angle.   (*true*)

(ii) $a = 0 \Leftrightarrow 2a^6 + a^4 + 3a^2 = 0$.   (*true*)

(iii) $f(x)$ is differentiable $\Leftrightarrow$ $f(x)$ is continuous   (*false*)

(iv) n divides a or b $\Leftrightarrow$ n divides ab .   (*false*)

Necessary and sufficient.   There are two verbal forms of $\Leftrightarrow$ which are in common use. We will mostly avoid them, but others do not, so you should know them. They are:

A if and only if B   (abbreviated: A iff B) ;

A is a necessary and sufficient condition for B (abbreviated: nasc).

Occasionally these are separated into their component parts:

$A \Rightarrow B$: A is a *sufficient* condition for B (if A is true, B follows);

$B \Rightarrow A$: A is a *necessary* condition for B (i.e., B can't be true unless A is also true, since B implies A).

The "if and only if" is also separated:

"A, if B" : $B \Rightarrow A$;   "A, only if B": $A \Rightarrow B$.

This last is the worst, since ordinary English usage is different—"only if" is considered the same as "if and only if":

"You can go only if you are invited"   (..."but if you <u>are</u> invited, you can go" is automatically implied, or one had better be prepared for insurrection).

Stronger and weaker.   If $A \Rightarrow B$ is true, but $B \Rightarrow A$ is false, we say:

A is a *stronger* statement than B; B is *weaker* than A.

Example A.1E

"$\triangle ABC$ is equilateral" is stronger than "$\triangle ABC$ is isosceles", since

$\triangle ABC$ is equilateral $\Rightarrow$ $\triangle ABC$ is isosceles.

The same terminology applies to entire "if-then" statements (theorems):

Example A.1F   The if-then statement

(4) $\triangle ABC$ is equilateral $\Rightarrow$ $\triangle ABC$ has two equal angles

can be made stronger in two different ways: make the hypothesis weaker:

(5) $\triangle ABC$ is isosceles $\Rightarrow$ $\triangle ABC$ has two equal angles;

or make the conclusion stronger:

(6) $\triangle ABC$ is equilateral $\Rightarrow$ $\triangle ABC$ has three equal angles.

Both (5) and (6) are stronger than (4) since they both imply (4): if you know (5) or (6) is true, then (4) follows, but not vice-versa.

> *Strengthen $A \Rightarrow B$ by making B stronger, or A weaker.*

Questions A.1   (Answers at end of Appendix A)

1. Write in the form $A \Rightarrow B$, with a preliminary sentence if appropriate; write the converse without using $\Rightarrow$; mark the original statement and its converse as true or false.

    (a) An integer n is divisible by 6, provided it is divisible by 2 and 3.

    (b) The derivative of x^2 is $2x$.

    (c) A quadrilateral whose diagonals are equal is a rectangle.

    (d) Let $\{a_n\}$ be an increasing sequence. If $\{a_n\}$ is bounded, it has a limit.

    (e) Two parallel lines make equal angles with a line intersecting them.

2. Which statement is stronger, which weaker?

    (a) $a \geq 0$; $a > 0$ (b) $\{a_n\}$ is bounded above; $\{a_n\}$ is bounded

3. Form all stronger-weaker pairs from the following statements; give the pairs by using the $\Rightarrow$ symbol. (If this gets confusing, the boxed statement above will be helpful.)

    (a) An increasing sequence which is bounded above has a limit.

    (b) A bounded increasing sequence has a limit.

    (c) A bounded monotone sequence has a limit.

A.2   Contraposition and indirect proof.

We turn now to discussing a style of mathematical proof which involves forming the negatives of statements.

Negation.   In general, if A is a statement, we will use either not–A or $\sim A$ to denote its negation. Often the word "not" doesn't appear explicitly in the negation. Here are three examples (in the first, a is a positive integer).

$\underline{A}$	not–A
a is prime	a is composite or $a = 1$
$a > 2$	$a \leq 2$
$4a^2 + 2 = 3b$	$4a^2 + 2 \neq 3b$

Contraposition.   In proving $A \Rightarrow B$, sometimes it is more convenient to use *contraposition*, i.e., prove the statement in its contrapositive form:

(7)   not–$B \Rightarrow$ not–A (*contrapositive* of $A \Rightarrow B$).

This means exactly the same thing as $A \Rightarrow B$: if you prove one, you've proved the other. We will give a little argument for this later; however you will probably be even more convinced by looking at examples. Below, the original statement is on the left, the contrapositive is on the right; they say the same thing.

$$f(x) = x^2 \;\Rightarrow\; f'(x) = 2x \qquad f'(x) \neq 2x \;\Rightarrow\; f(x) \neq x^2$$
$$a \geq 0 \;\Rightarrow\; \sqrt{a} \text{ real} \qquad \sqrt{a} \text{ not real} \;\Rightarrow\; a < 0$$

Example A.2A   Prove $2a^6 + a^4 + 3a^2 = 0 \;\Rightarrow\; a = 0$.

Solution.   We use contraposition (the last line is overkill):

$$a \neq 0 \;\Rightarrow\; a^2 > 0, \; a^4 > 0, \; a^6 > 0;$$
$$\Rightarrow\; 2a^6 + a^4 + 3a^2 \;>\; 0;$$
$$\Rightarrow\; 2a^6 + a^4 + 3a^2 \;\neq 0. \qquad\qquad \square$$

Indirect proof.   This has the same style as contraposition but is more general. To give an indirect proof that a statement S is true, we assume it is not true and derive a contradiction, i.e., show some statement C is both true and false. C can be anything.

Example A.2B   Prove that $\sqrt{2}$ is irrational.

Solution. (Indirect proof). Suppose it were rational, that is,

$$\sqrt{2} \;=\; \frac{a}{b} \;;$$

we may assume the fraction on the right is in lowest terms, i.e., a and b are integers with no common factor. (Call this last clause "statement C".)

If we cross-multiply the above and square both sides, we get

$$2b^2 \;=\; a^2 \;;$$

the left side is even, so the right side is even, which means a itself is even (since the square of an odd number is easily seen to be odd). Thus we can write $a = 2a'$, where a' is an integer. If we substitute this into the above equation and divide both sides by 2, we get

$$b^2 = 2a'^2 \; ;$$

by the same reasoning as before, b is even. Since we have shown both a and b are even, they have 2 as a common factor; but this contradicts the statement C. □

The above (attributed to Pythagoras) is probably the oldest recorded indirect proof. Note how the statement C to be contradicted just appears in the course of the proof; it's not part of the statement of the theorem.

The above example is a little atypical for us, in that almost always in this book the statement S to be proved will be an if-then statement $A \Rightarrow B$. To prove it indirectly, we have to derive a contradiction from the assumption that $A \Rightarrow B$ is false, i.e., that A does not imply B: in other words, A can be true, yet B be false. So we can now formulate

Indirect proof for if-then statements.

To prove $A \Rightarrow B$ indirectly, assume A true but B false, and derive a contradiction: C and not-C are both true.

Our earlier *proof by contraposition* is just the special type of indirect proof where $C = A$. Namely, to prove $A \Rightarrow B$ by contraposition, we

(a) assume A true and B false (i.e., not-B true);

(b) prove not-$B \Rightarrow$ not-A (the contrapositive).

It follows that not-A is true, which contradicts our assumption that A is true.

To confuse you a little further, we illustrate the difference between the two styles of proof by giving two proofs of a simple proposition.

Proposition. $a^2 = 0 \Rightarrow a = 0$.

Proof by contraposition. $a \neq 0 \Rightarrow a > 0$ or $a < 0$;

$$\Rightarrow a^2 > 0 \; ;$$
$$\Rightarrow a^2 \neq 0. \qquad\qquad □$$

Indirect proof.   Suppose the conclusion is false, that is,

$$a^2 = 0, \qquad \text{but} \quad a \neq 0 \; .$$

Since $a \neq 0$, we can divide both sides of the above equation by a; this gives

(8) $a = 0$,

which contradicts our supposition that $a \neq 0$.   □

Why not just stop the proof at line (8) — it says $a = 0$ and isn't that what we were supposed to prove?

This would be wrong; the last line of the proof is absolutely essential. We only got to line (8) by making a false supposition: that $a \neq 0$. Therefore (8) has no validity in itself; it is only a line in a bigger argument whose ultimate goal is to produce a contradiction.

The advantage of contraposition over the more general type of indirect proof is that since we know at the outset the statement A that is going to be contradicted, what has to to be proved (not–B $\Rightarrow$ not–A) becomes a direct statement that we hope can be proved by a direct argument.

The general argument against all indirect proofs is that they require you to focus on a false statement (A is true but B is not), and derive from it other false statements until finally you get a contradiction. Such a proof requires you to read one wrong thing after another, until a moment arrives when the author proudly announces, "This contradicts statement C!" but you can't remember that C was ever mentioned.

The other problem with indirect proofs (this applies to contraposition also) is that they require you to form the negation of statements, which is not always so easy to do in analysis, since many of the common statements are linguistically rather complicated. Directions for negating a statement are given in Appendix B, but students are well-advised not to read it until they have read most of the book; those who eat the apple of negation tend to fall into the habit of trying to prove even the simplest statements indirectly (and generally incorrectly).

In this book, we will avoid indirect proofs—even proofs by contraposition—whenever possible; if an indirect proof must be given, negation will be treated informally, which is how most professional mathematicians handle it.

Questions A.2

1. Formulate the negative statement without using "not":

   (a) In the plane, lines L and M are parallel.
   (b) Triangle ABC is isosceles.
   (c) There are infinitely many prime numbers.

2. Write the converse and contrapositive, using $\Rightarrow$, and mark T or F:

   The square of an odd integer a is odd.

3. Prove the following by contraposition (the a_i are real numbers):

   (a) if $a_1 a_2 < 0$, exactly one of the $a_i < 0$.
   (b) if $a_1 + \ldots + a_n = n$, at least one $a_i \geq 1$.

A.3   Counterexamples.

Some statements in mathematics are particular, i.e., they assert that something is true for some definite numbers, or other objects. For example.

$$3^2 + 4^2 = 5^2; \qquad \triangle ABC \text{ is isosceles}; \qquad \text{there is a number} \geq 22.$$

Other statements are general; they assert something about a whole class of numbers or other objects. For example:

(i) if a and b are numbers satisfying $a^2 = b^2$, then $a = b$;

(ii) a triangle with three equal sides has three equal angles;

(iii) every positive integer n is the sum of four squares of integers:
$$n = a_1^2 + a_2^2 + a_3^2 + a_4^2 \; ;$$

(iv) if a, b, c are numbers satisfying $ab = ac$, then $b = c$.

These respectively assert that something is true about any numbers satisfying $a^2 = b^2$, all equilateral triangles, all positive integers, any numbers satisfying $ab = ac$.

As it happens, statements (ii) and (iii) are true — (iii) is hard to prove — while (i) and (iv) are false. The problem we consider is:

How does one show a general statement like (i) or (iv) is false?

Since a general statement claims something is true for every member of some class of objects, to show it is false we only have to produce a single object in that class for which the general statement fails to hold. Such an object is called a **counterexample** to the general statement. For example a counterexample to (i) would be the pair $a = 3$, $b = -3$. (What would be a counterexample to (iv)?)

Example A.3 (uses Prop. 2.4)

(a) Prove: if the sequence $\{a_n\}$ is bounded, then $\{a_n^2\}$ is bounded.

(b) In part (a), can "bounded" be replaced by "bounded above"?

Solution.

Part (a). By hypothesis, there is a B such that (see Prop. 2.4)
$$|a_n| < B \quad \text{for all } n.$$
Squaring both sides, and using the law $|ab| = |a||b|$,
$$|a_n^2| < B^2 \quad \text{for all } n.$$
Therefore $\{a_n^2\}$ is bounded, again by Prop. 2.4.

Part (b). **Solution No. 1** (by our good student)

Yes, "bounded" can be replaced by "bounded above"; to prove it, just drop the absolute value signs from the above proof:

(9)
$$\begin{aligned} \{a_n\} \text{ is bounded above} &\Rightarrow a_n < B \quad \text{for all } n; \\ &\Rightarrow a_n^2 < B^2 \quad \text{for all } n; \\ &\Rightarrow \{a_n^2\} \text{ is bounded above} . \quad \square?? \end{aligned}$$

(STOP; don't continue until you have spotted the error!)

Solution No. 2   (by our better student)

No, "bounded" cannot be replaced by "bounded above", since in the above argument which tries to prove the amended statement, (9) can fail if $a_n < 0$; thus the proof doesn't work, so the amended statement must be false.   □??

Solution No. 3   (by our best student, and correct)

The amended statement is false; the sequence $\{-n\}$ is a counterexample, since it is bounded above by 0, but $\{n^2\}$ is not bounded above.   □

Students in general seem to dislike counterexamples and are reluctant to produce them. Asked to show some general statement S is false, like our "better" student above they usually produce some reasoning which tries to prove S, but fails. Then they point out that the proof doesn't work, and conclude the statement must be false. One can sympathize with this psychologically, but mathematically it's nonsense.

> *The failure of your attempted proof of S doesn't show S is false, since someone else might come along with a different argument which succeeds in proving it.*

Produce a counterexample to S instead; then no one can ever prove it.

Like students, mathematicians dislike counterexamples (particularly to their own theorems!) and producing them is a sure way to lose friends and alienate people. There seems to be something unsporting about demolishing a whole edifice of theorems by a single counterexample. Alas, it happens; worst of all in Ph.D. orals: a committee member wonders aloud how the candidate's theorem would apply to a favorite example, and a minute later the example has turned into a counterexample.

Of course you don't want to see a counterexample to your theorem: you want to know where your reasoning went astray. But even your best friend won't tell you; you have to figure it out yourself.

Questions A.3

1. Decide which of these statements is false; for each such, prove that it is false.

   (a) Every positive integer is the sum of three squared integers (some of which can be zero).

   (b) Three lines in the plane, no two of which are parallel, determine a unique triangle whose sides lie on the lines.

   (c) (It is conjectured there are an infinity of "twin primes": that is, pairs $(n, n + 2)$ where both numbers are prime, like $(5, 7)$, $(11, 13)$, and $(17, 19)$.)

   In a twin prime pair, the number between the two primes is divisible by 3.

A.4   Mathematical induction.

This is a way of proving a proposition whose statement involves all positive integers $n \geq$ some n_0. Some examples (note the different values of n_0 used):

(a) $1 + x + \ldots + x^n = \dfrac{1 - x^{n+1}}{1 - x}, \quad n \geq 0$

(b) A positive integer n is the product of one or more primes, if $n \geq 2$.

(c) The sum of the interior angles of an n-sided polygon is $(n - 2)\pi, \quad n \geq 3$

(d) $\displaystyle\int \dfrac{dx}{x^n} = \dfrac{x^{1-n}}{1 - n}, \quad n \geq 2.$

We will denote the proposition by $P(n)$, to show its dependency on n. Though most think of it as a single proposition, proof by induction depends on thinking of it as a whole sequence of propositions, one for each value of n.

Proof by Mathematical Induction   To prove $P(n)$, $n \geq n_0$,

(a) *prove $P(n_0)$*   *(the basis step)*;

(b) *prove $P(n+1)$; in the proof you are allowed to use $P(n)$,*
and if necessary, $P(k)$ for any lower values, $n_0 \leq k \leq n$, as well

(the induction step).

The best way to approach proof by induction is to see a variety of examples. As you will see, the boxed statement above has to adapted in various small ways for the different examples, but it is a good place to start.

Example A.4A   Prove: $1^2 + 2^2 + 3^2 + \ldots + n^2 = \dfrac{n(n + 1)(2n + 1)}{6}, \quad n \geq 1.$

Solution.   By induction. The basis step $P(1)$ says $1^2 = (1 \cdot 2 \cdot 3)/6$, which is true. Here is the induction step.

$$
\begin{aligned}
1^2 + 2^2 + \ldots + (n + 1)^2 &= (1^2 + 2^2 + \ldots + n^2) + (n + 1)^2 \\
&= \frac{n(n + 1)(2n + 1)}{6} + (n + 1)^2, && \text{using } P(n); \\
&= \frac{(n + 1)\big[n(2n + 1) + 6(n + 1)\big]}{6}, && \text{factoring out } n + 1; \\
&= \frac{(n + 1)(n + 2)(2n + 3)}{6}, && \text{after some algebra;}
\end{aligned}
$$

and this last is the right side of $P(n + 1)$.   □

The above shows that if $P(n)$ is true, then $P(n + 1)$ follows: $P(n) \Rightarrow P(n + 1)$. Since we know $P(1)$ is true, this shows $P(2)$ is true; this in turn shows $P(3)$ is true, and continuing in this way, $P(n)$ is shown to be true for any value of n — it's been likened to a row of falling dominos: you tip over the first one, which then tips over the second, and ultimately the n-th is reached.

Example A.4B   Prove $n! \geq 2^n$, for $n \geq$ some n_0.

Solution.   The induction step runs:

$$
\begin{aligned}
(n+1)! &= n!\,(n+1) \\
&\geq 2^n(n+1), && \text{using } P(n); \\
&\geq 2^{n+1}, && \text{if } n+1 \geq 2 .
\end{aligned}
$$

So we have proved $P(n) \Rightarrow P(n+1)$, if $n \geq 1$, i.e., for all $n \in \mathbb{N}$.   □

But take $n = 2$: $2! \not\geq 2^2$. $P(2)$ is false!

Didn't we prove that $P(1) \Rightarrow P(2)$? Yes, but $P(1)$ is also false. So is $P(3)$.

The basis step is $P(4)$: $4! \geq 2^4$; the proposition is true only for $n \geq 4$.   □□

The basis step is trivial or obvious most of the time, so students tend to skip it or ignore it, focussing on the harder induction step. But this can get you into trouble, as illustrated above.

Sometimes people argue over what the basis step should be. In mathematics, empty sums are assigned the value 0, and empty products are assigned the value 1 (as for example: $a^0 = 1$, $0! = 1$). Thus the basis step for Example A.4A could also be taken as $n = 0$.

In addition to proof by induction, there is also *inductive definition* or as it is also called, *recursive definition*, in which the terms of a sequence $\{a_n\}$, $n \geq n_0$ are defined by expressing them in terms of lower values of n; as the basis, a starting value a_{n_0} must also be given.

Example A.4C   Let $a_n = a_{n-1} + \dfrac{1}{n(n+1)}$, $a_0 = 0$. Find a formula for a_n.

Solution.   We have
$$
\begin{aligned}
a_1 &= a_0 + 1/(1 \cdot 2) = 1/2 , \\
a_2 &= a_1 + 1/(2 \cdot 3) = 2/3 , \\
a_3 &= a_2 + 1/(3 \cdot 4) = 3/4 ,
\end{aligned}
$$

so we guess $a_n = \dfrac{n}{n+1}$.

Taking this last statement as $P(n)$, we prove it by induction. It is true for a_0; as the induction step, we get

$$
\begin{aligned}
a_n &= a_{n-1} + \frac{1}{n(n+1)} \\
&= \frac{n-1}{n} + \frac{1}{n(n+1)} , && \text{using } P(n-1), \\
&= \frac{n}{n+1} , && \text{by algebra,}
\end{aligned}
$$

which completes the proof by induction.   □

Notice that here we proved $P(n)$, using $P(n-1)$. This just amounts to a change of variable in the boxed method on the previous page, and is therefore equally valid.

Example A.4D   Let $a_0 = 1$, $a_1 = 2$, and $a_{n+2} = 2a_{n+1} - a_n$, $n \geq 0$.
Prove that $a_n = n + 1$ for $n \in \mathbb{N}$.

Solution. We use as the induction step the proof of $P(n+2)$:

$$
\begin{aligned}
a_{n+2} &= 2a_{n+1} - a_n, & n \geq 0; \\
&= 2(n+2) - (n+1), & \text{using } P(n+1) \text{ and } P(n); \\
&= n+3 .
\end{aligned}
$$

$\square$

Here we proved $P(n+2)$, using in the proof not just $P(n+1)$, but $P(n)$ as well. When one uses in the proof of $P(n)$ not just the preceding value but lower values of n as well, the proof method is generally referred to as **strong** or **complete** induction; in this style of induction, often more than one value of n is needed for the basis step.

> In strong induction, the basis step consists of all $P(n)$ not covered by the argument in the induction step, i.e., for which there are no lower $P(k)$ available to imply $P(n)$.

In the example above, the proof of $P(n)$ uses the two previous values of n; therefore $P(1)$ and $P(0)$ are not covered by this proof, since these cases do not have two previous values — they must be proved separately, as the basis step. (In this case, both are trivial, since we are given $a_0 = 1$ and $a_1 = 2$.)

Example A.4E   Prove that the sum of the interior angles of a convex polygon with n sides is $(n-2)\pi$.

Solution.   A convex polygon is one where any line segment joining two vertices lies inside the polygon.

We give a proof by strong induction. To prove $P(n+1)$ we are given a convex polygon C_{n+1} with $n+1$ sides. Draw a line connecting two non-adjacent vertices. It divides C into two convex polygons C_{k+1} and C_{l+1} as shown, having respectively $k+1$ and $l+1$ sides. We have

$$
\begin{aligned}
\text{angle sum of } C_{n+1} &= \text{angle sum of } C_{k+1} + \text{angle sum of } C_{l+1} \\
&= (k-1)\pi + (l-1)\pi , & \text{by strong induction;} \\
&= (k+l-2)\pi = (n-1)\pi, & \text{since } k+l = n+1.
\end{aligned}
$$

The case not covered by the above is when no such dividing line segment exists; in this case C must be a triangle, and therefore $P(3)$ is the basis step, which is true: the sum of the angles of a triangle is π. $\square$

Here one could regard the basis step as the hardest step, since probably more students could give the above induction argument than remember how to prove the statement about a triangle.   The induction in this example is strong induction, since we don't know how many sides each of the smaller polygons has — just that it's less than $n+1$. (One could also give a proof by regular induction: see Question 3, but read the comments in the Answers since this work is often mishandled.)

Example A.4F   Prove that every integer $n \geq 2$ is the product of primes.

Solution.   Here is a case where you can only use strong induction, since there is no relation between the prime factorizations of n and $n + 1$.

> Keeping in mind the freshman in one class, who on seeing the above proposition on the board, yelled triumphantly, "False!! Five is not a product of primes!" I hasten to add that in higher mathematics, a sum is allowed to have just one term, and a product just one factor. They can even have none, if you believe in the empty sums and products we mentioned earlier.

If n is prime, we are done. If not, it factors into the product of two smaller positive integers, both ≥ 2 (since the factorization is not the trivial one $n = n \cdot 1$):

$$n = n_1 \cdot n_2, \qquad 2 \leq n_1, n_2 < n;$$
$$= (p_1 p_2 \cdots p_k)(q_1 q_2 \cdots q_l), \qquad p_i, q_j \text{ primes},$$

since by strong induction, we can assume the smaller numbers n_1 and n_2 factor into the product of primes.   □

Remarks.   Question 3 asks for a proof that the method of induction works; it is a good example of indirect proof.

The problem many beginners have with proof by induction is, of course, the apparent circularity: "How can you assume and use $P(n)$ in the proof, since that's what you're trying to prove?" The answer is, it's $P(n + 1)$ that you're trying to prove.

The same problem appears in recursive definitions: in the briefest and most efficient form, the definition of $n!$ is

$$n! = n \cdot (n - 1)!, \qquad 0! = 1.$$

The definition looks circular, but because the factorial on the right is for a smaller n, the definition makes sense.

In many mathematical arguments, induction is concealed by such phrases as "similarly" or "continuing in the same way, we see that". This is a genial tradition that makes for easier reading; the preference in this book is for such informal stratagems. Induction is used explicitly only for arguments that can't be made without it. Example A.4A is a good one for that: no calculation proving the formula for $\sum_1^n i^2$ for just the first few values of n will convince anyone that the general formula for the sum is correct; only induction will do that.

Questions A.4

1. In Example A.4B, proving $n! \geq 2^n$, show that $P(0)$ is true, according to the conventions about empty products described in the subsequent remarks. If $P(0)$ is true, why can't we take it as the basis step?

2. Prove that $n < 2^n$ for all $n \in \mathbb{N}_0$.

3. Prove the method of regular induction works: that is, if $P(n_0)$ is true and $P(n+1)$ is true whenever $P(n)$ is, for $n \geq n_0$, then $P(n)$ is true for all $n \geq n_0$.

(Hint: consider the set $S = \{n \geq n_0 : P(n)$ is false$\}$; it has a least element.)

4. Prove Example A.4E using regular induction.

5. In Example A.4F, what is the basis step?

Exercises

A.4

1. Prove by induction that $1 + x + \ldots + x^n = \dfrac{1 - x^{n+1}}{1 - x}$, $n \geq 0$.

2. Find a formula for $1 + 3 + 5 + \ldots + (2n - 1)$ and prove it by induction.

3. Let D denotes differentiation with respect to x. From the product rule for differentiation, $D(uv) = uDv + vDu$, and the fact that $Du = 0$ if $u(x)$ is a constant function, prove by induction that $D(x^n) = nx^{n-1}$, if $n \in \mathbb{N}_0$.

4. The coefficients of a series $\displaystyle\sum_0^\infty a_n x^n$ are given by $a_{n+2} = \dfrac{n+1}{n+3} a_n$.

(a) Find the power series if $a_0 = 1$ and $a_1 = 0$; prove it by induction.

(b) Find the power series if $a_0 = 0$ and $a_1 = 1$; prove it by induction.

5. The terms of a sequence $a_0, a_1, a_2, \ldots$ are given by the recursive relation

$$a_{n+1} = 2a_n - a_{n-1} + 2, \qquad a_0 = 0, \; a_1 = 1.$$

Find a formula for a_n, and prove it.

6. Fermat's Little Theorem in number theory is the basis of the RSA algorithm, currently the most secure encryption algorithm. The theorem says:

if p is prime, then $n^p - n$ is an integer multiple of p, for all $n \in \mathbb{N}$.

(a) Prove this by induction if $p = 3$.

(b) Prove it if $p = 5$.

(c) Show it is false if $p = 4$ (cf. Section A.3).

7. With an unlimited supply of three-cent and seven-cent stamps, it is possible to make any integral postage n, when $n \geq n_0$. Find the smallest n_0 for which this is true, and prove it by strong induction. What is the basis step?

8. (a) Prove that $1^3 + 2^3 + \ldots + n^3 < n^4$, if $n \in \mathbb{N}$, $n > 1$.

(b) Prove the sum in part (a) is $< n^4/2$, if $n > 2$ (a bit harder).

9. Fix an $a \in \mathbb{N}$. Prove by strong induction (regular induction is clumsier here) that any $n \in \mathbb{N}_0$ can be written in the form below; what is the basis step?

$$n = qa + r, \qquad \text{where } q \in \mathbb{N}_0, \quad 0 \leq r < a .$$

10. Prove $n^2 < 2^n$ for $n \geq n_0$. (If stuck, Question 2 might be helpful.)

Answers to Questions

A.0

1. (a) $1.12323\ldots = 1.1 + (.1)(.2323\ldots)$;
     $.2323\ldots = .23(1 + 10^{-2} + 10^{-4} + \ldots) = .23(100/99)$, by 4.2, (2).

    (b) $3/7 = .428571428571\ldots$; repetition is inevitable.

2. (a) $\mathbb{R}^+ \cup \{0\}$, or $[0, \infty)$ (b) $\mathbb{Q} \cap [a, b]$ (c) $(a, b) \times (c, d)$

    (d) $\mathbb{Q} \cap \mathbb{R}^+$ (e) $[0, \infty)$ (f) $\mathbb{R} \setminus \mathbb{Z}$

3. $(0, 1) \subseteq S$: if $x \in (0, 1)$, then $x > 0$, $1 - x > 0$, so $x(1 - x) > 0$.

    $S \subseteq (0, 1)$: if $x(1-x) > 0$, either both factors are positive or both factors are negative. If both are positive, $x > 0$ and $1 - x > 0$, so $x \in (0, 1)$. If both are negative, $x < 0$ and $1 - x < 0$, i.e., $x < 0$ and $x > 1$, which is impossible.

A.1

1. (a) n is divisible by 2 and 3 $\Rightarrow$ n is divisible by 6. (T)
    Converse: A number divisible by 6 is divisible by 2 and 3. (T)

    (b) $f(x) = x^2 \Rightarrow f'(x) = 2x$. (T)
    Converse: If $f'(x) = 2x$, then $f(x) = x^2$. (F)

    (c) Q is a quadrilateral with equal diagonals $\Rightarrow$ Q is a rectangle. (F)
    Converse: The diagonals of a rectangle are equal. (T)

    (d) Let $\{a_n\}$ be increasing; $\{a_n\}$ bounded $\Rightarrow$ $\{a_n\}$ has a limit. (T)
    Converse: if the increasing sequence $\{a_n\}$ has a limit, it is bounded. (T)

    (e) L, M are two lines, N a third line intersecting them.
         L, M parallel $\Rightarrow$ angle LN = angle MN. (T)
    Converse: If two lines make equal angles with a third line intersecting them, they are parallel. (T)

2. (a) $a > 0$ is stronger   (b) $\{a_n\}$ bounded is stronger

3. (a) $\Rightarrow$ (b),   (c) $\Rightarrow$ (b) are the stronger-weaker pairs.

A.2

1. (a) L and M intersect in one point. (Why specify "one point"?)

    (b) Triangle ABC has sides of three different lengths.

    (c) There are only a finite number of primes.

2. Converse: a^2 odd $\Rightarrow$ a odd. (T)

    Contrapositive: a^2 even $\Rightarrow$ a even. (T)

3. (a) $a_1 \geq 0$ and $a_2 \geq 0 \Rightarrow a_1 a_2 \geq 0$; $a_1 < 0$ and $a_2 < 0 \Rightarrow a_1 a_2 > 0$.

    (b) All $a_i < 1 \Rightarrow a_1 + a_2 + \ldots + a_n < n$.

A.3

1. All are false.

    (a) 7 is a counterexample;

    (b) any three lines meeting in a point are a counterexample;

(c) the pair $3, 5$ is a counterexample (the only one, actually).

A.4

1. We have $0! = 2^0$, since both sides are 0; thus $P(0)$ is true. This is not the basis step, however, since the proof of $P(n + 1)$, assuming $P(n)$, only works when $n \geq 1$. So the basis integer n_0 must be at least 1 to get the induction going, and as we have seen, the basis step is actually $P(4)$.

2.
$$
\begin{aligned}
n + 1 \ &< \ 2^{n+1}, \qquad \text{using } P(n); \\
&< \ 2^n + 2^n, \qquad \text{if } n \geq 1; \\
&= \ 2^{n+1}.
\end{aligned}
$$

The basis step is $P(1)$, which is true; we cannot use $P(0)$ as the basis, since the above proof doesn't show $P(0) \Rightarrow P(1)$: it requires $n \geq 1$. So we have shown $P(n)$ is true for $n \geq 1$. Nonetheless, $P(0)$ is also true (by a separate argument), so $P(n)$ is true for all $n \geq 0$.

3. We prove by indirect argument that S is empty, i.e., that $P(n)$ is true for all $n \geq n_0$.

If S is non-empty, then it contains a smallest integer $m \geq n_0$, and $P(m)$ is false. Look at the number just before m:

$$
\begin{aligned}
m - 1 \geq n_0, \qquad &\text{since } m \geq n_0, \text{ and } m \neq n_0 \text{ (for } P(n_0) \text{ is true);} \\
m - 1 \notin S, \qquad &\text{since } m \text{ is the smallest number in } S.
\end{aligned}
$$

Therefore $P(m - 1)$ is true; but since $P(n) \Rightarrow P(n + 1)$ for $n \geq n_0$, it follows that $P(m)$ is true, contradiction. $\qquad\qquad\square$

(Note: The self-evident fact we used:

a non-empty set of positive integers has a smallest element

is known as the *well-ordering property* of $\mathbb{N}$.)

4. The proof is almost the same: the basis step is still $P(3)$, and the only difference is that the line segment must be drawn so it divides C_{n+1} into a triangle C_3 and a C_n. Then

$$
\begin{aligned}
\text{angle sum of } C_{n+1} &= \text{angle sum of } C_n + \text{angle sum of } C_3 \\
&= (n - 2)\pi + \pi, \qquad \text{using } P(n) \text{ and } P(3); \\
&= (n - 1)\pi \ .
\end{aligned}
$$
$\qquad\qquad\square$

Students who think of the induction step as $P(n) \Rightarrow P(n + 1)$ (rather than "prove $P(n + 1)$ by using $P(n)$") will often start by drawing a C_n, then adding a vertex to make it a C_{n+1}. This is illegal — if you're proving $P(n + 1)$, you must start with whatever C_{n+1} is given to you — you can't make up your own! In many areas of discrete mathematics, graph theory for instance, the $P(n) \Rightarrow P(n + 1)$ approach to induction proofs produces immediate disaster. Remember: start with $P(n + 1)$, if that's what you are supposed to prove.

5.   The statements $P(p)$ for p prime are the basis step, though this is concealed the way the proof is given; it is worded so that the basis step is included in the proof of $P(n)$.

Appendix B

Quantifiers and Negation

This appendix can be read almost any time, but is probably best postponed until late; students who read it too early tend to embrace it too closely and start writing unreadable proofs that turn out to be incorrect as well.

Prerequisites: Chapter 3 for Sections B.1, B.2; Chapter 10 for Section B.3

B.1 Introduction. Quantifiers.

Though the basic ideas of analysis are generally agreed upon, the way they are expressed can vary from one book to another. You need to be familiar with these different styles if you want to be able to consult other references in analysis, and study from more advanced texts and monographs.

A good example is the definition of the limit of a sequence $\{a_n\}$:

$$(1) \qquad\qquad \text{given } \epsilon > 0, \quad a_n \underset{\epsilon}{\approx} L \text{ for } n \gg 1 \; ;$$

this is the abbreviated form we have been using. If we expand it by replacing the phrase "for $n \gg 1$" by its definition, then (1) becomes

$$(2) \qquad \text{given } \epsilon > 0, \quad a_n \underset{\epsilon}{\approx} L \quad \text{for } n \geq \text{ some } N_\epsilon \text{ (depending on } \epsilon \text{)};$$

or being still more formal, we replace the word "some" by its meaning:

$$(3) \qquad \text{given } \epsilon > 0, \text{ there is an } N_\epsilon \text{ such that } \quad a_n \underset{\epsilon}{\approx} L \quad \text{for } n \geq N_\epsilon.$$

The general format of (3) is the most widely accepted way to write the definition of the limit of a sequence. There is some latitude in the exact choice of words, as well as the order of the last two clauses. You can mix and match from the following lists (the symbol : stands for "such that"):

for all (any) $\epsilon > 0$,	there exists an N :	for all $n \geq N$, $\|a_n - L\| < \epsilon$
for every $\epsilon > 0$,	there is an N_ϵ :	$\|a_n - L\| < \epsilon$ for all $n \geq N_\epsilon$
given (any) $\epsilon > 0$,	for some N_ϵ,	$a_n \underset{\epsilon}{\approx} L$ for all $n \geq N_\epsilon$

The top options give the "standard" choice:

$$(4) \quad \text{for all } \epsilon > 0, \text{ there exists an } N \text{ such that, for all } n \geq N, \ |a_n - L| < \epsilon.$$

You notice that, as is customary, we have omitted the subscript ϵ on N. The dependence of N on ϵ is expressed by its placement in the sentence: once the ϵ is introduced in the first clause, everything coming later is understood to depend on it. If we wanted N to be independent of ϵ, we would place it first in the sentence. Let's do this and see how it changes the meaning:

(5)   there exists an N such that, for all $\epsilon > 0$, $a_n \underset{\epsilon}{\approx} L$ for $n \geq N$.

What does this mean? Look at the two central clauses:

$$\text{for all } \epsilon > 0, \quad a_n \underset{\epsilon}{\approx} L \;.$$

They are equivalent to saying $a_n = L$, since two numbers which are arbitrary close must be equal. Thus (5) can be rewritten

(5′)   there exists an N such that $a_n = L$ for $n \geq N$;

In other words, the sequence a_n becomes constant and equal to L when you go far enough out. This illustrates how changing the order changes the meaning.

> Note how the phrasing (1) conceals these issues by not mentioning the N explicitly. It makes sense to do so, since outside of the earliest examples of proofs that a given sequence has a particular limit, one rarely needs to introduce an N. The above does explain however why in (1) you must put "given $\epsilon > 0$" first; if you begin instead with "for n large", what you are saying is (5) or the equivalent (5′).

Quantifiers.   Phrases such as "for all"and "there exists" are called *quantifiers*, and there is a symbolic notation for them that is sometimes used:

   universal quantifier: ∀   (for all, for any, for every, given any, given);

   existential quantifier: ∃   (there exists, there is, for some) .

Using these, along with the "such that" symbol : , definition (5) of the limit of the sequence a_n reads

(6) $\forall \epsilon > 0, \quad \exists N : |a_n - L| < \epsilon$ for $n \geq N$.

The last clause "for $n \geq N$" also uses a quantifier; we put it last to make the sentence less awkward and more readable, but in formal writing these considerations are overridden (we will see why) in favor of placing all the quantifiers first; the final version thus reads:

(6a) $\forall \epsilon > 0, \; \exists N : \forall n \geq N, \; |a_n - L| < \epsilon.$

> As in a German sentence, one has to wait until the very end to see what is really being said—all the previous quantification has the purpose of picking and naming the variables (in the right order), and telling you for what values of the variables the last clause will be true. Note that the order is rigid: first the ϵ, then N because it depends on ϵ, then the third clause which uses N, and finally the main statement, which uses the preceding three clauses.

> Too much symbolic notation can make mathematical analysis look as if it's written in shorthand: formidable and hard to read. For this reason many mathematicians and many analysis books (like this one) avoid entirely the use of ∀ and ∃. As we will see, however, they are useful in forming correctly the negation of a mathematical statement.

Note that $\exists$ is regularly followed by : ("such that"). Also $\forall$ can be followed by : if it is necessary to restrict the range of the variable in some way; thus "for all positive x" could be rendered in any of the ways:

$$\text{for all } x > 0; \qquad \forall\, x > 0; \qquad \forall\, x : x > 0\, .$$

The first two would be considered somewhat colloquial; the last is a bit pompous, but the only one which is grammatically correct.

A rough but valuable guide to the complexity of a mathematical sentence or definition is the number of quantifiers it requires. From this point of view, it is no wonder that students struggle with the definition of limit, which requires three quantifiers. A lot of the terminology in this book has been adopted in order to reduce the number of quantifiers you have to deal with at one time, sweeping them under the rug if they aren't really needed for the argument or definition.

To get more practice with the above ideas, let us consider examples using fewer quantifiers.

Example B.1 Define formally, making the quantifiers explicit:

"the sequence $\{a_n\}$ is bounded above".

Then interchange the order of the quantifiers, and interpret the resulting statement, using informal mathematical prose.

Solution. The definition of bounded above is

(7)   there is a number B such that $a_n \leq B$ for all $n \geq 0$.

Putting the quantifiers in standard order, this becomes

(7a)   there exists a B such that, for every $n \geq 0$, $a_n \leq B$.

(7b) $\exists B : \forall n \geq 0,\ a_n \leq B$.

Interchanging the order of quantifiers gives

(8a) $\forall n \geq 0,\ \exists B : a_n \leq B$;

(8b)   given $n \geq 0$, there is a B such that $a_n \leq B$.

The important thing here is that since B is now in the second place, it is allowed to depend on n; we could have been kind and added a subscript or a phrase to make this explicit (writing B_n, or "there is a B depending on n such that..."), but one isn't required to: the position in the sentence says it all, mathematically speaking.

As for the new statement (8b), it is automatically true for any sequence at all since we can take $B = a_n$. Thus reversing the position of the quantifiers leads to a trivial statement that asserts nothing at all about a sequence.

On the other hand, reversing the order of two adjacent quantifiers when it makes sense and both of them are $\forall$ or both are $\exists$ generally has no effect on the statement, even though it may formally look a little different. The Questions and Exercises give some examples.

Questions B.1

1. How many quantifiers does the phrase "for n large" add to a statement?

2. For each of the following,

(a) write the statement formally, either in prose or in symbols, putting the quantifiers first and in the correct order;

(b) if there are two or more quantifiers, reverse the first two if it makes sense, and interpret the resulting statement.

   (i)   the sequence a_n, $n \geq 0$, is strictly increasing.

   (ii) the sequence a_n is strictly increasing for $n \gg 1$.

   (iii) the sequence a_n, $n \geq 0$, has a maximal element.

   (iv) $m + n$ is positive, if m and n are positive (integers)

   (v) $m - n$ is positive, if m, n are integers such that $m > n$.

   (vi) the sequence a_n is bounded above for $n \gg 1$.

B.2   Negation.

When giving a proof of $P \Rightarrow Q$ by contraposition or by contradiction, the hypothesis will be "not Q". To make such proofs therefore, one must be able to form the negation of statements. For simple statements, this is no problem; for instance (we use here $\Leftrightarrow$ for equivalent statements, and the symbol $\sim Q$ for the negation of Q),

$$\sim (x > 0) \quad \Leftrightarrow \quad x \leq 0;$$

but for a statement involving several quantifiers, putting its negation in a useful form is not so easy. It is partly to avoid this that this book tries to use direct proofs whenever possible, rather than proofs by contradiction.

There is however a systematic procedure for negating statements with quantifiers which we will describe now. If you study it, please don't go berserk using it; direct proofs are usually simpler and better than proofs by contradiction, and they keep your mind focussed on true things rather than false ones.

First of all one should note the elementary rules for negating compound sentences—those involving the conjunctives "and" and "or". Letting A and B represent simple statements,

(9) $\sim (A \text{ and } B) \quad \Leftrightarrow \quad \sim A \text{ or } \sim B$; $\sim (A \text{ or } B) \quad \Leftrightarrow \quad \sim A \text{ and } \sim B.$

For example,

$\sim (\{a_n\} \text{ and } \{b_n\} \text{both converge}) \quad \Leftrightarrow \quad$ either $\{a_n\}$ diverges or $\{b_n\}$ diverges;

$\sim (\text{either } \{a_n\} \text{ or } \{b_n\} \text{ is bounded}) \quad \Leftrightarrow \quad$ both $\{a_n\}$ and $\{b_n\}$ are unbounded.

We proceed now to the negation of statements involving a single quantifier, by way of example. Consider

(10) $$a_n > 0 \quad \text{for all } n;$$

what is its negation? Many students say: $a_n \leq 0$ for all n, but this is not correct; the negation is: $a_n \leq 0$ for some n, i.e.,

$$\sim (a_n > 0 \text{ for all } n) \quad \Leftrightarrow \quad \text{there is an } n \text{ such that } a_n \leq 0.$$

In symbols, we can write it:

$$\sim (\forall n, \ a_n > 0) \quad \Leftrightarrow \quad \exists n : a_n \leq 0.$$

Moreover, if we negate both sides of the above, we get

$$\sim (\exists n : a_n \leq 0) \quad \Leftrightarrow \quad \forall n, \ a_n > 0,$$

or more informally in words:

$$\text{there is no } n \text{ for which } a_n \leq 0 \quad \Leftrightarrow \quad a_n > 0 \ \text{ for all } n.$$

We can state this as a first general rule: suppose we have a statement $P(n)$ depending on n, with one quantifier in front. Then we negate it according to the rules:

(12) $\quad \sim (\forall n, P(n)) \quad \Leftrightarrow \quad \exists n : \sim P(n); \qquad \sim (\exists n : P(n)) \quad \Leftrightarrow \quad \forall n, \sim P(n).$

Treating it as a formal symbol-manipulation, what we do is move $\sim$ to the right; as it passes over $\forall$ it changes it to $\exists :$, and vice-versa.

In applying the rule (12), if $P(n)$ is further quantified, the rule still applies and so the process can be continued. To illustrate, here is an example. Before reading the solution, you are urged to try doing it intuitively first, and then use (12) as the solution does, and see if your answers agree.

Example B.2A   Express in readable mathematical prose the negation of

(13) $$\text{the sequence } \{a_n\} \text{ is positive for large } n.$$

Solution.   First write it formally, placing the quantifiers in front, and then (if you wish), change to symbolic notation:

(13a) $$\text{there exists an } N \text{ such that, for all } n \geq N, \ a_n > 0$$

(13b) $$\exists N : \forall n \geq N, \ a_n > 0.$$

To negate it, place the $\sim$ in front, and keep applying the rule (12):

$$\sim (\exists N : \forall n \geq N, \ a_n > 0) \quad \Leftrightarrow \quad \forall N, \ \sim (\forall n \geq N, \ a_n > 0)$$
$$\Leftrightarrow \quad \forall N, \ \exists n \geq N : \sim (a_n > 0)$$
$$\Leftrightarrow \quad \forall N, \ \exists n \geq N : \ a_n \leq 0 \ ;$$

so finally, the negation of (13) is in non-symbolic mathematical prose,

$$\text{given any } N, \quad a_n \leq 0 \ \text{ for some } n \geq N \ . \qquad \qquad \square$$

A formulation which is mathematically equivalent, though not formally, (i.e., one has to prove they are equivalent), would be

$$a_n \leq 0 \quad \text{for infinitely many } n \ .$$

Negation rule   To negate a quantified statement,

(a) *write the statement with all the quantifiers placed first, in the correct order, and put the "not" symbol $\sim$ at the left side;*

(b) *move the $\sim$ to the right; as it passes over each quantifier, change $\exists :$ to $\forall$ and vice-versa; stop after you have passed over all the quantifiers (or before, if you wish).*

Questions B.2

1. For each of the following negations,

(a) try writing the statement formally with the quantifiers in the right order, just using your intuition;

(b) then do it by first writing the corresponding positive assertion formally, and then following the negation rules given in the text.

(i) The sequence a_n is not bounded above.

(ii) The sequence a_n is not constant for large n.

B.3   Examples involving functions.

We give a few more examples of the use of quantifiers and their role in forming the negation of statements, this time involving properties of functions, and limits of functions. The ideas are the same as in our earlier work with sequences (which are after all just a special type of function).

Example B.3A   For the concept

(14)   "$f(x)$ is increasing on I",

(a) write it formally, exhibiting the quantifiers in the right order;

(b) express "$f(x)$ is not increasing on I" more explicitly.

Solution. (a) $\forall x \in I, \ \ \forall y \in I : y > x, \ \ f(x) \leq f(y)$.

(Note that for the sake of clarity here, we have to use the "such that" construction with the second quantifier.)

(b) Using the negation rules, we get

(14') $\exists x \in I, \ \ \exists y \in I : y > x : \ f(x) > f(y)$.

or less formally: "$f(x)$ is not increasing on I" means there are two points x, y in I such that $x < y$, but $f(x) > f(y)$.

Example B.3B   Carry out the program of the previous example for the statement "$f(x)$ is periodic", and then show $x \sin x$ is not periodic.

Solution.   Note that we are not told in advance what the period c is.

   (a) $\exists c > 0 : \forall x, \; f(x+c) = f(x)$.   "$f(x)$ is periodic";

   (b) $\forall c > 0, \; \exists x_c : \; f(x_c + c) \neq f(x_c)$.   "$f(x)$ is not periodic".

Here we have put the subscript on x_c to emphasize that it depends on the choice of c.

For the function $f(x) = x \sin x$, we show that it is non-periodic by showing that (b) is satisfied:

   (i) if $c \neq n\pi$, $n \in \mathbb{Z}$, take $x_c = 0$: then $c \sin c \neq 0$;

   (ii) if $c = n\pi$, take $x_c = \pi/2$: $(\pi/2 + n\pi) \sin(\pi/2 + n\pi) \neq (\pi/2) \sin \pi/2$,
since the two sin factors have 1 as their value.

Example B.3C   Carry out the program of Example B.3A for

(15) $$\lim_{x \to a} f(x) = 0 \; .$$

Solution   (a) Since this involves more quantifiers, we'll do it one step at a time:

(15a) $$\text{given } \epsilon > 0, \quad |f(x)| < \epsilon \quad \text{for } x \approx a;$$

(15b) $$\text{given } \epsilon > 0, \text{ there is a } \delta > 0 : |f(x)| < \epsilon \text{ for } x \underset{\delta}{\approx} a.$$

(15c) $$\forall \epsilon > 0, \; \exists \delta > 0 : \; \forall x \underset{\delta}{\approx} a, \quad |f(x)| < \epsilon \; .$$

   (b) The negation is often rendered incorrectly by students as
$$\text{"}\lim_{x \to a} f(x) = L, \text{ where } L \neq 0\text{"};$$
this is wrong because it fails to take into account the possibility that $f(x)$ may have no limit at all. Proceeding instead systematically from (15c), using the negation rules, we get

(15′) $$\exists \epsilon > 0 : \; \forall \delta > 0, \; \exists x \underset{\delta}{\approx} a : \; |f(x)| \geq \epsilon \; .$$

In words, there is a positive number ϵ such that, no matter how small a positive number δ is given, we will have $|f(x)| \geq \epsilon$ for some x within δ of a.

Questions B.3

   1. Carry out the program in Example A for "$f(x)$ has a maximum on I"; then prove that x^2 has no maximum on $(0,1)$.

   2. Carry out the program in Example A for "$f(x) > 0$ for $x \approx a$"; then prove that $\sin(1/x)$ is not positive for $x \approx 0^+$.

Exercises

B.1

1. A *tail* of a sequence $\{a_n\}$ is the sequence that remains after removing the first N terms of $\{a_n\}$. Write formally, with the quantifiers in the right order; interchange the order of the first two quantifiers if possible, and interpret.

> *Every tail of $\{a_n\}$ has a maximal element.*

2. Follow the same instructions as in 1. for

> *The set S has a maximal element.*

3. Write formally, putting the quantifiers in the right order:

> *The sequence a_n converges.*

4. $\forall n,\ \exists m :\ a_m > a_n$. Put this into prose, by completing this sentence in five words or less: "the sequence $\{a_n\}$ ... "

B.2

1. Write these negations with the quantifiers in the right order. Follow the program: first write formally the corresponding positive assertion, then use the rules given in the text to form its negation.

    (a) The sequence a_n has no minimum.

    (b) The sequence a_n has a tail with no minimum; (cf B.1/1).

    (c) The set S is not bounded above; (cf. B.1/2).

    (d) The sequence a_n diverges; (cf. B.1/3).

B.3

1. Carry out the program in Example B.3A for "$f(x)$ is bounded above on the interval I", and then use it to show that $1/x$ is not bounded above on $(0, 1]$.

2. Carry out the program in Example B.3A for "$f(x)$ is defined on I". (Use the graph G_f of $f(x)$.)

3. Carry out the program in Example B.3A for "$f(x)$ has a (finite) limit as $x \to \infty$".

4. Carry out the program in Example B.3A for "$f(x)$ is continuous on I".

5. Carry out the program in Example B.3A for "$f(x)$ is uniformly continuous on I".

Answers to Questions

B.1

1. two: $\exists N : \forall n \geq N$ ...

2. (i) $\forall n \geq 0$, $a_{n+1} > a_n$.

(ii) $\exists N : \forall n \geq N$, $a_{n+1} > a_n$. Interchanging the order of quantifiers makes no sense, since N has to be introduced before it can be used.

(iii) $\exists n_0 : \forall n \geq 0$, $a_n \leq a_{n_0}$. Interchanging the first two quantifiers gives: $\forall n \geq 0$, $\exists n_0 : a_n \leq a_{n_0}$. This is trivial (i.e., it holds for any sequence), for since n_0 is allowed to depend on n, we can simply take $n_0 = n$.

(iv) $\forall m > 0$, $\forall n > 0$, $m + n > 0$. Interchange of the quantifiers does not change the statement.

(v) $\forall n > 0$, $\forall m > n$, $m - n > 0$. Interchange is not possible, since n must be named before it can be used.

(vi) $\exists B$, $\exists N_B : \forall n \geq N_B$, $a_n \leq B$. (N depends on the choice of B). Interchange gives: $\exists N$, $\exists B_N : \forall n \geq N$, $a_n \leq B_N$. (Here B depends on the choice of N).

The two statements are formally different, because the dependency of one variable on the other changes, but they make the same assertion: a pair of numbers (B, N) exists such that $a_n \leq B$ when $n \geq N$.

B.2

1. (Part (b) only: congratulations if your guess for part (a) coincides with part (b), but note that only two quantifiers are involved; as the number of quantifiers increases, the difficulty of forming a correct negative just by unsing intuition increases dramatically.)

(i) a_n bounded above: $\exists B : \forall n \geq 0$, $a_n \leq B$.

a_n is not bounded above: $\forall B$, $\exists n \geq 0 : a_n > B$.

(ii) a_n is constant for $n \gg 1$: $\exists N : \forall n > N$, $a_n = a_N$.

a_n is not constant for $n \gg 1$: $\forall N$, $\exists n > N : a_n \neq a_N$.

B.3

1. $\exists x_0 \in I : \forall x \in I$, $f(x) \leq f(x_0)$.

Negation: $\forall x_0 \in I$, $\exists x \in I : f(x) > f(x_0)$.

For the function x^2, given $x_0 \in (0, 1)$, choose any x such that $x_0 < x < 1$; then $x^2 > x_0^2$.

2. $\exists \delta > 0 : \forall x \underset{\delta}{\approx} a$, $f(x) > 0$.

Negation: $\forall \delta > 0$, $\exists x_0 \underset{\delta}{\approx} a : f(x_0) \leq 0$;   (x_0 depends on δ).

For the function $f(x) = \sin(1/x)$, given any $\delta > 0$, choose a positive integer k such that $1/k\pi < \delta$; take $x = 1/k\pi$, then $f(1/k\pi) = 0$ and $1/k\pi \underset{\delta}{\approx} 0$.

Appendix C

Picard Iteration

Picard's method for finding roots of equations is an excellent application of many of the theorems of continuity and differentiation; it shows the ideas in action. It is also a good preliminary to studying Picard's method for proving the existence and uniqueness theorem for solutions to differential equations.

Prerequisite: Chapter 15

C.1   Introduction.

According to the Intersection Principle 12.2, solving an equation of the form

$$(1) \qquad\qquad h(x) \; = \; g(x)$$

means finding the x-coordinates of the points where the graphs of g and h intersect. The "cobweb" picture at the right suggests an iterative procedure for doing this. We draw alternately horizontal and vertical line segments joining the two graphs, getting a sequence of points (x_n, y_n) which converge to the intersection point (c, d), that is, the point for which

$$h(c) \; = \; g(c), \qquad c = \lim x_n.$$

The picture shows that two successive steps are related by the equation

$$h(x_{n+1}) \; = \; g(x_n) \, ,$$

so that the recursion formula for the successive x_n is

$$(2) \qquad\qquad x_{n+1} = \; h^{-1}\big(g(x_n)\big) \, .$$

The form of (2) suggests that to analyze this process, without losing generality we can simplify things by replacing the two functions h and g by the single function $f = h^{-1} \circ g$, and the equation (1) to be solved by

$$(3) \qquad\qquad x \; = \; f(x) \, .$$

Following the procedure in (2), therefore, we propose finding the roots of (3) as the limit of the sequence defined recursively by

$$(4) \qquad\qquad x_{n+1} = \; f(x_n) \, .$$

The questions before us are therefore
  ⋆ *existence:*   does (3) have a solution?
  ⋆ *uniqueness:*   does (3) have a unique solution?
  ⋆ *convergence:*   does the sequence $\{x_n\}$ defined by (4) converge to a solution of (3)?

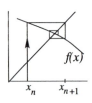 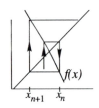 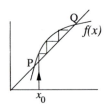

C.2   The Picard iteration theorems.

The two right-hand figures at the top show that things can sometimes go wrong with the cobweb picture; they suggest that we need assumptions about the function $f(x)$ if our questions are to have positive answers. We state them now.

The Picard conditions.   Let $I = [a, b]$ be a compact interval.

   Pic-1 f is continuous on I, and $f(I) \subseteq I$.

   Pic-2 $f'(x)$ exists on I, and there is a constant K such that

(5) $$|f'(x)| \leq K < 1, \qquad \text{for } x \in I.$$

Theorem C.2A   Existence theorem.

   If f satisfies Pic-1, then $x = f(x)$ has a solution in I.

Proof.   Since $f(I) \subseteq I$, it follows that $f(a), f(b) \in [a, b]$, so that

$$f(a) \geq a, \qquad f(b) \leq b .$$

If equality holds at either endpoint, we have a solution in I. If not, then the strict inequality holds at both endpoints, which shows the function $x - f(x)$ changes sign on $[a, b]$. Since the function is continuous, by Bolzano's Theorem (12.1) it has a zero $c \in I$, and this c is a solution to $x = f(x)$.   □

Theorem C.2B   Uniqueness theorem.

   If f satisfies Pic-2, then $x = f(x)$ has at most one solution in I.

Proof.   If x and x' are any two points in I, the Mean-value Theorem 15.1 together with (5) shows that

(6) $$|f(x) - f(x')| \leq K|x - x'| , \qquad \text{where } K < 1 .$$

   Suppose c and c' are solutions in I, so that $c = f(c)$, $c' = f(c')$. By (6),

$$|c - c'| \leq K |c - c'| .$$

This shows that $c = c'$, for otherwise we could divide the inequality through by $|c - c'|$, getting the contradiction $K \geq 1$.   □

Theorem C.2C   Convergence theorem.   *If f satisfies Pic-1 and Pic-2 on I, then starting with any $x_0 \in I$, the resulting sequence $\{x_n\}$ defined by (4) converges to the unique solution $c \in I$ of the equation $x = f(x)$.*

Proof.   Since $x_0 \in I$, all the succeeding $x_n \in I$, by Pic-1. By (6),

$$|f(x_n) - f(c)| \leq K |x_n - c| , \qquad K < 1 \, n \geq 0;$$

so that by (4), and since $f(c) = c$,

$$|x_{n+1} - c| \leq K |x_n - c| , \qquad K < 1 ;$$

using this in turn for $n = 0, 1, 2, \ldots$ (or by induction), we get

(7) $$|x_n - c| \leq K^n |x_0 - c| .$$

Since $K < 1$, we have $K^n \to 0$, which shows that the sequence $\{x_n\} \to c$, by the Squeeze Theorem for limits of sequences 5.2. □

> You should at this point look over the earlier graphs to see how the condition Pic-2 determines the behavior of Picard iteration.

In trying to use the Picard iteration method, if it turns out that Pic-2 fails because $|f'(x)| \geq 1$, one should not give up too easily. Often it is possible to transform the equation into an equivalent one for which the derivative condition Pic-2 will be satisfied.

In particular, if it looks like $f'(x) > 1$ near the solution c and f has an inverse function over some suitable interval I containing c, try changing $f(x) = x$ to the equivalent equation obtained by applying f^{-1} to both sides:

$$x = f^{-1}(x) ,$$

which has the same solutions in I. According to Theorem 14.2C,

$$\frac{d}{dx} f^{-1}(x) \Big|_{x=c} = \frac{1}{f'(c)} < 1, \quad \text{if } f'(c) > 1.$$

Thus f^{-1} satisfies Pic-2 near c, and we are in business once more. (Of course, f^{-1} might be impossible to calculate, or there might be trouble identifying a suitable interval.)

Example C.2   Solve $x^5 + x = 1$ by Picard's method.

Solution.   Let $f(x) = 1 - x^5$; try first the form
$$x = f(x) .$$
Drawing the graphs and experimenting numerically,
we see that $f(x) - x$ changes sign on $[.7, 1]$, so the root is in this interval. Unfortunately, Pic-2 fails since

$$|f'(x)| = 5x^4 > 1.2 \quad \text{on } [.7, 1].$$

Let us therefore try using the function $g = f^{-1}$ instead, which we calculate (cf. Section 9.4) by solving the equation $y = 1 - x^5$ for x; this gives

$$x = g(y) = \sqrt[5]{1 - y} ;$$

Then the new equation to be solved is $x = g(x)$, that is
$$x = \sqrt[5]{1 - x} .$$

To get the graph of g, flip the picture around the diagonal, or simply switch the labels on the x and y axes. The intersection point still occurs on the interval $[.7, 1]$; moreover, since the slope of $f(x)$ at the intersection with the diagonal had a value > 1.2, the slope of $g(x)$ at this point will be $< 1/1.2 \approx .8$.

Thus Pic-2 is satisfied for $x = g(x)$, and the Picard method is applicable to this equation. Starting with say .7, it leads to a zero at $\approx .76$. □

As the example shows, you don't usually bother to find an exact interval where the conditions are satisfied; instead you determine the approximate location of the solution c somehow, check that the derivative there is < 1, pick a nearby starting point, and calculate the sequence of successive approximations.

The method requires more iterations than does Newton's method, but it avoids calculating the derivative, often a crucial consideration if $f(x)$ is complicated. Comparing it with the bisection method of Chapter 12, the method avoids having to decide at each stage which of the two subintervals to select—for Picard's method on a pocket calculator, you just keep pressing over and over again the programmed function key which calculates $f(x)$ and watch the display converge. It's fun; try it.

Questions C.2

1. The proof of the Existence Theorem C.2A avoids using the Intersection Principle 12.2, at the cost of having to repeat its proof. Rewrite and streamline it by using the principle.

2. Write out the justification for the lines: (i) (6);   (ii) (7).

3. Solve the equation $\ln x = 4 - x^2$ by first sketching the curves and identifying a suitable interval containing the solution; then put it in the form $x = f(x)$ two different ways and select one suitable for Picard's method.

C.3   Fixed points.

Up to this point we have visualized the Picard method using the graphs. We now get some new insight by thinking of f from the mapping viewpoint. Pic-1 says we are given a map of $[a, b]$ to itself:

$$(8) \qquad\qquad\qquad f : [a, b] \rightarrow [a, b] \, .$$

How shall we interpret a solution c to the equation $x = f(x)$?

Definition C.3A   A **fixed point** of the map (8) is a point c such that $f(c) = c$.

From this point of view, the Picard method uses an iteration procedure to search for fixed points of the mapping f. Analyzing the proof of the three preceding theorems shows that Pic-2 is only needed to prove (6), and then (6) is what the subsequent proof of uniqueness and convergence requires. So for more generality and clarity, we make (6) the key property.

Definition C.3B   The map f is called a **contraction mapping** on $[a, b]$ if it satisfies the following two conditions:

Con-1 $f : [a, b] \rightarrow [a, b]$;

Con-2   there is a constant $K < 1$ such that
$$|f(x) - f(x')| \, \le \, K \, |x - x'|, \qquad \text{for all } x, x' \in [a, b] \, .$$

Since $K < 1$, the condition Con-2 shows that f shrinks the distance between any two distinct point of $[a, b]$; this explains the word "contraction". We need not as in Pic-1 add the hypothesis that f is continuous, since a contraction map is always continuous (see the Exercises). Thus our previous work may be reinterpreted as

Theorem C.3   Picard fixed-point theorem.

Let f be a contraction mapping on $[a, b]$. Then

(i) f has a unique fixed point on $[a, b]$;

(ii)   starting with any $x_0 \in [a, b]$, the sequence $\{x_n\}$ defined recursively by $x_{n+1} = f(x_n)$, $n = 0, 1, 2, \ldots$, converges to this unique fixed point.

The iteration procedure can be generalized: instead of constructing sequences of points that converge to the solution of an equation, one can use it to construct sequences of functions that converge to the solution of a differential equation (see Appendix E on the Existence and Uniqueness Theorem for differential equations).

From there it can be further generalized to a method for finding solutions to partial differential equations and integral equations, by interpreting a solution as a fixed point of a mapping. In this form it plays a central role in modern analysis.

Questions C.3

1. Let $I = [a, b]$ and suppose $f : I \to I$; also assume that on I, the function $f(x)$ is continuously differentiable and $|f'(x)| < 1$.

Prove f is a contraction map.

Exercises

C.2

1. Use Picard's method and a calculator to solve $xe^x = 1$ to three decimal places. Verify the hypotheses of the method for a suitable interval.

2. Use Picard's method to find to two decimal places the smallest positive solution to the equation $\sin x = x \cos x$. Be careful.

How many iterations would be required to get the answer to within 10^{-n}, and why?

3. What is wrong with this proof of the Convergence Theorem C.2C?

Let $\lim x_n = L$; then $\lim x_{n+1} = L$ also. Since f is differentiable, it is continuous. Therefore it follows by the Sequential Continuity Theorem 11.5 that $\lim f(x_n) = f(L)$, and so as $n \to \infty$,

$$x_{n+1} = f(x_n) \quad \Rightarrow \quad L = f(L).$$

This shows L is a solution to the equation $x = f(x)$; since by Theorem C.2B, the solution is unique, it follows that $L = c$, and therefore $\lim x_n = c$.

4. Identify an interval of length .2 on which $x^3 + x - 1$ has a zero. Two ways of rewriting the equation so as to use Picard's method are

$$x = 1 - x^3, \quad \text{and} \quad x = \frac{1}{1 + x^2}.$$

Choose one and find the zero.

C.3

1. Prove that a contraction mapping is continuous on the interval on which it is defined.

Problems

C-1 (a) Show that Newton's method for finding a solution of $f(x) = 0$ can be interpreted as the Picard method applied to a certain function $F(x)$ derived from $f(x)$.

(b) Assuming that $f'(x) > 0$ on $[a, b]$, what hypotheses on $f(x)$ and its first two derivatives guarantee that Pic-1 and Pic-2 will be satisfied for $F(x)$?

C-2 Let us call a function $f : [a, b] \to [a, b]$ a "pseudo-contraction" if

$$|f(x) - f(x')| < |x - x'| \qquad \text{for all } x, x' \in [a, b], \quad x \neq x'.$$

(a) Prove a pseudo-contraction is continuous.

(b) Prove it has a unique fixed point on $[a, b]$.

(c) Show that $\sin x$ is a pseudo-contraction on $[-\frac{\pi}{2}, \frac{\pi}{2}]$, but not a contraction.

(d) Try using a calculator to find the fixed point of $\sin x$ by the Picard iteration method, starting with some small positive number. What answer do you get? (Give as many decimal places as your calculator displays.)

(e) Prove that, despite the experimental evidence in (d) to the contrary, the sequence produced by the Picard method from $\sin x$ really does converge, and to the fixed point of the mapping $\sin x$.

(Hint: try drawing the graphical picture.)

Answers to Questions

C.2

1. After the first part of the proof as given, we may assume $f(a) > a$ and $f(b) < b$. This shows the graphs of x and $f(x)$ change their relative position over the interval $I = [a, b]$. Since both functions are continuous, it follows from the Intersection Principle 12.2 that the two graphs intersect over I, i.e., there is a point $c \in I$ such that $c = f(c)$.

2. (i) By the Mean-value Theorem,

$$|f(x) - f(x')| = |f'(c)||x - x'|, \quad \text{for some } c \in I.$$
$$|f'(c)| \le K < 1, \quad \text{by Pic-2, since } c \in I.$$

Therefore $|f(x) - f(x')| \le K|x - x'|.$

(ii) To prove (7), we use induction (Appendix A.4): the case $n = 0$ is trivial, and the case $n + 1$ follows from the case n since

$$|x_{n+1} - c| \le K|x_n - c|$$
$$\le K \cdot K^n|x_0 - c|, \qquad \text{by induction.}$$

3. A picture shows that the two graphs intersect over the interval $[1, 2]$.

The equation can be put in the form $x = f(x)$ either by solving the left-hand side for x, or the right-hand side. This leads to the two forms

$$x = e^{4-x^2}, \quad \text{and} \quad x = \sqrt{4 - \ln x}\ .$$

The first is no good, since calculation shows $|f'(x)| > 1$ on $[1, 2]$; for the second, however, $|f'(x)| < 1$ on the interval; thus Picard iteration can be used on the function $\sqrt{4 - \ln x}$: starting with either 1 or 2 leads in a couple of steps to 1.8, and then to as many more decimal places as you want.

C.3

1. Since $|x|$ is continuous and $f'(x)$ is continuous on I, the composite function $|f'(x)|$ is also continuous on I. Therefore it has a maximum point $x_0 \in I$, by the Maximum Theorem 13.3; letting $|f'(x_0)| = K$, we thus have

$$|f'(x)| \le |f'(x_0)| = K < 1, \quad \text{for all } x \in I, \text{ by hypothesis;}$$

from this the condition Con-2 follows by the Mean-value Theorem, as in (6).

Appendix D

Applications to Differential Equations

This appendix shows how the main theorems about sequences and functions (especially those involving continuity and differentiability) in the first half of the book are used to study the solutions of second-order linear differential equations. It will give you a better understanding of the significance of these theorems.

Prerequisites: chapters 1-15. The relevant information about differential equations is presented as needed; however, it would help to have studied in a calculus, physics, or engineering class the second-order equation with constant coefficients — it models oscillations in an RLC-circuit or a spring-mass-dashpot system.

D.1   Introduction.

Many physical systems are modeled by the second-order linear homogeneous differential equation, with coefficients continuous on an interval I:

$$(1) \qquad y'' + p(x)y' + q(x)y = 0, \qquad p, q \text{ continuous on } I.$$

An important question is the following: where do the zeros of a solution to (1) lie? We need to know, because often the physical problem leading to (1) will require us to find a solution $y(x)$ satisfying homogeneous boundary conditions:

$$(1a) \qquad y(x_0) = 0, \quad y(x_1) = 0; \qquad x_0, x_1 \in I, \ x_0 < x_1.$$

To do this, one must know which solutions have zeros, and where they are located, approximately. The example below illustrates.

Of course, the equation (1) always has the solution with too many zeros: $y = 0$ for all x. We will refer to this as the *trivial* solution, and the main theorems will always exclude it, by speaking of *non-trivial* solutions. (It has an irritating sound, but calling them "non-zero" solutions would be misleading, since they do have zeros; calling them "not-identically-zero" solutions would be a lot worse.)

Example D.1   For the differential equation (k is a positive constant)

$$(2) \qquad\qquad y'' + k^2 y = 0 \ ,$$

determine for what values of k the boundary conditions (1a) can be satisfied.

Solution.   The solutions are the two-parameter family of functions

$$y = c_1 \sin kx + c_2 \cos kx, \qquad c_1, c_2 \in \mathbb{R}.$$

Using trigonometric identities, the family can be written more expressively as

$$y = A \sin k(x + \phi) , \qquad A, \phi \in \mathbb{R}.$$

In this form, one sees that every non-trivial solution $y(x)$ is a translated sine curve, oscillating with frequency k; its zeros therefore are regularly spaced at intervals of π/k.

It follows that we can find a solution satisfying the boundary conditions (1a) if and only if for some positive integer n, we have

$$x_1 - x_0 = \frac{n\pi}{k}, \qquad \text{i.e.,} \qquad k = \frac{n\pi}{x_1 - x_0}. \qquad \square$$

With this example in mind of the usefulness of knowing where the zeros of solutions to (1) lie, our aim in this appendix is to get qualitative information about these zeros, without having to find explicit formulas for the solutions (which are usually unavailable). For example, under what conditions can we be sure a solution to (1) will have infinitely many zeros? If it does, can the zeros have a cluster point? If there is none, can we estimate how far apart the zeros are?

We will summarize the general theorems about differential equations of the form (1) as we need them; we begin with the question about a cluster point.

D.2   Discreteness of the zeros.

The most essential general theorem about (1) is the following; all the other general results we need follow from it.

Theorem D.2A   Existence and uniqueness theorem.

There is one and only one function $y(x)$ having a second derivative in the interval I and satisfying both (1) and the initial conditions (3) below:

(1) $y'' + p(x)y' + q(x)y = 0,$ $p(x), q(x)$ continuous on I) ,

(3) $y(x_0) = y_0, \quad y'(x_0) = y_0',$ $x_0 \in I$; y_0, y_0' constants.

A function having a second derivative on I and satisfying the differential equation (1) is called a *solution* to (1) on I. If it also satisfies (3), it is called a *solution to the initial-value problem.*

Corollary D.2A   Zero-solution theorem.

Let $c \in I$ and $y(x)$ be a solution to (1) on I.

$$y(c) = 0, \quad y'(c) = 0 \;\; \Rightarrow \;\; y(x) = 0 \quad \text{for all } x \in I.$$

Proof. The function 0 is a solution to (1) satisfying the same initial conditions; therefore $y(x) = 0$ on I since by Theorem D.2A the solution to the initial-value problem is unique. $\square$

Geometrically, the Corollary says that the graph of a non-trivial solution to (1) cannot be tangent to the x-axis at any point of I. We shall make frequent use of the Corollary.

We can now prove the remarkable fact that the zeros of a solution to (1) cannot have a cluster point.

Theorem D.2B   Discreteness of the zeros.

A non-trivial solution to (1) has only a finite number of zeros in any finite interval $[a, b] \subseteq I$.

Proof. By contraposition.

Let $y(x)$ be a solution with infinitely many zeros in $[a, b]$; select an infinite sequence of these zeros, without repetitions. Since the interval $[a, b]$ is finite and closed, it has the sequential compactness property (13.1), so from the infinite sequence of zeros we get a subsequence $\{x_n\}$ converging to a point $c \in [a, b] \subseteq I$:

$$(4) \qquad \lim x_n = c, \qquad c \in I, \qquad y(x_n) = 0 \text{ for all } n.$$

We now show that $y(c) = 0$ and $y'(c) = 0$; by the Zero-solution Theorem (Cor. D.2A), it then follows that $y = 0$ on I, which completes the proof by contraposition.

Since $y(x)$ is differentiable on I, it is continuous at $c \in I$. Using (4) and the Sequential Continuity Theorem 11.5, we see that

$$(5) \qquad y(c) = \lim_{n \to \infty} y(x_n) = 0 \ .$$

To show similarly that $y'(c) = 0$, we have by the definition of derivative

$$y'(c) = \lim_{x \to c} \frac{y(x) - y(c)}{x - c} \ ;$$

$$= \lim_{x_n \to c} \frac{y(x_n) - y(c)}{x_n - c}, \qquad \text{by Theorem 11.5A;}$$

$$= \lim_{x_n \to c} \frac{0 - 0}{x_n - c} = 0, \qquad \text{by (4) and (5).} \quad \square$$

Remarks.

There can be twice differentiable functions which have infinitely many zeros in a finite interval, for example the function

$$f(x) = x^3 \sin(1/x), \quad x \neq 0, \qquad f(0) = 0,$$

on any interval of positive length containing 0. (The factor x^3 is put in to make it twice differentiable at 0.) What the theorem is saying is that such a function cannot be a solution to a second-order differential equation of the form (1) in such an interval.

When a function has only a finite number of zeros in any finite interval, it makes sense to speak of the *next* zero after a given zero; i.e., starting at some zero x_0, the zeros to the right of it, in their natural order, form a sequence. By contrast, this is not so for the function $f(x)$ above — since 0 is a cluster point for both its positive and its negative zeros, there is no "next" zero after 0; neither the positive nor the non-negative zeros form, in their natural order, a sequence.

D.3   The alternation of zeros.

Our goal in this section is to prove that if you have two solutions of (1) with different zeros, then their successive zeros "alternate": a zero of one is followed by a zero of the other. The equation (2) illustrates: the zeros of two translated sine curves alternate. Our main tool will be the Wronskian.

Definition D.3A   Two functions y_1 and y_2 are called **linearly independent** on an interval I if neither is a constant multiple of the other on I, or equivalently,

$$a_1 y_1 + a_2 y_2 = 0 \quad \text{on } I \quad \Rightarrow \quad a_1 = 0, \ a_2 = 0,$$

Otherwise, they are **linearly dependent** — i.e., one is a constant multiple of the other, or equivalently, there is a nontrivial relation $a_1 y_1 + a_2 y_2 = 0$ on I.

An important theoretical way of expressing the linear independence of two functions is by using their Wronskian.

Definition D.3B   We define the **Wronskian** of two functions y_1 and y_2 which are differentiable on I to be

$$W(y_1, y_2) \ = \ \begin{vmatrix} y_1 & y_2 \\ y_1' & y_2' \end{vmatrix} \ = \ y_1 y_2' - y_1' y_2 \ .$$

From the definition, it is clear that

$$y_1, y_2 \text{ linearly dependent on } I \quad \Rightarrow \quad W(y_1, y_2) = 0 \text{ on } I,$$

since one column of the Wronskian will be a constant multiple of the other. We are however primarily interested in the converse of this — i.e., in using the Wronskian to show linear dependence. For this, a special result holds, provided y_1 and y_2 are solutions to (1) and not just arbitrary functions: either their Wronskian is always zero, or it is never zero.

Theorem D.3A   Wronskian vanishing theorem.

If y_1 and y_2 are two solutions to (1) on I, then on I either

$$W(y_1, y_2) = 0 \quad \text{for all } x \in I \quad \text{and } y_1, y_2 \text{ are linearly dependent, or}$$

$$W(y_1, y_2) \neq 0 \quad \text{for all } x \in I \quad \text{and } y_1, y_2 \text{ are linearly independent.}$$

Proof.   As a preliminary, we remark that since y_1 and y_2 are solutions to (1), one sees by substituting into (1) that $a_1 y_1 + a_2 y_2$ is also a solution, if $a_1, a_2 \in \mathbb{R}$ (the "superposition principle" for linear differential equations).

For the theorem, it suffices to prove that if $W(y_1, y_2) = 0$ at some point $c \in I$, then y_1 and y_2 are linearly dependent on I (so that their Wronskian is identically zero). To show this, we have by linear algebra,

$$\begin{vmatrix} y_1(c) & y_2(c) \\ y_1'(c) & y_2'(c) \end{vmatrix} = 0 \quad \Rightarrow \quad \begin{cases} a_1 y_1(c) + a_2 y_2(c) = 0 & \text{has a non-trivial} \\ a_1 y_1'(c) + a_2 y_2'(c) = 0 & \text{solution } (\bar{a}_1, \bar{a}_2). \end{cases}$$

The corresponding function $\bar{a}_1 y_1 + \bar{a}_2 y_2$ is thus a solution to (1) which is zero at c and has zero derivative at c. By the Zero-solution Theorem D.2B, we have $\bar{a}_1 y_2 + \bar{a}_2 y_2 = 0$ on I, showing y_1 and y_2 are linearly dependent on I.   □

Theorem D.3B   Sturm oscillation theorem.   *On an open interval I, let $u(x)$ and $v(x)$ be two linearly independent solutions of (1). Then between two successive zeros a, b of $v(x)$ on I lies exactly one zero c of $u(x)$: that is, $u(c) = 0$ for some c such that $a < c < b$.*

Since $u(x)$ and $v(x)$ are interchangeable, between two successive zeros of $u(x)$ lies exactly one zero of $v(x)$; this shows the zeros of the two solutions alternate.

We need not say the solutions are non-trivial: since u and v are linearly independent, neither of them can be identically zero. And to repeat an earlier remark, "successive zeros" only makes sense because we proved in the previous section that the set of zeros has no cluster points.

The proof needs the following simple but useful lemma.

Lemma D.3   Up-down lemma.   *If $f(x)$ is positive on (a, b), and differentiable and zero at a and b, then $f'(a^+) \geq 0$ and $f'(b^-) \leq 0$.*

The proof is left as an Exercise (apply the Limit Location Theorem to the difference quotient which defines the derivative).

Proof of the Sturm Theorem.

Let a, b be two successive zeros of $v(x)$ on I.

The plan is to find zeros of $u(x)$ on $[a, b]$ by using Bolzano's Theorem 12.1; this means we need to know the signs of $u(a)$ and $u(b)$; this will come from information about $v(x)$ and the Wronskian of the two functions.

We first study the Wronskian of u and v on $[a, b]$:
$$W(u, v) = uv' - u'v .$$
It is continuous on $[a, b]$ and never zero (by the Wronskian Vanishing Theorem), so it has the same sign at a and b, by Bolzano's Theorem. But since $v = 0$ at a and b, this implies

$(*)$ uv' *has the same sign at a and b .*

We now study the sign of $v'(x)$ at the two points. Since $v(x)$ is differentiable, it is continuous. Either $v(x) > 0$ or $v(x) < 0$ on (a, b), for otherwise by Bolzano's Theorem $v(x)$ would have a zero on (a, b), and then a and b would not be successive zeros.

So say $v(x) > 0$.

By the Up-down Lemma, $v'(a) > 0$, $v'(b) < 0$; neither can be zero, since v would then be identically 0 on I, by the Zero-solution Theorem D.2B. Thus v' changes sign on $[a, b]$; it follows from $(*)$ that $u(x)$ does also, and therefore it has a zero on (a, b), by Bolzano's Theorem.

The function $u(x)$ cannot have two zeros on (a, b), since then (reversing the roles of u and v), it would follow that v would have a zero between these two zeros of u; but this would contradict our assumption that a and b are successive zeros of v.   □

D.4   Reduction to normal form.

In analyzing the qualitative behavior of solutions to

(1) $$y'' + p(x)y' + q(x)y = 0, \qquad p, q \text{ continuous on } I ,$$

it is convenient to make a change in the dependent variable y so as to eliminate the term in y'. This may be done by setting

(6) $$y(x) = u(x)h(x); \qquad h(x) = e^{-\frac{1}{2} \int p(x)\, dx} .$$

Straightforward substitution shows that (1) is changed into the *normal form*

(7) $$u'' + r(x)u = 0; \qquad r = q - \frac{p'}{2} - \frac{p^2}{4} .$$

(In doing this, use is made several times of the relation $2h' = -ph$.)

Of course, (1) and (7) are different equations, but (6) tells us the relation between their solutions. In particular, since $h(x) > 0$ for all $x \in I$, two corresponding solutions have the same zeros:

$$y_1(c) = 0 \quad \Leftrightarrow \quad u_1(c) = 0, \qquad \text{where } y_1 = u_1 h, \text{ and } \begin{cases} y_1 \text{ solves (1)}, \\ u_1 \text{ solves (7)}. \end{cases}$$

Thus in our study of the existence and location of zeros, we can take the differential equation to be in the normal form (7), rather than (1).

The transformation $y = uh$ is the analogue for second-order linear differential equations of the algebraic transformation $x = u - a/2$ which eliminates the linear term in the second-degree equation $x^2 + ax + b = 0$.

If $r(x) > 0$, the differential equation (7) may be thought of as modeling the motion of an undamped spring-mass system, whose stiffness (Hooke's law) constant varies with the time x. Thus a plausible guess is that the solutions will oscillate for $x \gg 1$, but irregularly; they should have infinitely many zeros, but with irregular spacing. The equation (7) is therefore often called the *anharmonic oscillator*.

Example D.4   Change to normal form *Bessel's equation of order p:*

(9) $$x^2 y'' + xy' + (x^2 - p^2)y = 0, \qquad p \text{ constant}$$

Solution.   We first have to put it in the form (1):

$$y'' + \frac{1}{x} y' + \left(1 - \frac{p^2}{x^2}\right) y = 0, \qquad I = (0, \infty) .$$

Following (6), we get $h(x) = 1/\sqrt{x}$, and then either by letting $y = u/\sqrt{x}$ and substituting into (1), or using (7), we calculate

$$r(x) = 1 + \left(\frac{1 - 4p^2}{4x^2}\right) ;$$

thus $y = u/\sqrt{x}$ changes (9) into *Bessel's equation in normal form:*

(10) $$u'' + \left(1 + \frac{1 - 4p^2}{4x^2}\right) u = 0 \qquad \qquad \Box$$

Solutions to (9) are called "Bessel functions of order p". We will study their zeros by comparing (10) with the simpler equation $u'' + k^2 u = 0$.

D.5   Comparison theorem for zeros.

An important reason for putting the second-order linear differential equation in normal form is that it allows us to make comparisons between different second-order equations. The theorem which allows us to do this is another theorem of Sturm:

Theorem D.5   Sturm comparison theorem.

Suppose that $u(x)$ and $v(x)$ are respectively solutions on an interval I of

$$(11) \qquad\qquad u'' + r(x)u = 0 \quad \text{and} \quad v'' + s(x)v = 0 \;,$$

where $r(x)$ and $s(x)$ are continuous on I and

$$r(x) > s(x) \quad \text{on } I.$$

Then between two successive zeros a and b of $v(x)$ lies at least one zero of $u(x)$, that is, $u(c) = 0$ for some c such that $a < c < b$.

Proof.   Just as for the Sturm Oscillation Theorem D.3, the idea is to relate the behavior of $u(x)$ to that of $v(x)$ by studying their Wronskian:

$$(12) \qquad\qquad W(u,v) \;=\; uv' - u'v \;;$$

we will also need its derivative, calculated using (11):

$$(13) \qquad\qquad \frac{dW}{dx} \;=\; uv'' - u''v \;=\; uv(r - s) \;.$$

Since a and b are successive zeros of $v(x)$, we see that $v(x) \neq 0$ on (a,b).

Reasoning indirectly, suppose that also $u(x) \neq 0$ on (a,b). Then since $u(x)$ and $v(x)$ are continuous (even twice-differentiable), according to Bolzano's Theorem 12.1 one of these cases must hold on (a,b):

$$u > 0, \; v > 0; \quad u > 0, \; v < 0; \quad u < 0, \; v > 0; \quad u < 0, \; v < 0.$$

The reasoning for all is similar — we will just do the first case.

Since $u > 0, \; v > 0$, and by hypothesis, $r - s > 0$ on (a,b), we get by (13), and Theorem 15.2,

$$(*) \qquad\qquad \frac{dW}{dx} > 0 \quad \text{on } (a,b): \quad W \text{ is strictly increasing on } (a,b).$$

On the other hand, by the Up-down Lemma D.3,

$$v > 0 \text{ on } (a,b) \;\Rightarrow\; v'(a) \geq 0, \; v'(b) \leq 0;$$
$$\Rightarrow\; v'(a) > 0, \; v'(b) < 0,$$

since $v'(a) \neq 0, v'(b) \neq 0$ by the Zero-solution Theorem D.2A (remember that $v(a) = v(b) = 0$).

Finally, since $u(a) \geq 0$ and $u(b) \geq 0$ by continuity, (12) shows that

$$W(a) = u(a)v'(a) \geq 0 \;; \qquad W(b) = u(b)v'(b) \leq 0 \;,$$

which contradicts the fact (*) that $W(x)$ is strictly increasing on (a,b) □

If $r(x)$, $s(x) > 0$, think of (11) as modeling two undamped spring-mass systems, with the spring constants r, s varying with time x, but one spring always stiffer than the other. In this situation, the theorem says roughly that the mass at the end of the stiffer spring passes through the equilibrium point more often — it oscillates more rapidly. Of course, the theorem also applies when r and s are negative, or when there are no oscillations.

Corollary D.5A   *If $r(x) < 0$ on I, then any non-trivial solution $u(x)$ of the equation $u'' + r(x)u = 0$ has at most one zero on I.*

Proof.   Apply the Theorem D.5 to $u'' + ru = 0$ and $v'' + 0v = 0$.   By contradiction, if $u(x)$ had two zeros on I, then by the Comparison Theorem the solution $v(x) = 1$ to the second equation would have a zero between them. $\square$

Corollary D.5B   *If $r(x) > k > 0$ on an infinite interval I, then a solution $u(x)$ to $u'' + r(x)u = 0$ has infinitely many zeros on I.*

Proof.   We compare $u'' + r(x)u = 0$ with $v'' + kv = 0$, which has a solution $\sin \sqrt{k}x$ having an infinity of zeros on the infinite interval I. Between any two of them lies a zero of $u(x)$. $\square$

Example D.5A   Let $u_p(x)$ be a non-trivial solution to the Bessel equation

(10) $$u'' + (1 + \frac{1 - 4p^2}{4x^2}) u = 0, \qquad x > 0 .$$

Prove that $u_p(x)$ has an infinity of zeros on $(0, \infty)$, for any p.

Solution.   Use Corollary D.5B. By the Function Location Theorem (11.3D),

$$\lim_{x \to \infty} 1 + \frac{1 - 4p^2}{4x^2} = 1 \quad \Rightarrow \quad 1 + \frac{1 - 4p^2}{4x^2} > \frac{1}{2} \text{ for } x \gg 1, \text{ say for } x > a ;$$

then by the Corollary $u_p(x)$ has an infinity of zeros on (a, ∞). $\square$

Example D.5B   Continuing with the previous example, let x_n and x_{n+1} be two successive positive zeros of $u_p(x)$. Prove that

(a) $x_{n+1} - x_n < \pi$,   if $0 \le p < 1/2$;

(b) $x_{n+1} - x_n > \pi$,   if $p > 1/2$.

Solution.   We compare the solutions of (10) with those of

(14) $$v'' + v = 0; \qquad \text{solutions: } A \sin(x - a).$$

Note that if $[a, a + \pi]$ is an arbitrary interval of length π in $(0, \infty)$, then the function $v(x) = \sin(x - a)$ is a solution of (14) having successive zeros at a and $a + \pi$.

(a) We observe that if $0 \le p < 1/2$, then $1 + \dfrac{1 - 4p^2}{4x^2} > 1$ on $(0, \infty)$. Reasoning indirectly, suppose $x_{n+1} > x_n + \pi$ for some n. Then there would be no zero of $u_p(x)$ between the successive zeros x_n and $x_n + \pi$ of $\sin(x - x_n)$, which contradicts the Sturm Comparison Theorem.   □

(b) This time $p > 1/2$, so that $1 > 1 + \dfrac{1 - 4p^2}{4x^2}$ on $(0, \infty)$. Reasoning indirectly, suppose $x_{n+1} < x_n + \pi$ for some n. Then there would be no zero of $\sin(x - x_n)$ between the successive zeros x_n and x_{n+1} of $u_p(x)$, which contradicts the Sturm Comparison Theorem.   □

Discussion.

Since $1 + \dfrac{1 - 4p^2}{4x^2} \to 1$ as $x \to \infty$, it is tempting to conjecture that as $x \to \infty$, the spacing of the successive zeros of a solution $u_p(x)$ to (10) approaches the constant spacing π of the zeros to a solution of (14), i.e,

(15) $$x_{n+1} - x_n \to \pi, \qquad \text{as } n \to \infty.$$

This is true, and left for the Problems, in order to give you some practice with the Comparison Theorem.

Some feel that since $r(x) \to 1$ as $x \to \infty$, the result (15) should follow automatically. The problem with this is that we do not know whether the solutions to $u'' + r(x)u = 0$ vary continuously as $r(x)$ varies, nor do we know whether the zeros of a solution vary continuously as the solution varies continuously. In fact, we have not even formulated what it would mean for a solution to "vary continuously"; this is a topic for higher-level analysis courses.

To refine Corollary D.5B, what if $r(x)$ is positive, but cannot be bounded away from 0? For example, what about $u'' + (1/x)u = 0$? Do its solutions oscillate? Think of it as modeling an aging spring — one losing its stiffness as time x passes; does such a spring oscillate indefinitely, or does it give up?

The answer depends on how rapidly $r(x) \to 0$ as $x \to \infty$. A useful comparison equation is the following one, because it can be solved explicitly:

(16) $$v'' + \frac{k}{x^2} v = 0.$$

Its solutions are found from the roots of $t^2 - t + k = 0$. The results are:

(i) if $k < 1/4$, the roots are real numbers t_1 and t_2, and the general solution is non-oscillating:
$$v = c_1 x^{t_1} + c_2 x^{t_2};$$

(ii) if $k > 1/4$, the roots are complex, $a \pm bi$, and the general solution is
$$v = x^a (c_1 \cos(b \ln x) + c_2 \sin(b \ln x)),$$

which oscillates infinitely often, but more and more slowly with increasing x.

This is needed in the Exercises, which give further examples and applications.

Exercises

D.2

1. Suppose that $u(x)$ and $v(x)$ are two solutions to the second-order linear differential equation (1) on the interval I, and $u(c) = v(c) = 0$ for some $c \in I$. Prove that one of the functions is a constant multiple of the other on I.

2. For the theorem on the discreteness of the zeros, give a different proof of the statement $y'(c) = 0$, based on Rolle's Theorem (Lemma 15.1).

3. Let $u = x$ and $v = x^2$, and let I be an interval containing 0.

(a) Find an equation of the form (1) having u and v as solutions, as follows: write $y = c_1 x + c_2 x^2$, calculate y' and y'', and eliminate c_1 and c_2 from these three equations so as to get a single equation connecting y, y' and y''.

(b) Prove using the theorems of this section that there cannot be any equation of the form (1) having u and v as solutions on I.

(c) Reconcile parts (a) and (b).

4. Find, by calculating the derivatives and fooling around, a second-order linear homogeneous differential equation satisfied by

$$y = x^3 \sin(1/x), \quad y(0) = 0 ,$$

which is a function having an infinity of zeros on $(-1, 1)$. Why does this not contradict the theorem on the discreteness of the zeros?

D.3

1. Prove the Up-down Lemma.

2. The functions e^{ax} and e^{bx} satisfy an equation of the form (1) having constant coefficients p and q.

(a) Prove using the Sturm Oscillation Theorem that a non-trivial function of the form

$$y = c_1 e^{ax} + c_2 e^{bx}, \quad a \neq b$$

has at most one zero.

(b) Prove the same thing directly.

3. Prove that the Sturm Oscillation Theorem does not generalize to solutions of say third-order linear homogeneous differential equations with continuous coefficients, by producing a counterexample; use an equation with constant coefficients.

4. Prove that if one non-trivial solution to the equation (1) has an infinity of zeros, then all solutions have an infinity of zeros.

D.4

1. Derive the normal form (7) from (1), using the method indicated.

2. Carry out the derivation of the normal form for Bessel's equation:

(a) by using the results in (7);

(b) from scratch, by making the substitution $y = u/\sqrt{x}$.

D.5

1. Which of the following equations has, on some interval (a, ∞), solutions which have an infinity of zeros? Indicate reasoning.

(a) $xu'' + u = 0$;   (b) $(x^2 \ln x) u'' + u = 0$;   (c) $u'' + \sqrt{\dfrac{x+1}{x^5 - 1}}\, u = 0$.

2. Nothing was said about what happens with Bessel's equation if $p = 1/2$. You can solve equation (9) exactly in this case; what is the spacing of the zeros?

3. How would the conclusion of the Sturm Comparison Theorem be modified if the hypothesis were weakened to $r(x) \geq s(x)$ on I? Trace through the proof, indicating what modifications to make.

4. Give the argument proving the Sturm Comparison Theorem for the case $u < 0$, $v > 0$ on (a, b).

5. Give a direct proof of Corollary D.5A (i.e., without using the Sturm Comparison Theorem) along the following lines.

Suppose a, b are two successive zeros of a solution $u(x)$; say $u(x) > 0$ on (a, b). The differential equation that $u(x)$ satisfies shows that $u'(x)$ is strictly increasing on (a, b), but this is easily contradicted.

6. Chapter 15 is a prerequisite for this appendix, but actually the Mean-value Theorem is only needed (implicitly) once. Where?

Problems D

D-1 Let $0 \leq p < 1/2$, and let $u_p(x)$ be a non-trivial solution to Bessel's equation (10) in normal form.

Let x_n and x_{n+1} be two successive zeros of $u_p(x)$. Prove by going back to the definition of the limit of a sequence that

$$\lim_{n \to \infty} x_{n+1} - x_n = \pi \ .$$

You can use the results of Example D.5B .

D-2 By definition the Bessel function $J_p(x)$ is the unique series solution to Bessel's equation in standard form (9) which starts (assume p is a non-negative integer, to avoid explanations),

$$\frac{x^p}{2^p p!} \left(1 - \frac{x^2}{2^2 1!(p+1)} + \frac{x^4}{2^4 2!(p+1)(p+2)} - \cdots \right).$$

From the series, it is not difficult to see using term-by-term differentiation that

$$\frac{d}{dx}(x^{-p} J_p) = -x^{-p} J_{p+1} \ .$$

Assuming the above facts, prove that the zeros of $J_p(x)$ and $J_{p+1}(x)$ alternate (like the zeros of $\sin x$ and $\cos x$).

D-3 If a_1 is the first positive zero of $J_p(x)$ (assume $p > 0$), prove $a_1 > p$.

(Hint: let c be the maximum between 0 and a_1. Show $c > p$, by substituting $x = c$ into Bessel's equation (9).)

Appendix E

Existence and Uniqueness Theorems for Solutions to Differential Equations

This is one of the classic theorems of analysis. Though it lies at the foundation of any differential equations course, it is not often proved in such courses. But the proof is not difficult, and it gives a fine review of all of the major techniques in the book, using the theorems about continuous functions, differentiable functions, integration, uniform convergence of series and their term-by-term differentiation.

Prerequisites: Chapters 20, 22, 24 (just the theorems in 24.7); Appendix C is helpful as motivation, but not necessary—it is summarized briefly below.

E.1 Picard's method of successive approximations.

Appendix C discusses the method of Picard for finding a solution to equations of the form

$$(0) \qquad\qquad x = f(x) \ ,$$

where we assume that $f(x)$ is differentiable and satisfies, over some open x-interval containing the desired solution to (0), the Picard condition:

$$|f'(x)| \le K, \quad \text{for some constant } K < 1 \ .$$

To summarize how the solution is found: form the sequence x_n defined by

$$(1) \qquad\qquad x_{n+1} = \ f(x_n) \ .$$

This sequence converges; if we set $\lim x_n = c$, then by taking the limit as $n \to \infty$ of both sides of (1), we get $c = f(c)$, which shows that c is a solution to (0). It is also easy to see that c is the unique solution in the interval. If one thinks of f as a mapping, the equation $c = f(c)$ shows the solution c can be thought of as a *fixed point* of the mapping.

The above procedure has been enormously generalized in modern analysis to a method of showing existence and uniqueness for solutions to all sorts of equations — ordinary and partial differential equations and integral equations, as well as the algebraic and transcendental equations discussed in Appendix C. We illustrate here with the first order differential equation with initial condition:

$$(2) \qquad\qquad y' = f(x,y), \qquad y(a) = b \ .$$

As it stands, the connection between (0) and (2) is rather obscure. To show their relation, the idea is to convert (2) into the equivalent *integral equation*

(3)
$$y(x) = b + \int_a^x f(t, y(t))\, dt \ ,$$

"equivalent" meaning that a solution of $y' = f(x, y)$, $y(a) = b$ is a solution of (3) and vice-versa — the exact formulation and proof of equivalence is in Question E.1, and is an easy exercise in the two fundamental theorems of calculus.

The point to converting (2) into (3) is that we can then look on the right-hand side of (3) as an integral operator T which takes a function $y_n(x)$ into a function $y_{n+1}(x)$, according to the rule

(4)
$$y_{n+1}(x) = T(y_n(x)) = b + \int_a^x f(t, y_n(t))\, dt \ .$$

From this point of view, the solution $y(x)$ to (3) — and therefore also to (2) — will be a *fixed point* for the operator T, i.e., a function $y(x)$ for which
$$y(x) = T(y(x)) \ .$$

Thus T plays the role that f played in Picard's method for solving (0). The analogy suggests that to solve (3), we start with any continuous function $y_0(x)$ whose graph contains (a, b) and apply the operator T to it over and over. This will give us a sequence of functions $y_n(x)$ which we may hope will converge to the desired solution $y(x)$. This procedure is called *Picard's method of successive approximations* for solving $y' = f(x, y)$, $y(a) = b$; we illustrate it with a simple example (which is normally solved by separating variables).

Example E.1   Solve $y' = 2xy$, $y(0) = 1$, by Picard's method.

Solution.   The choice of starting function $y_0(x)$ is up to us; we pick $y_0 = 1$, since it at least satisfies the initial condition. The recursion formula (4) for the sequence of functions which approximate the solution is then
$$y_{n+1}(x) = 1 + \int_0^x 2t\, y_n(t)\, dt \ .$$

Using it, we get successively
$$y_1(x) = 1 + \int_0^x 2t\, dt = 1 + x^2 \ ;$$
$$y_2(x) = 1 + \int_0^x 2t(1 + t^2)\, dt = 1 + x^2 + x^4/2 \ ;$$
$$y_3(x) = 1 + \int_0^x 2t(1 + t^2 + t^4/2)\, dt = 1 + x^2 + x^4/2 + x^6/3! \ ;$$
$$y_4(x) = 1 + \int_0^x 2t(1 + t^2 + t^4/2 + t^6/3!)\, dt = y_3(x) + x^8/4! \ .$$

From this, one might guess that
$$y_n(x) = 1 + x^2 + \frac{x^4}{2!} + \frac{x^6}{3!} + \frac{x^8}{4!} + \ldots + \frac{x^{2n}}{n!} \ ,$$
and this can be confirmed by induction.

We get therefore as our solution the function $y(x) = \lim y_n(x)$, i.e., the function defined by the infinite series having $y_n(x)$ as its n-th partial sum; it is easy to see that this series sums to

$$y(x) \; = \; e^{x^2} \; ;$$

and it is easy to check by differentiation that this indeed is the solution to the initial-value problem $y' = 2xy, \; y(0) = 1$.

If we had started with a different initial function, the resulting sequence of functions would be different, but would have the same limit. Even a foolish starting choice like 2 will lead to $y = e^{x^2}$.

Picard's method becomes more difficult to carry out explicitly with non-linear first order equations, like $y' = y^2$ or $y' = 1/(x^2 + y^2)$, since the successive integrations required are usually impossible to perform using elementary functions.

Questions E.1

1.   Prove that (2) and (3) are equivalent in the following sense: let $y(x)$ be differentiable on an open interval I centered around a, and suppose $f(x, y)$ is continuous on $I \times J$, where J is an open interval centered around b; show that, for $x \in I$,

(i) $y(x)$ satisfies (2) $\Rightarrow$ $y(x)$ satisfies (3);

(ii) $y(x)$ satisfies (3) $\Rightarrow$ $y(x)$ satisfies (2).

2.   Calculate the first three approximations in Example E.1 using $y_0 = 2$ as the silly starting function. Watch what happens.

E.2   Local existence of solutions to $y' = f(x, y)$.

We show now that under suitable hypotheses, Picard's method leads to a solution of the first order ODE in some neighborhood of the point (a, b).

Theorem E.2   The local existence theorem.

Suppose $f(x, y)$ and $f_y(x, y)$ exist and are continuous in an open rectangle S containing the point (a, b). Let the starting function $y_0(x)$ satisfy $y_0(a) = b$ and be continuous for $x \approx a$, and set

(5) $$y_{n+1}(x) \; = \; b + \int_a^x f(t, y_n(t)) \, dt, \qquad n = 0, 1, 2, \dots \; .$$

Then for $x \approx a$, the sequence $y_n(x)$ converges to a function $y(x)$ which satisfies

(2) $$y' = f(x, y), \qquad y(a) = b \; , \qquad \text{for } x \approx a \; .$$

Proof.

   A.   Two preliminary steps.   We first get some control over the successive approximations by producing an interval of the form $\{x : |x - a| \leq h\}$ on which they will all be defined.

   We take a closed rectangle R centered at (a, b) and lying entirely inside the open rectangle S; since $f(x, y)$ is continuous on R, by Theorem 24.7A it is bounded on R, that is, for some constant M,

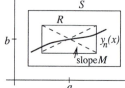

(6) $$\left| f(x, y) \right| \leq M, \qquad \text{for } (x, y) \in R .$$

By shrinking R in its vertical or horizontal direction if necessary, we may proportion it so that its two diagonals have slope $\pm M$; letting $2h$ and $2k$ be its horizontal and vertical dimensions, we can write this:

(7) $$R = \{(x, y) : |x - a| \leq h, \ |y - b| \leq k\}, \quad \text{where } k = Mh .$$

   We claim now that if $y_0(x)$ is continuous and its graph lies in R for $x \underset{h}{\approx} a$, then the same will be true for all the succeeding $y_n(x)$; namely, if it is true for $y_n(x)$, it is true also for $y_{n+1}(x)$:

$$|y_{n+1}(x) - b| \leq \int_a^x \left| f(t, y_n(t)) \right| dt \qquad \text{for } x \underset{h}{\approx} a, \qquad \text{by (5);}$$

$$\leq M |x - a|$$

by (6) and our assumption about $y_n(x)$;

$$\leq k \qquad \text{for } x \underset{h}{\approx} a, \qquad \text{by (7);}$$

moreover, $y_{n+1}(x)$ exists and is continuous (by the Second Fundamental Theorem 20.2A) since $y_n(t)$ and $f(t, y_n(t))$ are, on the respective intervals $x \underset{h}{\approx} a$ and $t \underset{h}{\approx} a$.

   Another preliminary remark: since $f_y(x, y)$ is by hypothesis continuous on the compact rectangle R, it is bounded by some constant K:

(8) $$\left| f_y(x, y) \right| \leq K \qquad \text{for } (x, y) \in R .$$

By the Mean-value Theorem 15.1, given two points $(x, c), (x, d) \in R$,

$$\left| f(x, c) - f(x, d) \right| = \left| f_y(x, \bar{y}) \right| |c - d|$$

for some $\bar{y}$ between c and d; thus

(9) $$\left| f(x, c) - f(x, d) \right| \leq K |c - d|, \quad \text{for } (x, c), (x, d) \in R.$$

   B.   Proof of convergence.

   We now prove the sequence $y_n(x)$ of successive approximations converges uniformly for $x \underset{h}{\approx} a$.

   The most convenient way to prove uniform convergence is with the M-test (Theorem 22.2), but for this, we need a series rather than a sequence. So we use the standard method of converting the sequence into a series having the $y_n(x)$ as its successive partial sums:

(10) $$y_0(x) + \big(y_1(x) - y_0(x)\big) + \ldots + \big(y_n(x) - y_{n-1}(x)\big) + \ldots .$$

To prove this series converges uniformly via the M-test, we must estimate its terms. Since $y_0(x)$ and $y_1(x)$ both lie in R for $x \underset{h}{\approx} a$,

(11) $\quad \left| y_1(x) - y_0(x) \right| \ \leq \ 2k, \qquad$ if $|x - a| \leq h$;

$\qquad \left| y_2(x) - y_1(x) \right| \ \leq \ \pm \int_a^x \left| f(t, y_1(t)) - f(t, y_0(t)) \right| dt, \quad$ by (5);

$\qquad \qquad \qquad \quad \ \leq \ \pm K \int_a^x \left| y_1(t) - y_0(t) \right| dt, \quad$ by (9);

(the minus sign is needed in the above two lines when $x < a$). We thus get

(12) $\quad \left| y_2(x) - y_1(x) \right| \ \leq \ 2kK|x - a|, \qquad$ by (11).

In the same way,

$\qquad \left| y_3(x) - y_2(x) \right| \ \leq \pm K \int_a^x \left| y_2(t) - y_1(t) \right|, \quad$ by (5) and (9);

$\qquad \qquad \qquad \quad \ \leq \ \pm 2kK^2 \int_a^x |t - a|\, dt \ = 2kK^2 \frac{|x - a|^2}{2}, \quad$ by (12).

Continuing in the same way, we get in general

(13) $\qquad \left| y_{n+1}(x) - y_n(x) \right| \ \leq \ 2kK^n \frac{|x - a|^n}{n!} , \qquad$ for $x \underset{h}{\approx} a$.

Since $|x - a| \leq h$, and the series $\sum \dfrac{K^n h^n}{n!}$ converges for all K and h, the conditions for the M-test (22.2C) are satisfied; this means the series (10) converges uniformly for $x \underset{h}{\approx} a$ to its sum $y(x)$:

$$y(x) \ = \ \lim \Big(y_0(x) + \big(y_1(x) - y_0(x) \big) + \ldots + \big(y_n(x) - y_{n-1}(x) \big) \Big)$$

$$\qquad = \ \lim_{n \to \infty} y_n(x), \text{ uniformly.}$$

C. Proof that $y(x)$ solves the problem (2).

Finally, we show $y(x)$ satisfies the differential equation with initial condition (2), for $x \underset{h}{\approx} a$; we do this by differentiating (10) term-by-term. We have

$$y_{n+1}(x) \ = \ b + \int_a^x f(t, y_n(t))\, dt, \qquad \text{for } x \underset{h}{\approx} a, \qquad \text{by (5)};$$

therefore, by the Second Fundamental Theorem 0.2A we get

(14) $\qquad y'_{n+1}(x) \ = \ f(x, y_n(x))$.

Using this, we can differentiate (10) term-by-term and apply the M-test to show the differentiated series converges uniformly. To estimate its terms, we have

$$\left| y'_{n+1}(x) - y'_n(x) \right| \ = \ \left| f(x, y_n) - f(x, y_{n-1}) \right|, \qquad \text{by (14)};$$

$$\qquad \qquad \leq \ K \left| y_n - y_{n-1} \right|, \qquad \text{by (9)};$$

$$\qquad \qquad \leq 2kK \frac{(Kh)^{n-1}}{(n-1)!}, \qquad \text{by (13), since } |x - a| \leq h \ .$$

Using this estimation, the M-test shows that the series obtained by differentiating (10) term-by-term converges uniformly.

Thus we have verified the hypotheses for Theorem 22.5 on term-by-term differentiation of series, so from

$$(10) \quad y(x) \; = \; y_0(x) + \big(y_1(x) - y_0(x)\big) + \big(y_2(x) - y_1(x)\big) + \ldots \; = \; \lim_{n \to \infty} y_n(x).$$

we get using term-by-term differentiation,

$$y'(x) \; = \; \lim_{n \to \infty} y_n'(x), \qquad \text{for } x \underset{h}{\approx} a;$$

$$= \; \lim_{n \to \infty} f(x, y_{n-1}(x)), \qquad \text{by (14)};$$

$$= \; f(x, y(x)),$$

where the last step used the Sequential Continuity Theorem 11.5; in applying that theorem, think of x as fixed, so that y_n is a sequence of constants and f is a function of y alone.

This shows that $y(x)$ satisfies the differential equation (2) for $x \underset{h}{\approx} a$; it also satisfies the initial condition, since $y(a) = \lim_{n \to \infty} y_n(a) = b$, by (5). $\qquad\square$

E.3   The uniqueness of solutions.

We now show there is only one solution to the differential equation with initial condition (2) in a neighborhood of a.

Theorem E.3   Local uniqueness theorem.

With the hypotheses and notations of the Local Existence Theorem E.2, the solution $y(x)$ is unique in R. That is,

if $\bar{y}(x)$ is a solution to (2) for $x \underset{h}{\approx} a$, then $y(x) = \bar{y}(x)$ for $x \underset{h}{\approx} a$.

Proof.   We show first that the graph of $\bar{y}(x)$ lies in R for $x \underset{h}{\approx} a$.

By contradiction, suppose for example that $\bar{y}(x) > b + k$ somewhere on the interval $[a, a+h]$. By the Intermediate Value Theorem (Cor.12.1), there is a point x' (see the picture) such that

$$\bar{y}(x') \; = \; b + k, \qquad a < x' < a + h .$$

Then by the above and the inequality laws, together with (7),

$$(15) \qquad \frac{\bar{y}(x') - b}{x' - a} \; > \; \frac{k}{h} \; = \; M ;$$

but the Mean-value Theorem 15.1 shows that for some c, where $a < c < x'$,

$$\frac{\bar{y}(x') - b}{x' - a} \; = \; \bar{y}'(c) \; = \; \big|f(c, \bar{y}(c))\big|, \qquad \text{by (2)};$$

$$\leq \; M , \qquad \text{by (6)},$$

which contradicts (15).

Now that we know $\bar{y}(x)$ lies in R for $x \underset{h}{\approx} a$, we can show $\bar{y}(x) = y(x)$.

Since both $\bar{y}(x)$ and $y(x)$ satisfy the integral equation (3) and lie in R,

$$\left|\bar{y}(x) - y(x)\right| \;\le\; \int_a^x \left|f(t, \bar{y}(t)) - f(t, y(t))\right| dt \quad \text{for } x \underset{h}{\approx} a \; ;$$

$$\le\; K \int_a^x \left|\bar{y}(t) - y(t)\right| dt \qquad \text{by (9);}$$

$$\le\; K\,h \max_I \left|\bar{y}(t) - y(t)\right| , \qquad I = [a - h, a + h] \; .$$

This shows that

(16) $$\max_I \left|\bar{y}(t) - y(t)\right| \;\le\; K h \max_I \left|\bar{y}(t) - y(t)\right| \; .$$

If the rectangle R is small enough so that

(∗) $$h < 1/K,$$

this completes the proof, since (16) shows then that

$$\max_I \left|\bar{y}(t) - y(t)\right| \;=\; 0 \; ;$$

(for otherwise one could divide (16) through by this maximum, getting $1 \le Kh$, which contradicts (∗)), and it follows therefore that

$$\bar{y}(x) \;=\; y(x) \qquad \text{for } x \in I, \quad \text{i.e., for } x \underset{h}{\approx} a \; . \qquad \square$$

If one is willing to accept the Uniqueness Theorem for the smaller rectangle pictured below, having a width $2h'$, where $h' < 1/K$, we are done. We can however finish the argument for the original R as follows.

Suppose by contradiction that $\bar{y}(x) \ne y(x)$ in R; say the two solutions diverge somewhere to the right of $a + h'$, as in the picture. Let x' be the point at which they diverge, i.e.

$$x' \;=\; \inf\{x : \bar{y}(x) \ne y(x), \quad a + h' \le x \le a + h\}.$$

Then we have

$$\bar{y}(x) \;=\; y(x) \qquad \text{for } a \le x < x'.$$

By continuity also

$$\bar{y}(x') \;=\; y(x') \; .$$

On the other hand,

$$\bar{y}(x) \;\ne\; y(x)$$

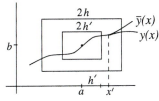

for infinitely many points $x > x'$ but arbitrarily close to x' (otherwise x' would not be the infimum).

But this last is impossible, since by the first part of the proof, there is a small rectangle centered around the point $(x', y(x'))$ in which $\bar{y}(x) = y(x)$. $\square\square$

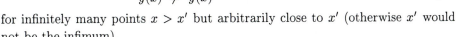

E.4   Extending the existence and uniqueness theorems.

We give four directions in which the theorems may be extended.

A.   Lipschitz conditions.

Just as for the Picard method for solving $x = f(x)$ that we gave in Appendix C, analysis of the argument for existence and uniqueness shows that we used the continuity of $f_y(x, y)$ only to get the estimation

$$(9) \qquad \left| f(x, c) - f(x, d) \right| \leq K \left| c - d \right|, \qquad \text{for any points } (x, c), (x, d) \in R.$$

The inequality in (9) is called a *Lipschitz condition* for $f(x, y)$ in R. It is weaker than assuming that $f_y(x, y)$ is continuous in R. So we get the stronger statement that local existence and uniqueness for solutions to (2) holds if $f(x, y)$ is continuous and satisfies a Lipschitz condition (9) in the open rectangle S.

B.   Complete solutions.

By a procedure analogous to that used in the end of the uniqueness proof above, one can extend a solution obtained in the rectangle R to one valid in a larger rectangle. We call a solution *complete* if it cannot be further extended. For example,

$$y' = 2xy^2, \quad y(0) = 1$$

has the solution

$$y = \frac{1}{1 - x^2}, \quad -1 < x < 1,$$

which is complete on the open interval $(-1, 1)$, but not on any smaller subinterval.

C.   Theorems for systems.

The existence and uniqueness proof given above works also for *systems* of first-order ODE's with initial conditions; a 2×2 system would be:

$$\begin{aligned} y_1' &= f_1(x, y_1, y_2) & y_1(a) &= b_1 \ ; \\ y_2' &= f_2(x, y_1, y_2) & y_2(a) &= b_2 \ ; \end{aligned}$$

or in vector form, setting $\quad \mathbf{y} = (y_1, y_2), \quad \mathbf{f} = (f_1, f_2), \quad \mathbf{b} = (b_1, b_2),$

$$\mathbf{y}' = \mathbf{f}(x, \mathbf{y}), \qquad \mathbf{y}(a) = \mathbf{b}.$$

In this vector notation, the same proof of existence and uniqueness goes through with almost no changes. The correct hypotheses are that the f_i and the partial derivatives $\partial f_i / \partial y_j$ should all be continuous in some open rectangle $S \subseteq \mathbb{R}^3$ containing the point $(a, \mathbf{b})$.

When given in vector form, the above argument generalizes immediately to an $n \times n$ system of differential equations.

D.   Existence and uniqueness for higher order ODE's.

The theorems for a single higher order ODE can be obtained without further argument by converting such an equation to an equivalent system. To illustrate on a general second-order equation,

$$y'' = f(x, y, y'), \qquad y(a) = b_1, \ y'(a) = b_2,$$

gets converted to a system by setting $y = y_1$ and $y' = y_2$, giving the system

$$
\begin{aligned}
y_1' &= y_2, & y_1(a) &= b_1; \\
y_2' &= f(x, y_1, y_2), & y_2(a) &= b_2.
\end{aligned}
$$

A similar procedure changes an n-th order ODE to an equivalent $n \times n$ system.

Exercises

E.1

1. Use Picard's method of successive approximations to find the solution of the following differential equations with initial condition; starting functions $y_0(x)$ are also given.

(a) $y' = x - y$, $y(0) = 0$ (i) $y_0 = 0$ (ii) $y_0 = 1$ (silly);

(b) $y' = 1 + y^2$, $y(0) = 0$;   take $y_0 = 0$ and find $y_3(x)$,

then solve the equation exactly and relate $y_3(x)$ to the exact solution somehow.

E.2

1. Prove that under the hypotheses of the Local Existence Theorem, Picard's method still produces a sequence converging to the solution if the starting function $y_0(x)$ is continuous and has its graph in the rectangle R, but the graph does not go through the point (a, b). (This is easy.)

E.3

1. Consider $y' = y^{1/3}$, $y(a) = 0$.

(a) Show it does not satisfy the hypotheses of the Local Uniqueness Theorem.

(b) Find an infinity of differentiable solutions to this initial value problem, valid on the whole x-axis.

(Two are easily found; if you can find a third, you can find an infinity.)

E.4

1. Show that $y^{1/3}$ does not satisfy a Lipschitz condition in any rectangle centered at $(a, 0)$. (Do this directly, not by using E.3/1b.)

2. Prove that the function $f(x, y) = x\,|y|$ satisfies a Lipschitz condition in a rectangle centered at $(a, 0)$, but it does not satisfy the hypotheses given in the book for the Local Existence and Uniqueness Theorems in any such rectangle. (This gives an example where having weaker hypotheses is useful.)

3. Consider the linear differential equation with initial conditions, with form
$$y'' + p(x)y' + q(x)y = r(x), \qquad y(a) = b, y'(a) = b' ;$$
by using the remarks in this this section, state hypotheses under which it will have a unique solution in an interval $[x_0, x_1]$ containing a, and prove they are sufficient. (Reduce it to a system.)

Problems

E-1 In the proof of the Local Existence Theorem, instead of using term-by-term differentiation to show that $y(x)$ satisfies (2), prove instead that $y(x)$ satisfies the integral equation (3).

(Take limits of both sides of (5) and justify the steps carefully.)

Answers to Questions

E.1

1. (i) Since $y(x)$ is a solution, we have (after changing the independent variable from x to t),
$$y(t)' = f(t, y(t)), \quad t \in I.$$
The right side is continuous for $t \in I$, since f is continuous in $I \times J$; also, $y(t)$ is continuous on I, and $y(t) \in J$ for $t \in I$. Therefore the left side is also continuous; integrating both sides and applying the First Fundamental Theorem 20.1 to the left side gives
$$y(x) - y(a) = \int_a^x f(t, y(t)) \, dt ;$$
since $y(a) = b$, this proves (3). □

(ii) Starting with (3), the integrand is continuous by the argument of part (i); therefore the Second Fundamental Theorem 20.2A can be applied to the integral; it gives (2):
$$y(x)' = f(x, y(x)), \quad x \in I .$$

2.
$$y_1(x) = 1 + \int_0^x 2t \cdot 2 \, dt = 1 + 2x^2 ;$$
$$y_2(x) = 1 + \int_0^x 2t(1 + 2t^2) \, dt = 1 + x^2 + x^4 ;$$
$$y_3(x) = 1 + \int_0^x 2t(1 + t^2 + t^4) \, dt = 1 + x^2 + x^4/2 + x^6/3 ;$$

The effect of using a poor starting function is shown in the incorrect highest power of x, which keeps getting pushed further and further to the right as the successive approximations are made; presumably it gets excreted at infinity.

Index